UNIVERSITY OF ROCHESTER LIBRARIES
3 9087 01866490 6

D1604645

BIOMEDICAL CHEMISTRY

BIOMEDICAL CHEMISTRY
APPLYING CHEMICAL PRINCIPLES TO THE UNDERSTANDING AND TREATMENT OF DISEASE

Edited by

Paul F. Torrence
Department of Chemistry
Northern Arizona University
Flagstaff, Arizona

A John Wiley & Sons, Inc., Publication

New York • Chichester • Weinheim • Brisbane • Singapore • Toronto

This book is printed on acid-free paper. ♾

Published simultaneously in Canada.

For ordering and customer service, call 1-800-CALL-WILEY.

Library of Congress Cataloging-in-Publication Data:

Biomedical chemistry/applying chemical principles to the understanding and treatment of disease / edited by Paul F. Torrence.

p. cm.

Includes index.

ISBN 0-471-32633-X (cloth : alk. paper)

1. Pharmaceutical chemistry. 2. Drugs–Design. 3. Clinical biochemistry. I. Torrence, Paul F.

RS403 .B53 2000

615′.19 21–dc21 99-043428

Printed in the United States of America

10 9 8 7 6 5 4 3 2 1

This volume is dedicated to my mentors who made chemistry come alive for me and for many others: Roy Adams, Paul Wunz, Howard Tieckelmann, and Bernhard Witkop.

P. F. TORRENCE

CONTENTS

PREFACE

Dramatic advances in the new fields of robotics, nanotechnology, microfluidics, combinatorial chemistry, photolithography, high-throughput screening, genomics, proteomics, microchip arrays, transgenics, and bioinformatics are driving a revolution that will cause major changes in the structure of the world's economy.[1] Functional genomics and the other components of the molecular revolution will impact medicine, agriculture, biotechnology, and the environmental and energy sciences. A novel synergy is arising from the interplay of genomics, biology, biotechnology, and *chemistry.*

As powerful as these enabling techniques are, they play on a few basic refrains, originally generated by relatively simple observations and experiments. In their refinement, they sometimes obscure the fundamentals from which they arose. For scientists to remember, for policymakers to realize, and for students to understand the intimate relation of fundamental discovery in biomedicine to simple axioms of chemistry, it is sometimes necessary to return to examples in which the linkage is still visible. We train students well to be effective in the present information-producing age, but they need to recall that information is meaningless unless tied to the creative scientific process, which itself relies on core scientific facts and principles.

The central theme of this volume is biomedical chemistry; specifically, the application of chemical concepts to the challenges of drug discovery and determination of disease etiology. This treatise does not pretend to cover serendipitous discoveries, various structure–activity, screening, or computer-driven approaches, but rather focuses on new applications of chemical principles in the understanding and treatment of disease. One goal of this book is to emphasize the importance of central chemical concepts in the generation of new approaches in medicine.

This treatise aims to fill several voids. First, in the explosion of biotechnology and molecular biology, scientists, physicians, administrators, and politicians tend to forget the critical importance of fundamental chemical concepts and studies in the advancement of the scientific enterprise. One excellent example of this, as related to biotechnology and the manipulation of genetic expression, has been the development of the essential chemistry required to synthesize DNA. Once, DNA and RNA synthesis were esoteric enterprises, practiced by a few stubborn academics who took years to put together a few short nucleotide sequences. Now so much has evolved in the chemistry of nucleic acid synthesis, thanks to these pioneering efforts, that almost anyone can synthesize a long DNA sequence, without even understanding the underlying chemistry. In this regard, I strongly recommend a commentary[2] by Dan Brown entitled "A Brief History of Oligonucleotide Synthesis." Commercially available "gene machines" power the heart of present-day biotech and molecular biology. Oligonucleotide synthesis plays a pivotal role today as DNA "chip" technology enables rapid identification of expressed genes.

This book attempts to show how chemists can begin with a basic chemical idea, often one learned by chemistry students early in their education, and develop a novel approach to complicated biological, biochemical, and/or medical challenges. By example, this book will show that the progress in the major problems of medicine is still very dependent on the richness of the basic science of chemistry and its creative application by the imaginative mind.

Biomedical Chemistry deals with topics in medicine that have a widespread impact on humans and that are in the forefront of research endeavors, including AIDS and other diseases of viral etiology, cancer, alcoholism, Parkinson's disease, glaucoma, trypanosomiasis, and malaria. The subject matter is interdisciplinary, dealing with the application of chemistry to drug discovery and development, pharmacology, the origins of disease, and future directions in biotechnology.

Because the theme of the book is the application of central chemical principles, the editor believes that it will be attractive to a variety of audiences including undergraduate organic and biochemistry students. The contributions herein can be a source of illustration of the application of chemistry to solve societal problems. They may serve as a well-spring to relieve what is often student-perceived as dry inanimate tedium. Students need to feel the energy and power of the foundations of chemistry in an arena they can see as germane. Some of the chemical principles that the authors employ in approaches to diseases include topics such as nucleophilic addition, hydrogen bonding, base-pairing in DNA and RNA, mechanism-based enzyme inhibitors, redox reactions, free-radical chemistry, magnetism, photochemical reactions, protecting groups, carbocation chemistry, and the biochemical effects of fluorine and sulfur substitution.

REFERENCES

1. Enriquez, J., *Science* 1998 **281**, 925–926.
2. Brown, D. M., in *Protocols for Oligonucleotides and Analogs*, S. Agrawal, ed., Human Press, Towata, NJ, 1993, pp 1–17.

P. F. TORRENCE

CONTRIBUTORS

James K. Bashkin, Monsanto Company R3A, 800 North Lindbergh Blvd., St. Louis, MO 63167 USA

Ronald T. Borchardt, Department of Pharmaceutical Chemistry, The University of Kansas, Lawrence, KS 66047 USA

Terrence R. Burke, Jr., Laboratory of Medicinal Chemistry, Division of Basic Sciences, National Cancer Institute, National Institutes of Health, Bethesda, MD 20892 USA

Fu-Chou Cheng, Department of Chemistry and Biochemistry, University of Oklahoma, Norman, OK 73019 USA

Jared N. Cumming, Department of Chemistry, Johns Hopkins University, 3400 North Charles St., Baltimore, MD 21218 USA

Eugene G. DeMaster, Department of Pharmacology, University of Minnesota, Duluth, MN 55812 USA

John S. Driscoll, Laboratory of Medicinal Chemistry, Division of Basic Sciences, National Cancer Institute, National Institutes of Health, Bethesda, MD 20892 USA

Glenn Dryhurst, Department of Chemistry and Biochemistry, University of Oklahoma, Norman, OK 73019 USA

Jacques Fahy, Division de Chimie Medicinale V, Centre de Recherche Pierre Fabre, 17 Avenue Jean Moulin, 81106 Castres, France

James M. Gallo, Department of Pharmacology, Fox Chase Cancer Center, Philadelphia, PA 19111 USA

Yang Gao, Laboratory of Medicinal Chemistry, Division of Basic Sciences, National Cancer Institute, National Institutes of Health, Bethesda, MD 20892 USA

Francis Garnier, Laboratoire des Materiaux Moleculaires, CNRS, rue Dunant 94320, Thiais, France

Gilles Gosselin, Laboratoire de Chimie Biorganique, UMR CNRS 5625, Case Courrier 008, Université Montpellier II, Sciences et Techniques du Languedoc, Place Eugene Bataillon, 34095 Montpellier, France

Jilin Han, Department of Chemistry and Biochemistry, University of Oklahoma, Norman, OK 73019 USA

Jean-Louis Imbach, Laboratoire de Chimie Biorganique, UMR CNRS 5625, Case Courrier 008, Université Montpellier II, Sciences et Techniques du Languedoc, Place Eugene Bataillon, 34095 Montpellier, France

Marco d'Ischia, Department of Organic and Biological Chemistry, University of Naples Federico II, Via Mezzocannone 16, I-80134 Naples, Italy

Jean-Claude Jacquesy, Laboratoire de Synthese et Reactivité des Substances Naturelles, UNR CNRS 6514, Faculté des Sciences, Université de Poitiers, 40 avenue du Recteur pineau, 86022 Poitiers, France

George L. Kenyon, College of Pharmacy, University of Michigan, USA

Kenneth L. Kirk, Laboratory of Bioorganic Chemistry, National Institute of Diabetes and Digestive and Kidney Diseases, National Institutes of Health, Bethesda, MD 20892 USA

Mikhail Krasavin, Department of Chemistry, Johns Hopkins Unviersity, 3400 North Charles St., Baltimore, MD 21218 USA

Hong Li, Department of Chemistry and Biochemistry, University of Oklahoma, Norman, OK 73019 USA

Susan M. Ludeman, Duke Comprehensive Cancer Center, and Department of Medicine, Duke University Medical Center, Durham, NC 27710 USA

Herbert T. Nagasawa, Medical Research Laboratories, DVA Medicinal Center and Department of Medicinal Chemistry, University of Minnesota, Minneapolis, MN 55417 USA

Peter E. Nielsen, Center for Biomolecular Recognition, Department of Medical Biochemistry & Genetics, Biochemical Laboratory B, The Panum Institute, Blegdamsvej 3c, 2200 Copenhagen N Denmark

Todd A. Osiek, Department of Chemistry, Washington University, St. Louis, MO 63130-4899 USA

Christian Périgaud, Laboratoire de Chimie Biorganique, UMR CNRS 5625, Case Courrier 008, Université Montpellier II, Sciences et Techniques du Languedoc, Place Eugene Bataillon, 34095 Montpellier, France

Gary H. Posner, Department of Chemistry, Johns Hopkins Unviersity, 3400 North Charles St., Baltimore, MD 21218 USA

Giuseppe Prota, Department of Organic and Biological Chemistry, University of Naples Federico II, Via Mezzocannone 16, I-80134 Naples, Italy

Sharon K. Pulfer, Department of Pharmacology, Fox Chase Cancer Center, Philadelphia, PA 19111 USA

William C. Putnam, Department of Chemistry, Washington University, St. Louis, MO 63130-4899 USA

Xue-Ming Shen, Department of Chemistry and Biochemistry, University of Oklahoma, Norman, OK 73019 USA

Frances N. Shirota, Medical Research Laboratories, DVA Medicinal Center and Department of Medicinal Chemistry, University of Minnesota, Minneapolis, MN 55417 USA

Patrick M. Woster, Department of Pharmaceutical Sciences, Wayne State University, Detroit, MI 48202 USA

Xiaoda Yang, Department of Pharmaceutical Chemistry, The University of Kansas, Lawrence, KS 66047 USA

Zhaoliang Yang, Department of Chemistry and Biochemistry, University of Oklahoma, Norman, OK 73019 USA

Zhu-Jun Yao, Laboratory of Medicinal Chemistry, Division of Basic Sciences, National Cancer Institute, National Institutes of Health, Bethesda, MD 20892 USA

Dan Yin, Department of Pharmaceutical Chemistry, The University of Kansas, Lawrence, KS 66047 USA

Chong-Sheng Yuan, Tanabe Research Laboratories USA, San Diego, CA 92121 USA

PART I

DRUG DISCOVERY AND DEVELOPMENT LEADS FROM FUNDAMENTALS TO ENZYME TRANSFORMATIONS AND MECHANISMS

Receptors are those biological entities with which a drug interacts to produce a therapeutic effect. John Langley first suggested this concept, but it was Paul Ehrlich who originated and expanded the idea with his *receptor sidechain theory* of toxin–antibody interactions. *Pharmocodynamics* involves the interaction of drug with its receptor. Receptors with catalytic properties are *enzymes*. In this section, we see how an understanding of the enzyme and its catalytic mechanism can contribute to drug discovery and development.

Sleeping sickness or *trypanosomiasis* causes the suffering of more than 66 million people in sub-Sahara Africa. The disease is caused by *Trypanosoma brucei*, which is able to evade the host's immune system by a process known as *antigenic variation*. In South America, another trypanosomal parasite, *T. cruzi*, causes *Chagas disease*. In Chapter 1 Woster details one approach to developing a new chemotherapeutic agent for trypanosomes by targeting an enzyme of the polyamine transport system: *S-adenosylmethionine decarboxylase*.

Virus-caused diseases, such as *Ebola fever, Lassa fever, parainfluenza*, and *measles*, have no available chemotherapy. In Chapter 2, Yin, Yang, Yuan, and Borchardt describe efforts to provide mechanism-based inhibitors of *S-adenosyl-*L-*homocysteine hydrolase*. This enzyme occurs in normal uninfected cells and may provide the basis for a broad-spectrum antiviral agent that would be effective against a number of different viruses.

Alcoholism and related problems cost the United States alone tens of thousands of lives and billions of dollars, not to mention the incalculable loss of human potential. We know that part of the risk for development of alcoholism lies in an individual's genes. To help people recover from the disease and prevent relapse, neuroscientists and medicinal chemists are evolving new medications. Nagasawa, Shirota, and DeMaster have been utilizing our knowledge of alcohol biochemistry, in particular the role of *alcohol dehydrogenase*, to find alcohol deterrents, and describe their research in Chapter 3.

As many as 900,000 Americans may be infected with *human immunodeficiency virus* (HIV), and approximately 30.6 million individuals worldwide were infected with HIV at the end of 1997. *AIDS dementia* is a debilitating syndrome affecting 30% of adult and 75% of pediatric infections. Although remarkable progress has been made in arresting the progress of AIDS through the use of drug "cocktails," these agents do not penetrate the normally protective *blood–brain barrier.* Thus HIV can replicate and do damage in the brain in relative "sanctuary." In Chapter 4 Driscoll describes his research aimed at disguising anti-HIV agents in a *lipophilic* form that can cross this blood–brain-barrier and then be unmasked by a naturally occurring enzyme *adenosine deaminase*. Illustrating why there are other reasons to synthesize lipophilic drugs, in Chapter 5 Perigaud, Gosselin, and Imbach narrate their work on *phosphotriester* derivatives of nucleosides that will be enzymatically converted to biologically active mononucleotides. This research is often called the "eye of the needle" in determining the effectiveness of an anti-HIV drug; namely, the cellular *kinase* phosphorylation of the nucleoside drug itself to yield a highly polar monophosphate derivative that is far too polar to enter the cell as such.

CHAPTER 1

Sleeping Sickness: Irreversible Inhibitors of *S*-Adenosylmethionine Decarboxylase as Antitrypanosomal Agents

PATRICK M. WOSTER

Department of Pharmaceutical Sciences, Wayne State University

INTRODUCTION

During the past 20 years (at the time of writing), remarkable advances have been made toward understanding the cellular roles of the natural polyamines spermidine and spermine.[1] In addition, a wide variety of analogs targeted to specific enzymes in the polyamine metabolic pathway have been synthesized, and their effects on the pathway and on cellular metabolism have been partially elucidated. Despite these advances, polyamine-related drug discovery efforts have resulted in only a handful of useful analogs, some of which are used as experimental tools, and a select few as therapeutic agents. However, recent studies indicating the role of polyamines in protection of DNA from damage due to reactive oxygen species,[2,3] the finding that certain alkylpolyamine analogs can induce apoptosis in tumor cells,[4–7] and the discovery of alkylpolyamines that affect tubulin polymerization[8] have provided new challenges in the area of drug discovery. At the present time, the most promising targets for therapeutic intervention are the polyamine cellular transport system, spermidine/spermine-N^1-acetyltransferase (SSAT, the rate-limiting step in the polyamine back-conversion pathway), and *S*-adenosylmethionine decarboxylase (AdoMet-DC), one of the regulatory enzymes in the polyamine biosynthetic pathway. The modulation of SSAT by terminally alkylated polyamine analogs has been reviewed.[1] The present chapter outlines the design, synthesis, and therapeutic effects of agents that inhibit the AdoMet-DC.

Biomedical Chemistry: Applying Chemical Principles to the Understanding and Treatment of Disease, Edited by Paul F. Torrence
ISBN 0-471-32633-x

MAMMALIAN POLYAMINE METABOLISM

The mammalian polyamine biosynthetic pathway is shown in Figure 1.1.[9,10] Ornithine is converted to putrescine by the action of the enzyme ornithine decarboxylase (ODC). This enzyme is a typical pyridoxal-phosphate-requiring amino acid decarboxylase, and has been studied quite extensively.[11] ODC is known to be one of the control points in the pathway, producing a product that is committed to polyamine biosynthesis. The synthesis and degradation of ODC are controlled by a number of factors, and the active enzyme has a half-life of about 10 min.[9,10] Putrescine is next converted to spermidine via an aminopropyltransferase known as *spermidine synthase*. A second closely related but distinct aminopropyltransferase, spermine synthase, then adds an additional aminopropyl group to spermidine to yield spermine, the longest polyamine occurring in mammalian systems. The by-product for the spermidine and spermine synthase reactions is 5′-methylthioadenosine (MTA), a potent product inhibitor for the aminopropyltransfer process. In

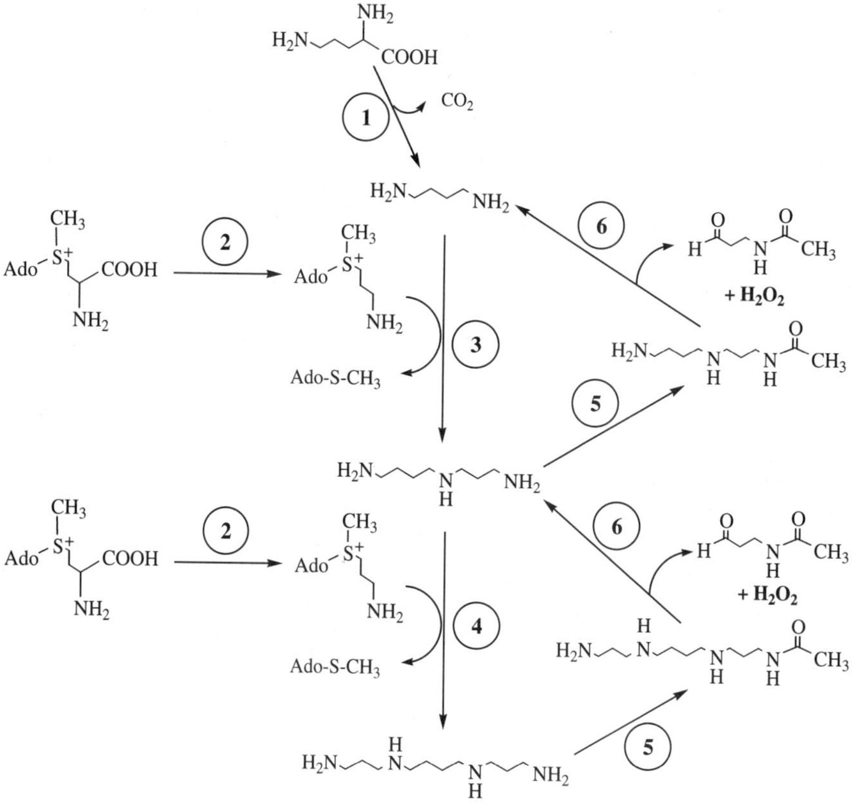

FIGURE 1.1 The biosynthesis of mammalian polyamines (1 = ornithine decarboxylase; 2 = *S*-adenosylmethionine decarboxylase; 3 = spermidine synthase; 4 = spermine synthase; 5 = spermidine/spermine acetyltransferase; 6 = polyamine oxidase; Ado = adenosyl).

mammalian systems, MTA is rapidly hydrolyzed by the enzyme MTA-phosphorylase, and the components are converted to adenosine and methionine via salvage pathways. The aminopropyl donor for both aminopropyltransferases is decarboxylated *S*-adenosylmethionine (dc-AdoMet), produced from AdoMet by the action of AdoMet-DC. AdoMet-DC, like ODC, is a highly regulated enzyme in mammalian cells, and also serves as a regulatory point in the pathway.[9,10] However, unlike ODC, AdoMet-DC belongs to a class of pyruvoyl enzymes that do not require pyridoxal phosphate as cofactor.

In addition to the enzymes mentioned above, cellular polyamine content is modulated by a pair of acetyltransferases. Spermidine in the cell nucleus is acetylated on the four carbon end by spermidine-N^8-acetyltransferase, possibly altering the compound's binding affinity for DNA.[12] A specific deacetylase can then reverse this enzymatic acetylation. Cytoplasmic spermidine and spermine serve as substrates for spermidine/spermine-N^1-acetyltransferase (SSAT), resulting in acetylation on the three carbon end of each molecule (Fig. 1.1).[1b] The acetylated spermidine or spermine then acts as a substrate for polyamine oxidase (PAOX), which catalyzes the formation of 3-acetamidopropionaldehyde and either putrescine or spermidine, respectively. SSAT and PAOX together serve as a reverse route for the interconversion of polyamines. Thus, ODC, AdoMet-DC, SSAT, and polyamine transport[13] act together to provide an exquisite level of control over the cellular levels of the individual polyamines.

The polyamine pathway represents a logical target for chemotherapeutic intervention, since depletion of polyamines results in the disruption of a variety of cellular functions, and ultimately in cell death.[14] It has now been well established that polyamines are absolutely required for normal cell division, and that this effect is especially pronounced in rapidly growing cell types.[14] Inhibitors of the polyamine pathway, therefore, have traditionally been developed as potential antitumor or antiparasitic agents. Such inhibitors also play a critical role as research tools to elucidate the cellular functions of the naturally occurring polyamines, especially if these agents are specific for a single enzyme in the pathway. The ability to selectively deplete individual polyamines by selective inhibition of the individual enzymes in the pathway would prove invaluable for the study of the role of polyamines in cellular metabolism. Inhibitors have now been developed for ODC,[11,15] AdoMet-DC,[16–20] and each of the two aminopropyltransferases,[21,22] but various problems (toxicity unrelated to polyamine biosynthesis, lack of specificity, resistance, etc.) have limited their development as chemotherapeutic agents and as research tools. Therefore, there is an ongoing need for novel inhibitors of the polyamine pathways. The most promising inhibitors that are currently available are briefly reviewed in the following sections. The structural aspects and molecular biology of the enzymes involved in polyamine biosynthesis have been reviewed.[2,11,19]

Inhibitors of Ornithine Decarboxylase (ODC)

Mammalian ODC, a dimeric enzyme with a molecular weight of about 80,000, is a highly unstable protein, and cellular levels of ODC depend on rates of synthesis and degradation.[9,11] For this reason, competitive inhibitors of ODC have proven to be of limited value, since the synthesis of new protein occurs very rapidly. The mechanism of ODC involves formation of a Schiff base between the amino group of ornithine and the pyridoxal phosphate cofactor, which is tightly bound to ODC.[19] The most useful inhibitor of ODC to date, α-difluoromethylornithine (DFMO), takes advantage of this aspect of the mechanism.[15] Formation of a Schiff base between DFMO and ODC results in the generation of a latent electrophile, and ODC is rapidly and irreversibly deactivated. The discovery of DFMO has provided an enormous stimulus to the field of mammalian polyamine biology. In addition, DFMO has been marketed as a treatment for *Pneumocystis carinii* secondary infections in AIDS and AIDS-related complex (ARC) patients, and has been shown to be somewhat effective in curing infections of *Trypanosoma brucei brucei* in limited clinical trials.[23,24] Another promising compound, (2*R*,5*R*)-6-heptyne-2,5-diamine (*R,R*-MAP), has been shown to possess a K_i of 3 μM, and appears to penetrate mammalian cells relatively well. This compound is currently undergoing clinical trials.

Inhibitors of Spermidine Synthase and Spermine Synthase

Selective inhibition of the individual aminopropyltransferases has proved to be a significant problem because of the similarity of the reactions catalyzed by the two enzymes. Mammalian spermidine synthase and spermine synthase each consist of two subunits (M_r 35,000 each or 44,000 each, respectively) and require no cofactors.[25] Both mammalian enzymes have been isolated and purified to homogeneity. A variety of MTA analogs have been shown to act as inhibitors for the aminopropyltransferases (5′-ethylthioadenosine, 5′-isobutylthioadenosine (SIBA), 5′-methylthiotubercidin, etc.), but these agents cannot distinguish between the two enzymes. An exception to this situation is dimethyl-(5′-adenosyl)sulfonium perchlorate [$AdoS^+(CH_3)_2$],[26] which appears to act selectively at spermine synthase ($IC_{50} = 8$ μM). However, this compound also has inhibitory activity toward AdoMet-DC, limiting its usefulness as a specific inhibitor for biochemical studies. By far the most significant advance in the selective inhibition of the aminopropyltransferases is the development of *S*-adenosyl-1,8-diamino-3-thiooctane (AdoDATO)[21] and *S*-adenosyl-1,12-diamino-3-thio-9-azadodecane (AdoDATAD),[22] which are multi-substrate analog inhibitors for spermidine synthase and spermine synthase, respectively. AdoDATO inhibits spermidine synthase with $IC_{50} < 50$ nM while having no significant effect on spermine synthase. AdoDATAD, conversely, inhibits spermine synthase selectively ($IC_{50} = 20$ nM) while leaving spermidine synthase unaffected. However, AdoDATAD appears to be rapidly metabolized in L1210 cells by polyamine oxidase, resulting in the formation of AdoDATO and other cytotoxic metabolites.[27]

Modulators of Spermidine/Spermine-N^1-Acetyltransferase (SSAT)

Recently, the present author and others have focused on the development of polyamine analogs that function by mimicking some of the regulatory roles of the natural polyamines.[28–32] The basis for studying these polyamine analogs is founded on investigations that demonstrated that the natural polyamines have several feedback mechanisms that regulate their own synthesis.[33] In addition, the polyamines are also known to have a highly specific energy-dependent transporter.[13,34] Symmetrically substituted bis(alkyl)polyamine analogs such as bis(ethyl)spermine were among the first compounds to use the transport system to enter tumor cells and successfully exploit the desired regulatory pathways.[29,30,35] Most alkylpolyamine analogs are readily accumulated by tumor cells, where they decrease ODC and AdoMet-DC activity, increase polyamine efflux and breakdown of the natural polyamines, and produce a decrease in all three natural polyamines, generally resulting in cytostasis. Subsequently developed unsymmetrically substituted alkylpolyamines such as CPENSpm[28] produced particularly encouraging results in breast[36] and prostate[37] cancer systems. SSAT induction is an important feature in the toxicity profile of some analogs, but does not occur with every compound that produces cell-type-specific toxicity. Excellent examples of this are recent studies demonstrating that the induction of programmed cell death (PCD) in some responsive tumors is mediated, at least in part, in response to H_2O_2 production resulting from polyamine catabolism.[38] These studies are the first to directly implicate the polyamine catabolic pathway as a mediator of PCD. Further, it has recently been demonstrated that some alkylpolyamine analogs have striking and differential effects on tubulin polymerization, and produce a significant G_2/M cell cycle blockade.[8] These observations provide a number of new avenues for elucidating structure–function relationships for cytotoxic alkylpolyamines. As was mentioned above, a more comprehensive review of the use of alkylpolyamine analogs as therapeutic agents is available.[1]

AdoMet-DC Biosynthesis, Mechanism, and Regulation

Unlike ODC, AdoMet-DC belongs to a small class of proteins known as pyruvoyl enzymes. All the known forms of AdoMet-DC contain a covalently bound pyruvate prosthetic group that is required for activity, although the individual AdoMet-DC isozymes differ in their subunit structure and cation requirements. Because of this feature, AdoMet-DC is inhibited by $NaBH_4$[39] and $NaCNBH_3$.[40] The latter inhibits only in the presence of substrate, and during $NaCNBH_3$ reduction, the enzyme forms a covalently bond product, indicating that a Schiff's base is formed during the enzymatic reaction. The *Escherichia coli* form of AdoMet-DC is first synthesized as proenzyme, which is then cleaved post translationally into two smaller polypeptide subunits that are both components of the purified enzyme. The purified enzyme is now thought to be a tetramer consisting of 2α (M_r 19,000) subunits and 2β (M_r 14,000) subunits. The smaller β subunit has a free amino terminus, while the amino terminus of larger α subunit is blocked by the pyruvoyl group. AdoMet-DC from *Escherichia coli* has been shown to require Mg^{2+} for activity.[41] Mammalian AdoMet-DC, which is activated by putrescine rather than Mg^{2+}, is also formed from a proenzyme (M_r 38,000), which is cleaved and processed to the mature decarboxylase. The active form contains two pairs of subunits (M_r 32,000 and 6000), the larger of which contains the pyruvate prosthetic group.[42] The primary sequences and cDNA libraries for AdoMet-DC from variety of sources have been established,[43] but no crystal structures have yet been reported.

A small number of pyruvoyl enzymes have been discovered and purified. Among them, histidine decarboxylase from *lactobacillus 30a* has been studied most thoroughly, and is perhaps prototypical of this group of enzymes.[44] Active histidine decarboxylase is formed from a proenzyme that is composed of a single subunit of M_r 35,000 (designated the π chain), possessing an NH_2-terminal serine and a COOH-terminal tyrosine residue. The proenzyme form does not contain pyruvate. Proenzyme activation involves cleavage of the π chain at the amide bond between serine 81 and 82 to yield a β chain with serine 81 at its COOH terminus, and an α chain with a terminal pyruvate derived from serine 82. Most pyruvoyl enzymes produce their pyruvate prosthetic group by the same general mechanism, and as mentioned above, AdoMet-DC is also formed from a proenzyme form by this mechanism, as shown in Figure 1.2. On the basis of studies involving histidine decarboxylase processing, it was assumed that cleavage of the AdoMet-DC proenzyme would be at a serine–serine linkage. However, a subsequent study established that, in the case of AdoMet-DC from *E. coli*, the cleavage is at the Lys^{111}–Ser^{112} peptide bond, with the resulting polypeptides having molecular weights of 14,000 and 19,000.[45] The mammalian form of AdoMet-DC was first isolated from rat prostate,[42,46,47] and is a pyruvoyl enzyme that requires putrescine rather than a divalent cation for activation.[46] The proenzyme form of rat prostate AdoMet-DC (M_r 38,000) is processed into two subunits of M_r 32,000 and 6000, and the active enzyme contains two copies of each subunit ($\alpha_2\beta_2$).[45] In mammalian AdoMet-DC, the bond between glutamic acid 67 and serine 68 is the site of cleavage. The serine residue at position 68 of the proenzyme is thus converted to the pyruvate prosthetic group.[45]

Mammalian AdoMet-DC has been expressed in *E. coli*, and the specific residues mediating putrescine stimulation of activity and processing have now been identified.[48] During the processing reaction, the pyruvate cofactor is generated on the larger subunit. Pegg and co-workers have purified the human form of AdoMet-DC to homogeneity.[49] In this study, human AdoMet-DC was expressed in high yield in *E. coli* using the pIN-III(*lpp*$^{P-5}$) expression vector, and then purified using MGBG-Sepharose affinity chromatography. The human enzyme, in terms of sequence and subunit structure, is very similar to the rat prostate form, and exhibits only nine conservative changes in 334 amino acids. The proenzyme form of human AdoMet-DC (M_r 38,000) is autocatalytically processed into two subunits of M_r 30,700 (pyruvate-containing) and 7700, and the active enzyme contains two copies of each subunit ($\alpha_2\beta_2$). Again, serine 68 becomes the pyruvate prosthetic group. Site-directed mutagenesis studies have been used to identify a putrescine binding site that mediates both enzyme activation and stimulation of processing.[50] Despite its similarity to the rat isozyme, the human form of AdoMet-DC is quite different from the *E. coli* form, both in subunit structure, and in primary sequence, showing less than 10% sequence homology.[42] In addition, a trypanosomal form of

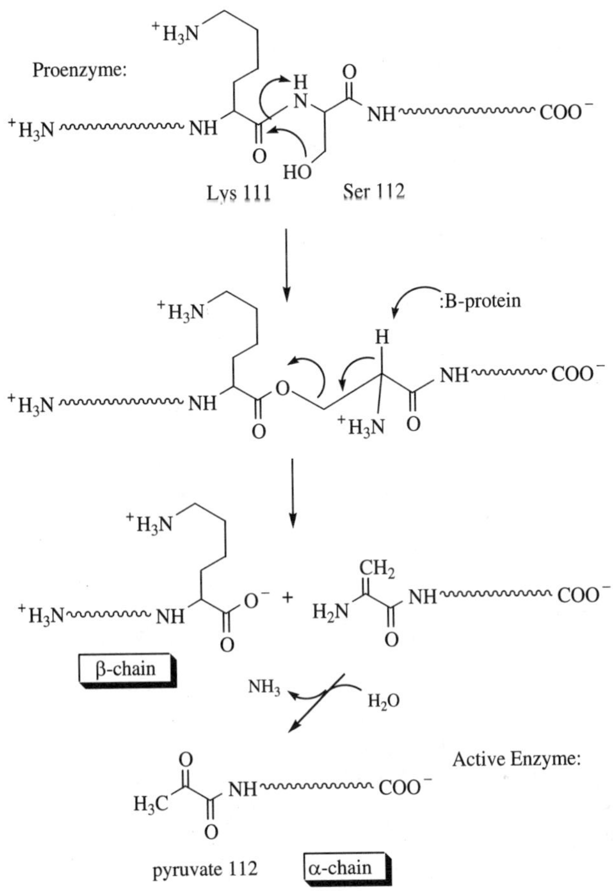

FIGURE 1.2 Origin of pyruvoyl residue of AdoMet-DC from *Escherichia coli.*

AdoMet-DC has been identified, and appears to exhibit significant differences in structure from both the human and bacterial forms of the enzyme.[51]

The differences in structure between the various forms of AdoMet-DC suggest that it may be possible to design isozyme-specific inhibitors for use as chemotherapeutic agents. The mechanism of *S*-adenosylmethionine (AdoMet) decarboxylation by AdoMet-DC has been studied, and the generally accepted mechanism is shown in Figure 1.3.[41,52] The pyruvoyl group of AdoMet-DC is thought to act like a pyridoxal phosphate cofactor by forming a Schiff base with the amino group of the substrate,[44] and then by serving as an electron sink to facilitate decarboxylation. The only clear mechanistic distinction between pyruvoyl-containing enzymes and enzymes that use pyridoxal phosphate as a cofactor is that the carbonyl of the pyruvoyl group does not form a Schiff base with an ε-amino group of a lysyl residue in the resting protein.

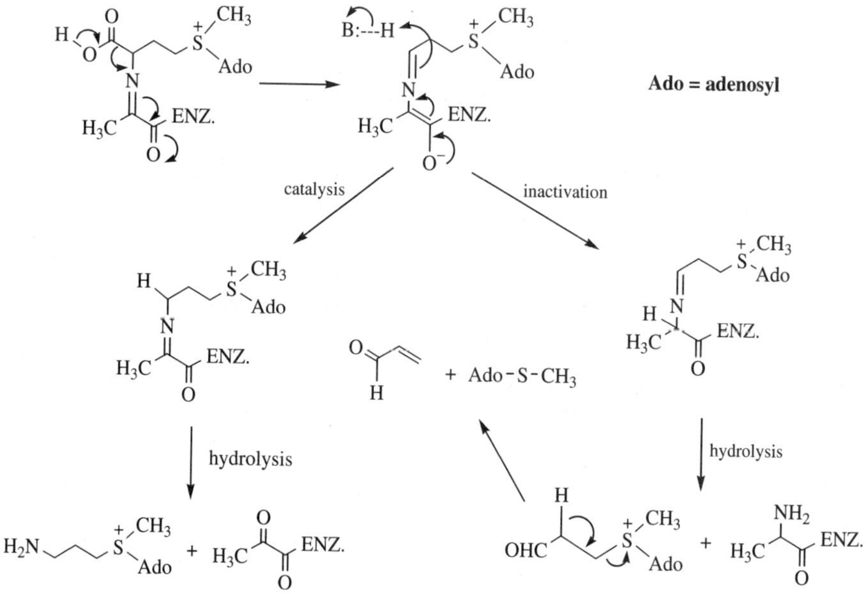

FIGURE 1.3 Mechanism of decarboxylation and substrate inactivation of AdoMet-DC.

A recent study has shown that *E. coli* AdoMet-DC undergoes time-dependent self-inactivation in the presence of substrate and Mg^{2+}. The rate of inactivation is dependent on the concentration of AdoMet, and the process is saturable. Once every 6000–7000 turnovers, the enzyme undergoes an inactivation step that results in the conversion of the pyruvate to an alanine residue *via* transamination between the pyruvate and decarboxylated *S*-adenosylmethionine (dc-Ado-Met).[52] This occurs when the normal decarboxylation step is followed by incorrect protonation of the enolate intermediate (Fig. 1.3). During this inactivation of AdoMet-DC, there is an additional reaction involving β elimination of methylthioadenosine to generate acrolein or an acrolein-like species, which stoichiometrically alkylates the cysteine residue in the second tryptic peptide from the NH_2 terminal of the α subunit of enzyme.[53] The modified residue is derived from Cys140 of the proenzyme and lies in the only sequence conserved between rat liver and *E. coli* AdoMet-DC. These observations suggest that the Cys residue modified during substrate inactivation is in the catalytic site, and could thus play a role in the enzyme-assisted decarboxylation of AdoMet.

Inhibitors of AdoMet-DC

Like ODC, AdoMet-DC is an inducible enzyme, and responds dramatically to polyamine depletion, or to elevation of spermidine or spermine. A number of reversible and irreversible inhibitors have been developed for the enzyme, as outlined below. The key to designing specific and effective inhibitors for any of the known forms of AdoMet-DC lies in exploiting the catalytic mechanism for the decarboxylation of AdoMet, which is shown in Figure 1.3. An enzyme-substrate complex results when the substrate, AdoMet, forms an imine linkage with the terminal pyruvate of the active-site-containing subunit. This mechanistic "handle" makes AdoMet-DC an excellent candidate for inhibition by rationally designed irreversible inhibitors.

Nonnucleoside Inhibitors of AdoMet-DC The finding that known antiparasitic agents such as pentamidine and berenil act as inhibitors of mammalian and parasitic AdoMet-DC[54] has spurred a renewed interest in the development of nonnucleoside inhibitors of the enzyme. The antileukemic agent methylglyoxal bis(guanylhydrazone) (MGBG) is a potent competitive inhibitor of the putrescine-activated mammalian enzyme, with $K_i < 1\ \mu M$.[55] However, MGBG is of limited use as a chemotherapeutic agent because of a wide variety of other effects on cells (induction of severe mitochondrial damage, interference with polyamine transport, induction of SSAT, etc.).[9] Thus, MGBG has never proved useful clinically, but has been used extensively as a research tool. A number of early attempts to modify the structure of MGBG resulted in active antitumor agents, but all of these MGBG analogs were as toxic or nonspecific as MGBG itself.[56–59] Interestingly, one of the newest developments in the area of AdoMet-DC inhibitors has been the discovery of a promising new series of MGBG analogs by Stanek et al.[60] All the analogs showing significant activity are conformationally restricted derivatives of MGBG in which the backbone assumes a partially fixed, all-*trans* conformation. Two of the most promising of these analogs are CGP 39937 and CGP 33829, shown in Figure 1.4, which inhibit AdoMet-DC with IC_{50} values of 6 and 36 nM, respectively, and show good antitumor activity in vitro.[61] Along these same lines, the conformationally restricted analog 4-amidinoindan-1-one 2′-amidinohydrazone shown in Figure 1.1 was synthesized and tested as an inhibitor of rat liver AdoMet-DC.[62]

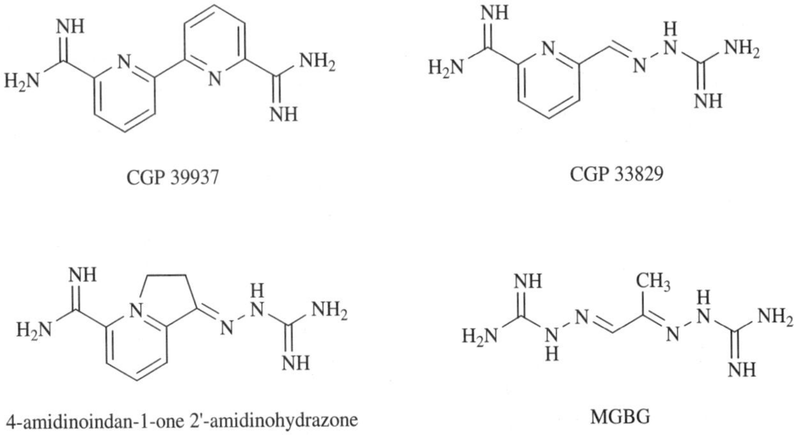

FIGURE 1.4 Analogs of MGBG that reversibly inhibit AdoMet-DC.

This analog is a remarkably potent, selective inhibitor of AdoMet-DC ($IC_{50} = 5$ nM), and also inhibits the growth of cultured T24 human bladder cancer cells with an IC_{50} of 0.71 μM. This compound and its congeners are currently being evaluated as antitumor agents. In addition, there is preliminary evidence that they may act as effective antiprotozoal agents, and may also be effective against *Pneumocystis carinii*.

Nucleoside-Based Inhibitors of AdoMet-DC Early attempts to develop nucleoside analogs as inhibitors of AdoMet-DC were concentrated on the synthesis of substrate and product analogs.[63–65] Although some of these analogs showed competitive, and in some cases noncompetitive inhibitory activity, none of them have proved to be useful clinically, and therefore none have been developed further. Interestingly, the α-difluoromethyl analog of AdoMet (analogous to DFMO) has been synthesized,[65] but has little useful activity against AdoMet-DC. The benefit of the early studies involving the synthesis of substrate and product analogs lies in the elucidation of critical structure–activity relationships. These studies indicated that both AdoMet and dc-AdoMet analogs could inhibit the enzyme, and that the methylsufonium center of AdoMet could be replaced by a charged nitrogen species. These data have proved critical for the development of second-generation AdoMet-DC inhibitors.

A number of structural analogs of AdoMet have been developed as potential inhibitors of AdoMet-DC, in which a nucleophilic amine surrogate has been appended to the molecule. S-(5′-Deoxy-5′-adenosyl)methylthioethylhydroxylamine (AMA; Fig. 1.5) has been shown to be an irreversible inhibitor of AdoMet-DC in L1210 cells with an IC_{50} concentration of 100 μm.[66–68] This compound acts presumably by forming a stable oxime linkage, rather than the usual imine, at the terminal pyruvate of AdoMet-DC, thus inactivating the enzyme. AMA has been reported to inhibit AdoMet-DC from rat liver at concentrations as low as 3 nM; however, this compound has not been studied adequately in purifed enzyme preparations, and no kinetic inhibitor constants are available. Secrist has described the synthesis of a series of nucleophilic AdoMet analogs which act as potent inhibitors of AdoMet-DC.[69] The two most promising inhibitors in this series were 5′-deoxy-5′-[(3-hydrazinopropyl)methylamino]adenosine (MHZPA; Fig. 1.5) and

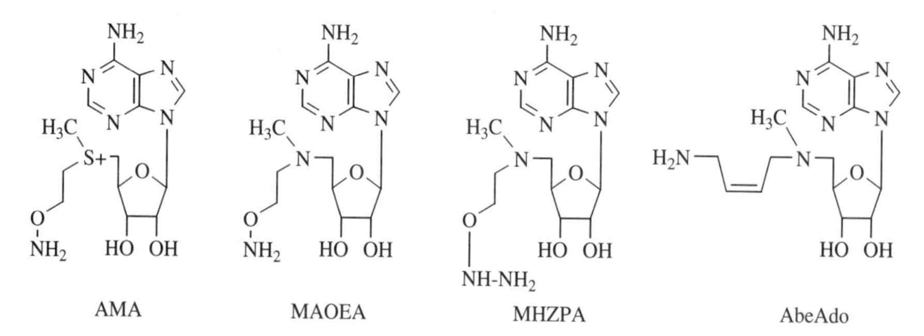

FIGURE 1.5 Structures of nucleoside-based, irreversible inactivators of AdoMet-DC.

5′-deoxy-5′-[(3-hydrazinopropyl)methylamino]adenosine (MAOEA; Fig 1.5). These derivatives are potent, irreversible inhibitors of the enzyme, with IC_{50} values of 7.5 and 40 nM against AdoMet-DC isolated from MRC5 cells, and I_{50} values of 400 and 70 nM against rat prostate AdoMet-DC,[17] respectively. The analogs act as effective growth inhibitors in L1210 cells, causing an elevation of putrescine and depletion of spermidine; however, the analogs appear to be rapidly metabolized in these cells, a fact that may limit their usefulness in vivo.[17] In a recent study, the adduct between MHZPA and the terminal pyruvate of human AdoMet-DC was characterized by isolating it from a LYS-C digest of the enzyme–inhibitor complex, confirming that MHZPA forms a stable hydrazone with the enzyme as predicted.[49]

Danzin has described the synthesis and enzymatic evaluation of 5′-{[(Z)-4-amino-2-butenyl]methylamino}-5′-deoxyadenosine (AbeAdo; Fig 1.5), a potent, irreversible inhibitor of AdoMet-DC.[18] This analog, which contains a nitrogen in place of the natural methylsulfonium center, is the first inactivator of AdoMet-DC that can be considered "enzyme-activated," since it was designed to rearrange to a latent electrophile in the catalytic site of AdoMet-DC, as shown in Figure 1.6. In theory, formation of an imine linkage between the primary amine of AbeAdo and the pyruvate cofactor of AdoMet-DC would activate the proton α to the imine. Deprotonation by general base catalysis would then lead to the formation of a latent electrophilic Michael acceptor following elimination of methylaminoadenosine

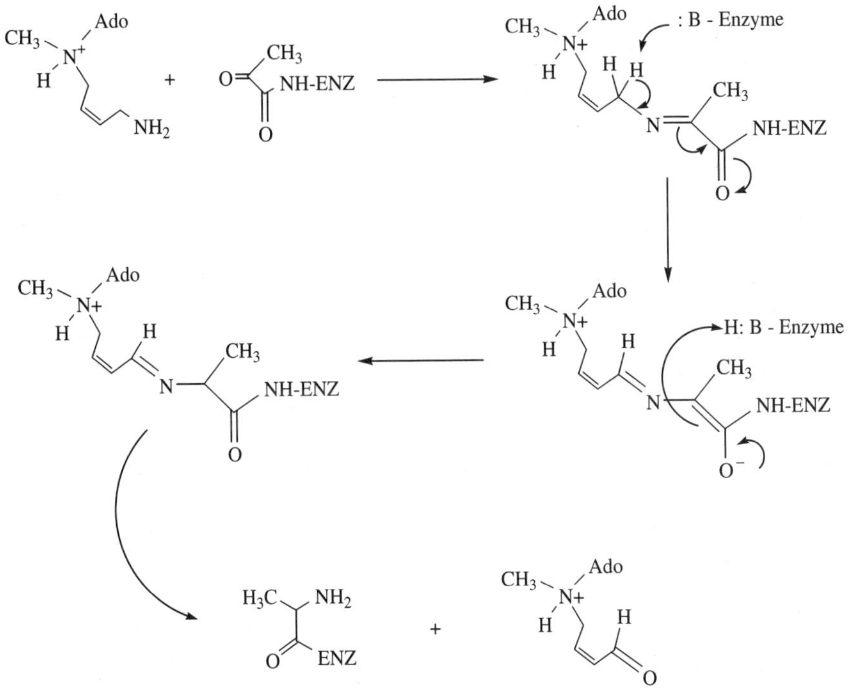

FIGURE 1.6 The mechanism of mammalian AdoMetDC inhibition by AbeAdo.

(MAA). Attack at this electrophilic center by a nucleophilic amino acid residue could then be envisioned, resulting in irreversible inactivation of the enzyme. As predicted, AbeAdo produces a rapid time-dependent loss of enzymatic activity, with a K_i of 0.3 μM and a K_{inact} of 3.6 min^{-1} against the *E. coli* form of the enzyme. The analog is also a potent inactivator of the rat liver form of AdoMet-DC ($K_i = 0.56$ μM, turnover number 1.5), and produces a long lasting, dose-dependent decrease in AdoMet-DC activity in vivo.[70] The *Z* isomer of AbeAdo is 100 times more potent against rat liver AdoMet-DC,[70] and 1000 times more potent against *E. coli* AdoMet-DC,[18] than is the corresponding *E* isomer. Interestingly, there is now evidence that AbeAdo inactivates *E. coli* and human AdoMet-DC by different mechanisms. Inactivation of bacterial AdoMet-DC by AbeAdo is accompanied by a quasi-stoichiometric production of the by-product, MAA.[18] However, inactivation of human AdoMet-DC by AbeAdo results in transamination of the terminal pyruvate moiety to an alanine. The transamination most likely proceeds by incorrect protonation of an enolate resulting from formation of the AbeAdo/AdoMet-DC adduct. Hydrolysis of this intermediate then produces the resulting alanine residue, and an aldehyde-containing aminoadenosine analog. The data available to date concerning the inhibition of AdoMet-DC by AbeAdo suggest that there are fundamental differences between the active sites of bacterial and human AdoMet-DC, and support the contention that it may be possible to develop isozyme-specific inhibitors of the various forms of the enzyme.

Trypanosomal Polyamine Metabolism

Polyamine biosynthesis in African trypanosomes[71] is somewhat similar to that occurring in mammalian cells, as shown in Figure 1.7. These parasites are known to have their own forms of ODC, AdoMet-DC, and spermine synthase. Trypanosomal ODC has a much longer half-life ($>$ 3 h) than the mammalian forms of the enzyme,[72] presumably because it lacks the PEST sequence that is present in a variety of rapidly degraded mammalian enzymes, including ODC. DFMO is an effective inhibitor of trypanosomal ODC,[73] and has found some use in treatment of trypanosomiasis. Little is known about trypanosomal spermidine synthase, which converts putrescine to spermidine.[71] Following the aminopropyl step, MTA is rapidly hydrolyzed by the enzyme MTA-phosphorylase, and the methionine portion is used to re-form AdoMet (see text below). Trypanosomes do not synthesize spermine, but incorporate spermidine into a glutathione-containing molecule known as *trypanothione*.[74] Trypanothione combines with reactive oxygen species (ROS) such as hydroxyl radical or superoxide, producing water and the oxidized form of trypanothione, and protecting the organism against oxidative stress. Trypanothione must then be reverted to the reduced state by the NADPH-dependent enzyme trypanothione reductase.[74]

In addition to the transformations mentioned above, trypanosomal polyamine metabolism depends on two additional enzymes: *S*-adenosylmethionine is formed from methionine and ATP by a trypanosomal form of *S*-adenosylmethionine synthase,[75] and the resulting AdoMet becomes a substrate for a variety of methyl-

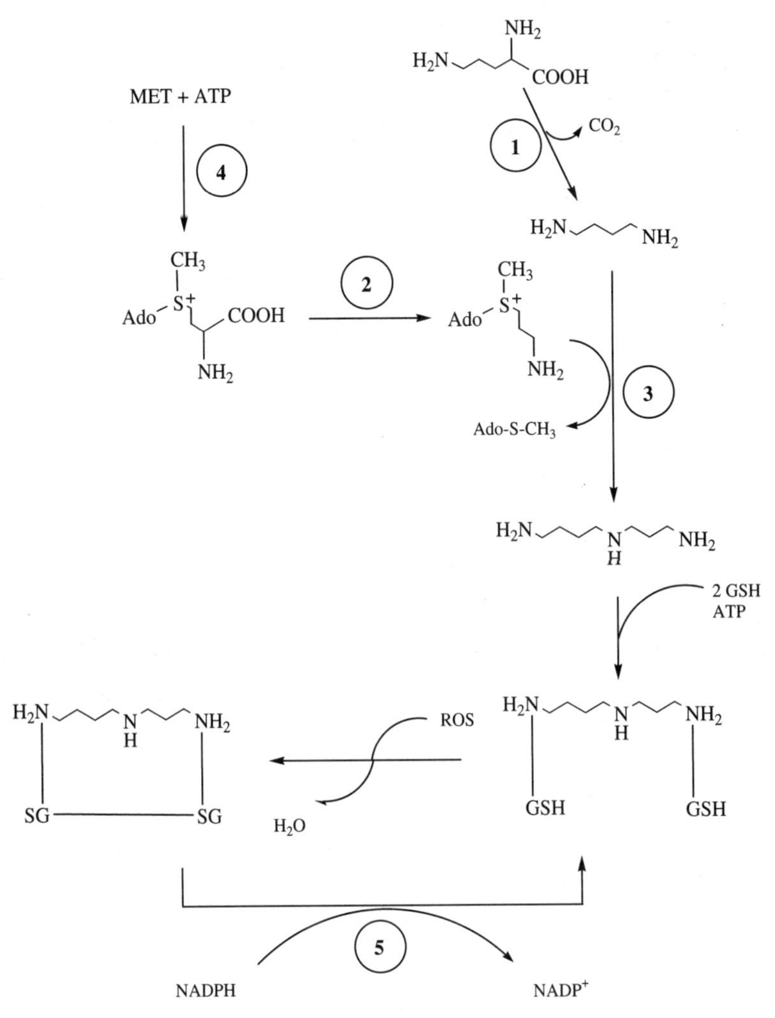

FIGURE 1.7 The polyamine biosynthetic pathway in African trypanosomes (1 = ornithine decarboxylase; 2 = *S*-adenosylmethionine decarboxylase; 3 = spermidine synthase; 4 = *S*-adenosylmethionine synthase; 5 = trypanothione reductase; Ado = adenosyl; ROS = reactive oxygen species).

ation reactions. A portion of AdoMet also becomes a substrate for trypanosomal AdoMet-DC,[76] producing dc-AdoMet for use in the synthesis of spermidine and trypanothione. Trypanosomal AdoMet-DC has been isolated as a single 34-kD subunit that is stimulated by putrescine and is extremely unstable in solution. Like mammalian AdoMet-DC, the trypanosomal enzyme contains a terminal pyruvoyl cofactor, and is inhibited by AbeAdo,[77] by other irreversible inhibitors of AdoMet-DC,[78,79] and by MGBG and its analogs.[80] Interestingly, trypanosomes do not possess a de novo synthetic pathway for purines, and as such rely on the import of exogenous purines from the host. There appear to be two trypanosomal purine transport systems, designated P1 and P2.[81] The P1 transporter preferentially transports inosine, while the P2 transporter imports adenosine, and is also the site of entry for melaminophenyl arsenical antitrypanosomal agents such as melarsoprol. The P2 transporter is not present in trypanosomal strains that are resistant to arsenicals, suggesting that the observed resistance is a transport phenomenon. Recent studies have shown that AbeAdo[82] and the trypanocidal adenosine analog HETA[83] enter trypanosomes using this purine transport system.

African trypanosomes are endemic to a number of third-world countries in Africa and South America, and are responsible for the disease commonly known as "sleeping sickness." One potential strategy for producing effective antitrypanosomal agents is to synthesize agents that specifically inhibit trypanosomal AdoMet-DC. Trypanosomal metabolism affords two distinct advantages in the design of such compounds: (1) the trypanosomal P1 and P2 transport systems can be utilized to import synthetic *S*-adenosylmethionine analogs, and (2) the terminal pyruvate of trypanosomal AdoMet-DC provides a site for attack by rationally designed mechanism-based inhibitors. Analogs that take advantage of these two characteristics have the potential to become parasite-specific chemotherapeutic agents.

AdoMac: An Enzyme-Activated, Irreversible Inhibitor of AdoMet-DC

As part of an ongoing program to develop restricted rotation analog inhibitors of the polyamine biosynthetic pathway, we designed the putative irreversible inhibitor *S*-(5′-deoxy-5′-adenosyl)-1-amino-4-thio-2-cyclopentene (AdoMac; Fig. 1.8),[84] a conformationally restricted analog of dc-AdoMet. We proposed that AdoMac

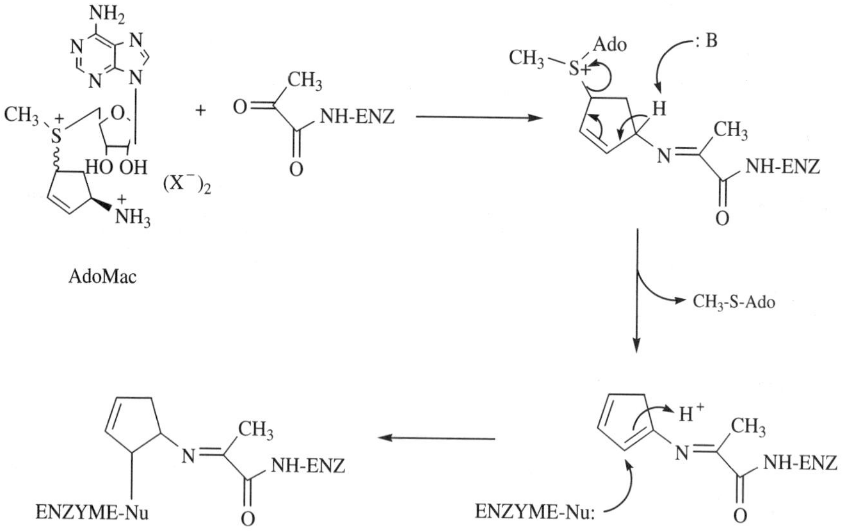

FIGURE 1.8 Structure of the AdoMet-DC inhibitor AdoMac and proposed mechanism of inactivation.

would bind to the enzyme-linked pyruvoyl residue of AdoMet-DC in the same manner as the natural substrate, and would then form a highly reactive Michael acceptor at the active site by general base catalysis, as shown in Figure 1.8. Attack by a nucleophile located on the surface of the enzyme would then lead to an irreversible binding of the inhibitor to the active site, resulting in a nonfunctional enzyme. In order to determine the feasibility of using AdoMac as an irreversible inhibitor of AdoMet-DC, we developed a synthetic route that afforded the target molecule in good overall yield, and that was adaptable to the ultimate production of each of the four possible diastereoisomers of AdoMac in pure form. The synthetic route leading to AdoMac is outlined in Scheme 1.1. (+)-1*R*,4*S*-*cis*-1-Acetoxy-4-hydroxy-2-cyclopentene was synthesized from the corresponding *meso*-diacetate[85,86] by porcine-lipase-mediated cleavage of the pro-*S* acetate.[87] The desired enantiomer was determined to be present in 97% enantiomeric excess (e.e.) by optical rotation, and by NMR analysis of the corresponding Mosher ester.[88] Substitution of phthalimide at the free hydroxyl under Mitsunobu conditions[89] produced the *N*-phthalimidoaminoester with 100% inversion of configuration. This compound was then subjected to hydrazinolysis to afford the corresponding aminoester. The amine was Boc protected (di-*tert*-butyldicarbonate[90]), followed by hydrolysis of the acetate (LiOH) to afford the free alcohol. The desired enantiomer of this alcohol was found to be present in 97% e.e. by NMR analysis of the Mosher ester,[88] confirming that no racemization had occurred during the synthesis. The *N*-Boc-protected aminoalcohol was then converted to the corresponding chloride (MsCl, LiCl[91]). During the conversion of this allylic alcohol to the corresponding chloride, the stereochemical control was lost at position 1, but retained at position 4. Thus the cyclopentenyl synthon was isolated as a 64 : 36 mixture of the *cis* and *trans* diastereoisomers, respectively, as determined by NMR analysis.

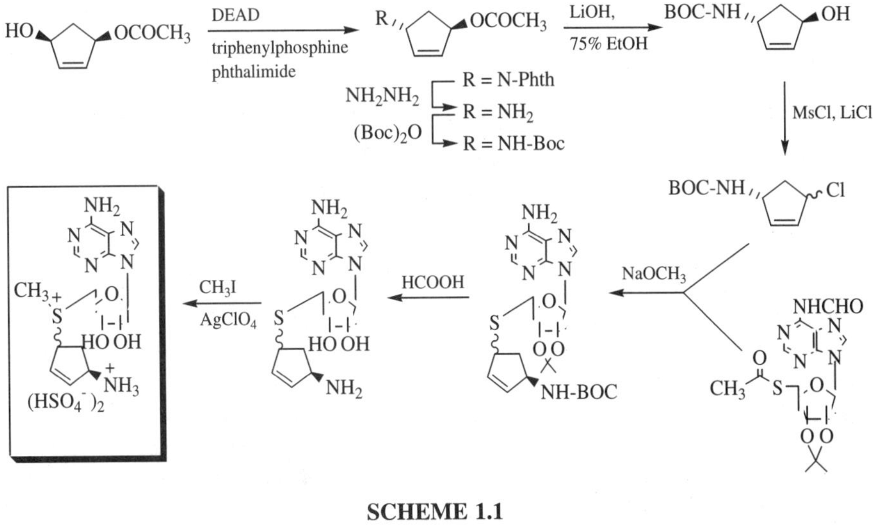

SCHEME 1.1

Coupling of the chlorinated cyclopentenyl ring to 5′-thioacetyl-2′,3′-isopropylidene- adenosine[92,93] was accomplished in a 50 : 50 mixture of methanol and DMF in the presence of sodium methoxide,[94] resulting in the formation of the fully protected thionucleoside. Rigorous exclusion of oxygen was ensured during this procedure by freezing and thawing the mixture 5 times under a vigorous stream of argon, since even a trace of oxygen catalyzed rapid dimerization of the thiolate. Simultaneous removal of both the *N*-Boc and 2′,3′-isopropylidene protecting groups (88% formic acid) yielded the thioether, which was methylated using a modification of the method of Samejima[95] (CH_3I, $AgClO_4$, AcOH/HCOOH) to provide the desired target compound AdoMac as a mixture of the 1*R*,4*S* and 1*R*,4*R* diastereomers.

A series of experiments were conducted to determine whether AdoMac, as a mixture of the *cis*-1*R*,4*S* and *trans*-1*R*,4*R* diastereomers, produced time-dependent inactivation of AdoMet-DC. The time-dependent decrease in enzyme activity was monitored at 7, 10, 20, and 30 μM concentrations of AdoMac over a period of 20 min. In each case, the decrease in activity was linear (Fig. 1.9), and a rate constant was derived from each line. The resulting data was then replotted using the Kitz–Wilson method[96] (Fig. 1.9). This plot indicated that AdoMac has a K_I value of 18.3 μM for AdoMet-DC, and exhibits a k_{inact} value of 0.133 min^{-1}. Two additional studies suggested that AdoMac is an active-site-directed, irreversible inhibitor of AdoMet-DC. When the enzyme was incubated with 100 μM MGBG for 10 min, then exposed to AdoMac for 10 min and assayed, 84.6% of the AdoMet-DC activity was retained. Under the same reaction conditions (except for the exclusion of MGBG), only 20% of the activity remained after 10 min, suggesting that AdoMac is an active-site-directed agent. The irreversibility of AdoMac was demonstrated by the inability to dialyze away the inhibitor following binding to AdoMet-DC. Finally, HPLC analysis of the product mixture after quenching the enzymatic reaction at various revealed the time-dependent appearance of a peak that coeluted with MTA, and the corresponding disappearance of the peak coeluting with AdoMac, suggesting that MTA is generated from AdoMac in the enzymatic reaction.

Synthesis and Evaluation of the Diastereomeric Forms of AdoMac

As was mentioned above, AdoMac was initially isolated as a mixture of the 1*R*,4*S* and 1*R*,4*R* diastereomers. Ironically, synthesis of the individual diastereomers in pure form was made possible by the loss of stereochemical control in the chlorination step mentioned above.[97] Coupling of the sidechain synthon to the nucleoside portion of the molecule, followed by removal of the protecting groups, afforded the unmethylated precursor to AdoMac as a mixture of isomers, as shown in Scheme 1.2. At this stage, the mixture of diastereomers was resolved by careful flash chromatography on silica gel (2 × 40-cm column length, $CHCl_3$: methanol : NH_4OH 14 : 14 : 1) to afford the 1*R*,4*R* and 1*R*,4*S* diastereomers of the unmethylated precursor in pure form. Each pure diastereomer was then individually methylated[95] to provide the desired 1*R*,4*R* or 1*R*,4*S*-AdoMac, isolated as their sulfate salts following recrystallization.

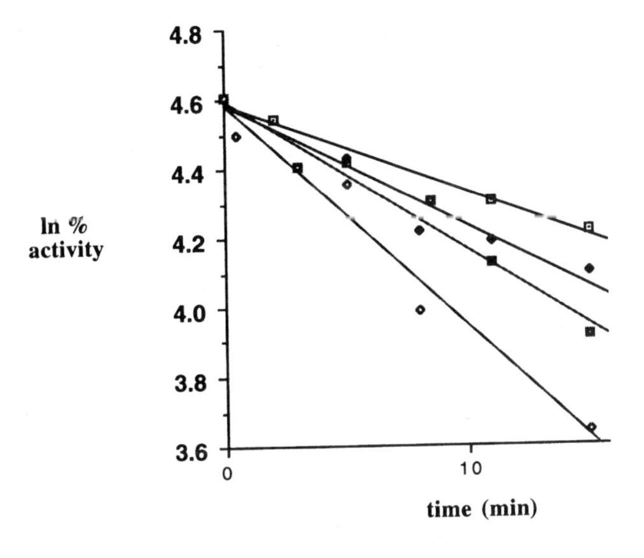

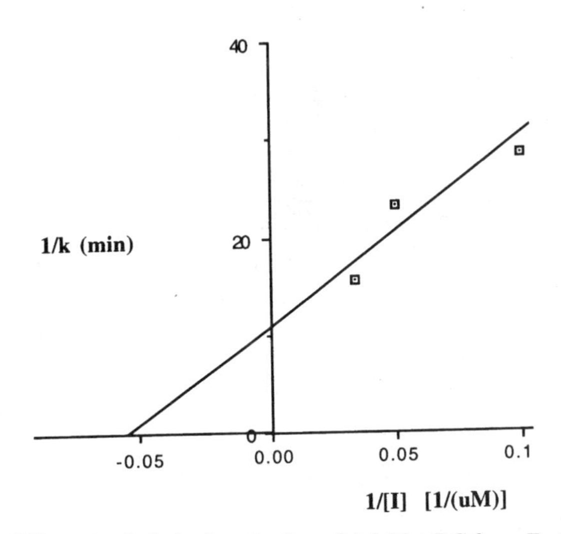

FIGURE 1.9 Kitz–Wilson Analysis for inactivation of AdoMet-DC from *Escherichia coli* by AdoMac. Top—time-dependent decay of enzyme activity at various concentrations; bottom—Kitz–Wilson replot. Each data point is the average of three determinations, which in each case differed by 5% or less.

The 1*S*,4*R* and 1*S*,4*S* diastereomers of AdoMac were accessed in similar fashion using the synthetic pathway shown in Scheme 1.3. Stereospecific hydrolysis of the pro-(*R*)-acetate of the mesodiacetate using pig liver esterase[98] afforded the corresponding 1*S*,4*R* hydroxyester (94% e.e.), which was elaborated and deprotected as described above to afford the unmethylated nucleoside precursor as a mixture of the 1*S*,4*R* and 1*S*,4*S* diastereomers. The mixture was then resolved by

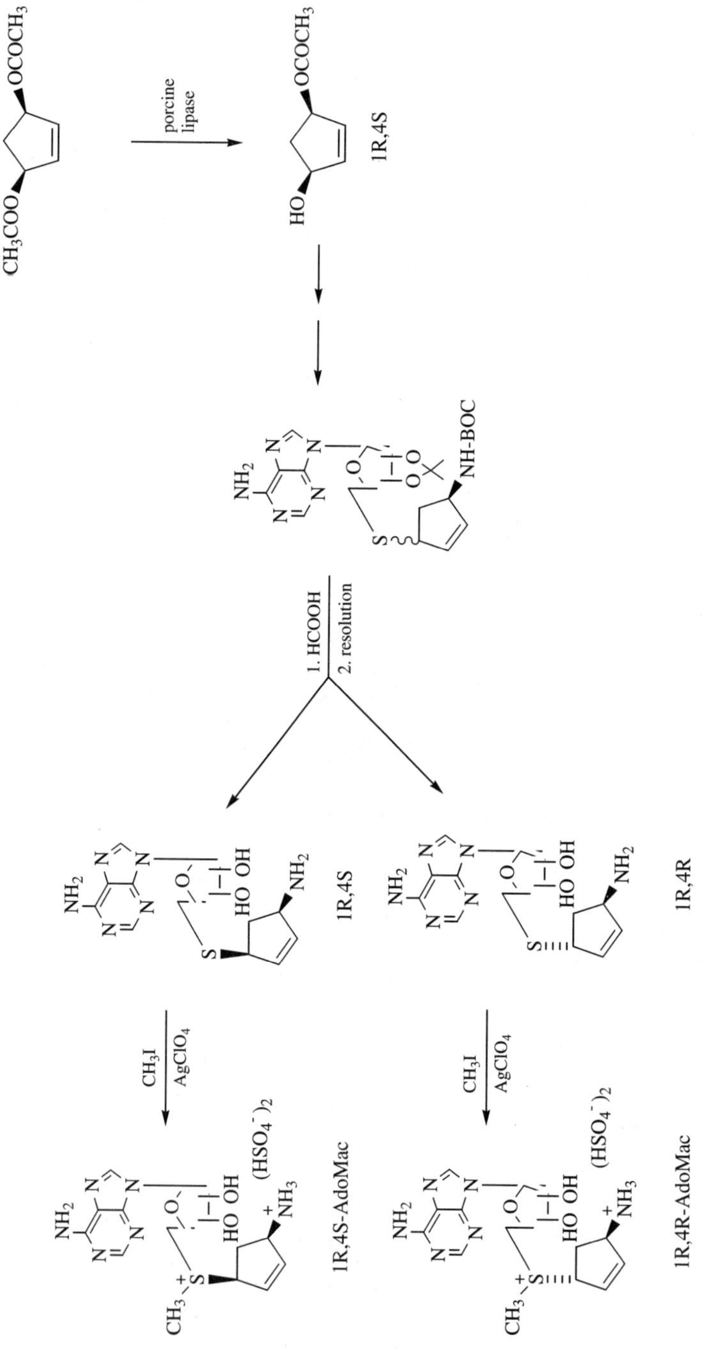

porcine lipase

1R,4S

1. HCOOH
2. resolution

1R,4S

1R,4R

CH_3I
$AgClO_4$

$(HSO_4^-)_2$

1R,4S-AdoMac

1R,4R-AdoMac

SCHEME 1.2

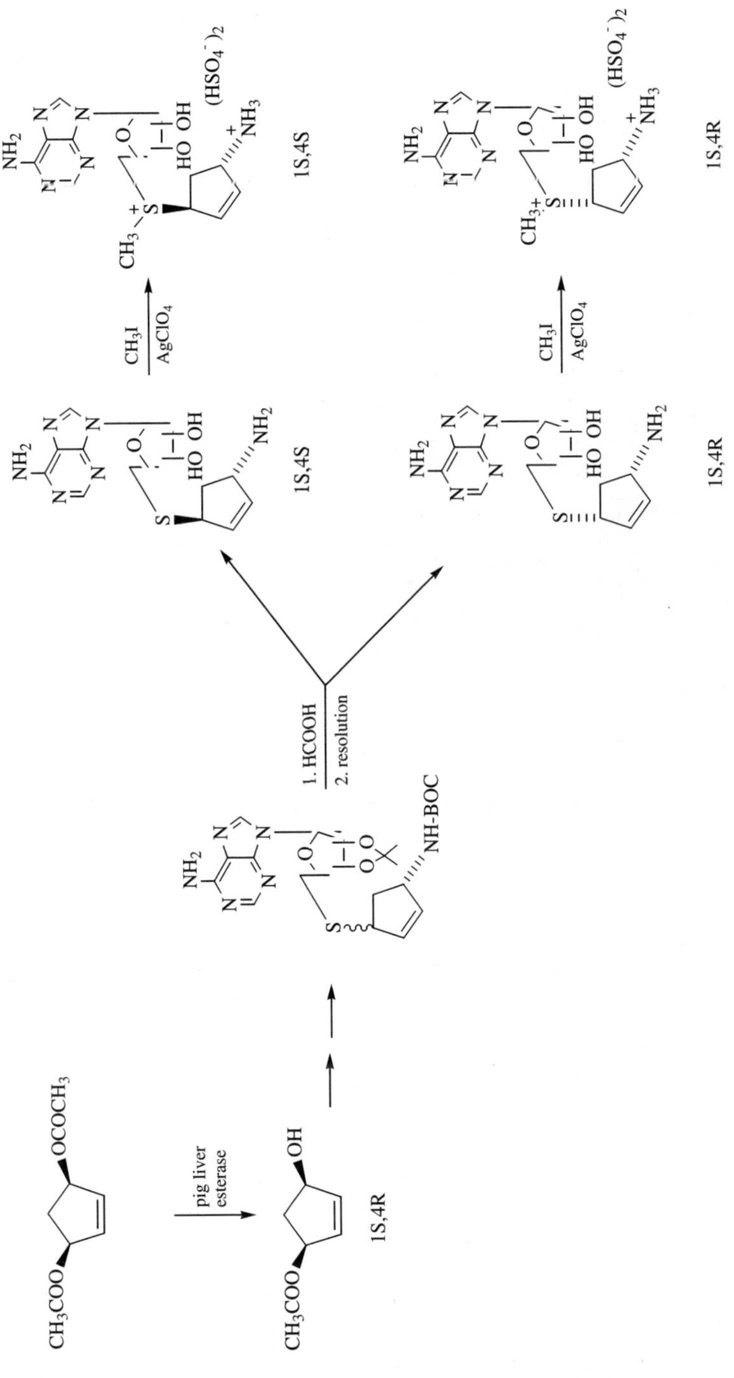

CH_3COO $OCOCH_3$

pig liver esterase

CH_3COO OH

1S,4R

NH_2

NH-BOC

1. HCOOH

2. resolution

NH_2

HO OH

NH_2

1S,4S

CH_3I

$AgClO_4$

NH_2

CH_3 S^+

HO OH

$(HSO_4^-)_2$

$\overset{+}{N}H_3$

1S,4S

NH_2

HO OH

NH_2

1S,4R

CH_3I

$AgClO_4$

NH_2

CH_3+ S

HO OH

$(HSO_4^-)_2$

$\overset{+}{N}H_3$

1S,4R

SCHEME 1.3

careful flash chromatography as described above to yield the pure 1*S*,4*R* and 1*S*,4*S* diastereomers, which were individually methylated to provide the desired 1*S*,4*R* and 1*S*,4*S*-AdoMac in diastereomerically pure form, isolated as the sulfate salts following recrystallization. Thus, all four of the possible diastereomers of AdoMac could be accessed in pure form using a common synthetic scheme.

To assess the configurational dependence of the binding of AdoMac to AdoMet-DC and the subsequent inactivation process, the pure diastereomers were individually evaluated as inhibitors of the *E. coli* form of AdoMet-DC using the method of Kitz and Wilson.[96] K_i and k_{inact} values were derived for each of the four diastereomers of AdoMac, as summarized in Table 1.1 and Figure 1.10. The K_i values observed, which range between 3.83 and 39.60 μM, demonstrate that the

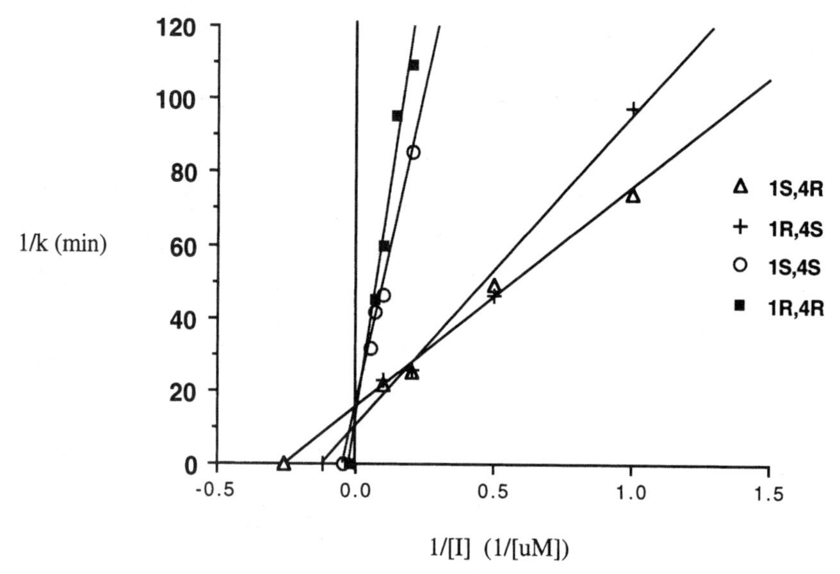

FIGURE 1.10 Determination of K_i and k_{inact} values for the individual pure diastereomers of AdoMac. Rate constants derived from the time inactivation studies were replotted according to the method of Kitz and Wilson, and kinetic constants were derived from the equation for the resulting straight line.

TABLE 1.1 Results of Inhibition Studies for Restricted Rotation AdoMet Analogs

Compound	K_i μM	k_{inact} min^{-1}	MTA Generation	Time-Dependence
cis-1*R*,4*S* + *trans*-1*R*,4*R*-AdoMac	18.3	0.133	Yes	Yes
cis-1*R*,4*S*-AdoMac	8.46	0.099	Yes	Yes
trans-1*R*,4*R*-AdoMac	39.60	0.079	Yes	Yes
cis-1*S*,4*R*-AdoMac	3.83	0.064	Yes	Yes
trans-1*S*,4*S*-AdoMac	23.81	0.068	Yes	Yes
cis-1*R*,4*S*-nor-Adomac	293.0	—	No	No
cis-1*S*,4*R*-dihydro-Adomac	93.0	—	No	No

binding of AdoMac to AdoMet-DC is significantly configuration-dependent. Interestingly, the k_{inact} values for the four diastereomers are remarkably constant, varying between 0.064 and 0.099 min^{-1} only.

As was the case with the diastereomeric mixture of AdoMac, AdoMet-DC was protected from inactivation by the pure diastereomers of AdoMac when the enzyme had been preincubated with the known competitive inhibitor MGBG, and when Mg^{2+} was omitted from the media. The inhibition produced by each diastereomer was irreversible, as demonstrated by the inability to dialyze away the inhibitor following binding to AdoMet-DC. HPLC product analysis of the enzymatic reaction mixture for each diastereomer of AdoMac revealed the time-dependent appearance of a peak that coeluted with MTA (retention time 17.08 min), and the corresponding disappearance of the peak coeluting with AdoMac. No other metabolites related to AdoMet could be detected under the assay conditions. To rule out the possibility of nonenzymatic generation of MTA from AdoMac under the assay conditions described, control experiments were carried out in the absence of AdoMet-DC, and in the presence of AdoMet-DC that had been inactivated by boiling at 100°C for 5 min. Under these conditions, no generation of MTA or any other metabolite of AdoMac was observed during the 20-min assay period.

In theory, the nonmethylated precursor to AdoMac, norAdoMac, and the saturated version of AdoMac, dihydroAdoMac (Fig. 1.11), should also form an imine linkage with the terminal pyruvate of AdoMet-DC. However, neither of these analogs possesses the driving force for elimination of MTA and formation of a latent electrophile within the catalytic site. Thus, these analogs should act as competitive inhibitors of the enzyme, and would thereby serve as suitable control compounds with respect to AdoMac. To test this hypothesis, compounds *cis*-1*R*,4*S*-nor-AdoMac and *cis*-1*S*,4*R*-dihydroAdoMac, each possessing the same absolute stereochemistry as the most potent inhibitor *cis*-1*R*,4*S*-AdoMac, were synthesized and evaluated for inhibitory activity against AdoMet-DC, as described above. As expected, compounds *cis*-1*R*,4*S*-norAdoMac and *cis*-1*S*,4*R*-dihydroAdoMac acted as weak

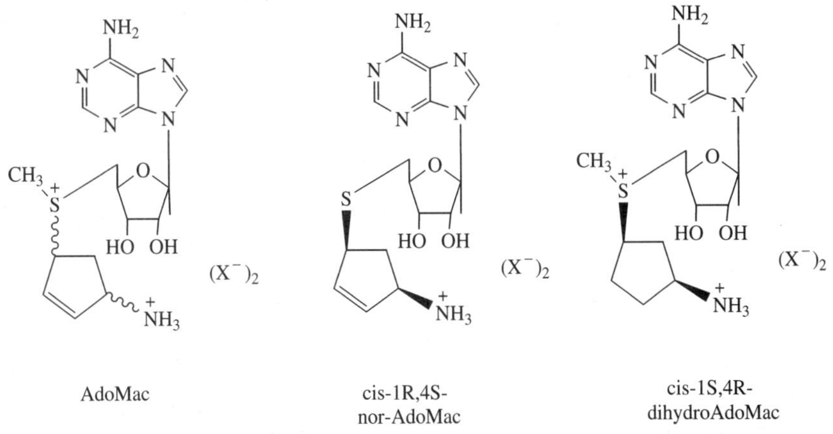

AdoMac

cis-1R,4S-nor-AdoMac

cis-1S,4R-dihydroAdoMac

FIGURE 1.11 Comparison of the structures of AdoMac, norAdoMac, and dihydroAdoMac.

competitive inhibitors of AdoMet-DC, with K_i values of 293.0 and 93.0 μM, respectively (Table 1.1). The inhibition observed in the presence of these analogs was not time-dependent, and no generation of MTA was detected in the enzymatic reaction mixture.

Previous studies have shown that there is little sequence homology (about 10%), between the *E. coli* and human forms of AdoMet-DC,[42,49,50,52] and that they possess different subunit structure. These facts suggest that, apart from the requisite terminal pyruvate moiety and a few other critical residues, there may be distinct differences in the amino acid composition of the catalytic sites within the various forms of the enzyme. In addition, the allosteric control mechanisms are different for these two forms of the enzyme, since the bacterial enzyme is strongly activated by Mg^{2+}, while the human form is activated by putrescine. To determine whether topological differences between the bacterial and human catalytic sites exist, the pure diastereomeric forms of AdoMac were also evaluated as inhibitors of human AdoMet-DC.[99] As was observed for the inactivation of *E. coli* AdoMetDC, each pure diastereomer of AdoMac inactivated the human enzyme in a time- and concentration-dependent manner. In each case, a pseudo-first-order rate constant of inactivation, k_{obs}, was obtained graphically, and the resulting data were replotted using the Kitz–Wilson method.[96] The kinetic parameters thus obtained are summarized in Table 1.2, along with those previously determined for the isolated *E. coli* enzyme. The K_i values for the four diastereomeric forms of AdoMac ranged between 11 and 63 μM, while the corresponding k_{inact} values did not vary significantly, and were similar to those observed for the inactivation of the bacterial enzyme. As was the case for the bacterial enzyme, norAdoMac and dihydroAdoMac acted as weak competitive inhibitors. The value of the partition ratio (i.e., the ratio of k_{cat}/k_{inact}) for *cis*-1*R*,4*S*-AdoMac, the most potent diastereomeric inactivator of human AdoMet-DC, was evaluated indirectly by the titration method. Increasing amounts of *cis*-1*R*,4*S*-AdoMac were added to a known amount of enzyme, and the reaction was allowed to go to completion (24-h incubation time). The mixture was then assayed to determining the percent activity remaining. Figure 1.12 plots the ratio of moles of inactivator per mole of enzyme ([I]/[E]) against the percent enzymatic activity remaining, in which the [I]/[E] ratios were varied from 1 to 30. Product inhibition arising from the presence of MTA could be observed, as indicated by the deviation from linearity at higher concentrations of the inhibitor. However, at ratios of [I]/[E] of 8.0 or less, a linear relationship was observed. Extrapolation of this region by linear regression revealed a turnover number of 9.84 ± 1.0, and thus the partition ratio was determined to be 8.84. This observation then allowed for the calculation of the k_{cat} value for *cis*-1*R*,4*S*-AdoMac, which was found to be $0.645\ \text{min}^{-1}$.

As shown in Table 1.2, both the human and *E. coli* forms of AdoMet-DC are able to discriminate between the four diastereomers of AdoMac, and in each case the *cis* diastereomers (with respect to the cyclopentene ring) are significantly more potent than the *trans*. However, human AdoMet-DC prefers the *cis*-1*R*,4*S* diastereomer, while the bacterial enzyme preferentially binds to *cis*-1*S*,4*R*-AdoMac. When subjected to computer-assisted molecular mechanics analysis, these two molecules possess significantly different least-energy conformations, as shown in Figure 1.13.

TABLE 1.2 Comparison of the Kinetic Parameters for the Inactivation of Human and *Escherichia coli* AdoMet-DC by the Pure Diastereomeric Forms of AdoMac, *cis*-1*S*,4*R*-*H*$_2$-AdoMac, and *cis*-1*R*,4*S*-nor-AdoMac

Inhibitor	K_i, μM (Human)	k_{inact}, min^{-1} (Human)	K_i, μM (*E. coli*)	k_{inact}, min^{-1} (*E. coli*)	Time-Dependence	MTA Generation
cis-1*R*,4*S*-AdoMac	11	0.073	8	0.099	Yes	Yes
cis-1*S*,4*R*-AdoMac	17	0.068	4	0.064	Yes	Yes
trans-1*S*,4*S*-AdoMac	53	0.088	24	0.068	Yes	Yes
trans-1*R*,4*R*-AdoMac	63	0.082	40	0.079	Yes	Yes
cis-1*S*,4*R*-dihydroAdoMac	72	—	93	—	No	No
cis-1*R*,4*S*-norAdoMac	307	—	293	—	No	No

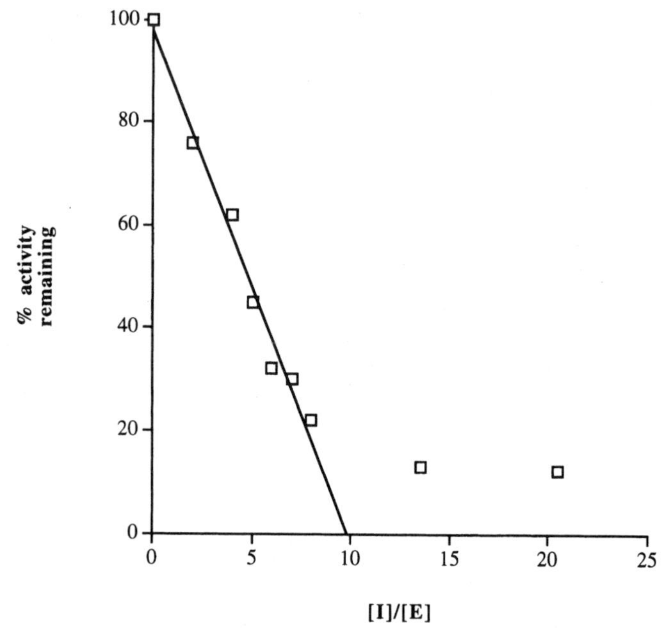

FIGURE 1.12 Partition ratio determination for the inactivation of human AdoMet-DC by *cis*-1*R*,4*S*-AdoMac. The partition ratio for the interaction of *cis*-1*R*,4*S*-AdoMac with human AdoMet-DC was determined by the titration method. Enzyme activity was measured following 24-h incubations at various inhibitor : enzyme ratios. Each data point is the average of two determinations, which in each case differ by 5% or less.

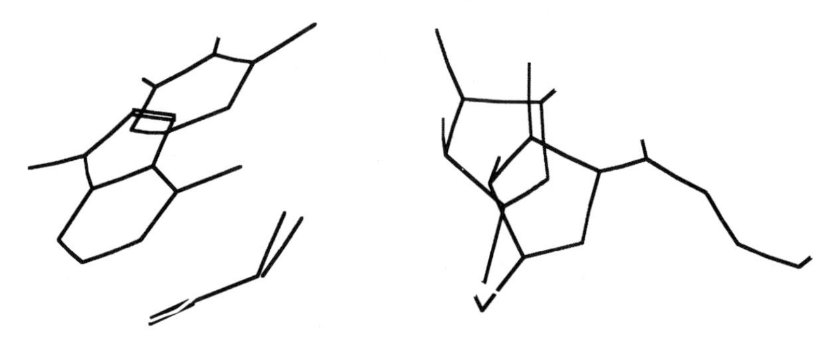

FIGURE 1.13 Results of RMS fitting experiments for the *cis*-1*R*,4*S* and *cis*-1*S*,4*R* diastereomers of AdoMac. The appropriate molecules were subjected to systematic conformational analysis, and then energy-minimized as described in the text. The resulting conformers were then compared using a standard RMS fitting routine. Hydrogens have been removed from the molecules for clarity. Left—normal view; right—orthogonal view.

The least-energy conformers for the two *cis* diastereomers were superimposed using the RMS fitting routine (SYBYL molecular modeling package, Tripos, Inc., St. Louis, MO) by specifying the atoms in the conformationally restricted aminopropyl sidechain. As seen in Figure 13, root-mean-square (RMS) fitting of these diastereomers results in relatively poor overlap, with a RMS distance of 0.557 Å. Since the charged ammonium and methylsulfonium moieties are reversed in the two diastereomers, the doubly bonded carbons in the cyclopentene ring lie on opposite sides of a plane containing these charged moieties. As a result, the 1*R*,4*S* and 1*S*,4*R*-methylene carbons at C5 are separated by a distance of 1.1 Å following RMS fitting. Overlap in the adenosyl portion of the molecule is even a poorer, with a separation of 1.6 Å between the ribose ether oxygens, and 3.8 Å between the adenosine N^6 amino groups. Thus, computer simulation suggests that there are significant conformational differences between the *cis*-1*R*-4*S* and *cis*-1*S*,4*R* diastereomers of AdoMac, which may account for the observed variation in the K_i values between the human and bacterial forms of the enzyme.

Second-Generation Enzyme-Activated Inhibitors of AdoMet-DC

As an extension of our studies involving AdoMac, the aminoxy cogener analog *S*-(5′-deoxy-5′-adenosyl)-1-aminoxy-4-methylsulfonio-2-cyclopentene (AdoMao; Fig. 1.14) was designed and synthesized. It was reasoned that the aminoxy amine–surrogate functional group would show enhanced nucleophilicity at the terminal pyruvate of AdoMet-DC, thus making AdoMao a more effective inhibitor. In addition, AdoMao would be expected to form a stable oxime linkage with the enzyme, rather than the more easily hydrolyzable imine formed by AdoMac. It is likely that AdoMet-DC would be irreversibly inactivated by AdoMao subsequent to this transformation.

During the synthesis of AdoMac, chlorination of the allylic alcohol (*trans*-1*R*,4*R*-1-hydroxy-4-{[(*tert*-butyloxy)carbonyl]amino}-2-cyclopentene) resulted in the formation of the corresponding chloride in a 64 : 36 ratio of *cis* to *trans* isomers, as determined by NMR analysis. The *cis* and *trans* diastereomeric forms for 1,4-substituted-2-cyclopentene derivatives can be readily discriminated by the ^{1}H NMR resonances of the diastereotopic protons at C5.[97] The intermediate *cis*-1*S*,4*R*-1-chloro-4-{[(*tert*-butyloxy)carbonyl]amino}-2-cyclopentene shows resonances for the C5 methylene protons, which are well separated (2.93 and 1.97 ppm), while the C5 protons in the corresponding *trans* derivative are more proximal and are shifted slightly upfield (2.64 and 2.05 ppm).[97] Synthesis of this chloride by the method of Meyers[100] and by the method of Corey[101] resulted in similar mixtures of the *cis* and *trans* diastereomers. When this chemistry was extended to the chlorination of the protected allylic aminoxy alcohol (Scheme 1.4) using the method of Corey[101] the desired allylic chloride was isolated *only* as the *cis*-1*S*,4*R* isomer, consistent with 100% inversion of configuration at the allylic center. The ^{1}H NMR spectrum for the *cis*-1*S*-4*R* chloride showed characteristic C5 resonances at 2.88 and 2.23, while no resonances appeared that could be attributed to *trans*-C5 methylene protons. Coupling of this chloride to the adenosyl portion of the molecule, followed by deprotection and methylation as described above, afforded only the 1*R*,4*R* diastereomer of AdoMao. The 1*S*,4*S* diastereomer of AdoMao was produced in similar fashion, beginning with porcine-lipase-mediated cleavage of the pro-*R* acetate of the starting mesodiacetate afforded the corresponding 1*S*,4*R* hydroxyester

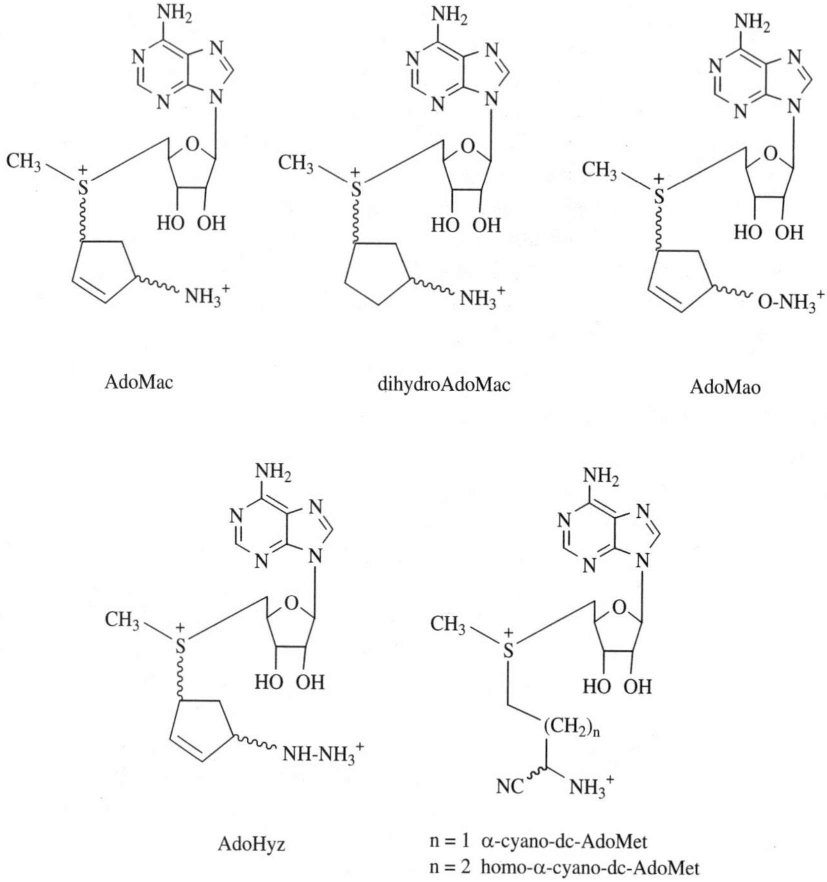

NH_2 N N N N O CH_3 $\overset{+}{S}$ HO OH NH_3^+

AdoMac

NH_2 N N N N O CH_3 $\overset{+}{S}$ HO OH NH_3^+

dihydroAdoMac

NH_2 N N N N O CH_3 $\overset{+}{S}$ HO OH O-NH_3^+

AdoMao

NH_2 N N N N O CH_3 $\overset{+}{S}$ HO OH NH-NH_3^+

AdoHyz

NH_2 N N N N O CH_3 $\overset{+}{S}$ HO OH $(CH_2)_n$ NC NH_3^+

n = 1 α-cyano-dc-AdoMet
n = 2 homo-α-cyano-dc-AdoMet

FIGURE 1.14 Structures of nucleoside-based reversible and irreversible inhibitors of AdoMet-DC.

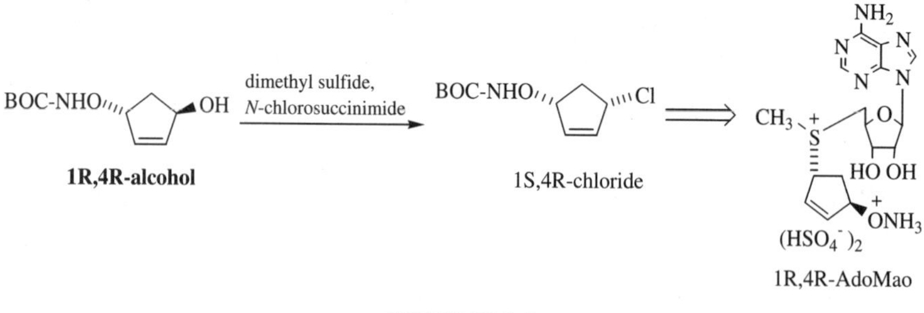

BOC-NHO OH

1R,4R-alcohol

dimethyl sulfide, *N*-chlorosuccinimide

BOC-NHO Cl

1S,4R-chloride

NH_2 N N N N O CH_3 $\overset{+}{S}$ HO OH $\overset{+}{O}NH_3$ $(HSO_4^-)_2$

1R,4R-AdoMao

SCHEME 1.4

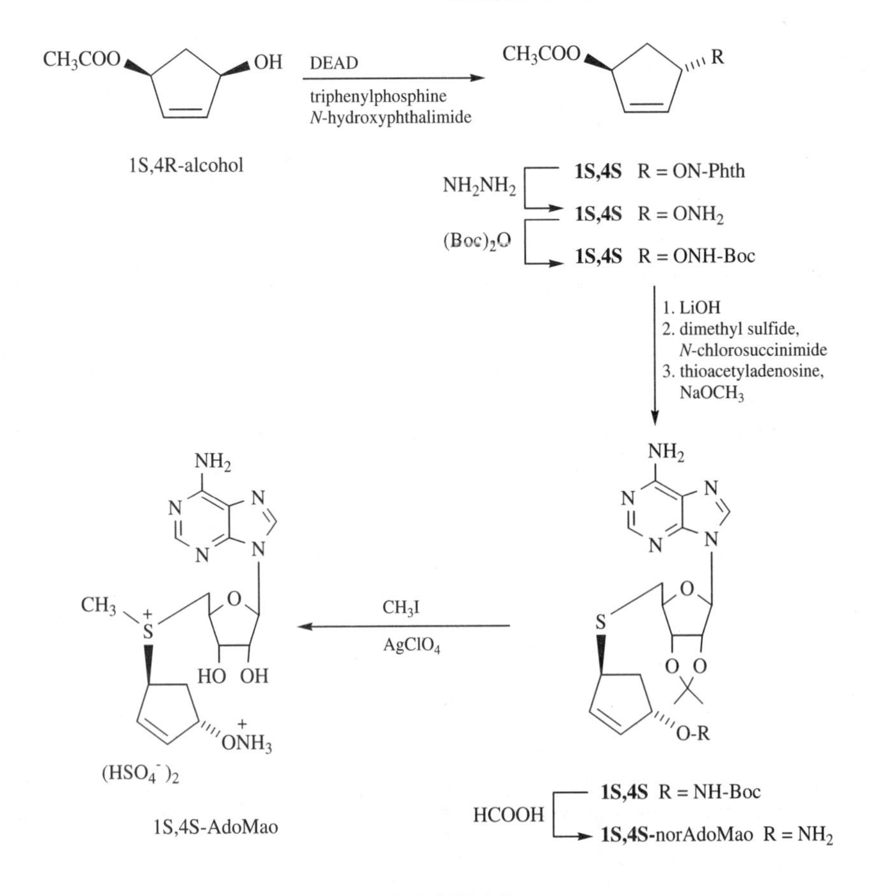

SCHEME 1.5

(Scheme 1.5), which was elaborated as described above to produce the 1*S*,4*S*-AdoMao as a single pure diastereomer.

The *trans*-1*R*,4*R* and *trans*-1*S*,4*S* diastereomers of AdoMao, all as well as the corresponding diastereomers of the unmethylated precursor molecule norAdoMao, all acted as time-dependent, irreversible inhibitors of AdoMet-DC from *E. coli*, exhibiting remarkably constant K_i values ranging between 20.6 and 23.7 μM. The fact that the unmethylated precursor norAdoMac was also an inactivator of AdoMet-DC suggests that base-catalyzed proton abstraction and elimination of MTA are not necessary for inhibition. These analogs also inactivated the human form of AdoMet-DC, although this form of the enzyme was able to discriminate between AdoMao (K_i values of 21.2 μM for the *trans*-1*R*,4*R* form and 19.6 μM for the *trans*-1*S*,4*S* form) and norAdoMao (K_i values of 95.2 μM for the *trans*-1*R*,4*R* form and 30.9 μM for the *trans*-1*S*,4*S* form). The observed differences in the inactivation of bacterial and human AdoMet-DC further support the contention that it is possible to design isozyme-specific inhibitors of AdoMet-DC.

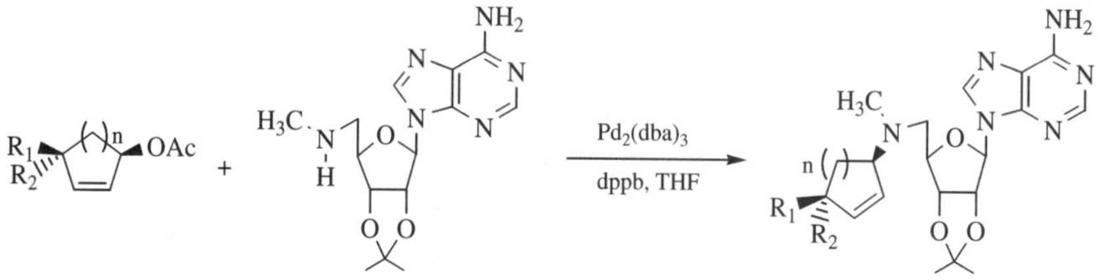

SCHEME 1.6 Palladium-catalyzed coupling of 5′-methylamino-2′,3′-isopropylideneadenosine (MAA) to cyclic allylic acetates.

Using synthetic routes analogous to those shown in Schemes 1.5 and 1.6, the putative AdoMet-DC inhibitor AdoHyz, in which the aminoxy group of AdoMao was replaced by a hydrazine moiety, was synthesized as the pure 1*S*,4*S*-diastereomer (Fig. 1.14). The author's preliminary enzymatic studies suggest that AdoHyz is an irreversible inactivator of human AdoMet-DC, with an IC_{50} value of 26 μM. The mechanism of inactivation, which likely involves formation of a stable hydrozone at the terminal pyruvate of AdoMet-DC, has yet to be verified, and the kinetic constants for inactivation have yet to be determined.

Differences in the catalytic sites of the isozymic forms of AdoMet-DC can also be exploited using AdoMet analogs that are not conformationally restricted. Wu et al. have described the synthesis of (5′-deoxy-5′-*S*-adenosyl)-2-amino-4-methylsulfonio-2-pentanenitrile (α-cyano-dc-AdoMet; Fig. 1.14), as well as an analog that contains one additional carbon in the sidechain, (5′-deoxy-5′-*S*-adenosyl)2-amino-4-methylsulfonio-2-pentanenitrile (homo-α-cyano-dc-AdoMet; Fig. 1.15).[102] The synthesis of these analogs was accomplished using the diphenylimino glycine equivalent shown in Scheme 1.7, which was appended to the appropriate TBDMS-protected alkyl iodide ($n = 1$ or 2). Conversion of this intermediate to the corresponding mesylate, followed by coupling to thioacetyladenosine, deprotection, and methylation then afforded the desired α-cyano analogs. As expected, α-cyano-

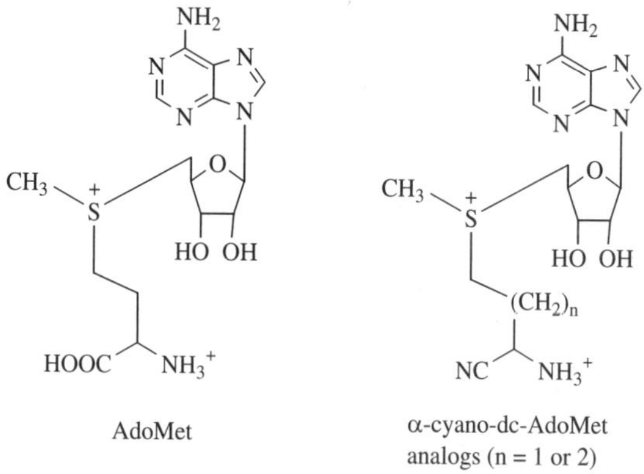

AdoMet

α-cyano-dc-AdoMet analogs (n = 1 or 2)

FIGURE 1.15 The structures of AdoMet and the proposed α-cyano-substituted dc-AdoMet analogs.

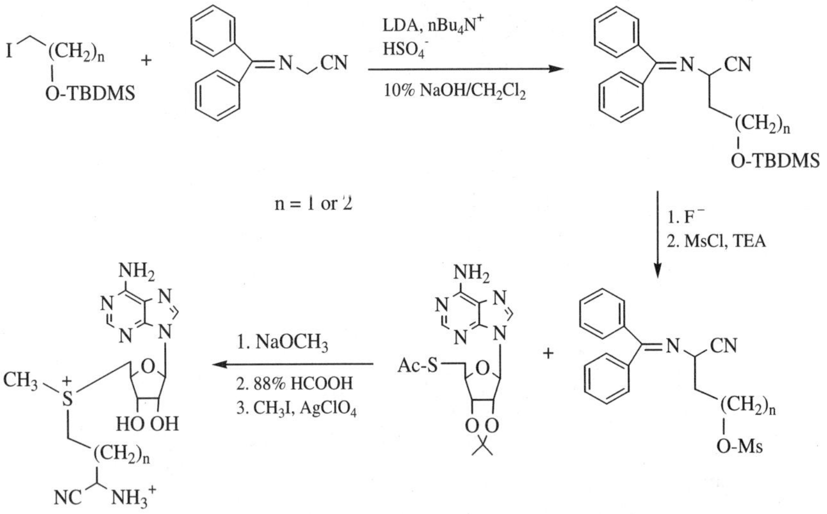

SCHEME 1.7

dc-AdoMet acted as an enzyme-activated, irreversible inhibitor of AdoMet-DC (see Fig. 1.16) from *E. coli*, with an IC_{50} value of 9 μM and a K_i value of 31 μM. The unnatural analog, homo-α-cyano-dc-AdoMet, was significantly less potent, exhibiting an IC_{50} value of 50 μM. Unexpectedly, the human form of AdoMet-DC showed a reverse preference for these two analogs, since homo-α-cyano-dc-AdoMet (K_i 7 μM) was considerably more active as an inhibitor than α-cyano-dc-AdoMet (K_i 247 μM). These data further support the hypothesis that there are significant structural differences between the catalytic sites of the human and *E. coli* forms of AdoMet-DC.

Third-Generation Irreversible AdoMet-DC Inhibitors

Because of the effectiveness of AbeAdo as an inhibitor of AdoMet-DC, the author undertook the synthesis and evaluation of a series of conformationally restricted, enzyme-activated inhibitors in which the sulfonium center had been replaced by a nitrogen.[103] The structures of these analogs are shown in Figure 1.17. Synthesis of these molecules proved to be unexpectedly difficult, but they were ultimately accessed using a modification of the palladium-catalyzed aminations developed by Miller[104] and Helquist.[105] As shown in Scheme 1.8, the readily synthesized nucleoside 5′-methylamino-2′,3′-isopropylidineadenosine could be directly coupled to the appropriate *N*-Boc-protected, cyclic allylic acetate in the presence of a palladium catalyst to afford the precursor to each desired analog. Deprotection in a single step then yielded the target analogs amino-AdoMac, amino-AdoHex, and amino-AdoHept, in high yield. The characterization of these analogs as irreversible inhibitors of AdoMet-DC has yet to be completed, and is an ongoing concern in the author's laboratories. However, these analogs have already been shown to be growth-inhibitory in cultured *Trypanosoma brucei brucei*, as described below.

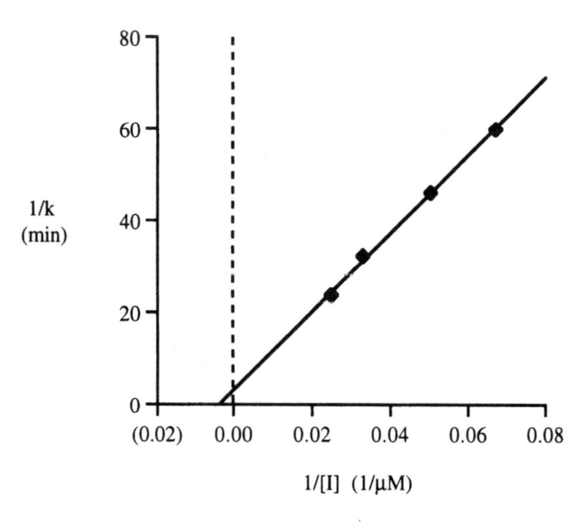

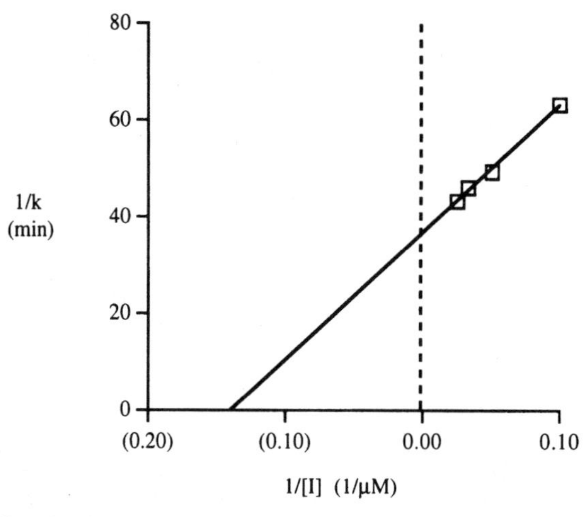

FIGURE 1.16 Inactivation of human AdoMet-DC by α-cyano-dcAdoMet analogs ($n = 1$ or 2).

Antiparasitic Activity of Inhibitors of AdoMet-DC

The author's initial interest in the use of AdoMet-DC inhibitors as antiparasitic agents stemmed from an early attempt to produce restricted rotation analogs of AdoMet and dc-AdoMet as potential antitumor agents.[106,107] The analogs shown in Figure 1.18 are representative of the compounds in this class. The author determined that these analogs had no effects on mammalian cell lines in culture (L1210 murine

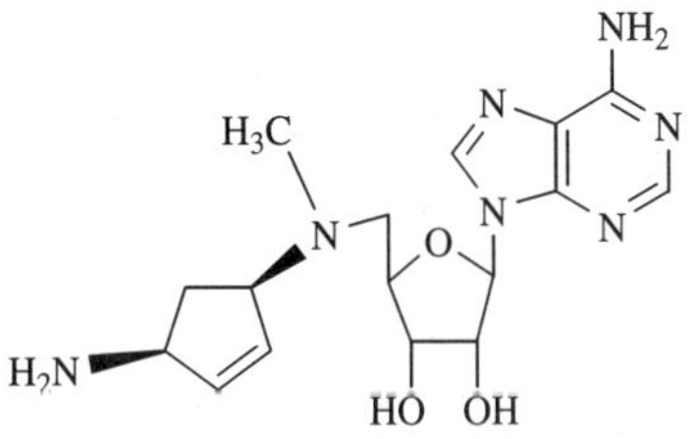

amino-AdoMac

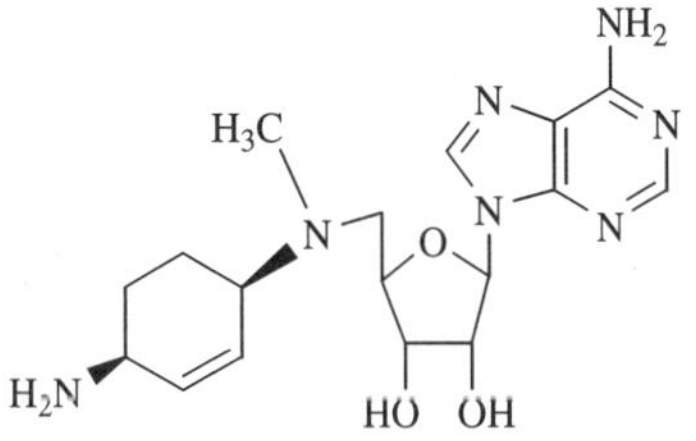

amino-AdoHex

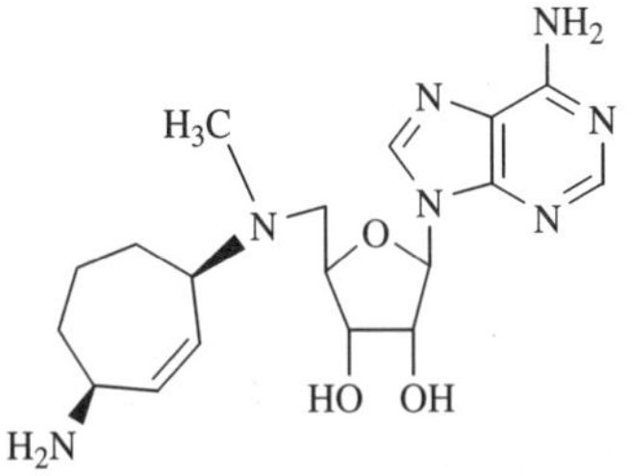

amino-AdoHept

FIGURE 1.17 AdoMet-DC inhibitors based on aminoadenosine.

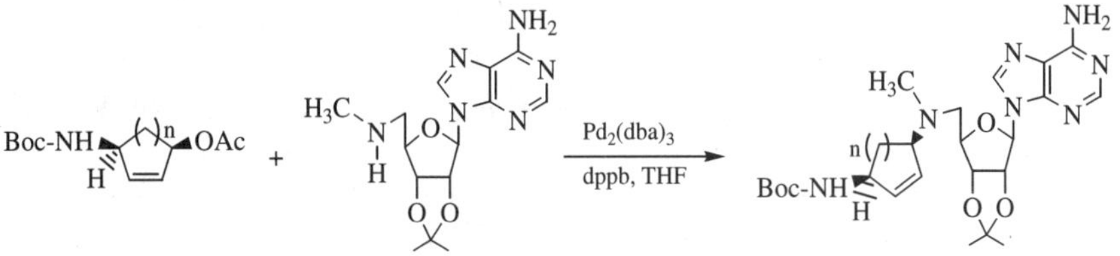

Pd$_2$(dba)$_3$

dppb, THF

SCHEME 1.8

lymphocytic leukemia, HL60 human promyelocytic leukemia, HT29 human colon carcinoma), and were very poor inhibitors of bacterial AdoMet-DC, with IC_{50} values ranging from 0.478 to 1.07 mM. However, as shown in Figure 1.18, these compounds were reasonable inhibitors of trypanosomal growth in culture,[107] even at micromolar concentrations. The author reasoned that these analogs must be substrates for the trypanosomal purine transport system mentioned above,[81] and that they were concentrated inside the organism at sufficient concentrations to cause growth inhibition. In all likelihood, the analogs shown in Figure 1.18 are unable to form an imine linkage with the terminal pyruvate of AdoMet-DC, a fact that would account for their weak inhibition of the bacterial enzyme. On the basis of this information, the irreversible inhibitors of AdoMet described above were designed and evaluated as antitrypanosomal agents.

Antitrypanosomal activity was determined on blood forms of *Trypanosoma brucei brucei* grown in vitro using a ^{3}H-hypoxanthine incorporation assay, or by

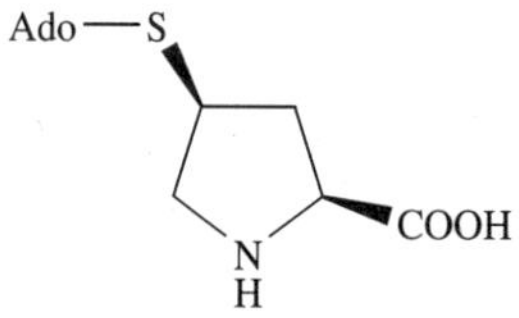

WLF-10-7
2S,4S-configuration
30% Inhibition
at 100 μM

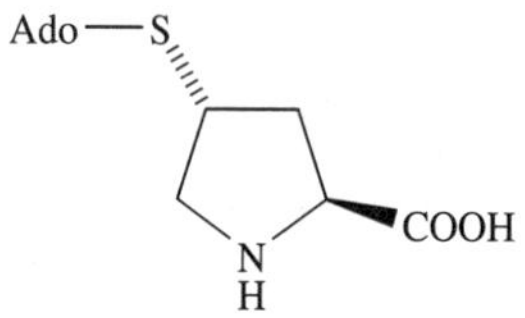

PMW-8-48
2S,4R-configuration
41% Inhibition
at 100 μM

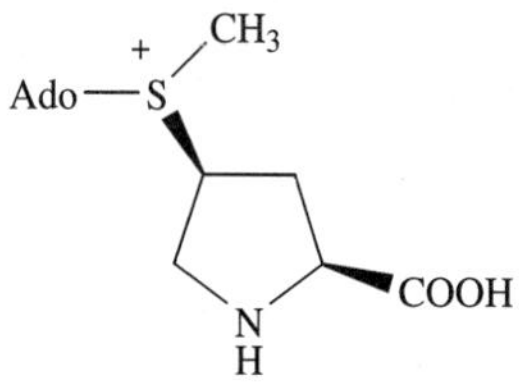

PMW-8-54
2S,4S-configuration
64% Inhibition
at 100 μM

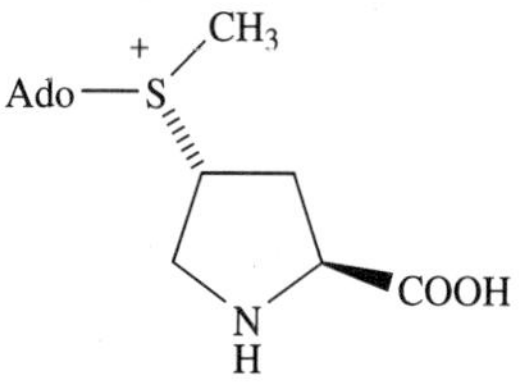

MMO-14-11
2S,4R-configuration
34% Inhibition
at 100 μM

FIGURE 1.18 Restricted rotation analogs of AdoMet containing 4-substituted proline sidechains, and their ability to inhibit the growth of *Trypanosoma brucei brucei* in culture.

direct cell count. Continuous cultures of blood forms were initiated in a feeder-layer-free system by inoculating wells of a 24-well culture dish (Falcon 3047) containing 1 mL of modified Iscove's medium[108] with 10^5 trypanosomes from mouse blood. Plates were incubated in 3% CO_2 in air at 37 °C. One half the volume of medium was replaced daily, and trypanosomes achieved peak densities of $5 \times 10^6\,mL^{-1}$ in 4 days. Inhibitor sensitivity tests were done by dissolving the agent in sterile medium and replacing one-half the volume daily with medium containing double-strength inhibitor. Cell counts were made daily and IC_{50} values calculated after 48 h of exposure. Alternatively, trypanosomes were incubated in modified Iscove's medium for 24 h with varying concentrations of inhibitor, washed, and then incubated for an additional 24 h with medium containing 1 μCi of [^3H]-hypoxanthine. The cells (5×10^6 per well) were then harvested by filtration, and the filters were scintillation counted. The results of the biological studies for the AdoMet-DC inhibitors that were evaluated are summarized in Table 1.3. Among the analogs tested, the most promising to date has been 1*R*,4*R*-AdoMao, which acts as a potent trypanocide in vitro. This analog was produced in large quantities for in vivo studies, and was found to have unremarkable in vivo activity. This may be due, in part, to the inability of AdoMet analogs to pass through biological barriers, since there were differences in the in vivo effects with different routes of administration. However, none of these routes of administration led to sufficient blood levels to cure trypanosomal infections in mice. Since the aminonucleoside analog AbeAdo can produce cures in animals,

TABLE 1.3 Antitrypanosomal Effects of Reversible and Irreversible AdoMet-DC Inhibitors

Compound	K_i Against Bacterial AdoMetDC, μM	Percent Growth Inhibition of *Trypanosoma brucei brucei* (Concentration)
WLF-10-7	478	30% (100 μM)
PMW-8-48	960	41% (100 μM)
PMW-8-54	1207	64% (100 μM)
MMO-14-11	1070	34% (100 μM)
1*R*,4*S* + 1*R*,4*R* AdoMac	18	50% (5.2 μM)
1*R*,4*R*-norAdoMao	20.6	50% (10.1 μM)
1*S*,4*S*-norAdoMao	21.3	50% (10.0 μM)
1*R*,4*R*-AdoMao	21.5	50% (3.0 μM)
1*S*,4*S*-AdoMao	34.9	50% (0.9 μM)
1*S*,4*S*-norAdoHyz	ND[a]	6% (100 μM)
1*R*,4*R*-AdoHyz	ND	39% (100 μM)
homo-α-cyano-dc-AdoMet	7.2	50% (7.6 μM)
aza-AdoMac	ND	50% (36.5 μM)
aza-AdoHex	ND	50% (10 μM)
aza-AdoHept	ND	50% (5.7 μM)

[a]No data.

we are actively pursuing the synthesis and evaluation of additional aminoadenosine analogs related to aza-AdoMac, aza-AdoHex, and aza-AdoHept. In addition, the data in Table 1.3 again suggest that there are significant differences in the active-site topology of the human, bacterial and trypanosomal forms of AdoMet-DC, and that these differences may be exploited to produce potent and target-specific inhibitors for trypanosomal AdoMet-DC. It should be pointed out that, by using a rational drug design strategy involving the application of chemical and mechanistic principles, the present author and other groups have been able to produce a relatively high number of active analogs, even though the overall number of compounds produced is low. Thus, such a rational approach may be considered a viable alternative to the currently popular methods of combinatorial synthesis and high-throughput screening. These methods for rapid drug discovery are quite useful, and do result in the discovery of new therapeutic agents. However, the optimization of leads discovered in this way will continue to depend on the chemical principles that govern binding to macromolecules and catalytic mechanisms. All the available avenues for bringing chemical principles to bear on drug discovery research are alive and well, and are currently being exploited in the author's laboratories.

REFERENCES

1. For more information on the function of the natural polyamines and the consequences of inhibition of the polyamine pathway, see: (a) "A Guide to the Polyamines," in S. Cohen,

ed., Oxford Univ. Press, New York, 1998; (b) Casero, R. A.; Pegg, A. E.; *FASEB J.* 1993, **7**, 653–661; Pegg, A. E., *Cancer Res.* 1988, **48**, 759–774.

2. Ha, H. C.; Sirisoma, N. S.; Kuppusamy, P.; Zweier, J. L.; Woster, P. M.; Casero, R. A., *Proc. Natl. Acad. Sci.* (USA) 1998, **95**, 11140–11143.
3. Ha, H. C.; Yager, J. D.; Woster, P. M.; Casero, R. A., *Bioch. Biophys. Res. Commun.* 1998, **244**, 298–303.
4. McCloskey, D. E.; Yang, J.; Woster, P. M.; Davidson, N. E.; Casero, R. A., *Clin. Cancer Res.* 1996, **2**, 441–446.
5. McCloskey, D. E.; Casero, R. A.; Woster, P. M.; Davidson, N. E., *Cancer Res.* 1995, **55**, 3233–3236.
6. Ha, H. C.; Woster, P. M.; Yager, J. D.; Casero, R. A., *Proc. Natl. Acad. Sci.* (USA) 1997, **94**, 11557–11562.
7. Ha, H. C.; Woster, P. M.; Casero, R. A., *Cancer Res.* 1998, **58**, 2711–2714.
8. Webb, H. K.; Wu, Z. Q.; Sirisoma, N.; Ha, H. C.; Casero, Jr., R. A.; Woster, P. M., *J. Med. Chem.* 1999, **42**, 1415–1421.
9. Pegg, A. E., *Bioch. J.* 1986, **234**, 249–262.
10. Coward, J. K., *Annu. Rep. Med. Chem.* 1982, **17**, 253–259.
11. McCann, P. P.; Pegg, A. E., *Pharm. Ther.* 1992, **54**, 195–215.
12. Marchant, P.; Dredar, S.; Manneh, V.; Alshabanah, O.; Matthews, H.; Fries, D.; Blankenship, J., *Arch. Bioch. Bioph.* 1989, **15**, 128–136.
13. Byers, T. L.; Pegg, A. E., *J. Cell. Physiol.* 1990, **143**, 460–467.
14. Sunkara, P. S.; Baylin, S. B.; Luk, G. D., in *Inhibition of Polyamine Metabolism: Biological Significance and Basis for New Therapies*, P. P. McCann, A. E. Pegg, and A. Sjoerdsma, eds., Academic Press, New York, 1987, pp. 121–140.
15. Metcalf, B. W.; Bey, P.; Danzin, C.; Jung, M. J.; Casara, P.; Vevert, J. P., *J. Am. Chem. Soc.* 1978, **100**, 2551–2553.
16. Williams-Ashman, H. G.; Pegg, A. E., in *Polyamines in Biology and Medicine*, D. R. Morris and L. J. Marton, eds., Marcel Dekker, New York, 1981, pp. 43–73.
17. Pegg, A. E.; Jones, D. B.; Secrist III, J. A., *Biochemistry* 1988, **27**, 1408–1415.
18. Casara, P.; Marchal, P.; Wagner, J.; Danzin, C., *J. Am. Chem. Soc.* 1989, **111**, 9111–9113.
19. Pegg, A. E.; McCann, P. P., *Pharm. Ther.* 1992, **56**, 359–377.
20. Bey, P.; Danzin, C.; Jung, M., in *Inhibition of Polyamine Metabolism: Biological Significance and Basis for New Therapies*, P. P. McCann, A. E. Pegg, and A. Sjoerdsma, eds., Academic Press, New York, 1987, pp. 1–31.
21. Tang, K.-C.; Mariuzza, R.; Coward, J. K., *J. Med. Chem.* 1981, **24**, 1277–1284.
22. Woster, P. M.; Black, A. Y.; Duff, K. J.; Coward, J. K.; Pegg, A. E., *J. Med. Chem.* 1989, **32**, 1300–1307.
23. Bacchi, C. J.; Yarlett, N.; Goldberg, B.; Bitonti, A. J.; McCann, P. P., in *Biochemical Protozoology*, G. H. Coombs and M. J. North, eds., Taylor & Francis, Washington, DC, pp. 469–481.
24. Bacchi, C. J.; Nathan, H. C.; Hutner, S. H.; McCann, P. P.; Sjoerdsma, A., *Science* 1980, **210**, 332–334.

25. Pegg, A. E.; Williams-Ashman, H. G., in *Inhibition of Polyamine Metabolism: Biological Significance and Basis for New Therapies*, P. P. McCann, A. E. Pegg, and A. Sjoerdsma, eds., Academic Press, New York, 1987, pp. 33–48.
26. Pegg, A. E.; Coward, J. K., *Biochem. Biophys. Res. Commun.* 1985, **133**, 82–89.
27. Pegg, A. E.; Wechter, R.; Poulin, R.; Woster, P. M.; Coward, J. K., *Biochemistry* 1989, **28**, 8446–8453.
28. Bergeron, R. J.; Neims, A. H.; McManis, J. S.; Hawthorne, T. R.; Vinson, J. R. T.; Bortell, R.; Ingeno, M. J., *J. Med. Chem.* 1988, **31**, 1183–1190.
29. Bergeron, R. J.; Hawthorne, T. R.; Vinson, J. R. T.; Beck, D. E., Jr.; Ingeno, M. J., *Cancer Res.* 1989, **49**, 2959–2964.
30. Porter, C. W.; Regenass, U.; Bergeron, R. J., in Anonymous *Falk Symposium on Polyamines*, Kluwer Press, Lancaster, U.K., 1992, pp. 301–322.
31. Saab, N. H.; West, E. E.; Bieszk, N. C.; Preuss, C. V.; Mank, A. R.; Casero, R. A.; Woster, P. M., *J. Med. Chem.* 1993, **36**, 2998–3004.
32. Bellevue, F. H.; Boahbedason, M. L.; Wu, R. H.; Casero, Jr., R. A.; Rattendi, D.; Lane, S.; Bacchi, C. J.; Woster, P. M., *Bioorg. Med. Chem. Lett.* 1996, **6**, 2765–2770.
33. Porter, C. W.; Sufrin, J., *Anticancer Res.* 1986, **6**, 525–542.
34. Mandel, J. L.; Flintoff, W. F., *J. Cell. Physiol.* 1978, **97**, 335–343.
35. Bergeron, R. J.; McManis, J. S.; Liu, C. Z.; Feng, Y.; Weimar, W. R.; Luchetta, G. R.; Wu, Q.; Ortiz-Ocasio, J.; Vinson, J. R. T.; Kramer, D.; Porter, C., *J. Med. Chem.* 1994, **37**, 3464–3476.
36. Davidson, N. E.; Hanm, H. A.; McCloskey, D. E.; Woster, P. M.; Casaro, R. A., *Endocrine-Related Cancer* 1999, **6**, 69–73.
37. McCloskey, D. E.; Prestigiacomo, L. J.; Woster, P. M.; Casero, R. A.; Davidson, N. E., *Proc. Am. Assoc. Cancer Res.* 1996, **37**, 400.
38. Ha, H. C.; Woster, P. M.; Yager, J. D.; Casero, R. A., *Proc. Natl. Acad. Sci.* (USA) 1997, **94**, 11557–11562.
39. Wickner, R. B.; Tabor, C. W.; Tabor, H., *J. Biol. Chem.* 1970, **245**, 2132–2139.
40. Markham, G. D.; Tabor, C. W.; Tabor, H., *J. Biol. Chem.* 1982, **257**, 12063–12068.
41. Anton, D. L.; Kutny, R., *J. Biol. Chem.* 1987, **262**, 2817–2822.
42. Pajunen, A.; Crozat, A.; Janne, O. A.; Ihalainen, R.; Laitinen, P. H.; Stanley, B.; Madhubala, R.; Pegg, A. E., *J. Biol. Chem.* 1988, **263**, 17040–17049.
43. Tabor, C. W.; Tabor, H., *J. Biol. Chem.* 1987, **262**, 16037–16040.
44. Recsei, P. A.; Snell, E. E., *Ann. Rev. Biochem.* 1984, **53**, 357–387.
45. Stanley, B. A.; Pegg, A. E.; Holm, I., *J. Biol. Chem.* 1989, **264**, 21073–21079.
46. Pegg, A. E.; Williams-Ashman, H. G., *J. Biol. Chem.* 1969, **244**, 682–693.
47. Shirahata, A.; Pegg, A. E., *J. Biol. Chem.* 1985, **260**, 9583–9588.
48. Stanley, B. A.; Shantz, L. M.; Pegg, A. E., *J. Biol. Chem.* 1994, **269**, 7901–7907.
49. Shantz, L. M.; Stanley, B. A.; Secrist III, J. A.; Pegg, A. E., *Biochemistry* 1992, **31**, 6848–6855.
50. Stanley, B. A.; Pegg, A. E., *J. Biol. Chem.* 1991, **266**, 18502–18506.
51. Tekwani, B. L.; Bacchi, C. J.; Pegg, A. E., *Molec. Cell. Biochem.* 1992, **117**, 53–61.
52. Anton, D. L.; Kutny, R., *Biochemistry* 1987, **26**, 6444–6447.

53. Diaz, E.; Anton, D. L., *Biochemistry* 1991, **30**, 4078–4081.
54. Karvonen, E.; Kauppinen, L.; Partanen, T.; Poso, H., *Biochem. J.* 1985, **231**, 165–169.
55. Williams-Ashman, H. G.; Pegg, A. E., in *Polyamines in Biology and Medicine*, D. R. Morris and L. J. Marton, eds., Marcel Dekker, New York, 1981, pp. 43–73.
56. Hibasami, H.; Tsukada, T.; Maekawa, S.; Sakurai, M.; Shirakawa, S.; Nakashima, K., *Cancer Chemother. Pharm.* 1988, **22**, 187–190.
57. Hibasami, H.; Maekawa, S.; Murata, T.; Nakashima, K., *Cancer Res.* 1989, **49**, 2065–2068.
58. Elo, H.; Mutikainen, I.; Alhonen-Hongisto, L.; Laine, R.; Janne, J., *Cancer Lett.* 1988, **41**, 21–30.
59. Nakashima, K.; Hibasami, H.; Tsukuda, T.; Maekawa, S., *Eur. J. Med. Chem.* 1987, **22**, 553–558.
60. Stanek, J.; Caravatti, G.; Capraro, H. G.; Furet, P.; Mett, H.; Schneider, P.; Regenass, U., *J. Med. Chem.* 1993, **36**, 46–54.
61. Regenass, U.; Caravatti, G.; Mett, H.; Stanek, J.; Schneider, P.; Muller, M.; Matter, A.; Vertino, P.; Porter, C., *Cancer Res.* 1992, **52**, 4712–4718.
62. Stanek, J.; Caravatti, G.; Frei, J.; Furet, P.; Mett, H.; Schneider, P.; Regenass, U., *J. Med. Chem.* 1993, **36**, 2168–2171.
63. Bey, P.; Vevert, J. P.; VanDorsselaer, V.; Kolb, M., *J. Org. Chem.* 1979, **44**, 2732.
64. Pankaskie, M.; Abdel-Monem, M. M., *J. Med. Chem.* 1980, **23**, 121–127.
65. Kolb, M.; Danzin, C.; Barth, J.; Claverie, N., *J. Med. Chem.* 1982, **25**, 550–556.
66. Artamonova, E. Y.; Zavalova, L. L.; Khomutov, R. M.; Khomutov, A. R., *Bioorg. Khim.* 1986, **12**, 206–212.
67. Khomutov, A. R.; Kritsky, A. M.; Artamonova, E. Y.; Khomutov, R. M., *Nucleosides Nucleotides* 1987, **6**, 359–362.
68. Kramer, D. L.; Khomutov, R. M.; Bukin, Y. V.; Khomutov, A. R.; Porter, C. W., *Biochem. J.*, 1989, **259**, 325–331.
69. Secrist III, J. A., *Nucleosides Nucleotides*, 1987, **6**, 73–83.
70. Danzin, C.; Marchal, P.; Casara, P., *Biochem. Pharmacol.* 1990, **40**, 1499–1503.
71. Bacchi, C. J.; Yarlett, N.; Goldberg, B.; Bitonti, A. J.; McCann, P. P., in *Biochemical Protozoology*, G. H. Coombs and M. J. North, eds., Taylor & Francis, Washington, DC, 1991, Chapter 43, pp. 469–481.
72. Phillips, M. A.; Coffino, P.; Wang, C. C., *J. Biol. Chem.* 1987, **262**, 8721–8727.
73. Yarlett, N.; Bacchi, C. J., *Molec. Biochem. Parisitol.* 1988, **27**, 1–10.
74. Walsh, C.; Bradley, M.; Nadeau, K., *TIBS* 1991, **16**, 305–309.
75. Yarlett, N.; Garofalo, J.; Goldberg, B.; Ciminelli, M. A.; Ruggiero, V.; Sufrin, J. R.; Bacchi, C. J., *Bioch. Bioph. Acta* 1993, **1181**, 68–76.
76. Tekwani, B. L.; Bacchi, C. J.; Pegg, A. E., *Molec. Cell. Biochem.* 1992, **117**, 53–61.
77. Bitonti, A. J.; Byers, T. L.; Bush, T. L.; Casara, P. J.; Bacchi, C. J.; Clarkson, A. B. Jr.; McCann, P. P.; Sjoerdsma, A., *Antimicrob. Agents Chemother.* 1990, **34**(8), 1485–1490.
78. Tekwani, B. L.; Bacchi, C. J.; Secrist III, J. A.; Pegg, A. E., *Biochem. Pharmacol.* 1992, **44**, 905–911.
79. Guo, J. Q.; Wu, Y. Q.; Rattendi, D.; Bacchi, C. J.; Woster, P. M., *J. Med. Chem.* 1995, **38**, 1770–1777.

80. Bacchi, C. J.; Brun, R.; Croft, S. L.; Alicea, K.; Buhler, Y., *Antimicrob. Agents Chemother.* 1996, **40**(6), 1448–1453.
81. Carter, N. S.; Fairlamb, A. H., *Nature* 1993, **361**, 173–176.
82. Byers, T. L.; Casara, P.; Bitonti, A. J., *Biochem J.* 1992, **283**, 755–758
83. Bacchi, C. J.; Goldberg, B.; Rattendi, D.; Gorrell, T. E.; Spiess, A. J.; Sufrin, J. R., *Biochem Pharmacol.* 1999, **57**. 89–96
84. Wu, Y. Q.; Woster, P. M., *J. Med. Chem.* 1992, **35**, 3196–3201.
85. Kaneko, C.; Sugimoto, A.; Tanaka, S., *Synthesis*, 1974, 876–877.
86. Johnson, C. R.; Penning, T. D., *J. Am. Chem. Soc.* 1988, **110**, 4726–4735.
87. Laumen, K.; Schneider, M. P., *J. Chem. Soc. Chem. Commun.* 1986, 1298–1299.
88. Dale, J. A.; Dull, D. L.; Mosher, H. S., *J. Org. Chem.* 1969, **34**, 2543–2549.
89. Mitsunobu, O., *Synthesis* 1981, 1–28.
90. Keller, O.; Keller, W. E.; van Look, G.; Wersin, G., *Org. Synth.* 1985, **63**, 160–171.
91. Collington, E. W.; Meyers, A. I., *J. Org. Chem.* 1971, **36**, 3044–3045.
92. Coward, J. K.; Anderson, G. L.; Tang, K.-C., *Meth. Enzymol.* 1984, **94**, 286–294.
93. Woster, P. M.; Black, A. Y.; Duff, K. J.; Coward, J. K.; Pegg, A. E., *J. Med. Chem.* 1989, **32**, 1300–1307.
94. Douglas, K. A.; Zormeier, M. M.; Marcolina, L. M.; Woster, P. M., *Bioorg. Med. Chem. Lett.* 1991, **1**, 267–270.
95. Samejima, K.; Nakazawa, Y.; Matsunaga, I., *Chem. Pharm. Bull.* 1978, **26**, 1480–1485.
96. Kitz, R.; Wilson, I. B., *J. Biol. Chem.* 1962, **237**, 3245.
97. Wu, Y. Q.; Woster, P. M., *Bioorg. Med. Chem.* 1993, **1**, 349–360.
98. Laumen, K.; Schneider, M., *Tetrahedron Lett.* 1984, **25**, 5875–5878.
99. Wu, Y. Q.; Woster, P. M., *Biochem. Pharmacol.* 1995, **49**, 1125–1133.
100. Collington, E. W.; Meyers, A. I., *J. Org. Chem.* 1971, **36**, 3044–3045.
101. Corey, E. J.; Kim, C. U.; Takeda, M., *Tetrahedron Lett.* 1972, **42**, 4339–4342.
102. Wu, Y. Q.; Lawrence, T.; Guo, J. Q.; Woster, P. M., *Bioorg. Med. Chem. Lett.* 1993, **3**, 2811–2816.
103. Sirisoma, N. S.; Woster, P. M., *Tetrahedron Lett.* 1998, **39**, 1489–1492.
104. Ghosh, A.; Ritter, A. R.; Miller, M. J., *J. Org. Chem.* 1995, **60**, 5808–5813.
105. Connell, R. D.; Akermark, B.; Helquist, P., *J. Org. Chem.* 1989, **54**, 3359–3370.
106. Douglas, K. A.; Zormeier, M. M.; Marcolina, L. M.; Woster, P. M., *Bioorg. Med. Chem. Lett.* 1991, **1**, 267–270.
107. Guo, J. Q.; Wu, Y. Q.; Douglas, K. A.; Farmer, W. L.; Garofalo, J.; Bacchi, C. J.; Woster, P. M., *Bioorg. Med. Chem. Lett.* 1993, **3**, 147–152.
108. Hirumi, H.; Hirumi, J., *J. Parisitol.* 1989, **75**, 985–989.

CHAPTER 2

Mechanism-Based S-Adenosyl-L-Homocysteine Hydrolase Inhibitors in the Search for Broad-Spectrum Antiviral Agents

DAN YIN, XIAODA YANG and RONALD T. BORCHART
Department of Pharmaceutical Chemistry, The University of Kansas

CHONG-SHENG YUAN
Tanabe Research Laboratories USA

INTRODUCTION

Broad-spectrum antiviral drugs offer many advantages over narrow-spectrum agents. It is often difficult in clinical diagnoses to identify a viral pathogen in a short time, and results often arrive too late for the choice of a specific antiviral drug. Especially in acute infections, viral chemotherapy must start as soon as the patient presents clinical symptoms. Thus, the development of broad-spectrum antiviral drugs is highly desired. The basis of most attempts at antiviral chemotherapy is the inhibition of viral replication.[1] Although different virus families adopt many diverse mechanisms of replication, most viruses of eukaryotes require 5′-capped, methylated structures on their mRNA for efficient translation of viral proteins.[2,3] *S*-Adenosyl-L-methionine (AdoMet) serves as a methyl donor for methylations of the cap structure of mRNA as well as other nucleic acids, proteins, phospholipids, and small molecules that are catalyzed by specific AdoMet-dependent methyltransferases.[4] All AdoMet-dependent transmethylation reactions produce *S*-adenosyl-L-homocysteine (AdoHcy), a product inhibitor of this class of reactions. In eukaryotes, the only known pathway for the catabolism of AdoHcy is its hydrolysis to adenosine (Ado) and homocysteine (Hcy) by AdoHcy hydrolase (EC 3.3.1.1.).[5] Because of the key role that AdoHcy hydrolase plays in regulating crucial

Biomedical Chemistry: Applying Chemical Principles to the Understanding and Treatment of Disease, Edited by Paul F. Torrence
ISBN 0-471-32633-x

AdoMet-dependent methylations in eukaryotic cells, this enzyme has become a target for the design of potential antiviral agents, and this approach has met with considerable success.[6–8] In this chapter, we do not intend to provide a comprehensive review of the vast primary literature concerning AdoHcy hydrolase and its inhibitors. Instead, we focus on the rationale for selecting AdoHcy hydrolase as a target for the design of antiviral agents, the application of the information about the catalysis mechanism, the structure, and mechanism of protype inhibitors of this enzyme in the design of more specific inhibitors, and the potential clinical utility of inhibitors of this enzyme as antiviral agents.

RATIONALE FOR TARGETING AdoHcy HYDROLASE FOR THE DESIGN OF ANTIVIRAL AGENTS

Methylation of the Viral mRNA Cap Structure in Viral Replication

Recent advances in the molecular biology of viral replication have led to more rational approaches to the design of antiviral chemotherapeutic agents. These approaches target specific stages of viral replication including (1) attachment of virus to host cell, (2) penetration and uncoating of virus, (3) transcription of viral mRNA (or cDNA in the case of the retroviruses), (4) replication of viral DNA or RNA, (5) viral mRNA capping or methylation, (6) translation of viral mRNA into protein, (7) posttranslational processing, (8) virus assembly, and (9) virus release.[9] Our laboratories, along with several others, have focused on the methylation of the cap structure on viral mRNA as one approach to the design of antiviral agents. This strategy is based on the observation that many viruses require 5′-methylated cap structures on their mRNA to promote the initiation of translation of viral proteins.[2,3,10] The appended cap structure of mRNA consists of a N^7-methylguanosine residue joined to the initial nucleoside of the mRNA via a 5′–5′ triphosphate bridge (Fig. 2.1). Most 5′-capped, methylated structures also contain a

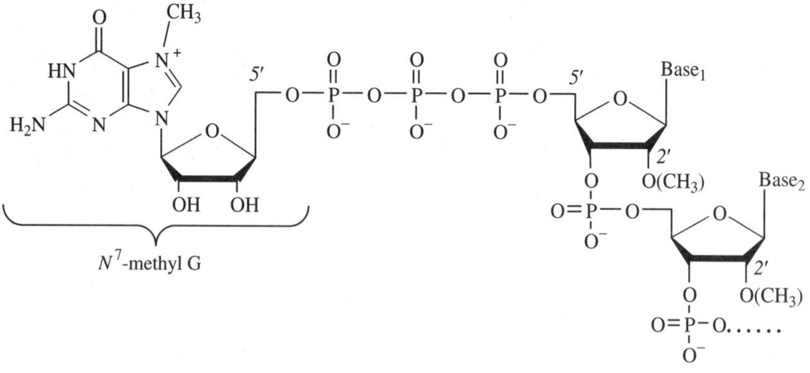

FIGURE 2.1 Structure of the 5′ cap of mRNA.

methyl group on the 2′-hydroxy group of the penultimate nucleotide having a general structure of $m^7G(5')ppp(5')N^m$.

Both capping and methylation of viral mRNA are enzymatically catalyzed in the cytoplasm, rather than the nucleus, where cellular mRNA is modified.[9] A number of viruses have the capping enzyme (RNA guanyltransferase) and the methylation enzymes (mRNA guanine-7 and nucleoside-2′ methyltransferases) within the viron (e.g., vaccinia virus, reovirus, vesicular stomatitis virus, Newcastle disease virus, and polyhedrosis viruses).[8] Other viruses (e.g., herpes) are presumed to use host cell capping enzymes.[8] The methylation of 5′-terminal cap plays an important role in protection of the integrity of mRNAs and efficient translation. The methylated cap structure protects mRNA from attack by phosphatases and digestion by 5′-end nucleases, thereby enhancing the stability of mRNA in the cytoplasm.[11] Methylation of the cap structure increases the affinity for ribosome binding to the 5′ end of the mRNA during formation of the translational initiation complex.[12] Studies of several viral replication systems have revealed a direct relationship between the presence of 5′-cap structures and completed viral mRNA translation.[13,14] Because uncapped or undermethylated viral mRNA are less effectively translated into viral proteins,[15] suppression of these 5′-capping and methylation reactions (e.g., through the inhibition of guanyltransferase, mRNA guanine-7 methyltransferase or nucleoside-2′ methyltransferase) could conceivably lead to inhibition of viral replication.

Approaches to Modulate AdoMet-Dependent Methylation

Biological methylation reactions are catalyzed by various methyltransferases that transfer a methyl group from a methyl donor to a methyl acceptor. In both eukaryotic and prokaryotic organisms, the major methyl donor for biological methylations is AdoMet.[4] It has been shown that viral-coded methyltransferases are also AdoMet-dependent.[16,17] Therefore, inhibition of AdoMet-dependent methyltransferases would lead to inhibition of viral mRNA methylation and, thus, viral replication. From Scheme 2.1, one might easily envisage that, metabolically, at least three approaches are available to inhibit AdoMet-dependent methyltransferases. These include use of (1) inhibitors that function directly on a particular methyltransferase (e.g., analogs of either substrate or product of enzyme **2** in transmethylation reaction), (2) inhibitors that function by inhibiting AdoMet biosynthesis (e.g., enzyme **1**, methionine adenosyltransferase), or (3) inhibitors that function by inhibiting AdoHcy metabolism [e.g., enzyme **3**, AdoHcy hydrolase (eukaryotes), or enzyme 7, AdoHcy nucleosidase (prokaryotes)] to elevate levels of intracellular AdoHcy, which is a potent product inhibitor of all AdoMet-dependent methyltransferases.

Utilization of analogs of the methyl acceptor substrates or the methylated products has been extensively exploited in the design of inhibitors for "small molecule" methyltransferases (e.g., catechol-*O* methyltransferase, phenylethanolamine methyltransferase, histamine-*N* methyltransferase, hydroxyindole-*O* methyltransferase, and indolethylamine methyltransferase).[18] This approach has not been used for the design of macromolecule methyltransferases inhibitors, probably

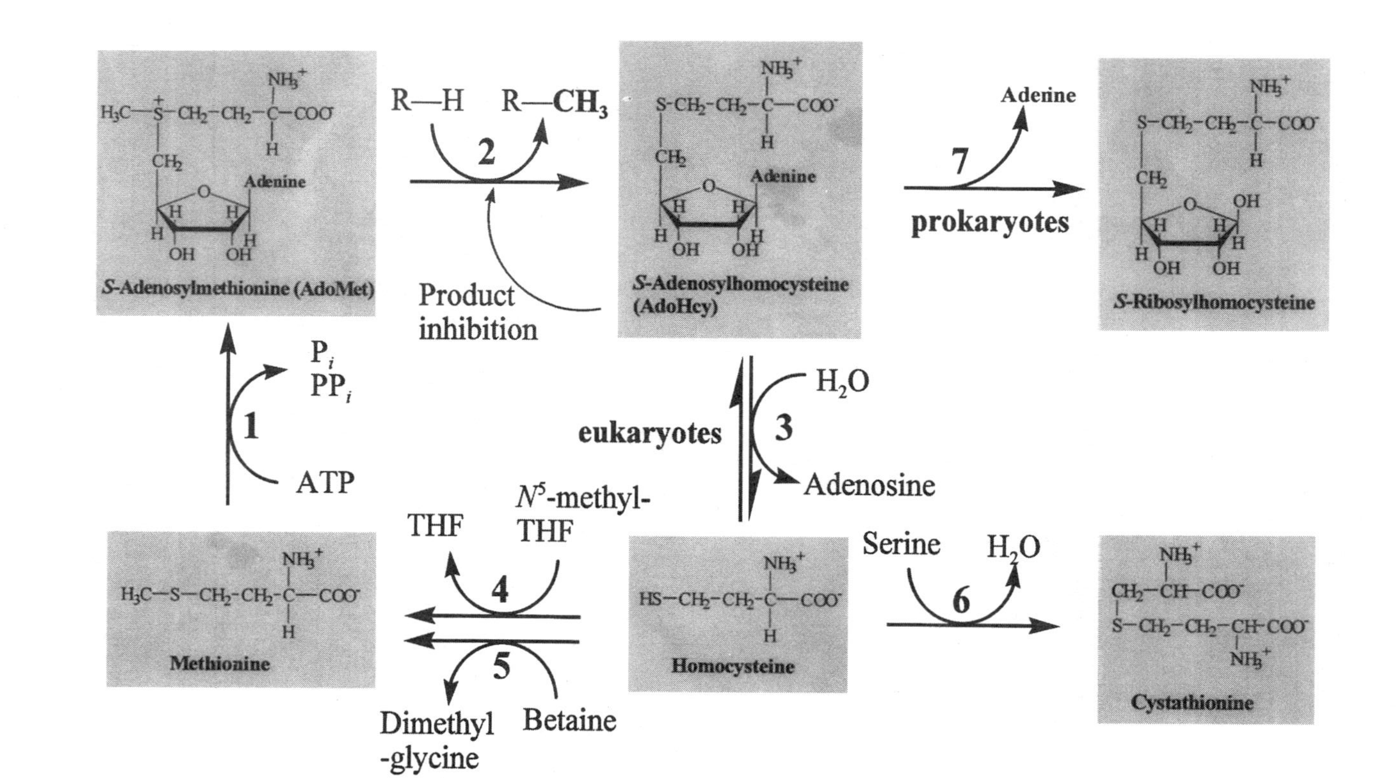

SCHEME 2.1 AdoMet-dependent methyltransferases and AdoHcy metabolism. Enzymes involved in the pathway: (**1**) methionine adenosyltransferase; (**2**) AdoMet-dependent methyltransferases; (**3**) AdoHcy hydrolase; (**4**) Hcy methyltransferase; (**5**) betaine-homocysteine methyltransferase; (**6**) cystathionine sythatase; (**7**) AdoHcy nucleosidase.

because of the difficulty in synthesizing such compounds, the potential problems with their cellular transport and metabolism, and the extremely high substrate specificity exhibited by different AdoMet-dependent methyltransferases. Each of the AdoMet-dependent methyltransferases utilizes only a single methyl acceptor substrate or a limited number of structurally related molecules. Analogs of the methyl donor substrate, AdoMet, have also been prepared by chemical procedures as possible inhibitors of AdoMet-dependent methyltransferases.[19] However, the analogs of AdoMet have poor substrate and/or inhibitor properties for methyltransferases. These results indicate that the enzymes have a very high specificity for the structural features of AdoMet. This disadvantage and problems associated with cellular transport of these compounds have diminished interest in this approach to the inhibition of methyltransferases.

A second approach that has been explored in an effort to inhibit AdoMet-dependent methyltransferases is that of altering the activity of methionine adenosyltransferase (Scheme 2.1, enzyme **1**).[20] The result of inhibiting this enzyme would be to decrease intracellular levels of AdoMet, thereby inhibiting all AdoMet-dependent enzymes. Numerous analogs of L-methionine have been prepared and evaluated as inhibitors of methionine adenosyltransferase.[20] Administration of some of the most potent methionine adenosyltransferase inhibitors to rodents in vivo resulted in the expected accumulation of L-methionine and depression of the levels of AdoMet in several tissues examined.[21] The apparent disadvantages of this approach include (1) the high structural specificity of methionine adenosyltransferase for L-methionine results in poor inhibitory activity of methionine analogs and (2) the fact that inhibition of this biosynthetic enzyme will result in a general inhibitory effect on all AdoMet-dependent methyltransferases, as well as polyamine biosynthesis.

Another approach to inhibiting AdoMet-dependent methyltransferases, which has attracted considerable attention and proved quite successful in vivo, is to focus on AdoHcy hydrolase as a primary target (Scheme 2.1, enzyme **3**).[6,8,22] AdoHcy is a competitive inhibitor of AdoMet-dependent methyltransferases. The rate of cellular methylation is regulated by existing intracellular ratios of AdoHcy/AdoMet.[23,24] Inhibition of AdoHcy hydrolase results in the intracellular accumulation of AdoHcy, causing a significant increase in the intracellular AdoHcy/AdoMet ratio and subsequent inhibition of AdoMet-dependent methylation reactions essential for viral replication.[6,8,25]

Inhibition of AdoHcy Hydrolase and Clinical Antiviral Potency

AdoHcy hydrolase inhibitors elevate the cellular levels of AdoHcy and AdoHcy/AdoMet ratios, which closely correlate with the extent of virus yield reduction.[8,26,27] De Clercq and Cools have demonstrated a strong relationship between the log IC_{50} values (inhibitor concentration at which viral replication reduced by 50%) against five viruses for a series of AdoHcy hydrolase inhibitors, including 3-deazaadenosine (C^3-Ado) and neplanocin A (NepA), and their log K_i values for inhibition of murine L929 cell or beef liver AdoHcy hydrolase.[28–30] Recently, they further confirmed the close correlations between inhibitory effects on murine L929 cell AdoHcy hydrolase activity and inhibitory effects on the replication of vaccinia virus and vesicular stomatitis virus with a series of acyclic and carbocyclic Ado analogs.[6] These correlations strongly suggest that AdoHcy hydrolase is the site of action for AdoHcy hydrolase inhibitors and that the subsequent inhibition of AdoMet-dependent methylation by elevated AdoHcy is the direct cause for inhibition of viral replication.

From a medicinal chemistry prospective, AdoHcy hydrolase is an attractive target because it can be specifically inhibited by a variety of nucleoside analogs. By increasing the specificity of inhibitors for AdoHcy hydrolase, cellular toxicity concerns have been reduced.[6,31] For example, NepA analogs including 9-(*trans*-2′-*trans*-3′-dihydroxycyclopent-4′-enyl)adenine (DHCeA) and 9-(*trans*-2′-*trans*-3′-dihydroxycyclopentanyl)adenine (DHCaA) retained inhibitory activity toward cellular AdoHcy hydrolase but were devoid of substrate properties for other cellular enzymes such as Ado kinase and Ado deaminase. These results suggested that rational drug design can be used to identify even more potent and more specific AdoHcy hydrolase inhibitors that might have clinical use as antiviral agents.

AdoHcy hydrolase inhibitors exert their antiviral activity through interaction with a targeted cellular enzyme (i.e., AdoHcy hydrolase), these compounds may be expected, on one hand, to be effective against a broad-spectrum of viruses and not to lead to the rapid development of virus–drug resistance. On the other hand, inhibitors of this enzyme may also affect the growth and metabolism of the normal host cells. AdoHcy hydrolase gene has been shown to be indispensable for the development of embryos, and deletion of this gene in mouse embryos results in death.[32] Complete inhibition of AdoHcy hydrolase could inevitably cause unwanted cytotoxicity. However, partial inhibition can be beneficial in suppressing viral replication. The current dogma does have its exceptions; interference with normal cellular metabolism does not preclude a clinically useful antiviral agent. In fact, to maintain antiviral activity but with reduced cytotoxicity, several approaches have been tried, some of which have been successful, including (1) partial or short-term inhibition, (2) low inhibitor concentrations that do not produce cellular toxicity but perturb viral replication;[33,34] (3) synergistic effects with other inhibitors such as ribavirin, an inhibitor of mRNA cap structure formation, or L-*cis*-AMB, an inhibitor of AdoMet biosynthesis;[35] and (4) designing AdoHcy hydrolase inhibitors that are sufficiently selective to exhibit functions specific to the virus-infected cells with acceptable toxic effects on uninfected cells. It is these possibilities that make AdoHcy hydrolase an attractive target for design of broad-spectrum antiviral agents.

CATALYSIS MECHANISM AND STRUCTURE OF AdoHcy HYDROLASE

Knowledge of the mechanism of catalysis of AdoHcy hydrolase has been useful in the design of specific mechanism-based inhibitors of the enzyme. These mechanism-based inhibitors can be converted to their inhibitory forms catalyzed by AdoHcy hydrolase. Also, the recently resolved three-dimensional structure of AdoHcy hydrolase showing exactly how the enzyme interacts with substrates and inhibitors provides a rational basis for the molecular design of potent inhibitors that will specifically act on AdoHcy hydrolase.[36]

Catalytic Mechanism of AdoHcy Hydrolase

By catalyzing the following reversible reaction, AdoHcy hydrolase regulates the intracellular levels of AdoHcy to modulate AdoMet-dependent methylations:[37]

$$\text{AdoHcy} + H_2O \rightleftharpoons \text{Ado} + \text{Hcy}$$

The equilibrium of the reaction favors the synthesis of AdoHcy with a K_{eq} of 10^{-6} M. Discrete steps in the mechanism of catalysis of AdoHcy hydrolase have been elucidated by Palmer and Abeles[38] and are depicted in Scheme 2.2. Like other members of the class of enzymes that employ NADH/NAD^+ as a prosthetic group, AdoHcy hydrolase binds the cofactor very tightly and utilizes its redox properties to enable catalysis of a bond scission that would not otherwise be within the catalytic capabilities available to enzymes.[39,40] The enzymatic synthesis of AdoHcy starts with the oxidation of the 3′-OH of Ado by an enzyme-bound NAD^+ (step **1**). A base at the active site removes the 4′ proton from the 3′-keto-Ado to form an α-carbanion (step **2**). After elimination of H_2O, 3′-keto-4′,5′-dehydro Ado is formed (step **3**). The latter then reacts with Hcy to from 3′-keto-AdoHcy α-carbanion (step **4**), which accepts a proton at the 4′ position to form 3′-keto-AdoHcy (step **5**). AdoHcy is formed after the reduction of 3′-keto AdoHcy by enzyme-bound NADH (step **6**). The hydrolysis of AdoHcy to Ado and Hcy is the reversal of steps 1–6.

The enzyme AdoHcy hydrolase possesses both "oxidative" and "hydrolytic" activities, utilizing general acid–base catalysis or low-barrier hydrogen bonds (LBHBs) to effect catalysis.[40,41] We have developed mechanism-based inhibitors that enable a clean separation of the "oxidative" and the "hydrolytic" activities and hence allow the catalytic acceleration factors for each activity to be studied separately.[8] *Type I* inhibitors (e.g., DHCeA) serve as substrates for the "oxidative" activity of the enzyme, producing the 3′-keto derivatives of the inhibitors and arresting the enzyme in the NADH form. These type I inhibitors have played pivotal roles in the successful determination of the structure of human AdoHcy hydrolase and the separation of the "oxidative" and "hydrolytic" activities of the enzyme catalytic process.[8] *Type II* inhibitors of AdoHcy hydrolase utilize the catalytic activity of the enzyme to generate electrophiles at the active site, which can then irreversibly inactivate the enzyme through covalent modification.[42–44]

Structures of AdoHcy Hydrolase

The amino acid sequences of AdoHcy hydrolase from over 30 different sources, including human placenta, rat liver, *Leishmania donovani*, and *Dictyostelium discoideum*,[45–48] have been deduced from their encoding cDNAs. The primary

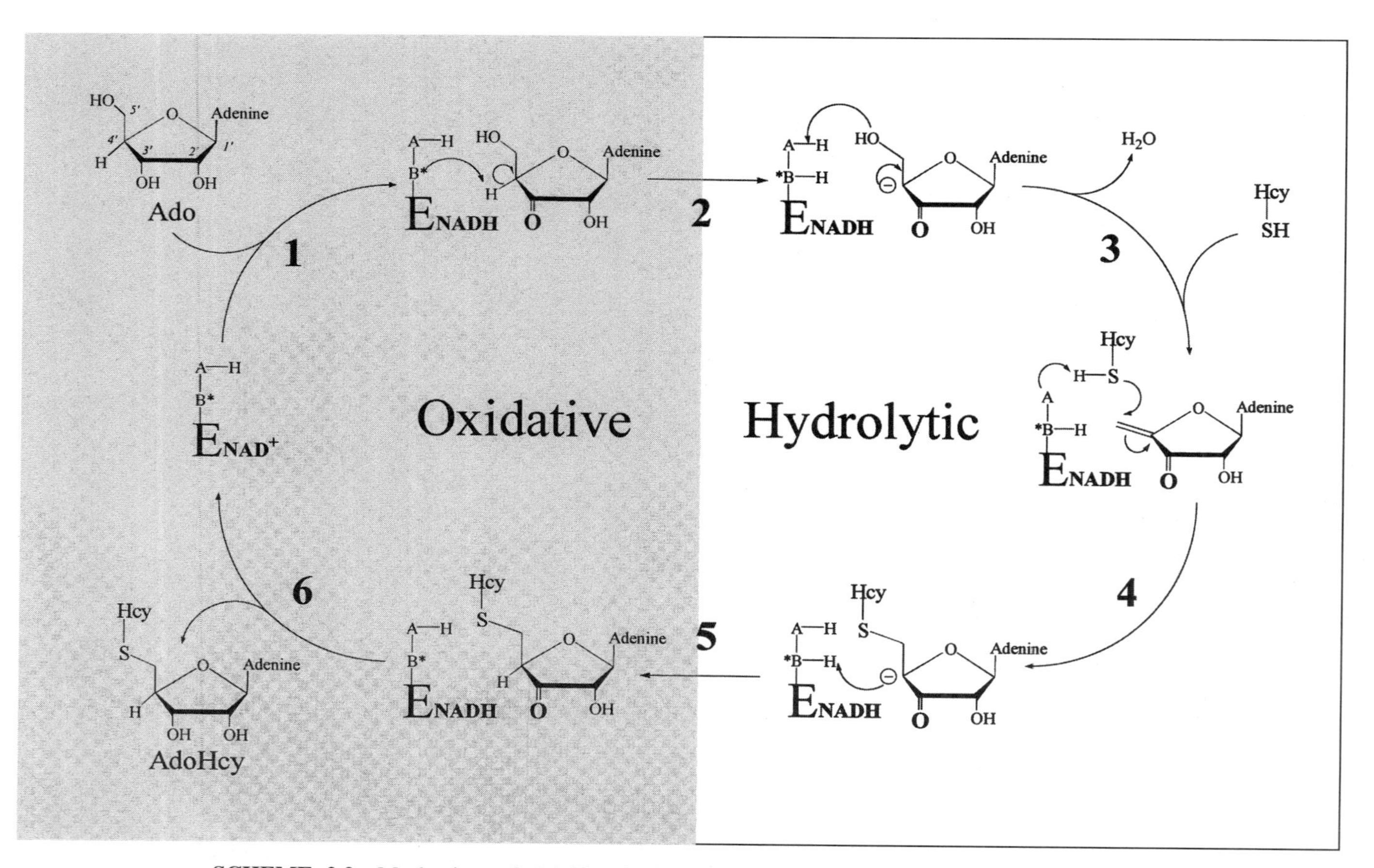

SCHEME 2.2 Mechanism of AdoHcy hydrolase–catalyzed interconversion of AdoHcy and Ado + Hcy.

structures of AdoHcy hydrolase are highly conserved (e.g., human placental and rat liver enzymes exhibit 97% homology). All the cloned AdoHcy hydrolases have a conserved NAD^+ binding site with a sequence of GYGDVGK. The mammalian AdoHcy hydrolase exists as a homotetramer of $\sim$48 kDa each subunit.[49] The crystal structure of human AdoHcy hydrolase complexed with NADH and an inhibitor 3′-keto-DHCeA was resolved by Turner et al.[40] Four nearly spherical subunits are arranged to form a flat, square tetramer with a hole in the center (Fig. 2.2*a*). The tetramer comprises two tightly associated dimers. Each subunit contains 432 amino acids, a cofactor NADH, and an inhibitor in the active site between two globular domains (Fig. 2.2*b*).

In addition to its unique catalytic capabilities, the structure of AdoHcy hydrolase offers potential insights into intersubunit regulation within the enzyme. Because it is a homotetrameric enzyme with an active site in each subunit, cooperative interactions are suggested on the basis of structural data of the human enzyme and on inhibitor studies.[40,44,50] The AdoHcy hydrolase monomer is structurally similar to NAD^+-dependent dehydrogenases with the substrate and cofactor binding sites located in a cleft between two domains (Fig. 2.3). However, cofactor binding and catalytic domains exhibit deviations from the classic dinucleotide binding fold. Unlike NAD^+-dependent dehydrogenases, the NAD^+ binding site of AdoHcy hydrolase requires interactions from two protein monomers. The *C* termini of the individual subunits appear to play crucial roles in facilitating monomer–monomer interactions within each dimer. Tyr430 and Lys426 in one monomer are involved in interactions with the NADH bound to the other monomer.[40,51,52]

Another interesting feature of the mechanism and structure of AdoHcy hydrolase is the role of water at the active site. Palmer and Abeles provided evidence for solvent exchange of the C4′H with $[H^3]$-H_2O (perhaps via an enolate intermediate shown in Scheme 2.2).[38] From the X-ray crystal structure of the human enzyme, the sequestered water molecule found at the active site is hydrogen bonded to His55, Asp131, and His301.[40] The water molecule appears to have a dual role in the catalytic mechanism. It not only is the sole candidate for the catalytic base responsible for the C4′ proton abstraction initiating Hcy cleavage but also probably adds to the intermediate 3′-keto-4′,5′-didehydro-5′-deoxyAdo in the formation of 3′-ketoAdo (Scheme 2.2).

Conformational Changes Associated with Catalysis

AdoHcy hydrolase undergoes a significant conformational change in alternating between the NAD^+ and NADH forms of the enzyme within each catalytic cycle.[53] Large differences in the hydrodynamic properties between the NAD^+ and NADH forms of the enzyme suggest that AdoHcy hydrolase adopts a more compact conformation on NAD^+ reduction and substrate oxidation.[54,55] As revealed by the crystal structure,[40] the inhibitor is completely buried in the active site located in a cleft between the two domains of each subunit in the NADH form (Fig. 2.3). The recently solved crystal structure of rat AdoHcy hydrolase in the NAD^+ form shows an open catalytic site in the absence of substrate.[56] Because domain II of each

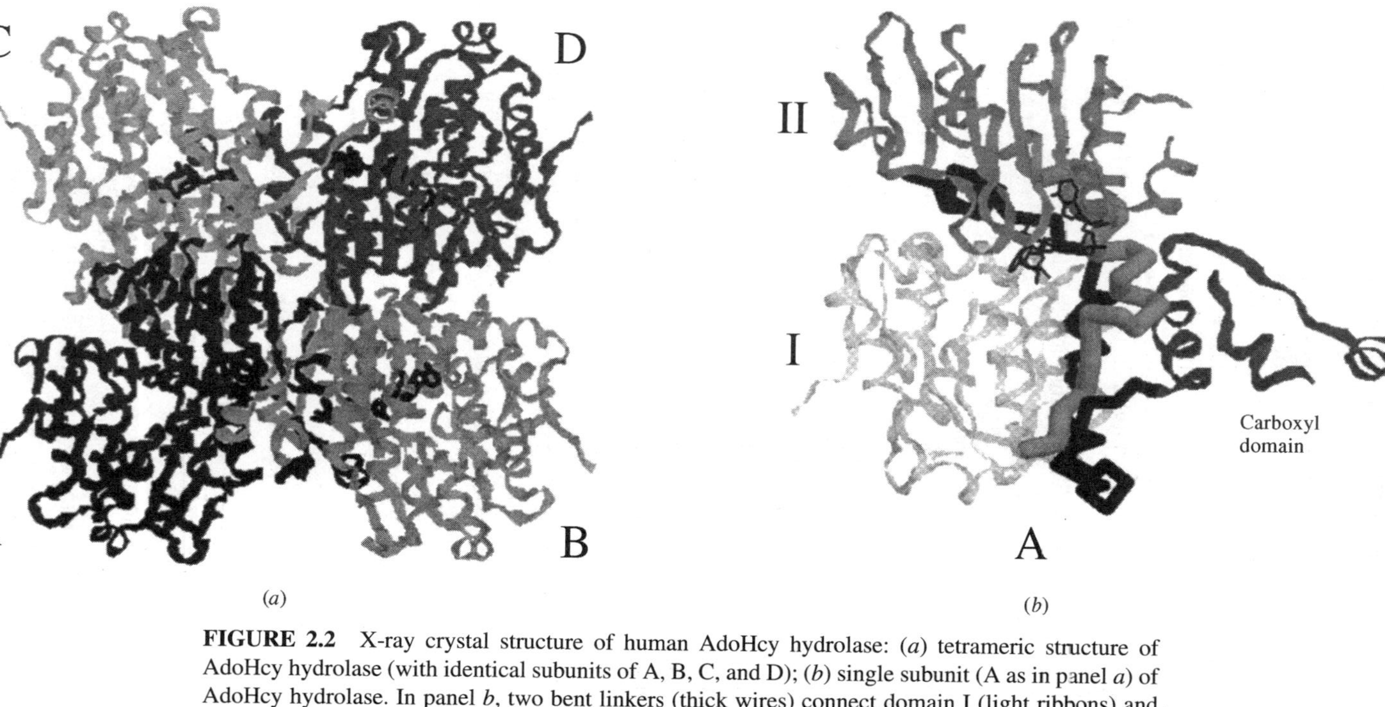

FIGURE 2.2 X-ray crystal structure of human AdoHcy hydrolase: (*a*) tetrameric structure of AdoHcy hydrolase (with identical subunits of A, B, C, and D); (*b*) single subunit (A as in panel *a*) of AdoHcy hydrolase. In panel *b*, two bent linkers (thick wires) connect domain I (light ribbons) and domian II (dark ribbons) of the subunit. Enzyme-bound cofactor NADH and the inhibitor 3′-keto-DHCeA in the catalytic site of AdoHcy hydrolase are represented by black sticks.

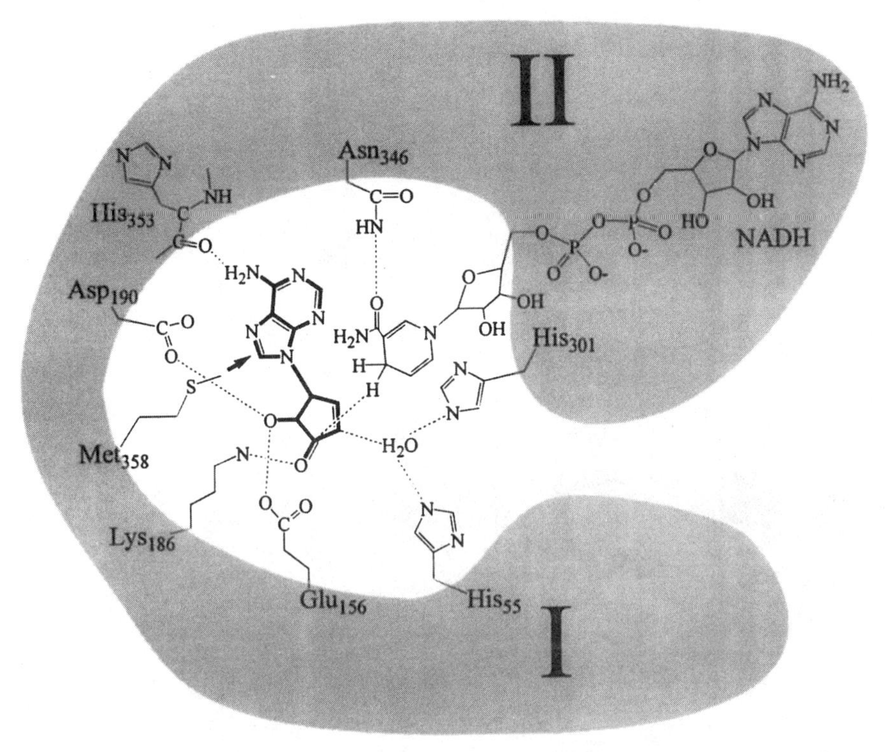

FIGURE 2.3 The interactions between AdoHcy hydrolase and the inhibitor, 3′-keto-DHCeA.

subunit in the enzyme is largely associated with the neighboring subunits, we have proposed that there is a hinged motion involving domain I of each subunit during the enzymatic reaction (Fig. 2.4).[54]

In the crystal structure of AdoHcy hydrolase complexed with the inhibitor, two bent linkers (linker I, residues 184–196; linker II, residues 350–360) connect the two domains in each subunit. Among all the known sequences of AdoHcy hydrolase, these two linkers ($VTK_{186}SKFD_{190}NLYGCR_{196}$ and $AMGH_{353}PSFVM_{358}SNS$) are highly conserved. Because residues Lys186, Asp190, His353, and Met358 within the two linkers are involved in binding to the oxidized substrate in the catalytic site (Fig. 2.3), it is possible that this binding induces bending of the two linkers, and subsequently swing the mobile domain I toward the stationary domain II. This hinged motion of domain I therefore closes up the catalytic site to engulf the substrate. In the "closed" conformation of the enzyme, His55 in domain I and His301 in domain II may act together for the enzyme hydrolytic activity.[40]

The model of alternative conformational changes between the "closed" and "open" forms of AdoHcy hydrolase is common among many other enzymes.[57–61] A classic example for this model is yeast hexokinase, whose crystal structures in the absence and presence of glucose have been resolved.[57] Another example is formate dehydrogenase (FDH), which shows striking structural homology with AdoHcy

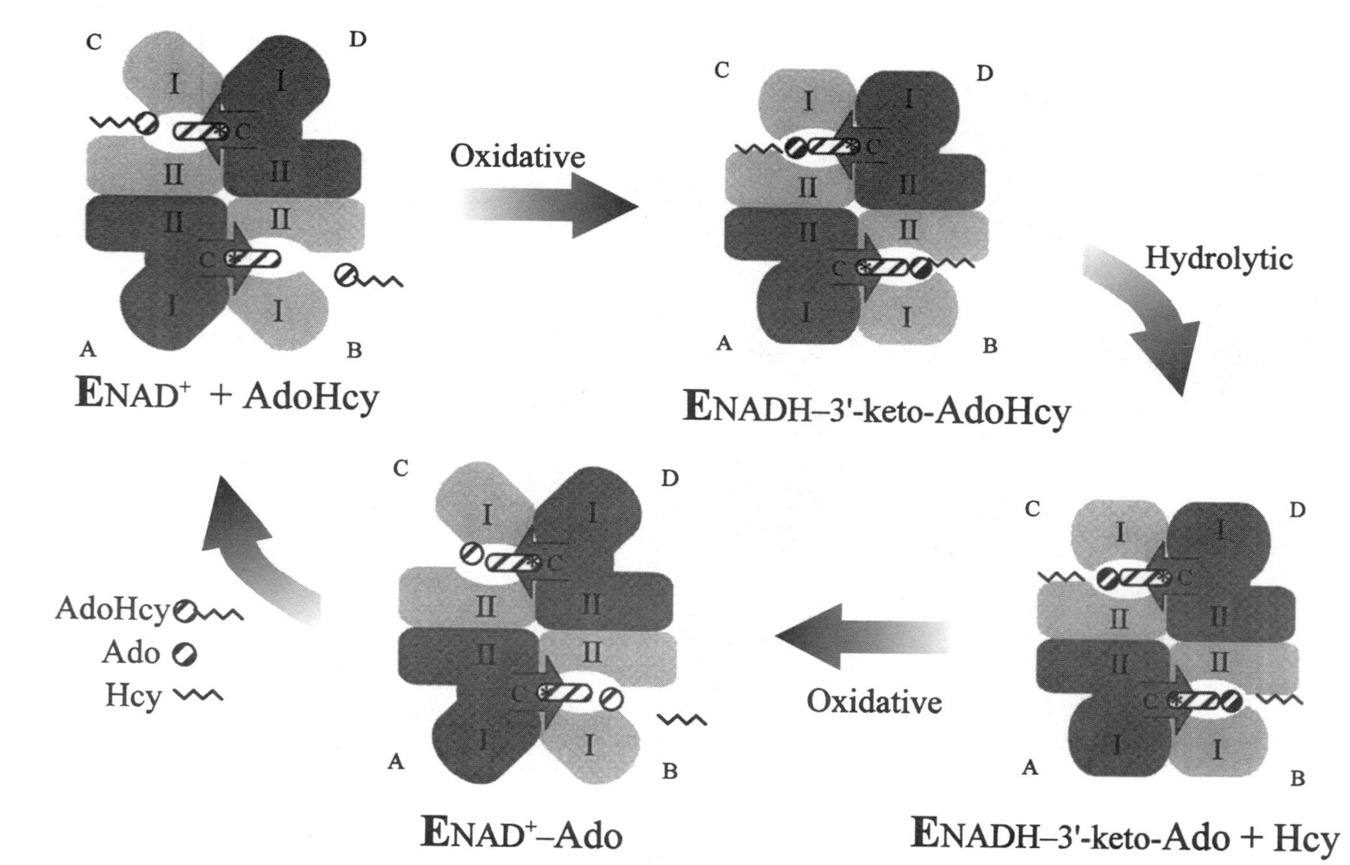

FIGURE 2.4 Structural changes of AdoHcy hydrolase associated with catalysis. Enzyme-bound cofactor is represented by hatched bar.

hydrolase.[58] From the structural information, it is possible to design specific inhibitors of the enzyme to restrain the conformational changes during enzymatic catalysis.[36]

AdoHcy HYDROLASE INHIBITORS: DESIGN AND MECHANISM OF INACTIVATION

Many inhibitors of AdoHcy hydrolase are compounds structurally similar to Ado (Fig. 2.5), a substrate of the enzyme. The inhibitors identified in the past 15 years (at the time of writing) can essentially be divided into three groups based on the history of their discovery. From the first to the third generation, the structural features and biochemical properties of these inhibitors have changed greatly. However, the general trend has been toward more potent and more specific inhibitors of AdoHcy hydrolase. The evolution of AdoHcy hydrolase inhibitors during this time period clearly reflects our increased understanding of the biochemical properties and mechanism of catalysis of AdoHcy hydrolase and its regulatory role in cellular metabolism.

First-Generation Inhibitors

The first generation of inhibitors of AdoHcy hydrolase was identified when AdoHcy hydrolase became an attractive target for the design of antiviral agents in the late 1970s and early 1980s.[62] Most of these compounds are naturally occurring carbocyclic Ado analogs [e.g., NepA, aristeromycin (Ari)] or synthetic cyclic or acyclic analogs of Ado [e.g., C^3-Ado, 9(*S*)-(2,3-dihydroxypropyl) adenine or (*S*)-DHPA, *D*-eritadenine, Ado dialdehyde] (Fig. 2.5). The mechanisms by which these compounds inhibit or inactivate AdoHcy hydrolase vary. Some of these compounds are competitive inhibitors [e.g., C^3-Ado, (*S*)-DHPA] that have K_i values ranging from 1 nM to 4 μM and bind at the same sites on the enzyme as does Ado.[8,63] However, they tend to have higher affinity for this site than do the natural substrates [K_M(Ado) $\approx$ 10 μM; K_M(AdoHcy) $\approx$ 70 μM].[64] Some of these compounds irreversibly inactivate AdoHcy hydrolase (e.g., NepA, D-eritadenine, and Ado dialdehyde).[29,65,66]

Among these first-generation inhibitors, NepA, which is a naturally occurring antibiotic, was of particular interest because of the unique mechanism by which it inactivated AdoHcy hydrolase. The inactivation of AdoHcy hydrolase by NepA was shown to be both time- and concentration-dependent. Scheme 2.3 illustrates the mechanism by which NepA inactivates the enzyme. After binding to the Ado binding site on the enzyme, NepA is oxidized by enzyme-bound NAD^+ to form 3′-keto-NepA, while NAD^+ is converted to NADH. The product, 3′-keto-NepA, has been isolated from the inactivated AdoHcy hydrolase under mild denaturing conditions, and the formation of enzyme-bound NADH has been verified by UV and fluorescence spectrometry.[67,68] Because the substrate activity of NepA results in conversion of the NAD^+ cofactor to its reduced form NADH, this process has been

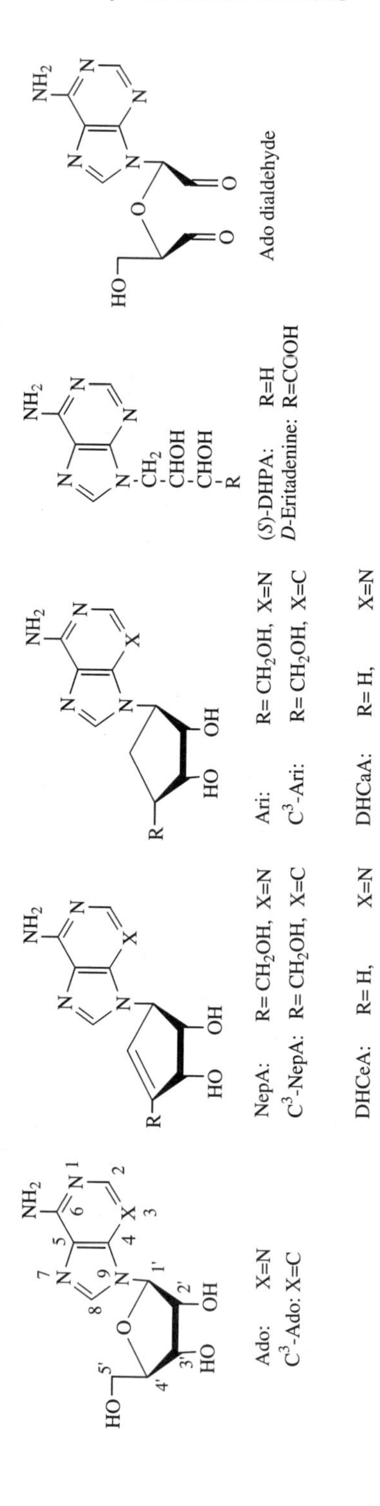

FIGURE 2.5 Structure of Ado and examples of first- and second-generation AdoHcy hydrolase inhibitors.

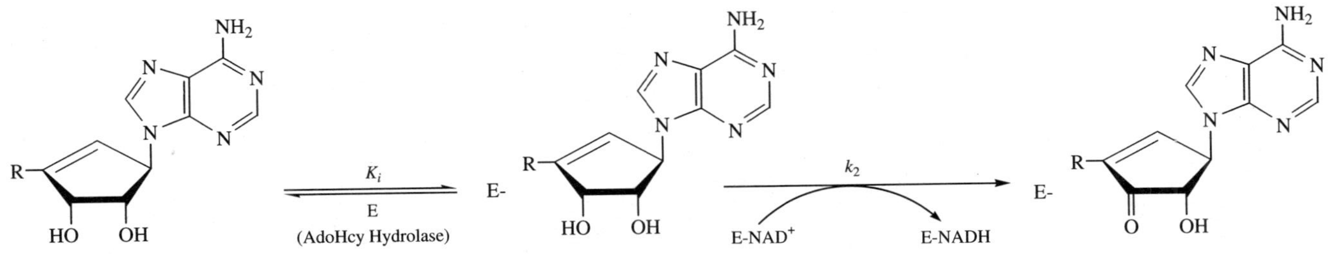

NepA $R=CH_2OH$
DHCeA $R=H$

K_i
E
(AdoHcy Hydrolase)

E-

k_2
E-NAD$^+$
E-NADH

E-

3'-keto-NepA $R=CH_2OH$
3'-keto-DHCeA $R=H$

SCHEME 2.3 Mechanism of inactivation of AdoHcy hydrolase by NepA and DHCeA, type I mechanism–based inhibitors.

referred to as a "cofactor depletion" mechanism, and compounds that inactivate AdoHcy hydrolase by this mechanism are referred to as *type I mechanism-based inhibitors* (see section titled "Catalysis Mechanism and Structure of AdoHcy Hydrolase," above). As discussed in the third subsection of the catalysis section, reduction of cofactor NAD^+ induces the closing of the catalytic site in AdoHcy hydrolase,[40,54,56] which traps the 3′-keto derivatives (e.g., 3′-keto-NepA) in its catalytic site and shuts off further enzymatic action.[69]

Of these first-generation inhibitors, NepA was one of the most potent antiviral agents.[29,30,70] However, the cellular toxicity of these first generation inhibitors of AdoHcy hydrolase precluded their advancement as clinically useful antiviral agents (see next section titled "AdoHcy Hydrolase Inhibitors: Antiviral Activity and Cytotoxic Effects," below). Nevertheless, by elucidating the mechanism by which these compounds produced their antiviral effects (e.g., inhibition of AdoHcy hydrolase) versus their cytotoxicity effects (e.g., substrate activity for Ado kinase and Ado deaminase), it was possible to design more potent AdoHcy hydrolase inhibitors that lacked the cytotoxicity of the first-generation inhibitors.

Second-Generation Inhibitors

Two different approaches were applied in an effort to design more potent and more specific inhibitors of AdoHcy hydrolase while precluding 5′-phosphorylation by Ado kinase and deamination by Ado deaminase. One approach involved replacing the adenine (Ade) moiety of NepA with 3-deazaadenine, resulting in C^3-NepA (Fig. 2.5). Another approach involved removing the 4′-hydroxymethyl substituent, producing DHCeA (Fig. 2.5).[8] Inhibition of AdoHcy hydrolase by C^3-NepA was reported to be reversible and competitive.[71–74] In contrast, DHCeA was shown to inactivate AdoHcy hydrolase irreversibly by the type I mechanism (Scheme 2.3).[75]

Third-Generation Inhibitors

The elucidation of the type I mechanism by which NepA and DHCeA cause inactivation of AdoHcy hydrolase has led several research groups to attempt to design type II mechanism–based AdoHcy hydrolase inhibitors, which can be catalytically activated by AdoHcy hydrolase and subsequently become covalently bound to the enzyme (see first subsection of the catalysis section, above). Initial efforts to prepare type II mechanism–based inhibitors exploited the oxidative activity of the enzyme to generate an electrophilic site on the inhibitor that could react with a protein nucleophile. For example, (Z)-4′,5′-didehydro-5′-deoxy-5′-fluoroadenosine (ZDDFA) was synthesized by Jarvi et al. in 1991 and shown to be a potent inhibitor of AdoHcy hydrolase (Fig. 2.6).[76] Enzyme inhibition was accompanied by reduction of NAD^+ to NADH and the release of fluoride ion, suggesting that ZDDFA might function by a type II mechanism (Scheme 2.4, pathways **b** and **b′**). However, subsequently it was shown that the mechanism of inactivation of AdoHcy hydrolase by ZDDFA involved rapid addition of water at the 5′ position of ZDDFA (hydrolytic activity) and elimination of fluoride ion, resulting

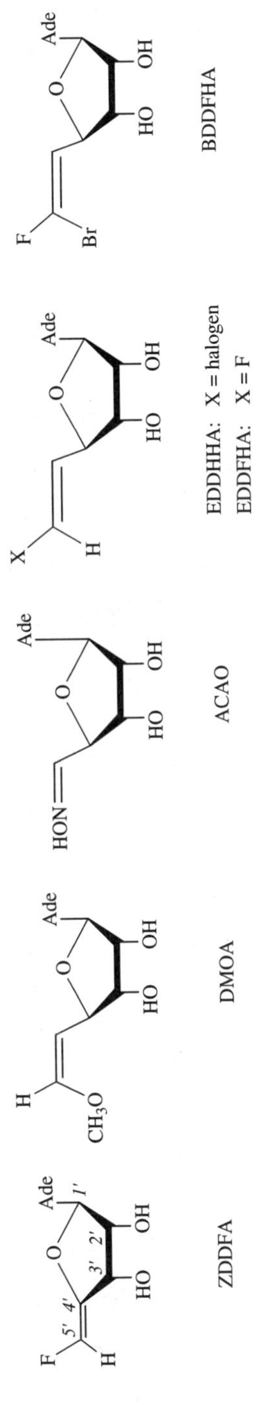

FIGURE 2.6 Structures of some inhibitors serving as substrates of the hydrolytic activity of AdoHcy hydrolase.

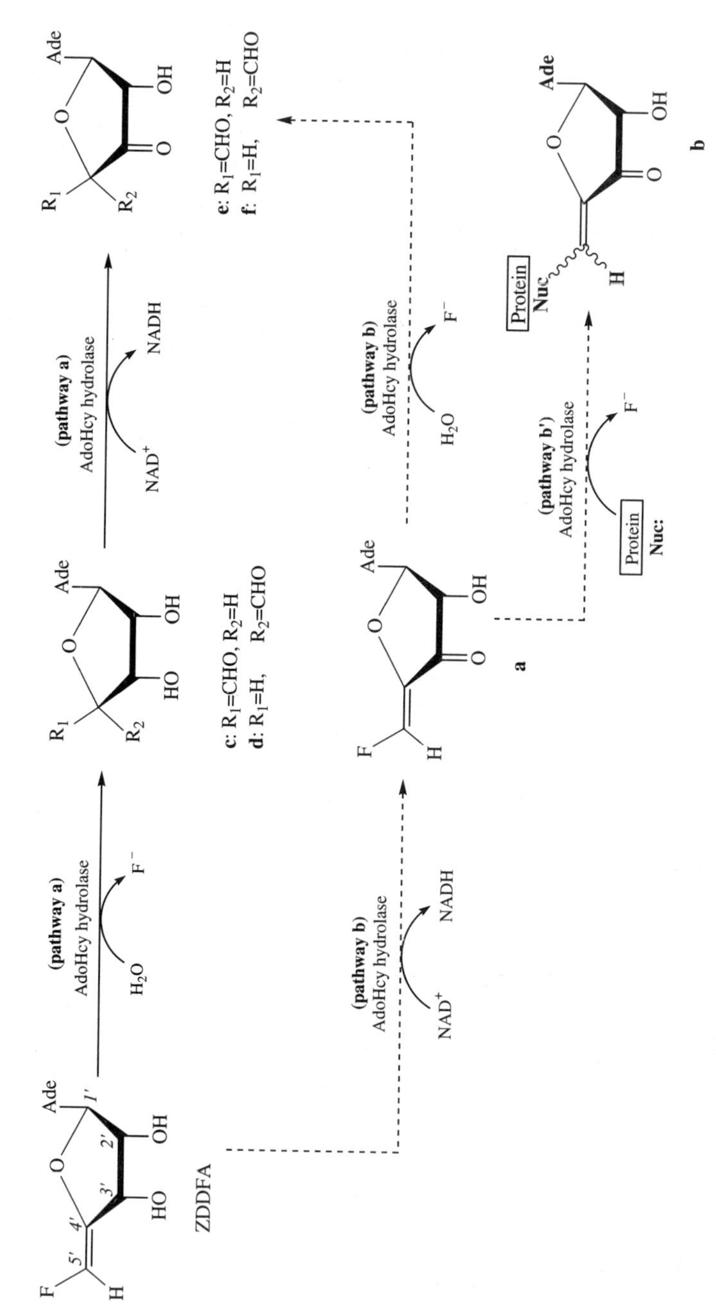

SCHEME 2.4 Mechanisms of inactivation of AdoHcy hydrolase by ZDDFA.

in the formation of Ado 5′-carboxaldehyde **c** and **d** (Scheme 2.4, pathway **a**).[8] The Ado 5′-carboxaldehydes were oxidized (oxidative activity) in a slower step to 3′-keto-5′-carboxaldehydes **e** and **f** by reduction of the enzyme-bound NAD^+ to NADH. Intermediate carboxaldehydes **c** and **d** were synthesized independently and proved to be potent type I mechanism–based inhibitors.[8] Carboxaldehyde c and ZDDFA have identical K_i values of 40 nM, but the k_{inact} value for carboxaldehyde **c** is 8 times greater than that for ZDDFA.[77]

These results clearly show that ZDDFA is simply a "pro inhibitor" for a type I mechanism–based inhibitor (Ado 5′-carboxaldehyde) of AdoHcy hydrolase. The unique aspect of this mechanism is that the conversion of the "pro inhibitor" (ZDDFA) to the "inhibitor" (Ado 5′-carboxaldehyde) actually occurs at the enzyme active site, utilizing the hydrolytic activity of the enzyme. These observations were particularly significant because they showed for the first time that the "hydrolytic activity" of the enzyme can function independently of the "oxidative activity." On the basis of the mechanism elucidated for the ZDDFA-induced inactivation of AdoHcy hydrolase, attempts have been made to use the hydrolytic activity of AdoHcy hydrolase to catalyze formation of type I mechanism–based inhibitors (e.g., Ado-5′-carboxaldehyde) using other nucleoside precursors or to catalyze the formation of strong electrophiles that could covalently modify the enzyme (type II mechanism–based inhibition). For example, like ZDDFA, 4′,5′-didehydro-5′-methoxy Ado (DMOA) and Ado 5′-carboxaldehyde oxime (ACAO) are potent inhibitors of AdoHcy hydrolase (Fig. 2.6).[78] The active species generated from DMOA or ACAO by the hydrolytic activity of the enzyme were determined to be Ado 5′-carboxaldehydes c and d (Scheme 2.4).

(*E*)-5′,6′-didehydro-6′-deoxy-6′-halohomoadenosines (EDDHHAs) (Fig. 2.6) are also substrates for the hydrolytic activity of AdoHcy hydrolase.[79] Here, the hydrolytic activity is defined as the ability of the enzyme to catalyze addition of water to the 5′,6′-bond of EDDHHAs. Scheme 2.5 shows the mechanism by which the fluorine derivative (EDDFHA) is processed by AdoHcy hydrolase. Incubation of EDDFHA with AdoHcy hydrolase produces a large molar excess of hydrolytic products [e.g., fluoride ion, Ade derived from chemical degradation of homoadenosine 6′-carboxaldehyde (HACA), and 6′-deoxy-6′-fluoro-5′-hydroxyhomoadenosine (DFHHA)] accompanied by a slow irreversible inactivation of the enzyme.[79] The enzyme inactivation is time-dependent, biphasic, and concomitant with the reduction of enzyme-bound NAD^+ to NADH. The reaction of EDDFHA with AdoHcy hydrolase proceeds by three pathways: pathway **a**, water attack at the 6′-position of EDDFHA and elimination of fluoride ion result in the formation of HACA, which degrades chemically to form Ade; pathway **b**, water attack at the 5′ position of EDDFHA results in the formation of DFHHA; and pathway **c**, oxidation of EDDFHA results in formation of the NADH form of the enzyme (inactive) and 3′-keto-EDDFHA, which could react with water at either the C5′ or C6′ position. The partition ratios among the three pathways were determined to be $k_{3'} : k_{6'} : k_{5'} = 1 : 29 : 79$, with one lethal event (enzyme inactivation) occurring every 108 nonlethal turnovers. The K_i and k_{inact} ($k_{3'}$) values for EDDFHA were determined to be 1.3 μM and 0.011 min^{-1}, respectively. The relatively large K_i and small k_{inact}

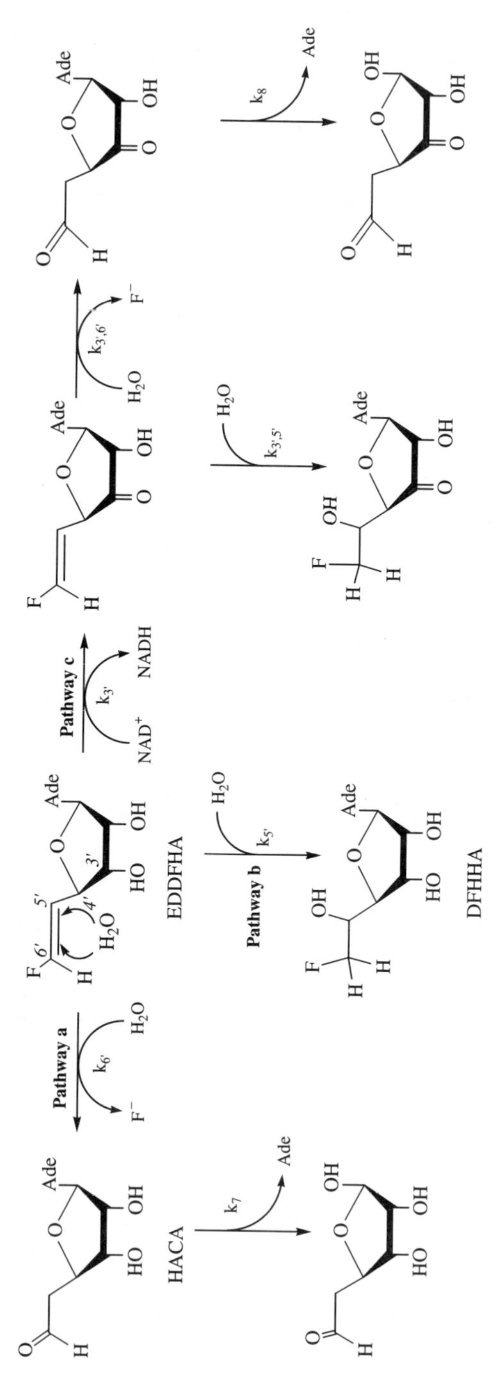

SCHEME 2.5 Mechanism of inactivation of AdoHcy hydrolase by EDDFHA.

values, which indicate that EDDFHA is a poor inhibitor of AdoHcy hydrolase, could be explained by the efficiencies of pathways **a** and **b**, which are nonlethal transformation of EDDFHA. In addition, the reaction products (Ade and DFHHA) from the hydrolytic activity of the enzyme effectively compete with EDDFHA for binding with AdoHcy hydrolase.

Similar results were obtained with the Cl, Br, and I analogs of EDDFHA.[79] The ratios of product of 5′ to 6′ hydrolytic activity are in the order of F > Cl > Br > I. While the partition ratios are in the order of I > Br > Cl > F. Although EDDFHA and its analogs are not potent inhibitors of AdoHcy hydrolase, these compounds suggested alternative ways by which the "5′/6′-hydrolytic activity" could be used to our advantage in designing inhibitors of this enzyme.

The first type II mechanism–based inhibitor of AdoHcy hydrolase that relies only on the hydrolytic activity of the enzyme for activation was characterized in 1998.[44] 6′-Bromo-5′,6′-didehydro-6′-fluorohomoadenosine (BDDFHA) was prepared by our laboratory (Fig. 2.6). Incubation of AdoHcy hydrolase with BDDFHA causes a maximum inactivation of 83%; however, the enzyme retains its original NAD^+/NADH content. The partial inactivation is concomitant with the release of both Br^- and F^- ions and formation of Ade. The enzyme can be covalently labeled with [8–^{3}H]BDDFHA, resulting in a stoichiometry of 2 mol of BDDFHA/per mole of the tetramer enzyme. Tryptic digestion and subsequent protein sequencing of the ^{3}H-labeled enzyme have revealed that Arg196 is the residue associated with the radiolabeled inhibitor. The partition ratio of Ade formation (nonlethal event) to covalent acylation (lethal event) is approximately 1 : 1.

From the experimental results described above, the mechanism by which BDDFHA caused inactivation has been proposed (Scheme 2.6).[44] BDDFHA functions as a substrate for the 6′-hydrolytic activity of the enzyme; water addition at the C6′ position of BDDFHA followed by elimination of Br^- ion results in the formation of homoAdo 6′-carboxyl fluoride (HACF). HACF then partitions in one of two pathways: (**a**) attack by a proximal nucleophile amino acid residue (Arg196) to form an amide bond after expulsion of F^B ion (lethal event) or (**b**) depurination to form Ade and hexose-derived 6-carboxyl fluoride, which is further hydrolyzed to hexose-derived 6-carboxylic acid and F^- ion (nonlethal event).

More recent results with the third-generation inhibitors (e.g., EDDFHA, BDDFHA) suggest that it should be possible to design highly specific and very potent inhibitors of AdoHcy hydrolase by exploiting the "hydrolytic activity" of the enzyme.

AdoHcy HYDROLASE INHIBITORS: ANTIVIRAL ACTIVITY AND CYTOTOXIC EFFECTS

The unique feature of AdoHcy hydrolase inhibitors as antiviral agents is their broad-spectrum antiviral activity[8] (Table 2.1); this encompasses: (1) some DNA viruses, whereas other DNA viruses such as herpetoviridae (herpes simplex virus, varicella zoster virus, and Epstein–Barr virus) are much less sensitive or insensitive to

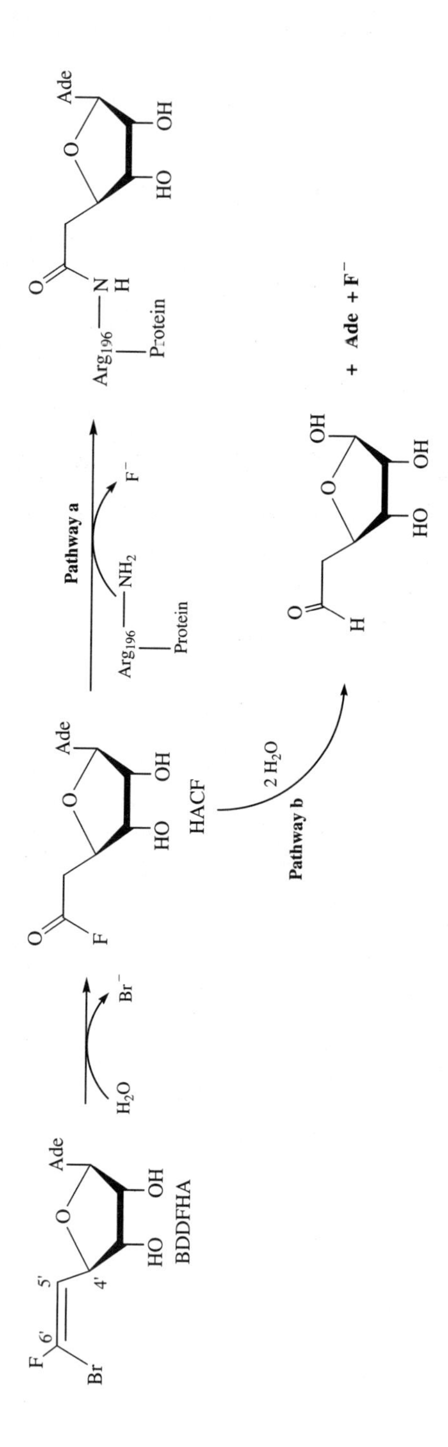

SCHEME 2.6 Mechanism of inactivation of AdoHcy hydrolase by BDDFHA, a type II mechanism-based inhibitor.

TABLE 2.1 Antiviral Activity Spectrum of AdoHcy Hydrolase Inhibitors

DNA viruses	Herpesviridae	Human cytomegalovirus (HCMV)
	Iridoviridae	African swine fever virus
	Poxviridae	Vaccinia virus
(−)RNA viruses	Paramyxoviridae	Parainfluenza virus, measles virus, mumps virus, espiratory syncytial virus (RSV)
	Arenaviridae	Junin virus, Tacaribe virus
	Rhabdoviridae	Vesicular stomatitis virus, rabies virus
	Filoviridae	Ebola virus
(±)RNA viruses	Reoviridae	Reovirus, rotavirus

AdoHcy hydrolase inhibitors; (2) (−)RNA viruses; and (3) (±)RNA viruses. In contrast, (+)RNA viruses [other than retroviruses, such as picomaviruses (poliomyelitis virus, Coxsackie virus) and togaviruses (Sindbis virus, Semliki forest virus)] are virtually insensitive to AdoHcy hydrolase inhibitors. Also, yellow fever virus, a (+)RNA virus belonging to the flavivirus family, is not affected by AdoHcy hydrolase inhibitors.

As described in the preceding section (on inhibitors), AdoHcy hydrolase inhibitors with increased potency, specificity, and reduced cytotoxicity have been designed in recent years. The first-generation AdoHcy hydrolase inhibitors are strikingly similar in their broad-spectrum antiviral activity. These compounds were observed to exhibit more selective activity toward DNA viruses (e.g., HCMV, vaccinia, and African swine fever virus), single-stranded (−)RNA viruses (e.g., vesicular stomatitis, rabies, parainfluenza and measles), and double-stranded (±)RNA virus (e.g., reovirus and rotavirus).[6] NepA is the most potent antiviral agent among the first-generation inhibitors. The IC_{50} values of NepA for different viruses vary from 0.01 to 0.3 fg/mL, which is about 2000 times lower than that of (*S*)-DHPA. It has been demonstrated that the inhibition of virus multiplication by NepA coincides with a rapid inhibition of AdoHcy hydrolase activity in the vaccinia-virus-infected cells and a subsequent 10-fold increase in the intracellular AdoHcy/AdoMet ratio.[80] In addition, the incorporation of [^{3}H]methyl groups into RNA isolated from vaccinia-virus-infected, Ado dialdehyde-treated cells was inhibited by approximately 30% in the cytoplasmic fraction and approximately 15% in the poly A^+-mRNA fraction compared to untreated controls. These data provide direct evidence in support of a mechanism involving inhibition of the methylation of the mRNA-capped structure, which results in suppressed translation of viral proteins essential for viral replication.[8]

Unfortunately, the first-generation inhibitors have a common problem of cellular toxicity, which precludes their clinical use as antiviral agents. Is their toxicity due to inhibition of AdoHcy hydrolase or interactions with other enzymes? Clues concerning the nature of this toxicity were derived from cellular metabolism studies on NepA (Scheme 2.7) and Ari. These studies showed that multifunctional

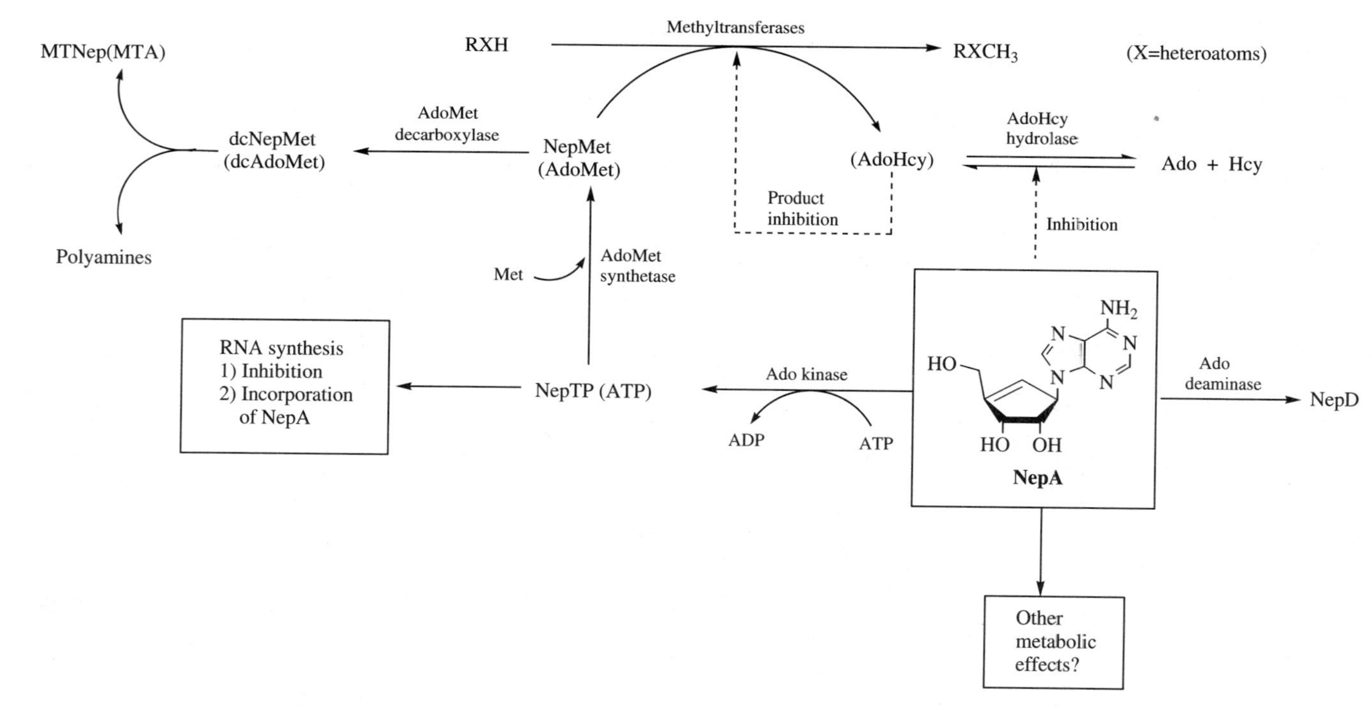

SCHEME 2.7 Metabolic pathways of NepA (normal Ado metabolites in parentheses).

metabolic activity is the major cause of toxicity. A concentration as low as 0.1 fM NepA was shown to inactivate 90% of the AdoHcy hydrolase in mouse L929 fibroblast cells, resulting in a marked increase in cellular AdoHcy/AdoMet levels.[80] While not a substrate for AdoHcy hydrolase, NepA was shown to serve as a substrate for Ado deaminase. NepA is converted by Ado deaminase to the biologically inactive NepD, the carbocyclic analog of inosine. This metabolic route does not seem to be important in mediating the toxic effects of NepA because administration of deoxycoformycin or EHNA (both Ado deaminase inhibitors) did not potentiate the effects of this carbocyclic nucleoside in several different cell lines.[8] However, this metabolic route may be important in vivo (e.g., NepA was not effective in protecting newborn mice against a lethal infection of vesicular stomatitis virus).[28] NepA was also shown to be a substrate for Ado kinase, leading to the formation of the 5′-triphosphate analog (NepTP), which is subsequently utilized as a substrate by methionine adenosyltransferase to generate the corresponding AdoMet derivative, *S*-neplanocylmethionine (NepMet).[8] Metabolism of NepA by this phosphorylation pathway has been proposed as the mechanism by which this agent produces cytotoxicity.[81]

Ari also exhibits significant cytotoxicity.[62] Its metabolism is somewhat similar to that of NepA. Kinases also metabolize Ari to its 5′-phosphate derivatives, and these nucleotides have been implicated in cellular toxicity. However, Ari apparently kills cells by different mechanisms in Ado kinase–deficient ($AdoK^-$) and normal ($AdoK^+$) cell lines.[82] In $AdoK^+$ cells, the phosphate metabolites of Ari are presumably responsible for the toxicity, while in the $AdoK^-$ cell, AdoHcy hydrolase inhibition may be the cause. Complete inhibition of AdoHcy hydrolase has been implicated in cytotoxicity; this could be due, at least in part, to Hcy depletion.[83] Another Ari metabolite, carbocylic GMP, is a possible cause of toxicity. The 5′-monophosphate of Ari serves as a substrate for AMP deaminase, converting it to the IMP analog of Ari, which is then transformed to phosphates of carbocyclic guanosine.[84] The carbocyclic analog of GMP is a good inhibitor of hypoxanthine–guanine phosphoribоxyltransferase, an important enzyme in the purine salvage pathway. This explains the complete blockade of the utilization of hypoxanthine and guanine by cells on treatment with Ari.

The second generation of inhibitors, which are NepA and Ari analogs, were designed to eliminate non-specific metabolic effects (e.g., through Ado deaminase and Ado kinase catalysis). It has been demonstrated that these second-generation AdoHcy hydrolase inhibitors exhibit broad-spectrum antiviral activity, whereas their cytotoxicity is considerably lower than that of the parent compounds (Table 2.2).[6] For example, NepA is a more potent inhibitor of vaccinia virus replication (IC_{50}) in murine L929 fibroblast cells than DHCeA by a factor of 3.5; however, DHCeA is 34 times less cytotoxic (measured as ID_{50} value, the concentration of drug that causes 50% inhibition of cellular replication). DHCeA is, therefore, a better antiviral agent than NepA by a 10-fold increase in the selectivity index ($SI = ID_{50}/IC_{50}$). By this criterion, the Ari analogs DHCaA and C^3-DHCaA also were much better antiviral agents than Ari by 670- and 330-fold increases in their SI values, respectively.[8]

TABLE 2.2 Comparison of Antiviral Potency, Cytotoxicity, and Selectivity of Second-Generation AdoHcy Hydrolase Inhibitors with their Parent Compounds against Vaccinia Virus (VV) Replication in Murine L929 Cells and African Swine Fever Virus (ASFV) Replication in CCL-81 Vero Cells[8]

	Antiviral Activity (IC_{50}, μM)[a]		IC_{50} (analog) over IC_{50} (parent)		Cytotoxicity (ID_{50}, μM)[b]		ID_{50} (analcg) over ID_{50} (parent)		Selectivity Index ID_{50}/IC_{50}	
Compounds	VV	ASFV	VV	ASFV	L929	CCL-81	VV	ASFV	VV	ASFV
NepA	0.08	0.57	1	1	0.5	38	1	1	6	66
DHCeA	0.28	2.15	3.5	3.8	17	858.4	34	22.6	61	400
C^3-DHCeA	0.95	4.31	11.9	7.6	56	858.4	112	22.6	59	200
Ari	6.62	—	1	—	4.3	—	1	—	0.64	—
DHCaA	0.17	1.29	0.03	—	73	858.4	17	—	429	667
C^3-DHCaA	0.13	0.34	0.02	—	27.4	858.4	6	—	221	2500

[a] Antiviral concentration required to effect a 50% reduction in plaque formation.
[b] Cytotoxic concentration inhibiting 50% of cell growth.

Antiviral activities of the second-generation inhibitors against the replication of African swine fever virus (ASFV) in Vero cells were also reported in Table 2.2,[85] the SI value for NepA is only 66, whereas DHCeA (SI = 400), C^3-DHCeA (SI = 200), DHCaA (SI = 667), and C^3-DHCaA (SI = 2500) have much higher SI values and thus are much less cytotoxic. The mechanism of antiviral action of the second-generation inhibitors has been demonstrated to be based on the inhibition of AdoHcy hydrolase, resulting in the elevation of intracellular AdoHcy/AdoMet ratio and subsequent suppression of methylation reactions needed for viral mRNA maturation.[62] Elimination of Ado kinase and Ado deaminase substrate activities apparently led to the reduced toxicity of these inhibitors.[30]

Huggins et al.[86] found that C^3-Ado and C^3-NepA were potent in vitro inhibitors of filoviruses Ebola and Marburg replications with EC_{50} values of 8 and 0.5 fg/mL, respectively. Immune-deficient SCID mice infected with Ebola virus (Mayinga strain, 100 LD_{50}) die uniformly with a mean time to death of 27 days, but without hemorrhagic disease. Twice daily prophylactic administration to mice of C^3-Ado or C^3-NepA increased mean time to death and suppressed virus replication in both circulation and major organs. C^3-NepA, the more potent inhibitor, was further evaluated in a 100% lethal African green monkey–Ebola primate model that closely mimics the clinical disease seen in humans. Prophylactic thrice daily (tid) dosing at the maximum tolerated dose (MTD) and $\frac{1}{3}$ MTD beginning 24 h prior to infection resulted in a significant increase in mean time to death in both groups (twofold with MTD/3) and reduction in the amount of viral antigen compared to placebo treatment.

Antiviral activity data for the third-generation AdoHcy hydrolase inhibitors are limited at present. ZDDFA was shown to be a potent antiviral agent against vaccinia virus.[33] Its IC_{50} and ID_{50} values were determined to be 0.05 and 15.6 fM, respectively, with a SI value of 312. ZDDFA has potential as an antiretroviral agent against Moloney leukemia virus (Molv).[87] The IC_{50} value of ZDDFA against Molv replication in murine fibroblasts (SC-1) cells was 0.18 fM, and the ID_{50} for SC-1 cells growth was 214 fM, which gives a SI value of 1189. The antiretroviral activity of ZDDFA has been demonstrated to correlate with its ability to inhibit AdoHcy hydrolase and maintain elevated intracellular AdoHcy levels.[86] ZDDFA is also an effective antiviral agent against ASFV.[85] The mechanism of the anti-ASFV action of ZDDFA is postulated to depend on the inhibition of AdoHcy hydrolase, which results in suppression of viral mRNA maturation. The IC_{50} value of ZDDFA against ASFV replication in Vero cells was estimated to be 0.077 fM, and the ID_{50} value was 52.1 fM, which gives a SI value of 677.

Comprehensive evaluation of the present third-generation AdoHcy hydrolase inhibitors as potential antiviral agents is under way. In general, among the three generations of AdoHcy hydrolase inhibitors, the second- and third-generation inhibitors are more specific and less toxic owing to their elimination of other metabolic activities. These compounds (e.g., C^3-NepA, DHCaA, ZDDFA, and BDDFHA) should be further explored for their potential in the treatment of various virus infections [e.g., pox (molluscum contagiosum, monkey pox), paramyxo (parainfluenza, measles), arena (Lassa, Machupo, Junin), rhabodo (rabies), Filo

(Ebola, Marburg), reo (rota)] that fall within the realm of the AdoHcy hydrolase inhibitors and for which there is currently no (chemo) therapy available. In particular, such inhibitors should be pursued as potential candidate drugs for the treatment of Ebola virus infections, as suggested by studies with C^3-NepA in SCID mice and African green monkeys infected with Ebola virus.[85] The filoviruses Ebola and Marburg cause the most severe viral hemorrhagic fevers with mortalities of 40–90% in sporadic human outbreaks, and no effective therapeutic agents are available at the present time to fight these viruses.

CONCLUSION

It is our hope that the recent success in the determination of the X-ray structure of human AdoHcy hydrolase[40] will open new avenues for the design of AdoHcy hydrolase inhibitors and for structure-based searches for potential non-nucleoside pharmacophores as inhibitors of the enzyme with higher selectivity and lower toxicity. Although AdoHcy hydrolase and its inhibitors have been studied since the late 1980s, understanding the structure and mechanism of this enzyme as well as the causes of cytotoxicity related to its inhibition holds promise for the control of many diseases from carcinogenesis to deadly viral infections and from inflammation to atherosclerosis.

REFERENCES

1. De Clercq, E., *Advances in Antiviral Drug Design*, JAI Press, Greenwich, CT, 1996, Vol. 2.
2. Narayan, P.; Rottman, F. M., *Advances in Enzymology*, A. Meister, ed., Wiley, New York, 1992, Vol. 65, pp. 255–285.
3. Dimmock, N. J.; Primrose, S. B., *Introduction to Modern Virology*, Blackwell Science, London, 1994, pp. 60–82.
4. Chiang, P. K.; Gordon, R. K.; Tal, J.; Zeng, G. C.; Doctor, B. P.; Pardhasaradhi, K.; McCann, P. P., *FASEB J.* 1996, **10**, 471–480.
5. Smolin, L. A.; Benevenga, N. J., *Absorption and Utilization of Amino Acids*, M. Friedman, ed., CRC Press, Boca Raton, FL, 1989, Vol. 1, pp. 157–187.
6. De Clercq, E., *Nucleosides Nucleotides* 1998, **17**, 625–634.
7. Chiang, P. K., *Pharmacol. Ther.* 1998, **77**, 115–134.
8. Yuan, C. S.; Liu, S.; Wnuk, S. F.; Robins, M. J.; Borchardt, R. T., in *Advances in Antiviral Drug Design*, E. De Clercq, ed., JAI Press, Greenwich, CT, 1996, Vol. 2, pp. 41–48.
9. Morrison, J. M., *Virus Induced Enzymes*, Wiley, New York, 1991, pp. 521–539.
10. Gallie, D. R., *Plant Molec. Biol.* 1996, **32**, 145–158.
11. Furuichi, Y.; LaFiandra, A.; Shatkin, A. J., *Nature* 1977, **266**, 235–239.
12. Both, G. W.; Furuichi, Y.; Muthukrishnan, S.; Shatkin, A. J., *Cell* 1975, **6**, 185–195.
13. Krug, R. M.; Broni, B. A.; Bouloy, M., *Cell* 1979, **18**, 329–334.
14. Banerjee, A. D.; Abraham, G.; Colonno, R. J., *J. Gen. Virol.* 1977, **34**, 1–8.

15. Ransohoff, R. M.; Narayan, P.; Ayers, D. F.; Rottman, F. M.; Nilsen, T. W., *Antiviral Res.* 1987, **7**, 317–327.
16. Barbosa, E.; Moss, B., *J. Biol. Chem.* 1978, **253**, 7692–7697.
17. Martin, S. A.; Moss, B., *J. Biol. Chem.* 1975, **250**, 9330–9335.
18. Borchardt, R. T., in *The Biochemistry of Adenosylmethionine*, F. Salvatore, E, Borek, V. Zappia, H. G. Williams-Ashman, and F. Schlenk, eds., Columbia Univ. Press, New York, 1977, pp. 151–170.
19. Zappia, V.; Zydek-Cwick, R.; Schlenk, F., *J. Biol. Chem.* 1969, **244**, 4499–4509.
20. Chou, T. C.; Coulter, A. W.; Lombardini, J. B.; Sufrin, J. R.; Talalay, P., in *The Biochemistry of Adenosylmethionine*, F. Salvatore, E. Borek, V. Zappia, H. G. Williams-Ashman, and F. Schlenk, eds., Columbia Univ. Press, New York, 1977, pp. 18–36.
21. Lombardini, J. B.; Talalay, P., *Molec. Pharmacol.* 1973, **9**, 542–560.
22. De Clercq, E., *Biochem. J.* 1982, **205**, 1–13.
23. Chiang, P. K.; Cantoni, G. L., *Biochem. Pharmacol.* 1979, **28**, 1897–1902.
24. Cantoni, L.; Budillon, G.; Cuomo, R.; Rodino, S.; Le Grazie, C.; Di Padova, C.; Rizzardini, M., *Scand. J. Gastroenterol.* 1990, **25**, 1034–1040.
25. Keller, B. T.; Borchardt, R. T., in *Biological Methylation and Drug Design*, R. T. Borchardt, C. R. Creveling, and P. M. Ueland, eds., Humana Press, Clifton, NJ, 1986, pp. 385–396.
26. Cools, M.; De Clercq, E., *Biochem. Pharmacol.* 1990, **40**, 2259–2264.
27. Ramakrishnan, V.; Borchardt, R. T., *Neurochem. Int.* 1987, **10**, 423–431.
28. De Clercq, E.; Cools, M., *Biochem. Biophys. Res. Commun.* 1985, **129**, 306–311.
29. De Clercq, E., *Biochem. Pharmacol.* 1987, **36**, 2567–2575.
30. Cools, M.; De Clercq, E., *Biochem. Pharmacol.* 1989, **38**, 1061–1067.
31. Wolfe, M. S.; Borchardt, R. T., *J. Med. Chem.* 1991, **34**, 1521–1530.
32. Miller, M. W.; Duhl, D. M.; Winkes, B. M.; Arredondo–Vega, F.; Saxon, P. J.; Wolff, G. L.; Epstein, C. J.; Hershfield, M. S.; Barsh, G. S., *EMBO. J.* 1994, **13**, 1806–1816.
33. Liu, S.; Wolfe, M. S.; Borchardt, R. T., *Antiviral Res.* 1992, **19**, 247–265.
34. Hasobe, M.; McKee, J. G.; Ishii, H.; Cools, M.; Borchardt, R. T.; De Clercq, E., *Molec. Pharmacol.* 1989, **36**, 490–496.
35. Ishii, H.; Hasobe, M.; in McKee, J. G.; Ault-Riche, D. B.; Borchardt, R. T., *Antiviral Chem. Chemother.* 1993, **4**, 127–130.
36. Air, G. M.; Luo, M., in *Structural Biology of Viruses*, W. Chiu, R. M. Burnett, and R. L. Garcea, eds., Oxford Univ. Press, New York, 1997, pp. 411–431.
37. de la Haba, G.; Cantoni, G. L., *J. Biol. Chem.* 1959, **234**, 603–608.
38. Palmer, J. L.; Abeles, R. H., *J. Biol. Chem.* 1979, **254**, 1217–1226.
39. Clarke, A. R.; Dafforn, T. R., in *Comprehensive Biological Catalysis*, M. L. Sinnott, ed., Academic Press, London, 1998, Vol. III, pp. 1–76.
40. Turner, M. A.; Yuan, C. S.; Borchardt, R. T.; Hershfield, M. S.; Smith, G. D.; Howell, P. L., *Nat. Struct. Biol.* 1998, **5**, 369–376.
41. Kumar, G. A.; McAllister, M. A., *J. Am. Chem. Soc.* 1998, **120**, 3159–3165.
42. Robins, M. J.; Wnuk, S. F.; Yang, X.; Yuan, C. S.; Borchardt, R. T.; Balzarini, J.; De Clercq, E., *J. Med. Chem.* 1998, **41**, 3857–3864.

43. Wnuk, S. F.; Mao, Y.; Yuan, C. S.; Borchardt, R. T.; Andrei, G.; Balzarini, J.; De Clercq, E.; Robins, M. J., *J. Med. Chem.* 1998, **41**, 3078–3083.
44. Yuan, C. S.; Wnuk, S. F.; Robins, M. J.; Borchardt, R. T., *J. Biol. Chem.* 1998, **273**, 18191–18197.
45. Coulter-Karis, D. E.; Hershfield, M. S., *Ann. Hum. Genet.* 1989, **53**, 169–175.
46. Ogawa, H.; Gomi, T.; Mueckler, M. M.; Fujioka, M.; Backlund, P. S., Jr.; Aksamit, R. R.; Unson, C. G.; Cantoni, G. L., *Proc. Natl. Acad. Sci.* (USA) 1987, **84**, 719–723.
47. Henderson, D. M.; Hanson, S.; Allen, T.; Wilson, K.; Coulter-Karis, D. E.; Greenberg, M. L.; Hershfield, M. S.; Ullman, B., *Molec. Biochem. Parasitol.* 1992, **53**, 169–183.
48. Kasir, J.; Aksamit, R. R.; Backlund, P. S., Jr.; Cantoni, G. L., *Biochem. Biophys. Res. Commun.* 1988, **153**, 359–364.
49. Richards, H. H.; Chiang, P. K.; Cantoni, G. L., *J. Biol. Chem.* 1978, **253**, 4476–4480.
50. Parry, R. J.; Muscate, A.; Askonas, L. J., *Biochemistry* 1991, **30**, 9988–9997.
51. Huang, H.; Yuan, C. S.; Borchardt, R. T., *Protein Sci.* 1997, **6**, 1482–1490.
52. Ault-Riche, D. B.; Yuan, C. S.; Borchardt, R. T., *J. Biol. Chem.* 1994, **269**, 31472–31478.
53. Yuan, C. S.; Yeh, J.; Squier, T. C.; Rawitch, A.; Borchardt, R. T., *Biochemistry* 1993, **32**, 10414–10422.
54. Yin, D.; Yang, X.; Squier, T. C.; Borchardt, R. T., *Biophys. J.* 1999, **76**, A169.
55. Yuan, C. S.; Ault-Riche, D. B.; Borchardt, R. T., *J. Biol. Chem.* 1996, **271**, 28009–28016.
56. Hu, Y.; Komoto, J.; Huang, Y.; Takusagawa, F.; Gomi, T.; Ogawa, H.; Takata, Y.; Fujioka, M., *Biochemistry* 1999, **38** 8323–8333.
57. Steitz, T. A.; Shoham, M.; Bennett, W. S., Jr., *Phil. Trans. Roy. Soc. Lond. B, Biol. Sci.* 1981, **293**, 43–52.
58. Lamzin, V. S.; Dauter, Z.; Popov, V. O.; Harutyunyan, E. H.; Wilson, K. S., *J. Molec. Biol.* 1994, **236**, 759–785.
59. Blake, C. C.; Rice, D. W.; Cohen, F. E., *Internatl. J. Pept. Protein Res.* 1986, **27**, 443–448.
60. Stillman, T. J.; Baker, P. J.; Britton, K. L.; Rice, D. W.; Rodgers, H. F., *J. Molec. Biol.* 1992, **224**, 1181–1184.
61. Ha, Y.; McCann, M. T.; Tuchman, M.; Allewell, N. M., *Proc. Natl. Acad. Sci.* (USA) 1997, **94**, 9550–9555.
62. Keller, B. T.; Borchardt, R. T., in *Antiviral Drug Development—A Multidisplinary Approach*, E. De Clercq and R. T. Walker, eds., Plenum Press, New York, 1988, pp. 123–138.
63. Guranowski, A.; Montgomery, J. A.; Cantoni, G. L.; Chiang, P. K., *Biochemistry* 1981, **20**, 110–115.
64. Porter, D. J.; Boyd, F. L., *J. Biol. Chem.* 1991, **266**, 21616–21625.
65. De Clercq, E.; Cools, M.; Balzarini, J., *Biochem. Pharmacol.* 1989, **38**, 1771–1778.
66. Houston, D. M.; Dolence, E. K.; Keller, B. T.; Patel-Thombre, U.; Borchardt, R. T., *J. Med. Chem.* 1985, **28**, 467–471.
67. Paisley, S. D.; Wolfe, M. S.; Borchardt, R. T., *J. Med. Chem.* 1989, **32**, 1415–1418.
68. Matuszewska, B.; Borchardt, R. T., *J. Biol. Chem.* 1987, **262**, 265–268.
69. Porter, D. J.; Boyd, F. L., *J. Biol. Chem.* 1992, **267**, 3205–3213.
70. Snoeck, R.; Andrei, G.; Neyts, J.; Schols, D.; Cools, M.; Balzarini, J.; De Clercq, E., *Antiviral Res.* 1993, **21**, 197–216.

71. Montgomery, J. A.; Clayton, S. J.; Thomas, H. J.; Shannon, W. M.; Arnett, G.; Bodner, A. J.; Kion, I. K.; Cantoni, G. L.; Chiang, P. K., *J. Med. Chem.* 1982, **25**, 626–629.

72. Glazer, R. I.; Knode, M. C.; Tseng, C. K.; Haines, D. R.; Marquez, V. E., *Biochem. Pharmacol.* 1986, **35**, 4523–4527.

73. Glazer, R. I.; Hartman, K. D.; Knode, M. C.; Richard, M. M.; Chiang, P. K.; Tseng, C. K. H.; Marquez, V. E., *Biochem. Biophys. Res. Commun.* 1986, **135**, 688–694.

74. Tseng, C. K. H.; Marquez, V. E., Fuller, R. W.; Goldstein, B. M.; Haines, D. R.; McPherson, H.; Parsons, J. L.; Shannon, W. M.; Arnett, G.; Hollingshead, M.; Driscoll, J. S., *J. Med. Chem.* 1989, **32**, 1442–1446.

75. Paisley, S. D.; Hasobe, M.; Borchardt, R. T., *Nucleosides Nucleotides* 1989, **8**, 689–698.

76. Jarvi, E. T.; McCarthy, J. R.; Mehdi, S.; Matthews, D. P.; Edwards, M. L.; Prakash, N. J.; Bowlin, T. L.; Sunkara, P. S.; Bey, P., *J. Med. Chem.* 1991, **34**, 647–656.

77. Yuan, C. S.; Yeh, J.; Liu, S.; Borchardt, R. T., *J. Biol. Chem.* 1993, **268**, 17030–17037.

78. Robins, M. J.; Wnuk, S. F.; Mullah, K.; Dalley, N. K.; Yuan, C. S.; Lee, Y.; Borchardt, R. T., *J. Org. Chem.* 1994, **59**, 544–555.

79. Yuan, C. S.; Wnuk, S. F.; Liu, S.; Robins, M. J.; Borchardt, R. T., *Biochemistry* 1994, **33**, 12305–12311.

80. Borchardt, R. T.; Keller, B. T.; Patel-Thombre, U., *J. Biol. Chem.* 1984, **259**, 4353–4358.

81. Glazer, R. I.; Knode, M. C., *J. Biol. Chem.* 1984, **259**, 12964–12969.

82. Bennett, L. L., Jr.; Bowdon, B. J.; Allan, P. W.; Rose, L. M., *Biochem. Pharmacol.* 1986, **35**, 4106–4109.

83. Kim, I. K.; Aksamit, R. R.; Cantoni, G. L., *J. Biol. Chem.* 1982, **257**, 14726–14729.

84. Bennett, L. L., Jr.; Brockman, R. W.; Rose, L. M.; Allan, P. W.; Shaddix, S. C.; Shealy, Y. F.; Clayton, J. D., *Molec. Pharmacol.* 1985, **27**, 666–675.

85. Villalon, M. D.; GilBFernandez, C.; De Clercq, E., *Antiviral Res.* 1993, **20**, 131–144.

86. Huggins, J. W.; Zhang, Z. X.; Davis, K.; Coulombe, R. A., *The 8 Internatl. Conf. Antiviral Research*; Santa Fe, NM, 1995; abstract, page A301.

87. Prakash, N. J.; Davis, G. F., Jarvi, E. T., Edwards, M. L.; McCarthy, J. R.; Bowlin, T. L., *Life Sci.* 1992, **50**, 1425–1435.

CHAPTER 3

Alcoholism: Aldehyde Dehydrogenase Inhibitors as Alcohol Deterrent Agents

HERBERT T. NAGASAWA

Medical Research Laboratories, DVA Medical Center and
Department of Medicinal Chemistry, University of Minnesota

FRANCES N. SHIROTA

Medical Research Laboratories, DVA Medical Center

EUGENE G. DeMASTER

Medical Research Laboratories, DVA Medical Center and
Department of Pharmacology, University of Minnesota

Despite recent (at the time of writing) evidence that alcohol consumption per capita is declining, alcoholism continues to be a major medical and socioeconomic problem in developed countries. In the United States alone, it is estimated that 7.4% or 13.7 million of the U.S. population either abuse or are addicted to alcohol.[1] Disruption of family harmony and the economic burden of lost workdays due to alcoholism are difficult to quantitate, but it is estimated that alcohol contributes up to 45% of the 45,000 annual deaths from traffic accidents. Traffic fatalities are the leading cause of death for the 1–36-year age group, with more than 25% of all "driving while intoxicated" (DWI) [also termed "driving under the influence" (DUI)] deaths in 1993 caused by persons between 16 and 24 years of age, even though this group represented only 15% of all licensed drivers.[1] Increasing alcohol consumption among high-school and college students are deeply troubling as this portends the future, and binge drinking by the latter group that often leads to death is of grave concern. Among adults, alcoholic liver cirrhosis ranked 11th as the leading cause of death and accounted for 45,000 deaths in 1993.

While psychological counseling and peer intervention by Alcoholics Anonymous groups may have some beneficial effects in the management of alcohol abuse, there

Biomedical Chemistry: Applying Chemical Principles to the Understanding and Treatment of Disease, Edited by Paul F. Torrence
ISBN 0-471-32633-x

is a clear need for drugs that either alleviate the craving for or antagonize the euphoric effects of alcohol as primary pharmacotherapeutic agents for the treatment of alcoholism. It is axiomatic that total abstinence from alcohol, or even moderate drinking within the socially and legally acceptable norms, can eradicate most, if not all, of the adverse consequences of alcohol misuse.

CENTRALLY ACTING ANTICRAVING AGENTS

Drugs of this type reduce drinking frequency by attenuating the neurochemical stimuli that incite the psychological craving for alcohol that leads to uncontrollable consumption. Although a number of these anticraving drugs have been studied, only two are in clinical use. Acamposate is used in Europe, while naltrexone (ReVia) has recently been approved for use in the United States. Both drugs prevent relapse of recovering alcoholics to some degree, but do not reinforce abstinence from alcohol consumption.[2]

Acamposate, the calcium salt of *N*-acetylhomotaurine, is thought to interfere with NMDA receptor-mediated glutamate neurotransmission. It has been shown to suppress alcohol craving during alcohol deprivation in a dose-dependent manner without affecting the pharmacokinetics and pharmacodynamics of alcohol. Naltrexone, an opioid receptor antagonist, presumably reduces the desire to drink alcohol by blocking the μ-, δ-, and possibly the κ-opioid receptors in the brain, and this drug is used as an adjunct together with psychological counseling and peer support for the treatment of alcoholism.

PERIPHERALLY ACTING ALCOHOL DETERRENT AGENTS

Therapeutically used alcohol deterrent drugs of this type are disulfiram (DSSD, tetraethylthiuram disulfide, Antabuse), which is available to some extent worldwide, and, cyanamide (Fig. 3.1), or more correctly, its citrated calcium salt, calcium carbimide (Temposil, Dipsan, Abstem), which is available for use in prescription form in Europe and Canada, but not in the United States.[3] Formulated aqueous solutions of cyanamide itself (cyanamide, Yoshitomi) are prescribed in Japan. The mechanism of action of these drugs will be elaborated in more detail later.

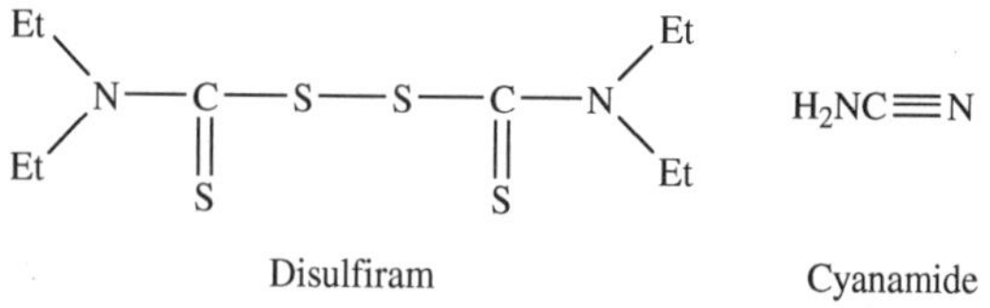

$$Et_2N-C(=S)-S-S-C(=S)-NEt_2 \qquad H_2NC\equiv N$$

Disulfiram   Cyanamide

FIGURE 3.1 Chemical structures of disulfiram and cyanamide.

BIOCHEMISTRY OF ETHANOL METABOLISM

Ethanol is converted in two metabolic steps to acetate, which then enters the tricarboxylic acid cycle to produce energy in the normal manner. Each step is catalyzed by a different enzyme and requires the coenzyme nicotinamide adenine dinucleotide (NAD^+), namely, *alcohol dehydrogenase* (ADH), which converts ethanol to acetaldehyde (AcH) reversibly, and *aldehyde dehydrogenase* (AlDH), which *irreversibly* transforms AcH to acetate:

$$CH_3CH_2OH \underset{NAD^+ \rightarrow NADH}{\overset{ADH}{\rightleftharpoons}} CH_3CHO \xrightarrow[NAD^+ \rightarrow NADH]{AlDH} CH_3COOH \quad (3.1)$$

A number of isoenzymes of ADH that catalyze the conversion of ethanol to AcH are known, while two major isozymes of AlDH are involved in AcH metabolism.[4] The low-K_m enzyme (AlDH2; K_m 3 μM) located in liver mitochondria is responsible for the bulk of the metabolism of AcH, since the concentration of AcH is normally low because of the reversibility of the ADH reaction [Eq. (3.1)], and the high-K_m AlDH (AlDH1; K_m 100 μM) located in the cytosol of liver is not saturated. It is generally accepted that the active-site residue of mammalian AlDH2 is Cys302, the only conserved cysteine among the AlDHs, whose sulfhydryl integrity, when compromised, leads to inhibition of the enzyme.[5] For example, replacement of Cys302 with Ala302 by site-directed mutagenesis gives an inactive enzyme. A metabolic block imposed on this enzyme by drugs such as DSSD or cyanamide, or a genetic defect of this enzyme, leads to the accumulation of AcH.

Yeast AlDH is also a mitochondrial enzyme that shows significant sequence identity with the mammalian AlDH2 near the *C*-terminal end where the active-site amino acid residues are believed to be located. Thus, Cys302 is conserved in yeast AlDH,[6] making it an ideal alternative to the liver enzyme for in vitro studies with potential inhibitors of AlDH2.

THE DISULFIRAM–ETHANOL REACTION (DER)

The causal relationship between ethanol hypersensitivity and exposure to DSSD was first described by Williams in 1937,[7] who observed that some workers in the rubber industry could not tolerate alcohol. DSSD is an industrial chemical, one of its uses being in the vulcanization of rubber. This disulfiram–ethanol reaction was later attributed to the effect of the accumulation of AcH.[8]

The overall mechanism underlying this alcohol sensitization by DSSD is currently understood to be as follows:[9,10] (1) DSSD inhibition of AlDH2 in vivo, (2) elevation of blood AcH as a result of this metabolic block, (3) AcH-mediated release of biogenic amines (catecholamines, histamine), and (4) resulting biogenic amine-mediated hemodynamic effects. Clinically, the DER is characterized by flushing of the face and trunk, headache, nausea, vertigo, palpitations, tachycardia, and (occasionally) hypotension. These unpleasant physiological effects brought on by DSSD on ingestion of ethanol-containing substances (including cough syrups) are presumed to act as deterrents to drinking alcohol.

GENETIC POLYMORPHISM OF ALDH2

It has long been recognized that certain individuals, particularly of Oriental origin, have a natural aversion to alcohol and sustain a DER manifested by a pronounced facial flush on consumption of only modest quantities of alcoholic beverages. These individuals lack a functional low-K_m hepatic mitochondrial aldehyde dehydrogenase

TABLE 3.1 Frequency of ALDH2 Isozyme Deficiency in Asian Mongoloids and Other Populations[11]

Population	Sample Size	Percent Deficient
Asian Mongoloids		
Japanese	184	44
Chinese		
Mongolian	198	30
Zhuang	106	45
Han	120	50
Korean (Mandschu)	209	25
Chinese living abroad	196	35
Koreans (South Korea)	75	27
Vietnamese	138	53
Indonesians	30	39
Filipinos	110	13
Ainu	80	20
Thais (north)	110	8
Other populations		
Germans	300	0
Egyptians	260	0
Sudanese	40	0
Kenyans	23	0
Liberians	184	0
Turks	65	0
Fangs	37	0
Israeli	177	0
Hungarians	179	0
Matyo	106	0
Romai	50	0
Asian Indians	50	0

Source: Reprinted with permission of S. Karger AG, Basel.

(AlDH2), the enzyme that is inhibited in vivo by DSSD and cyanamide as well as by other drugs known to interact with ethanol (see discussion below). This inborn enzyme deficiency observed in 30–50% of Orientals (Table 3.1)[11] has been shown to be due to a genetic defect in AlDH2, which is characterized by a point mutation in the codon for amino acid 487 involving a G/C–A/T (guanine/cytosine–adenine/thymine) transition.[12] The consequence is the substitution of the amino acid lysine (Lys) for glutamate (Glu) at this position resulting in an inactive enzyme. The consumption of ethanol by such individuals elevates blood AcH levels (Fig. 3.2)[13] and induces a flushing reaction reminiscent of the DER leading to alcohol avoidance and self-imposed sobriety.

Individuals with this genotypic, inactive AlDH2 (flushers) are considerably less prone to alcoholism and have a correspondingly low incidence of alcoholic liver disease (Table 3.2).[14,15] Conversely, individuals with normal AlDH2 (nonflushers) have a higher rate of alcoholism and a higher risk of developing alcoholic liver disease.[16] These genetic–epidemiological findings are also reflected in the low rate of alcoholism in Orientals compared to individuals of European descent[11] and offer compelling rationale for treating chronic alcoholics by pharmacological intervention with agents specifically designed to inhibit AlDH2, the same enzyme that is deficient in 30–50% of Orientals.

A thorough understanding of the biochemical and pharmacological mechanisms of documented drug-induced DERs should greatly aid in the design of second-generation drugs that can deter susceptible subjects from excessive consumption of alcohol. This theme is the focus of this chapter.

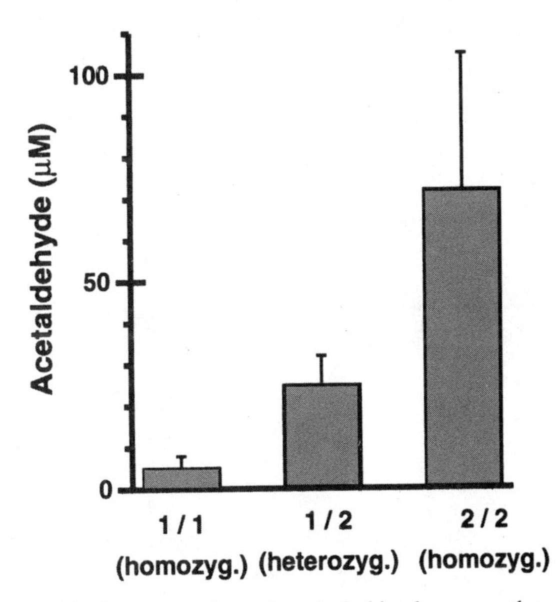

FIGURE 3.2 Blood AcH concentrations after alcohol intake among three different ALDH2 genotypes.[13] The 2/2 homozygotes lack AlDH2 while the 1/1 homozygotes have normal AlDH2. (Reprinted by permission of Wiley-Liss, Inc., a division of John Wiley & Sons, Inc.)

TABLE 3.2 AlDH2 Phenotypes, Alcoholism, and Alcoholic Liver Disease in Japanese Subjects

	AlDH2 Phenotype	
Subjects	Inactive (Flushers)	Normal (Nonflushers)
Healthy	43	62
Alcoholic	4	171
Alcoholic with liver disease	1	64

Source: Abstracted from Refs. 11 and 12.

DISULFIRAM (DSSD; TETRAETHYLTHIURAM DISULFIDE)

The only FDA-approved alcohol deterrent agent based on AlDH inhibition and marketed in the United Stated is disulfiram (Antabuse). However, the side effects associated with DSSD therapy—due to its nonspecific inhibition of a number of enzyme systems[10]—limits its clinical usefulness somewhat. Intact disulfiram does not affect the critical low-K_m enzyme (AlDH2), but does inhibit the high-K_m isozyme (AlDH1). More recent studies have shown that DSSD must first be metabolized in vivo to a reactive species before it can inhibit AlDH2. This requires

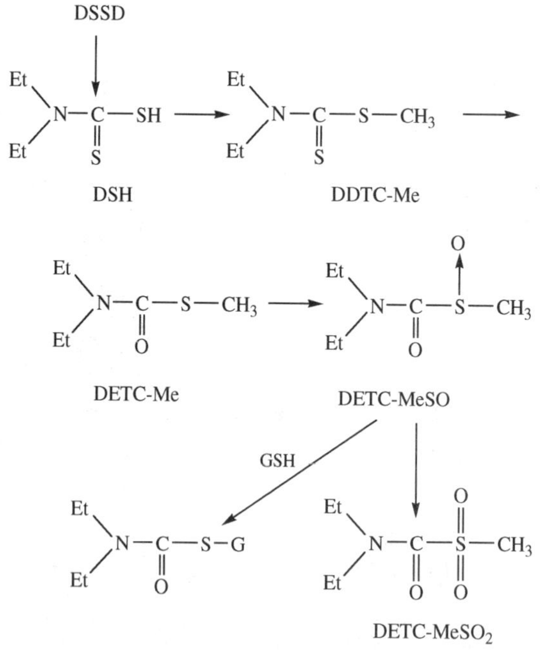

FIGURE 3.3 Metabolic activation pathway for disulfiram.

at least four sequential metabolic activation steps (Fig. 3.3).[17,18] Thus, DSSD administration leads to the rapid reduction of the disulfide bond to give diethyldithiocarbamate (DSH), which is then methylated to the thiol ester, DDTC-Me. Oxidative desulfurization of the thiono sulfur of DETC-Me gives the corresponding oxygen analog *S*-methyl-*N,N*-diethylthiolcarbamate (DETC-Me). Sulfoxidation of the latter produces the putative inhibitor of AlDH, specifically, DETC-MeSO. This metabolite is a specific inhibitor of hepatic AlDH2 in rats and raises blood AcH following an ethanol dose. Administration of DETC-MeSO not only inhibits AlDH2 but also leads to the excretion of the glutathione (GSH) conjugate, *S*-diethylcarbamoyl glutathione, in the bile,[19] suggesting that DETC-MeSO might be a carbamoylating agent capable of irreversibly carbamoylating the active site Cys302 of liver AlDH. Further oxidation of DETC-MeSO to the sulfone is also possible, but DETC-$MeSO_2$ is toxic because of its high reactivity.

CYANAMIDE

The use of cyanamide as an alcohol deterrent agent is therapeutically advantageous when other drugs are coadministered, since, unlike disulfiram, cyanamide does not inhibit enzymes such as dopamine-β-hydroxylase or cytochrome P-450. Cyanamide has a rapid onset of action that is short-lived with maximal pharmacological effect occurring at 1 h and lasting up to 24 h.[3] Cyanamide inhibits AlDH only in vivo and does not inhibit the partially purified AlDH1 or AlDH2 in vitro, suggesting that a *metabolite* of cyanamide is the active inhibitor of AlDH.[20,21]

Deitrich et al.,[20] using ^{14}C-labeled cyanamide, showed that it was rapidly converted in rats to a major (> 94%) urinary metabolite which was devoid of AlDH activity in vitro. They established that this metabolite (1) was acidic with a pK_a of 3.92, (2) exhibited a UV maximum at 219 nm in base, and (3) retained an intact *N*-cyano functional group. However, the chemical structure of the metabolite and the nature of this biotransformation were not elucidated.

This major metabolite of cyanamide, also found to be excreted in rabbit, dog, and human urine, has now been identified as *N*-acetylcyanamide.[22] Structural characterization of the isolated metabolite was based on the identity of its physiochemical properties with authentic *N*-acetylcyanamide and by the conversion of the metabolite and synthetic *N*-acetylcyanamide to *N*-benzyl-*N*-acetylcyanamide (or to *N*-(*p*-nitrobenzyl)-*N*-acetylcyanamide) and showing their identity by CI-MS:

$$CH_3C(=O)NHCN + X\text{-}C_6H_4\text{-}CH_2Br \xrightarrow{X = H,\ NO_2} CH_3C(=O)N(CN)CH_2\text{-}C_6H_4\text{-}X \quad (3.2)$$

N-Acetylcyanamide

The conversion of cyanamide to *N*-acetylcyanamide in vivo is catalyzed by an acetyl-*S*-CoA-dependent *N*-acetyltransferase, as demonstrated in vitro by the acetylation of [^{14}C]-cyanamide by acetyl-*S*-CoA in the presence of this enzyme from rabbit or dog liver. Acetylcyanamide did not inhibit the low-K_m AlDH2 found in isolated rat liver mitochondria, in agreement with the earlier report.[20] Thus, acetylation represented a detoxication/conjugation/elimination pathway for cyanamide and suggested that a second—albeit minor—pathway for cyanamide must exist that leads to an active inhibitor of AlDH.

Metabolic Activation of Cyanamide

In vitro studies[23,24] indicated that a rat liver mitochondrial fraction contained significant cyanamide activation activity. The subcellular localization, stability, and apparent lack of cofactor requirement, suggested that the cyanamide bioactivating enzyme was *catalase*.[25,26] This premise was tested by evaluating the effects of the catalase inhibitor, 3-amino-1,2,4-triazole (3-AT), on cyanamide activation. Administration of cyanamide to rats followed by ethanol raised blood AcH levels 90-fold compared to controls, a reflection of the inhibition of hepatic AlDH. 3-AT treatment prior to cyanamide and ethanol inhibited hepatic catalase activity by 90% and concomitantly lowered blood AcH levels by more than 90%. 3-AT alone had no significant effect on blood AcH. Purified bovine liver catalase also catalyzed the conversion of cyanamide to its active form, whose formation was monitored by determining the degree of inhibition of yeast AlDH. Inhibition of this enzyme required both cyanamide and catalase and increased with catalase concentration. Sodium azide, another catalase inhibitor, also effectively blocked this activation of cyanamide by bovine or rat liver mitochondrial catalase.

Incubation of [^{13}C]-cyanamide, prepared from [^{13}C]-thiourea [99 at% (atom percent) enrichment], with bovine liver catalase/glucose–glucose oxidase (to provide a continuous source of H_2O_2) and NMR analysis of the volatile ^{13}C-labeled product trapped in 0.10 N KOD/D_2O, showed unequivocal evidence for the presence of [^{13}C]HCN (as [^{13}C]KCN). [^{13}C]HCN was not produced when catalase or the H_2O_2 source was omitted from the incubation mixture.[27] Incubation of uniformly labeled [^{15}N]-cyanamide with this catalase system and analysis of the volatile gaseous product in the headspace by GC/MS gave [^{15}N]N_2O with m/z 46 (relative abundance 100%). With unlabeled cyanamide, the N_2O had the expected m/z of 44.

The results of these tracer studies were in accord with the postulate[28] that cyanamide was bioactivated by catalase/H_2O_2 to *N*-hydroxycyanamide, which spontaneously decomposed to nitroxyl (HNO) and cyanide [Eq. (3.3)]. Nitroxyl is known to dimerize to hyponitrous acid which, in turn, dehydrates to nitrous oxide [Eq. (3.4)]. Since cyanide in concentrations up to 5 mM did not inhibit AlDH, the conclusion was reached that *nitroxyl* produced in the oxidation of cyanamide was the inhibitor of this enzyme.

$$\underset{\text{Cyanamide}}{H_2NC{\equiv}N} \xrightarrow[H_2O_2]{\text{catalase}} \underset{N\text{-Hydroxycyanamide}}{\left[\begin{array}{c} HO \\ H \end{array} \!\!\! > N{-}C{\equiv}N \right]} \longrightarrow \underset{\text{Nitroxyl}}{HN{=}O} + HCN \tag{3.3}$$

$$2HN{=}O \longrightarrow \underset{\text{Hyponitrous acid}}{H_2N_2O_2} \longrightarrow N_2O + H_2O \tag{3.4}$$

In subsequent studies, it was shown that ^{13}C-labeled cyanamide gave rise not only to ^{13}C-labeled cyanide when incubated with catalase/glucose–glucose oxidase but also to ^{13}C-labeled CO_2.[29] Moreover, a time-dependent formation of *nitrite* was observed when unlabeled cyanide was incubated in this system. These results suggested that the initial product of cyanamide oxidation, namely, *N*-hydroxycyanamide [Eq. (3.3)], was being further oxidized by catalase/H_2O_2 to nitrosyl cyanide (O=NCN), which hydrolyzed to the four end products detected: nitroxyl, cyanide, nitrite, and CO_2. Direct proof required that authentic nitrosyl cyanide also yield these same products at physiological pH. Accordingly, both unlabeled and ^{13}C-labeled nitrosyl cyanide were prepared by nitrosylation of K-(18-crown-6)cyanide with nitrosyl tetrafluoroborate at −40 to −50°C. Solvolysis of this product in phosphate-buffered solution, pH 7.4, and analysis of the headspace by gas chromatography showed the presence of N_2O, the dimerization/dehydration product of nitroxyl. Cyanide was detected (colorimetrically) in the aqueous solution, while $[^{13}C]CO_2$ with *m/z* 45 was identified by GC/MS (gas chromatography—mass spectrometry).

An oxidative biotransformation pathway for cyanamide that accounts for all the products detected (NH_3 was not measured) is shown in Figure 3.4. The formation of these products can be rationalized by invoking *both* *N*-hydroxycyanamide and nitrosyl cyanide as tandem intermediates. Both can generate *nitroxyl*, the putative inhibitor of AlDH, while nitrosyl cyanide can also inhibit this enzyme by *S*-nitrosylating the active site cysteine directly. That nitrosylation of AlDH can lead to profound enzyme inhibition was indicated by the observation that isoamyl nitrite, a chemical nitrosylating agent, rapidly inhibited yeast AlDH when incubated with the enzyme.[29]

Acylated Prodrugs of Cyanamide

Although *N*-acetylcyanamide did not inhibit AlDH in vitro, doses of 1.0 mmol/kg administered to rats raised ethanol-derived blood AcH levels 25-fold over control

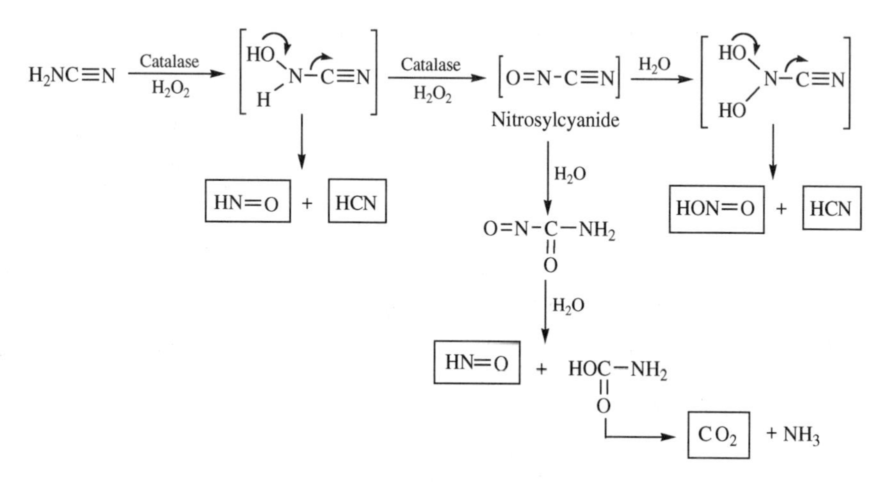

FIGURE 3.4 Catalase-mediated oxidation of cyanamide to the various end-products detected.

values, constituting pharmacological evidence that some cyanamide must have been liberated in vivo by deacetylation.[30] Indeed, administration of [1-^{14}C]-acetylcyanamide to rats gave rise to the expiration of 2% of the administered radioactive dose as [^{14}C]CO_2 within 8 h, providing corroborative biochemical evidence that finite deacetylation to cyanamide had taken place.

On the basis of these results, a number of acyl derivatives of cyanamide were prepared specifically as prodrug forms.[31] These included the sterically bulky pivaloylcyanamide and adamantoylcyanamide designed to retard the rate of enzymatic hydrolysis of the acylcyanamide linkage, and lipophilic long-chain fatty acyl derivatives such as palmitoylcyanamide and stearoylcyanamide and others of intermediate carbon chain to prevent rapid renal excretion. *N*-Protected α-aminoacyl and peptidyl derivatives of cyanamide were also synthesized and tested in rats for their ability to elevate blood AcH after treatment with ethanol. The rationale for preparing the latter compounds was based on the premise that the α-aminoacylcyanamide linkage would be cleaved in vivo by peptidase action.

The observation that most of these compounds, including the *N*-protected α-aminoacyl- and peptidylcyanamides, were nearly uniformly active in elevating ethanol-derived blood AcH in rats (Figs. 3.5 and 3.6) provided deductive evidence that finite quantities of cyanamide must have been liberated in vivo, thereby fulfilling their designation as prodrugs of cyanamide. Moreover, since the released cyanamide must still be further bioactivated before the inhibition of AlDH can be manifest (i.e., two sequential enzymatic steps are necessary to liberate nitroxyl), these derivatives of cyanamide can be considered to be *pro*-prodrugs of *nitroxyl.* The nitroxyl liberated need only be minimal; for example, even though $>94\%$ of the administered dose of cyanamide is excreted as the acetylated conjugate within 6 h,[20] cyanamide is the most potent in vivo AlDH inhibitor known. This suggests that the relatively small fraction of cyanamide that is bioactivated to nitroxyl is sufficient to exert a profound inhibitory action on AlDH.

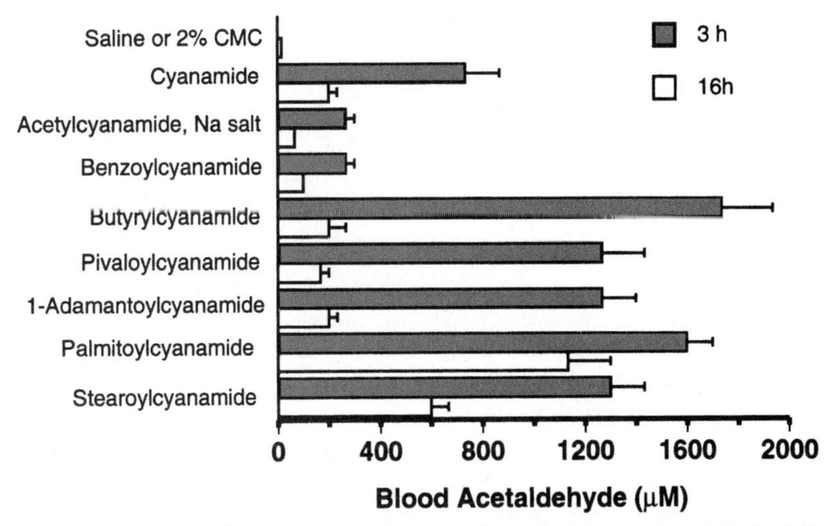

FIGURE 3.5 Effect of *N*-acylated derivatives of cyanamide on ethanol-derived blood acetaldehyde levels in rats. Dose: 1.0 mmol/kg; CMC = carboxymethylcellulose.

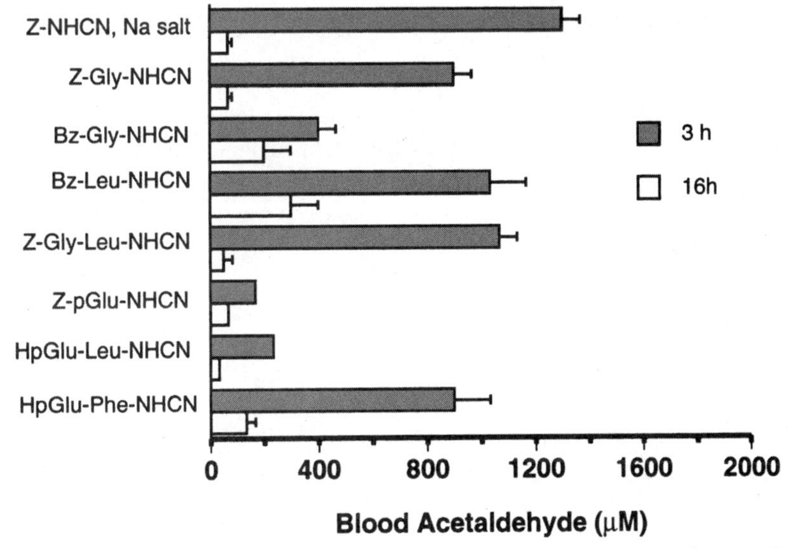

FIGURE 3.6 Effect of *N*-protected α-aminoacyl and peptidyl derivatives of cyanamide on ethanol-derived blood acetaldehyde levels in rats. Dose: 1.0 mmol/kg; Z = carbobenzoxy; Bz = benzoyl; HpGlu = pyroglutamyl.

S-Alkyl*iso*thioureas as Prodrugs of Cyanamide

The report[32] that thiobenzamide was oxidized by rat liver microsomes to thiobenzamide *S*-dioxide, which spontaneously decomposed to benzonitrile (Fig. 3.7), suggested that oxidation of thiourea by this enzyme system might lead to the

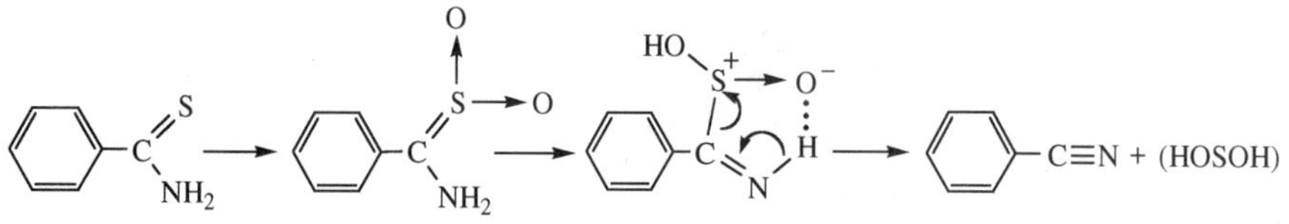

FIGURE 3.7 The oxidative metabolism of thiobenzamide to benzonitrile.

formation of cyanamide by an analogous mechanism (not shown). However, thiourea is toxic and known to be a suspect carcinogen and mutagen and hence could not be considered for development as a potential prodrug of cyanamide. On the other hand, the oxidative metabolic transformations of *S*-alkyl*iso*thioureas could be rationalized to give cyanamide.

Accordingly, *S*-methy*iso*thiourea would be expected to be oxidized to its *S*-oxide or the *S*-dioxide. These intermediate *S*-oxides should spontaneously decompose to cyanamide and methanesulfenic acid or methanesulfinic acid, respectively (Fig. 3.8). *S*-Methyl*iso*thiourea would, therefore, be expected to behave as a prodrug of cyanamide in vivo and to inhibit AlDH. In contrast, the corresponding oxygen analog, *O*-methyl*iso*urea (structure not shown), which cannot be oxidatively metabolized in the same manner, should have no effect on AlDH in vivo.

Administration of *S*-methyl*iso*thiourea to rats [0.5 mmol/kg, intraperitoneally (IP)], followed by an ethanol challenge (2.0 g/kg, IP) 5 h later, gave rise to a 119-fold increase in blood AcH levels at 1 h after ethanol administration compared to controls given ethanol alone (Fig. 3.9).[34] Persistent enzyme inhibition was reflected

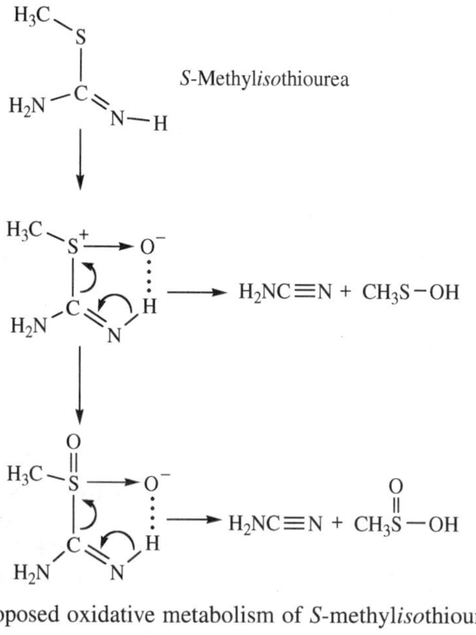

FIGURE 3.8 Proposed oxidative metabolism of *S*-methyl*iso*thiourea to cyanamide.

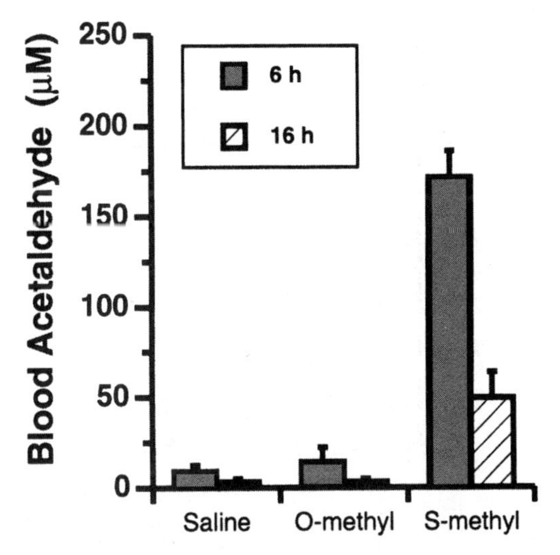

FIGURE 3.9 Blood AcH levels in rats treated with *S*-methyl*iso*thiourea (0.5 mmol/kg) and ethanol.

by the elevation of blood AcH when challenged again with ethanol 16 h after drug treatment. As predicted, blood AcH was *not* elevated above control levels with *O*-methyl*iso*urea.

Structural alteration of the *S*-alkyl group (i.e., replacement of the *S*-methyl group with *S*-*n*-butyl, *S*-*iso*butyl, or *S*-allyl groups) resulted in increased potency in inhibiting hepatic AlDH, as reflected by blood AcH levels that were 40–92% higher than that observed for the *S*-methyl compound (Fig. 3.10). The longer duration of action of the *S*-butyl compounds compared to cyanamide was evidenced by the fact that an ethanol challenge at 15 h still raised blood AcH levels. Blood *ethanol* levels were unaffected by these structural modifications, suggesting that the first step in ethanol metabolism—conversion of ethanol to AcH—was not affected.

The *S*-oxidation of thiobenzamide is mediated in large part by the hepatic microsomal cytochrome P-450 mixed-function oxidase system.[33] That the same enzyme(s) was (were) involved in the bioactivation of these *S*-alkyl*iso*thioureas to cyanamide was adduced by the 66–88% reduction in blood AcH levels in rats pretreated with 1-benzylimidazole (0.03 mmol/kg, IP), a known inhibitor of cytochrome P-450. The actual formation of cyanamide by cytochrome P-450 action was shown by its isolation following incubation of *S*-*n*-butyl*iso*thiourea with liver microsomes from rats pretreated with 3-AT (to inhibit catalase) in the presence of an NADPH-generating system. Cyanamide formation was minimal in the absence of this NADPH source, indicating the enzymatic nature of this reaction.

Thus, short-chain ($< C_4$) *S*-alkyl*iso*thioureas can be considered to be prodrug forms of the alcohol deterrent agent, cyanamide. These *S*-alkyl *iso*thioureas are bioactivated by the hepatic mixed function oxidase enzymes to sulfoxidized products that liberate cyanamide, which on further bioactivation by catalase-H_2O_2 gives *nitroxyl*, a potent inhibitor of AlDH.

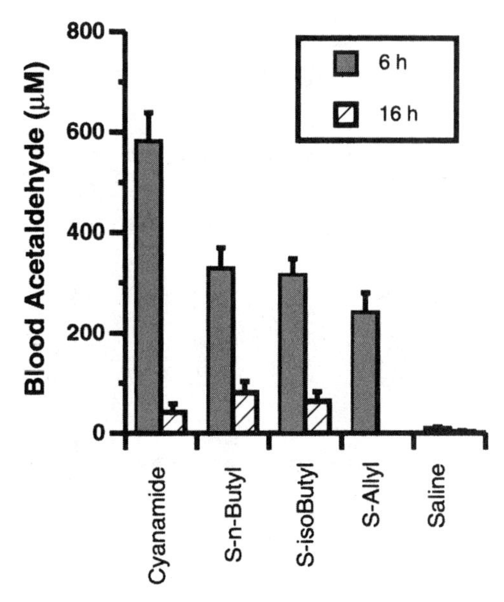

FIGURE 3.10 Blood AcH levels in rats treated with various *S*-alkylisothioureas (0.5 mmol/kg) and ethanol.

THE CHLORPROPAMIDE–ALCOHOL FLUSH (CPAF)

Approximately 30% of individuals with adult-onset diabetes, when treated with a sulfonylurea-type oral hypoglycemic agent such as chlorpropamide (CP), experience facial flushing, headache, breathlessness, nausea, giddiness, and tachycardia on consumption of only modest quantities of alcohol.[35] These symptoms are reminiscent of the DER, a consequence of the inhibition of hepatic mitochondrial AlDH2. Indeed, this chlorpropamide–alcohol flush (CPAF) was accompanied by high blood AcH levels in CP-induced flushers following ethanol ingestion, whereas blood AcH was not elevated in the nonflushers.[36,37] In rodent species given CP and ethanol, this CPAF was manifested by a significant dose-dependent inhibition of hepatic mitochondrial AlDH and a dramatic increase in blood AcH levels.[38,39]

The molecular mechanism of inhibition of AlDH by CP is yet unknown; however, metabolic activation in vivo must be involved, since CP is inactive as an AlDH inhibitor in vitro. Studies designed to elucidate the possible bioactivation mechanisms of CP suggested that oxidation or conjugation of the N^1-sulfonamide nitrogen must play a role, since hydroxylations on the *n*-propyl sidechain appeared to represent detoxication rather than bioactivation reactions.[39] In order to block direct metabolic oxidation or conjugation at this N^1 position of CP, an *ethyl* group was chemically substituted for H at this position. The product, N^1-ethylchlorpropamide (N^1-EtCP), proved to be several times more active than CP in elevating ethanol-derived blood AcH in rats.[40] Moreover, N^1-EtCP was devoid of any

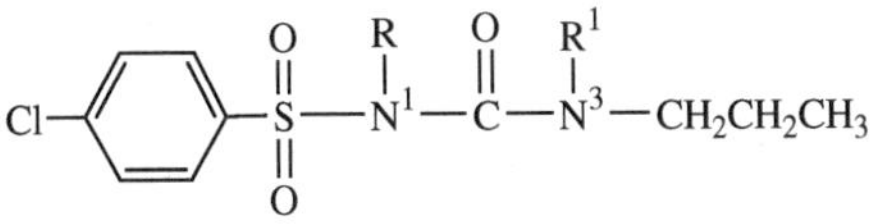

$$Cl-C_6H_4-S(=O)_2-N^1(R)-C(=O)-N^3(R^1)-CH_2CH_2CH_3$$

R	R'	
H	H	(Chlorpropamide)
CH_3	II	(N^1 MeCP)
C_2H_5	H	(N^1-EtCP)
n-C_3H_7	H	
i-C_3H_7	H	
n-C_4H_9	H	
i-C_4H_9	H	
t-C_4H_9	H	
$CH_2C(CH_3)_3$	H	
H	CH_3	(N^3-MeCP)

FIGURE 3.11 N^1- and N^3-substituted chlorpropamides.

hypoglycemic effect, making it a potential candidate for use in alcohol deterrent therapy.

To gain further insights on the mechanism of AlDH inhibition by N^1-EtCP *vis-a-vis* CP itself, a structure-activity study was initiated by (1) altering the alkyl substitutent at N^1 and (2) introducing a substitutent at N^3 (Fig. 3.11).[41] The results suggested that as the steric bulk of the N^1-substitutent was increased on CP, there was increased propensity for the resulting molecule to eliminate n-propylisocyanate (PrNCO)—*a potent inhibitor of AlDH*—by a nonenzymatic mechanism that required the presence of a free N-H group at N^3. Thus, when the N^3-H was replaced with a methyl group as in N^3-methyl-CP such that isocyanate elimination was blocked, the compound was devoid of any biological activity in vivo. The N^1-substituted analogs of CP may, therefore, be considered to be *prodrugs* of PrNCO that can release this enzyme inhibitor *non-enzymatically* in vivo.

It is of interest that metabolic *acetylation* on the N^1-sulfonamide nitrogen has been postulated to be a bioactivation mechanism for CP, since chemical acetylation of CP led to the spontaneous formation of PrNCO (Fig. 3.12).[42] Moreover, S-n-propylcarbamoyl-L-cysteine and S-n-propylcarbamoyl-L-glutathione, which are the L-cysteine and glutathione conjugates of PrNCO, respectively, also functioned as inhibitors of AlDH by generating PrNCO in vivo,[43] a consequence of the chemical reversibility of this reaction:

$$PrNH-C(=O)-S-R \rightleftharpoons PrNCO + RSH \tag{3.5}$$

R = cysteine and glutathione residues

These data lend some credence (but not direct proof) to the hypothesis that the mechanism of inhibition of AlDH by CP may be due to PrNCO release following

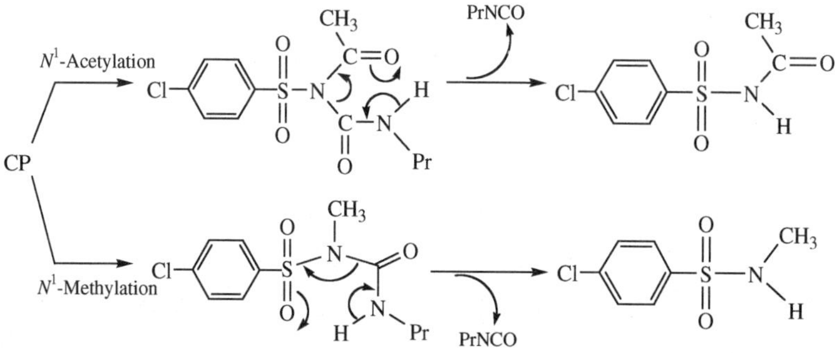

FIGURE 3.12 Proposed metabolic activation pathways for CP that liberate *n*-propyl*iso*-cyanate (PrNCO).

bioactivation by metabolic N^1-acetylation or N^1-methylation in vivo (Fig. 3.12). It is likely that the PrNCO liberated from these latent isocyanates is carbamoylating the active-site cysteine residue on AlDH in a manner similar to that for DETC-MeSO, the active metabolite of the alcohol deterrent agent DSSD.

LATENTIATED ALKYL ISOCYANATES AS INHIBITORS OF ALDH

On the basis of the results described above, a series of structurally analogous compounds with potential for releasing PrNCO in vivo, that is, *latentiated alkyl isocyanates,* were synthesized and tested for their ability to inhibit hepatic AlDH and thus raise ethanol-derived blood AcH in rats.[44] Some representative compounds are shown in Figure 3.13. The N^3-trifluoroacetyl-N^1-EtCP was designed to serve as a *pro*-prodrug of PrNCO specifically targeted to the liver, since PrNCO can be released only on hydrolysis of the trifluoroacetyl group by hepatic amidases or

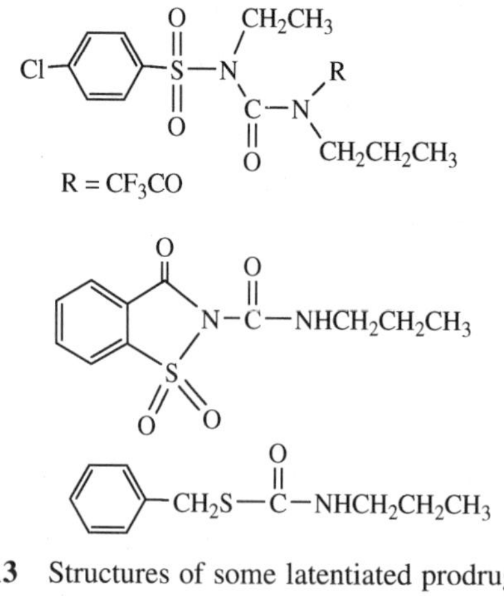

R = CF_3CO

FIGURE 3.13 Structures of some latentiated prodrugs of PrNCO.

esterases. The saccharin derivative, a functionally similar cyclic analog of N^1-acetyl-CP, was expected to be a more stable mimic of the latter, while the thiocarbamate ester represented a simplified analog of the glutathione and cysteine conjugates of PrNCO.

These latentiated alkyl isocyanates were all potent inhibitors of AlDH when administered to rats, and raised ethanol-derived blood AcH levels 1–2 times that elicited by N^1-EtCP. The activity of the *pro*-prodrug N^3 trifluoroacetylated N^1-EtCP equaled that for N^1-EtCP (blood AcH 425 μM), suggesting its precursor relationship to the latter.

PRODRUGS OF NITROXYL

To unequivocally prove the mechanism depicted for the decomposition mode of *N*-hydroxycyanamide, the intermediate in the bioactivation pathway from cyanamide to nitroxyl [Eq. (3.3)], a chemical synthesis of this compound was attempted (reaction of hydroxylamine with cyanogen bromide)—without success. However a *derivative* of *N*-hydroxycyanamide, namely, *N,O*-dibenzoyl-*N*-hydroxycyanamide (DBHC) was readily prepared.[46] It was rationalized that DBHC, after enzymatic hydrolysis of the benzoyl groups, should behave like *N*-hydroxycyanamide by liberating nitroxyl and inhibiting AlDH in vivo. In fact, only the *ester* moiety of DBHC need be hydrolyzed initially, as this should give the unstable *N*-benzoyl-*N*-hydroxycyanamide that can spontaneously decompose to yet another unstable intermediate: benzoylnitroxyl (PhCONO). The latter is known to readily hydrolyze to benzoic acid and nitroxyl[46] (Fig. 3.14).

When treated with dilute NaOH at 55 °C, DBHC gave rise to benzoic acid and cyanide. The gaseous product generated in a time-dependent manner was nitrous oxide (N_2O), the end product of nitroxyl dimerization and dehydration [Eq. (3.4)]. Since yeast AlDH, like the mammalian enzyme, is a sulfhydryl enzyme with intrinsic esterase activity, the action of this enzyme on DBHC should preferentially

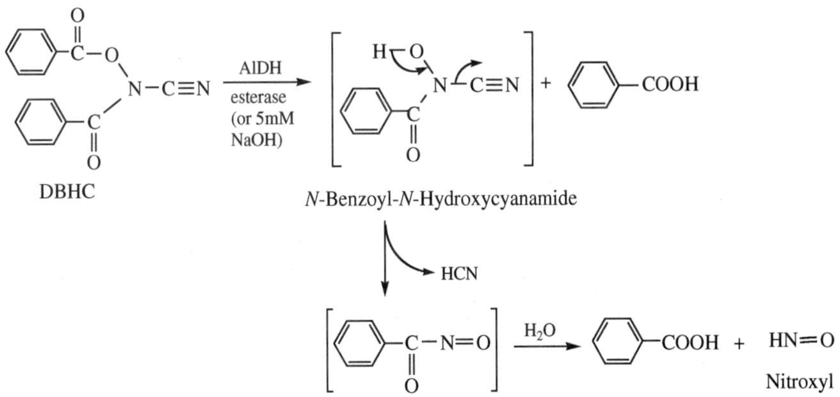

FIGURE 3.14 Esterase-mediated bioactivation of DBHC leading to nitroxyl.

hydrolyze the more labile benzoate ester group, leading to the ultimate generation of nitroxyl via the sequential reactions depicted in Figure 3.14. Therefore, it might be predicted, a priori, that DBHC would be an inhibitor of yeast AlDH, and indeed, DBHC inhibited this enzyme in a concentration-dependent manner with an IC_{50} of 25 μM. At a DBHC concentration of 0.1 mM, the enzyme was inhibited 97%.

These results are consistent with the postulate that nitroxyl is a potent inhibitor of AlDH, whether generated metabolically from cyanamide via *N*-hydroxycyanamide, or from a derivative of the latter as DBHC. Since DBHC (as well as cyanamide) can be considered to be a prodrug of nitroxyl, the design of other nitroxyl prodrugs—preferably *prodrugs that do not give rise to cyanide as a byproduct*—as inhibitors of AlDH presented a distinct challenge.

Nitroxyl Generation from N^1-Oxygenated CP Analogs

In search of other biologically relevant substituents at the N^1 position of CP, the *hydroxyl* group was considered, even though biological *N*-hydroxylation reactions on sulfonamides or on sulfonylureas are unknown. N^1-Hydroxy-CP could not be prepared (unstable) for pharmacological evaluation, but the corresponding oxygen-masked N^1-methoxy-CP was considered to be equivalent, since enzymatic *O*-demethylation in vivo would give the desired N^1-hydroxy-CP.[47] This *O*-methylated derivative of N^1-hydroxy-CP turned out to be a highly potent inhibitor of AlDH in rats (Fig. 3.15). However, since N^1-OMeCP can be construed to be an isostere of N^1-EtCP and can release PrNCO by the same mechanism, other N^1-OMe and N^1-Et analogs of CP were synthesized. The functionality of the molecule was also changed from a sulfonylurea to a sulfonylcarbamate (*O* in place of *N*-H at N^3) (Fig. 3.16). A

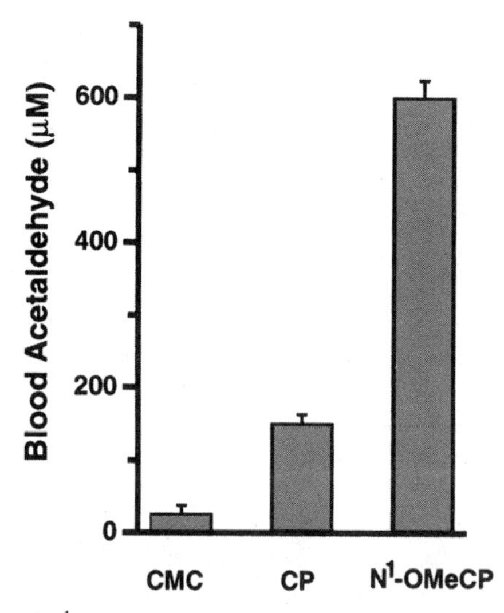

FIGURE 3.15 Effect of N^1-OMeCP (1.0 mmol/kg) on ethanol-derived blood acetaldehyde in rats. CMC = 2% carboxymethylcellulose vehicle control.

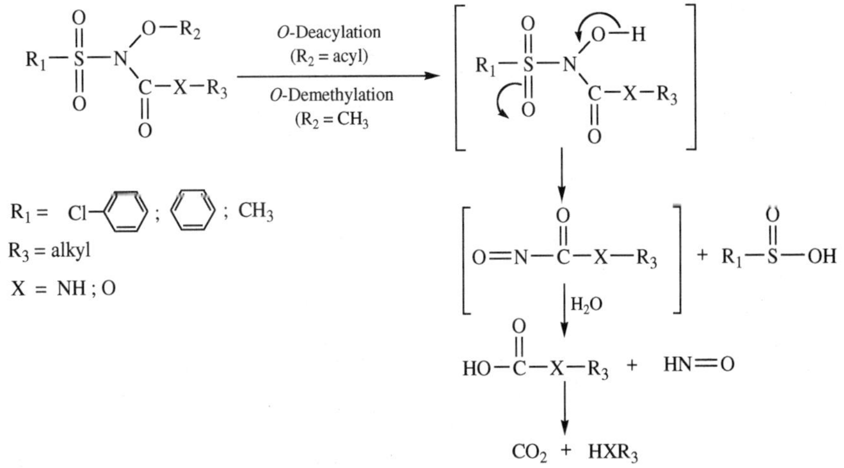

$R_1 = $ Cl–C$_6$H$_4$– ; C$_6$H$_5$– ; CH$_3$
$R_3 = $ alkyl
X = NH ; O

O-Deacylation (R_2 = acyl)
O-Demethylation ($R_2 = CH_3$)

$HO-C(=O)-X-R_3$ + HN=O

CO_2 + HXR_3

FIGURE 3.16 Enzyme-mediated bioactivation of Piloty's acid and MSHA derivatives, followed by a cascade of nonenzymatic reactions leading to nitroxyl.

detailed structure–activity study comparing the N^1-OMe and N^1-Et-substituted sulfonylureas with the corresponding sulfonylcarbamates revealed that the N^1-Et substituted sulfonyl *carbamates* were totally inactive whereas the corresponding N^1-OMe sulfonyl *carbamates* were reasonably active in vivo. Since the mechanism of action of the latter *cannot* be due to the release of PrNCO, an alternative mechanism was invoked: *the generation of nitroxyl* (Fig. 3.16).[49]

It was concluded that the 4-chlorobenzenesulfonylureas and sulfonylcarbamates with a free hydroxyl group on the sulfonamide nitrogen were unstable, disproportionating to 4-chlorobenzenesulfinic acid and an unstable acylnitroxyl intermediate that rapidly solvolyzed to nitroxyl (Fig. 3.16; $R_1 = 4\text{-Cl-}C_6H_4$). The coalescence of this mechanism with the mechanism of inhibition of AlDH by the alcohol deterrent drug, *cyanamide*, also postulated to be mediated by nitroxyl (see discussion above), was highly intriguing. The potency of the N^1-OMe analogs of CP must be due to dual mechanisms of action: nonenzymatic release of PrNCO and enzyme-mediated bioactivation to nitroxyl.

Cascade-Latentiated Prodrugs of Nitroxyl with Dual Mechanisms of Action

The preceding results suggested that derivatives of Piloty's acid (benzenesulfohydroxamic acid; *N*-hydroxybenzenesulfonamide) or of methanesulfohydroxamic acid (MSHA), which are (1) acylated on the sulfonamide nitrogen *and* (2) alkylated or acylated on the hydroxyl moiety, could serve as nitroxyl prodrugs by initial esterase mediated deacylation of the *O*-acyl group, or cytochrome P-450–mediated dealkylation of the *O*-alkyl moiety, followed by a cascade of nonenzymatic transformations leading to nitroxyl (Fig. 3.16).

The *N,O-bis*-carbethoxy derivatives of Piloty's acid and MSHA (Fig. 3.16: $R_1 = C_6H_5$ or CH_3; $R_2 = OCOEt$; $R_3 = Et$; $X = O$) with IC_{50} of 24 and 37 μM, respectively, were potent inhibitors of AlDH.[50,51] That esterase action was required to initiate the bioactivation cascade was supported by the detection of nitroxyl (as N_2O) when they were subjected to the action of porcine liver or rat plasma esterase. Additional studies not elaborated here suggested that the active-site sulfhydryl group of the enzyme was converted by these *O*-carbethoxy compounds to a *mono-thiolcarbonate ester* that would be expected to be much more stable to hydrolysis than a *thioester* group. A thioester is the penultimate chemical form following the dehydrogenase action on aldehydes by AlDH,[51] with subsequent hydrolysis of the thioester to the carboxylic acid. Thus, esterase action on these carbethoxy compounds would lead to an irreversibly bound *S*-ethoxycarbonyl group on the active-site Cys302 of the enzyme resulting in inactivation. The potent activities of these carbonate prodrugs are, like N^1-OMeCP, due to *dual action*: (1) irreversible enzyme carbethoxylation and (2) subsequent release of nitroxyl, which inactivates the remaining functional enzyme.

Chemistry of the Reaction of Nitroxyl with Thiols

This chemistry is briefly reviewed here to aid in the understanding of the mechanism of inhibition of AlDH by nitroxyl, as discussed in the next section. It is well established that nitroxyl reacts with thiols to give the corresponding disulfide and hydroxylamine according to Eqs. (3.6) and (3.7).[52–54] The *N*-hydroxysulfenamide intermediate can also undergo a molecular rearrangement to yield a sulfinamide [Eq. (3.8)]. The products formed are dependent both on the thiol and on reaction conditions; for example, at physiologic pH, the major product with reduced glutathione (GSH) is a sulfinamide,[55] whereas a cyclic disulfide is formed with dithiothreitol (DTT) at pH 9.0.[54]

$$\mathrm{RSH + HN{=}O \longrightarrow RSNHOH} \tag{3.6}$$

$$\mathrm{RSNHOH + RSH \longrightarrow RSSR + NH_2OH} \tag{3.7}$$

$$\mathrm{RSNHOH \longrightarrow\ \longrightarrow RS(O)NH_2\ (sulfinamide)} \tag{3.8}$$

The reaction of nitroxyl with thiols can be monitored by determining the effect of thiols on N_2O production. Thus, *N*-acetyl-L-cysteine (NAC) was shown to sequester the nitroxyl generated from cyanamide oxidation[56] or from Piloty's acid decomposition at alkaline pH,[57] while DTT effectively trapped nitroxyl produced by Angeli's salt.[54] Piloty's acid and Angeli's salt are chemical nitroxyl donors.[55] Excess GSH or DTT can, therefore, serve as competitive "traps" for nitroxyl and protect sulfhydryl enzymes such as AlDH from inactivation.

The major products from the reaction of nitroxyl with GSH and NAC are glutathione sulfinamide and NAC-sulfinamide, respectively. The identity of these products as sulfinamides was based on the similarities of their chromatographic properties with chemically synthesized glutathione sulfinamide and NAC-sulfinamide, prepared, respectively, by sodium cyanoborohydride reduction of *S*-nitrosoglutathione (GSNO) and by the reaction of NAC-sulfinyl chloride with NH_3. However, attempts to isolate these sulfinamides in crystalline form were unsuccessful because of their instability. With a simpler mercaptan, such as cyclohexylmercaptan, it was possible to identify the sulfinamide by GC/MS analysis.[55]

Mechanism of Inhibition of AlDH by Nitroxyl

The reaction of nitroxyl with enzyme thiols can lead to at least two types of thiol modifications, namely, one that can be reduced back to the free, active form of the enzyme by DTT, and a second form that is resistant to reduction by DTT. This is illustrated by the formation of a *reversibly* modified AlDH at pH ≤ 7.5 and an

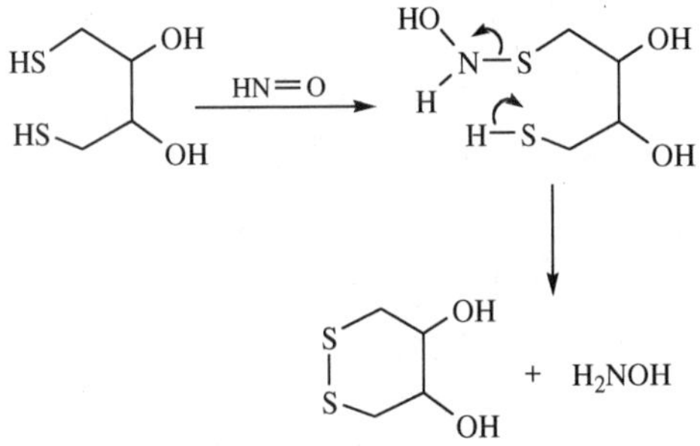

FIGURE 3.17 Reaction of nitroxyl with dithiothreitol, a model for the reversible inhibition of AlDH by nitroxyl.

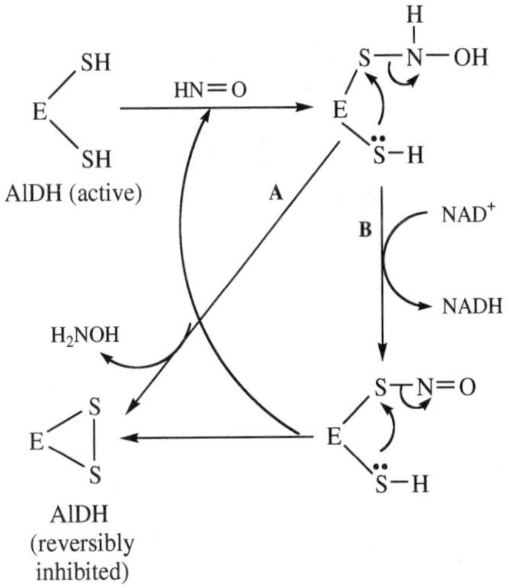

FIGURE 3.18 Proposed mechanism for the *reversible* inactivation of AlDH by nitroxyl.

irreversibly modified enzyme at pH ≥ 8.5.[56] Both of these thiol modifications proceed via an enzyme *N*-hydroxysulfenamide intermediate.

The *reversibly* inhibited or disulfide form of the enzyme is exemplified by the model reaction of nitroxyl with DTT, which yields hydroxylamine and the cyclic disulfide of DTT (Fig. 3.17). However, in the case of AlDH with cofactor-mediated dehydrogenase activity, two mechanisms for the oxidation of the proximal active-site thiols to a disulfide are possible (Fig. 3.18). In pathway **A**, the intramolecular expulsion of hydroxylamine by a neighboring sulfhydryl nucleophile produces the disulfide form of the inhibited enzyme, whereas in pathway **B**, the *N*-hydroxysulfenamide is first oxidized by NAD^+ to an *S*-nitroso intermediate that then can undergo intrasubunit disulfide formation with the elimination of nitroxyl that can recycle. This disulfide form of the nitroxyl-inhibited enzyme is likely identical to the inhibited form of this enzyme produced by nitric oxide,[54] and is readily reversible with DTT.

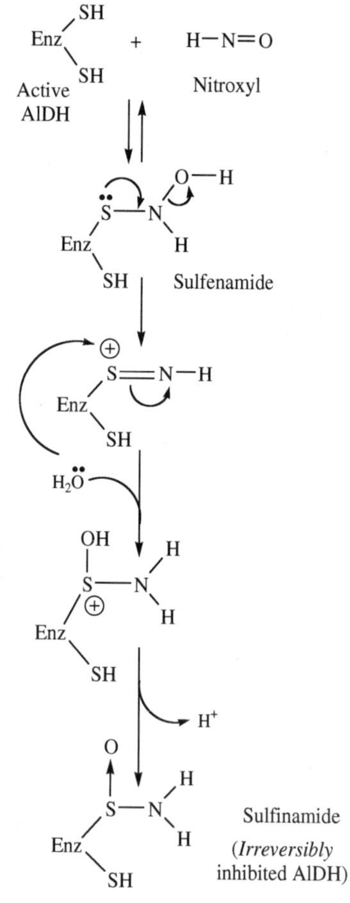

FIGURE 3.19 Proposed mechanism for the *irreversible* inactivation of AlDH by nitroxyl.

The mechanism for the *irreversible* inhibition of AlDH by nitroxyl (Fig. 3.19) is based, in part, on the chemistry of the reaction of aromatic nitroso compounds with thiols,[59] and, in part, on studies on the interaction of model thiol compounds with nitroxyl.[55,57] Accordingly, the *N*-hydroxysulfenamide formed when nitroxyl reacts with an active-site sulfhydryl group of AlDH can spontaneously rearrange via a sulfiminium ion intermediate to give the enzyme sulfinamide. The resulting active-site-modified enzyme would be expected to be resistant to reduction by DTT, thus accounting for the observed irreversible inhibition.

ACKNOWLEDGMENT

This work was supported by the Department of Veterans Affairs and, in part, by NIH/NIAAA Grant R01AA07317.

REFERENCES

1. U.S. Department of Health and Human Services, *Alcohol and Health*, 9th special report to the U.S. Congress, 1997, pp. 10–38.
2. Spangel, R.; Zieglgansberger, W., *TIPS* 1997, **18**, 54.
3. Peachey, J. E., *J. Clin. Psychopharmacol.* 1981, **1**, 368.
4. Weiner, H.; Wang, X., *Alcohol. Alcoholism.* 1994, Suppl. **2**, 141.
5. Weiner, H.; Farres, J.; Wang, T. T. Y.; Cunningham, S. J.; Zheng, C.-F.; Ghenbot, G., in *Enzymology and Molecular Biology of Carbonyl Metabolism*, H. Weiner, B. Wermuth, and D. W. Crabb, eds., Plenum Press, New York, 1991, Vol. 3, pp. 13–17.
6. Saigal D.; Cunningham, S. J.; Farres, J.; Weiner, H., *J. Bacteriol.* 1991, **173**, 3199.
7. Williams, E. E., *JAMA* 1937, **109**, 1472.
8. Hald, J.; Jacobsen, E., *Lancet* 1948, **255**, 1001.
9. Kitson, T. M., *J. Stud. Alc.* 1977, **38**, 96.
10. Gessner, P. K.; Gessner, T., in *Disulfiram and its Metabolite Diethyldithiocarbamate*, P. K. Gessner, and T. Gessner, eds., Chapman & Hall, London, 1962, pp. 167–203.
11. Goedde, Y. W.; Agarwal, D. P., *Enzyme* 1987, **37**, 29.
12. Yoshida, A.; Huang, I.-Y.; Ikawa, M., *Proc. Natl. Acad. Sci.* (USA) 1984, **81**, 258.
13. Harada, S., *Prog. Clin. Biol. Res.* 1990, **344**, 289.
14. Harada, S.; Agarwal, D. P.; Goedde, Y. W.; Tagaki, S.; Ishikawa, B., *Lancet* 1982, 827.
15. Takase, S.; Takada, A.; Yasuhara, M.; Tsutsumi, M., *Hepatology* 1989, **9**, 704.
16. Shibuya, A.; Yoshida, A., *Am. J. Hum. Genet.* 1988, **43**, 744.
17. Johansson, B., *Acta Psychiatr. Scand.* 1992, **86**, 15.
18. Yourick, J. J.; Faiman, M. D., *Biochem. Pharmacol.* 1991, **42**, 1361.
19. Jin, L.; Davis, M. R.; Hu, P.; Baille, T. A., *Chem. Res. Toxicol.* 1994, **7**, 526.
20. Deitrich, R. A.; Troxell, P. A.; Worth, W. S.; Irwin, V. G., *Biochem. Pharmacol.* 1976, **25**, 2733.
21. Kitson, T.; Crow, K. E., *Biochem. Pharmacol.* 1979, **28**, 2551.

22. Shirota, F. N.; Nagasawa, H. T.; Kwon, C.-H.; DeMaster, E. G., *Drug Metab. Disp.* 1984, **12**, 337.
23. DeMaster, E. G.; Kaplan, E.; Shirota, F. N.; Nagasawa, H. T., *Biochem. Biophys. Res. Commun.* 1982, **107**, 1333.
24. Svanas, G. W.; Weiner, H., *Biochem Pharmacol.* 1985, **34**, 1197.
25. DeMaster, E. G.; Shirota, F. N.; Nagasawa, H. T., *Biochem. Biophys. Res. Commun.* 1984, **122**, 358.
26. DeMaster, E. G.; Shirota, F. N.; Nagasawa, H. T., *Alcohol* 1985, **2**, 117.
27. Nagasawa, H. T.; DeMaster, E. G.; Redfern, B.; Shirota, F. N.; Goon, D. J. W., *J. Med. Chem.* 1990, **33**, 3120.
28. Shirota, F. N.; DeMaster, E. G.; Nagasawa, H. T., *Toxicol. Lett.* 1987, **37**, 7.
29. Shirota, F. N.; Goon, D. J. W.; DeMaster, E. G.; Nagasawa, H. T., *Biochem. Pharmacol.* 1996, **52**, 141.
30. Nagasawa, H. T.; Kwon, C.-H.; DeMaster, E. G.; Shirota, F. N., *Biochem. Pharmacol.* 1986, **35**, 129.
31. Kwon, C.-H., Nagasawa, H. T.; DeMaster, E. G.; Shirota, F. N., *J. Med. Chem.* 1986, **29**, 1922.
32. Hanzlik, R.; Cashman, J.; Traiger, G., *Toxicol. Appl. Pharmacol.* 1980, **55**, 260.
33. Hanzlik, R.; Cashman, J., *Drug. Met. Disp.* 1983, **11**, 201.
34. Shirota, F. N.; Stevens-Johnk, J. M.; DeMaster, E. G.; Nagasawa, H. T., *J. Med. Chem.* 1997, **40**, 1870.
35. Ohlin, H.; Jerntorp, P.; Bergstrom, B.; Almer, L.-O., *Br. Med. J.* 1982, **285**, 838.
36. Jerntorp, P.; Ohlin, H.; Ergstrom, B.; Almer, L.-O., *Diabetes* 1981, **30**, 788.
37. Barnett, A. H.; Gonzales-Auvert, C.; Pyke, D.; Saunder, J. B.; Williams, R.; Dickenson, C.; Rawlins, M. D., *Br. Med. J.* 1981, **283**, 939.
38. Little, R. G.; Peterson, D. R., *Toxicol. Appl. Pharmacol.* 1985, **80**, 206.
39. Nagasawa, H. T.; DeMaster, E. G.; Kwon, C.-H.; Fraser, P. S.; Shirota, F. N., *Alcohol* 1985, **2**, 123.
40. Nagasawa, H. T.; Elberling, J. A.; Shirota, F. N.; DeMaster, E. G., *Alcoholism Clin. Exp. Res.* 1988, **12**, 563.
41. Nagasawa, H. T.; Elberling, J. A.; DeMaster, E. G.; Shirota, F. N., *J. Med. Chem.* 1989, **32**, 1335.
42. Nagasawa, H. T.; Smith, W. E.; Kwon, C.-H.; Goon, D. J. W., *J. Org. Chem.* 1985, **50**, 4993.
43. Shirota, F. N.; Elberling, J. A.; Nagasawa, H. T.; DeMaster, E. G., *Biochem. Pharmacol.* 1992, **43**, 916.
44. Nagasawa, H. T.; Elberling, J. A.; Goon, D. J. W.; Shirota, F. N., *J. Med. Chem.* 1994, **37**, 422.
45. Nagasawa, H. T.; Lee, M. J. C.; Kwon, C. -H.; Shirota, F. N.; DeMaster, E. G., *Alcohol* 1992, **9**, 349.
46. Beckwith, A. L. J.; Evans, G. W., *J. Chem. Soc.* 1962, 130.
47. Lee, M. J. C.; Elberling, J. A.; Nagasawa, H. T., *J. Med. Chem.* 1992, **35**, 3641.
48. Lee, M. J. C.; Nagasawa, H. T.; Elberling, J. A ; DeMaster, E. G., *J. Med. Chem.* 1992, **35**, 3648.

49. Nagasawa, H. T.; DeMaster, E. G.; Goon, D. J. W.; Kawle, S. P.; Shirota, F. N., *J. Med. Chem.* 1995, **38**, 1872.
50. Conway, T. T.; DeMaster, E. G.; Lee, M. J. C.; Nagasawa, H. T., *J. Med. Chem.* 1998, **41**, 2903.
51. Dickenson, F. M.; Haywood, G. W., *Biochem. J.* 1987, **247**, 377.
52. Doyle, M. P.; Mahapatro, S. N.; Broene, R. D.; Guy, J. K., *J. Am. Chem. Soc.* 1988, **110**, 593.
53. Turk, T.; Hollocher, T. C., *Biochem. Biophys. Res. Commun.* 1992, **183**, 983
54. DeMaster, E. G.; Redfern, B.; Quast, B. J.; Dahlseid, T.; Nagasawa, H. T., *Alcohol* 1997, **14**, 181.
55. Wong, P. S.; Hyun J.; Fukuto, J. M.; Shirota, F. N.; DeMaster, E. G.; Shoeman, D. W.; Nagasawa, H. T., *Biochemistry* 1998, **37**, 5363.
56. DeMaster, E. G.; Redfern, B.; Nagasawa, H. T., *Biochem. Pharmacol.* 1998, **55**, 2007.
57. Shoeman, D. W.; Nagasawa, H. T.; DeMaster, E. G., *FASEB. J.* 1996, **10**, A177.
58. Feelish, M.; Stamler, J. S., in *Methods in Nitric Oxide Research.* M. Feelish and J. S. Stamler, eds., Wiley, New York, 1996, pp. 98–99.
59. Ellis, M. K.; Hill, S.; Foster, P. M. D., *Chem. Biol. Interact.* 1992, **82**, 151.

CHAPTER 4

AIDS: Adenosine Deaminase–Activated Prodrugs Designed for the Treatment of Human Immunodeficiency Virus in the Central Nervous System

JOHN S. DRISCOLL

Laboratory of Medicinal Chemistry, Division of Basic Sciences,
National Cancer Institute, National Institutes of Health

INTRODUCTION

The appearance and recognition of acquired immune deficiency syndrome (AIDS) as a major health problem in the early 1980s prompted an urgent search for useful therapeutic drugs. Shortly after human immunodeficiency virus (HIV) was identified as the causative agent for AIDS, the 2′,3′-dideoxynucleosides (ddN) and their analogs [e.g., 2′,3′-dideoxy-3′-azidothymidine (AZT), 2′,3′-dideoxycytidine (ddC), and 2′,3′-dideoxyinosine (ddI) (**2a**)] were soon discovered to inhibit reverse transcriptase (RT), a key enzyme used by HIV to replicate itself within cells.[1] With the recognition that the ddNs were fruitful lead compounds, a major effort was directed toward the discovery of additional members of this chemical family. This resulted in the successful development of additional drugs such as 2′,3′-didehydro-2′,3′-dideoxythymidine (d4T) and 3′-thia-2′,3′-dideoxycytidine (3TC). The design of a new drug for any disease is based on a marriage of medicinal chemistry and biochemical pharmacology. Each brings its own unique approach and knowledge base to the problem addressed, but neither alone is sufficient to maximize the probability for success. With the recent discovery of the beneficial effects of drug combinations ("cocktails") containing protease inhibitors as well as RT inhibitors, a significant reduction in AIDS-related deaths has occurred. While these combination treatments are not curative, they illustrate the power of an integrated chemistry/biochemistry/pharmacology approach to the discovery of useful new drugs.

Biomedical Chemistry: Applying Chemical Principles to the Understanding and Treatment of Disease, Edited by Paul F. Torrence
ISBN 0-471-32633-x

However, no current anti-HIV drug is problem-free. Deficiencies include toxicities, less than desirable potencies, and poor drug delivery to HIV disease sanctuaries. This chapter addresses one approach to dealing with the difficult problem of getting an active drug into the central nervous system (CNS) in order to treat HIV in that biological sanctuary.

BACKGROUND

The complications resulting from HIV infection of the CNS are major problems associated with AIDS, especially in advanced disease.[2] The observed neurological dysfunction, which includes a progressive, AIDS-related dementia, appears to be directly related to the presence of virus in the CNS, since HIV has many characteristics of the neurotropic lentiviruses.[3] Because the CNS can be a sanctuary for HIV as well as a site for significant physiologic damage, the need for effective anti-HIV drugs that cross the blood–brain barrier (BBB) has been recognized.[4] The BBB is a natural defense mechanism that protects the brain from toxins in the systemic circulation. Unfortunately, the BBB also can exclude beneficial drugs. Although AZT crosses the BBB to some extent, it possesses toxicities that limit both the amount of drug that can be administered and the duration of the treatment. Lipophilic, nonionic, low-molecular-weight materials generally appear to have the best passive diffusion properties for CNS penetration.[5] Our working hypothesis, therefore, is that anti-HIV agents that are more lipophilic than those currently available may be particularly useful for treatment of HIV sequestered in the CNS.

PROJECT APPROACH

The approach we have taken to test this hypothesis involves the synthesis of prodrugs of an active anti-HIV agent (F-ddI; **2b**) discovered earlier in this laboratory[6] which is structurally related to the clinically useful drug ddI (**2a**), as well as the 6-amino precursors **1a** and **1b**.

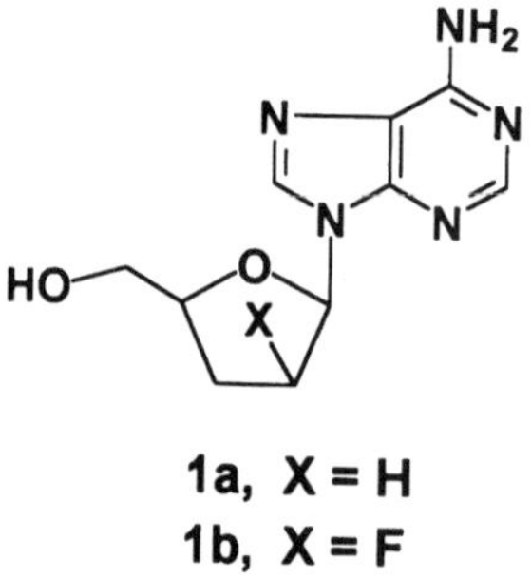

1a, X = H
1b, X = F

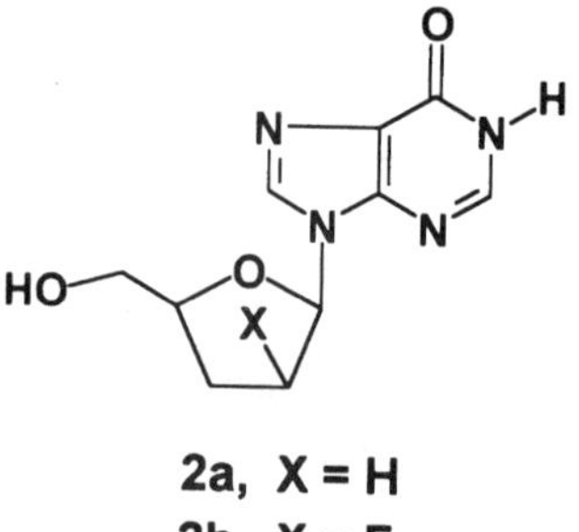

2a, X = H
2b, X = F

PRODRUGS

A *prodrug* is a derivative or analog of a known active compound modified to give it certain desirable properties relative to its active parent drug. In this work, the property we are hoping to improve is the transport of **2b** into the CNS. Normally a prodrug is a biologically inactive compound that is converted to an active drug by a chemical or biochemical reaction in vivo. It is anticipated that once our prodrug has entered the CNS, the activation step, which will convert the inactive prodrug to the anti-HIV agent, F-ddI (**2b**), is an enzymatic hydrolysis reaction catalyzed by the naturally occurring enzyme adenosine deaminase (ADA) (Scheme 4.1).

Prodrug Considerations

A number of drug design factors can determine whether a newly synthesized compound has a chance to become a useful drug. Some of the chemical factors can be estimated prior to organic synthetic work (e.g., lipophilicity, acid stability), but most of the biological factors (anti-HIV activity and potency, enzymatic anabolism and catabolism rates, systemic elimination, and in vivo toxicity) cannot be predicted with certainty. These properties must be determined experimentally after preparing the compounds.

Acid Stability Purine 2′,3′-dideoxynucleosides are notoriously unstable under acidic conditions.[7] This makes the oral administration of these compounds very difficult since much of the drug is decomposed to inactive products under the acidic conditions found in the human stomach. We have previously described the effect of sugar fluorine substitution, which renders purine ddNs acid-stable.[8] This effect is illustrated in Figure 4.1. Because of its stabilization the glycosidic bond towards acid-catalyzed cleavage, 2′-fluorosugar substitution was always included in the design of the prodrugs.

Lipophilicity The design of compounds with quantitatively predicted lipophilic properties is not difficult given the general principles and lipophilic substituent constants generated by Hansch and others in earlier quantitative structure–activity

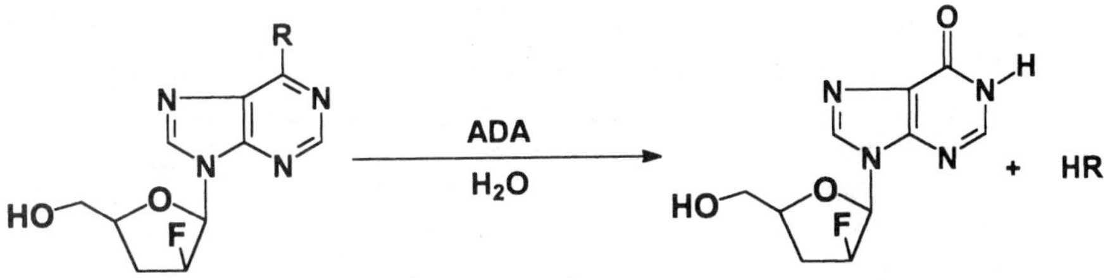

ADA

H_2O

+ HR

R = alkylamino, alkoxy, halo

F-ddI (2b)

SCHEME 4.1

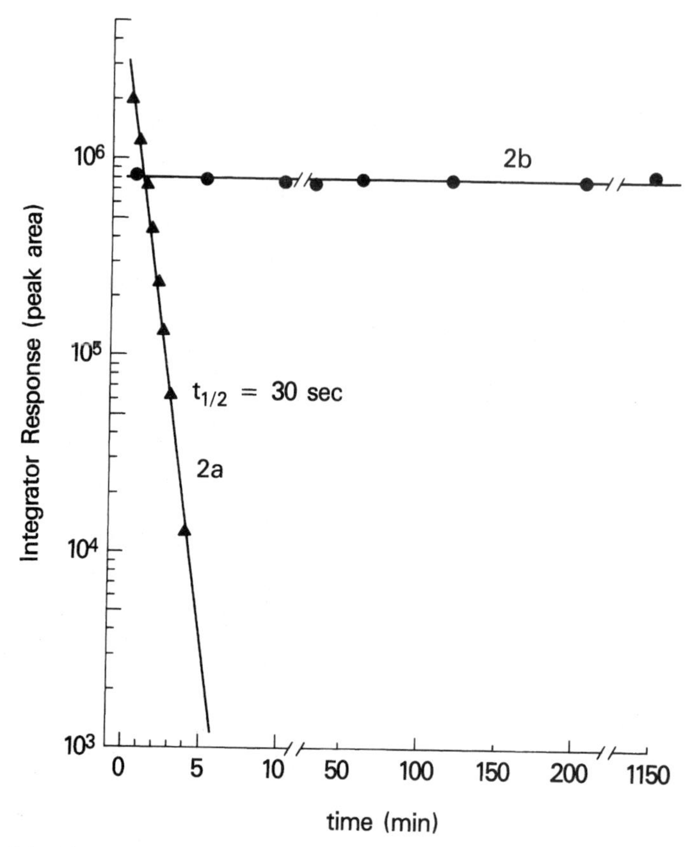

FIGURE 4.1 Concentration vs time profile for 2′,3′-dideoxyinosine (**2a**, ddI) and 2′-fluoro-2′,3′-dideoxyinosine (**2b**, F-ddI) at pH 1.0 and 37°C. The timescale for the first 10 min is greatly expanded. Initial concentrations of structures **2a** and **2b** were 10.2 and 8.4 μg/mL, respectively.

relationship (QSAR) studies.[9] A more uncertain part of drug design is whether anti-HIV activity will be retained after structural modifications are made to an active molecule. The anti-HIV activity of dideoxynucleosides is critically dependent on a series of enzymatic events (e.g., sequential kinase activation steps, catabolic events, RT interaction), any one of which can be adversely affected by a structural change (Fig. 4.2). For this reason, we began our test of the lipophilicity/CNS penetration hypothesis by making the most minor structural changes possible consistent with increasing the lipophilic character of prodrugs of **2b**. Hansch and co-workers have pioneered the use of octanol/water partition coefficients (*P*) to provide a relative measure of the lipophilic character of a compound, and to correlate compound structure with CNS penetration. *P* values (usually expressed as $\log_{10}$, i.e., log *P*) can be calculated prior to synthesis and these values approximate how a compound might distribute itself between aqueous and fat-like compartments in the body.

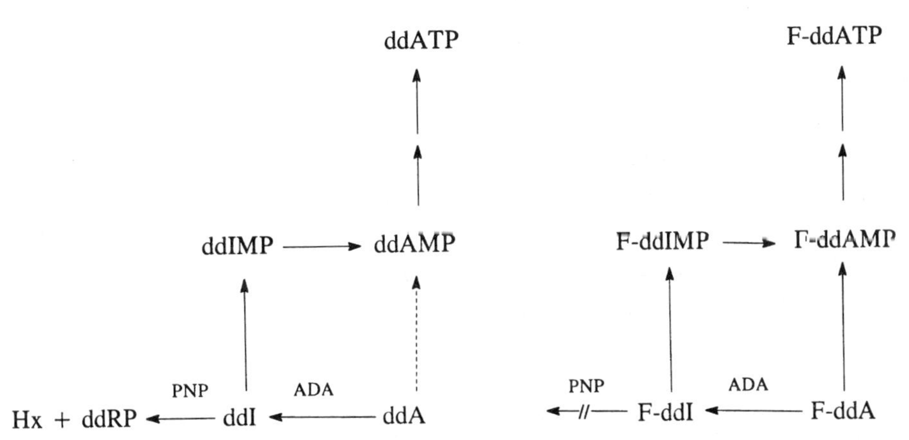

FIGURE 4.2 Metabolic pathways for ddA (**1a**) and F-ddA (**1b**). (*Abbreviations*: MP and TP, mono- and triphosphate, respectively; PNP, purine nucleoside phosphorylase; Hx, hypoxanthine; ddRP, 2,3-dideoxyriboside-1-monophosphate.)

Initially, the approach taken involved the addition of methyl groups to various positions of the 6-aminopurine ddN, **1b** (Table 4.1). Such an addition should increase a log P value by approximately +0.5 (a factor of 3) for each methyl group.[10]

Purine nucleosides were chosen over pyrimidines since they are more lipophilic to begin with, and thus easier to design in the log P of range 0.0–2.0, which we felt might be optimum. Ultimately, a number of additional 6-substituted FddNs were prepared and evaluated (Table 4.2). Preclinical pharmacologic studies on ddN suggest that a relationship may exist between partition coefficients and CSF/plasma ratios.[9,11,12] Azidothymidine (AZT), which has a log P value of 0.05, is one of the more lipophilic of the clinically useful ddNs. AZT is about 15 times more lipophilic than either 2′,3′-dideoxycytidine (ddC) or dideoxyinosine (ddI), and enters the CNS

TABLE 4.1 Methylated 2′-Fluoro-2′,3′-dideoxypurine Prodrugs

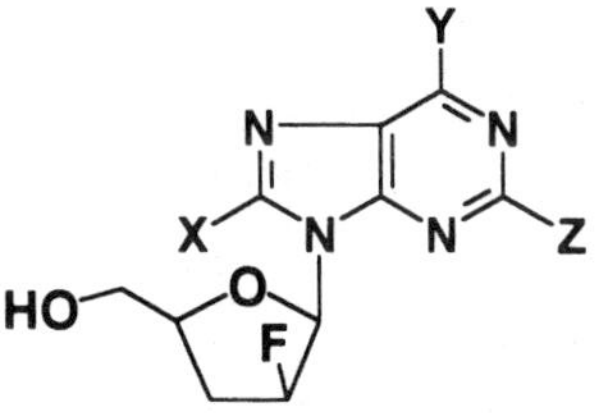

Compound	X	Y	Z
3	CH_3	NH_2	H
4	H	NH_2	CH_3
5	H	$NHCH_3$	H
6	H	$NHCH_3$	CH_3

TABLE 4.2 6-Substituted 2′-Fluoro-2′-3′-dideoxypurine Prodrugs

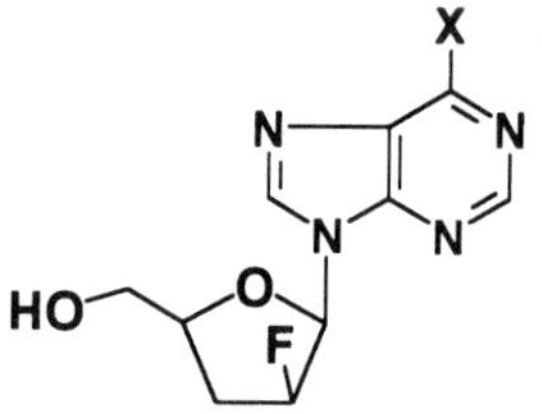

Compound	X
7	$N(CH_3)_2$
8	$NHCH_2CH_3$
9	NHOH
10	$NHOCH_3$
11	$NHOCH_2C_6H_5$
12	$NHNH_2$
13	NO_2
14	$NHCOC_6H_5$
15	H
16	Cl
17	Br
18	I
19	F
20	OCH_3
21	OCH_2CH_3

better than either of the latter two compounds. The poor CNS penetration of ddC and ddI was correlated with their highly hydrophilic nature (log P values of -1.33 and -1.24, respectively).

Hydrolysis by Adenosine Deaminase The role that ADA plays in the experimental process is as a prodrug activator once the compound has crossed the BBB and entered the CNS (Scheme 4.1). ADA serves as a naturally occurring biological catalyst for a hydrolysis reaction that converts an inactive prodrug (chosen for its good transport properties) into the anti-HIV-active ddN, F-ddI (**2b**) (Scheme 4.1). ADA is a ubiquitous catabolic enzyme present in almost all animal and human tissues.[13] It is present in the brain as well as the systemic circulation and is reported to increase in the CNS during HIV infection.[14] ADA is important in the proper functioning of the immune system and its regulatory function appears to involve T cells, which are also important in HIV infection. While the enzyme's name implies that it changes the 6-amino group of adenosine to a 6-hydroxyl group generating inosine, ADA also is capable of catalyzing an identical hydrolysis reaction with a large number of 6-substituted purine nucleosides. Early studies on the properties of ADA showed that this enzyme converted compounds such as the 6-

alkylamino, 6-alkoxy, and 6-halo analogs of adenosine to inosine at varying rates.[15,16] For this reason, certain 6-substituted ddNs were chosen for their ability to serve as ADA substrates and the ultimate generation of the anti-HIV agent F-ddI (**2b**), as well as their lipophilic properties.

Prodrug Design

As mentioned above, factors that might be addressed in F-ddI prodrug design before beginning organic synthetic work include a prediction of acid stability, and lipophilic character. The only biological property that one can attempt to predict qualitatively for these prodrugs is their possible utility as an ADA substrate. Other metabolic characteristics, as well as biological activities (e.g., anti-HIV activity and toxicity), must be determined experimentally after the prodrugs are prepared. As previously described (Fig. 4.1), acid stability was built into all the prodrugs by including a 2′-fluoro group in the nucleoside structures (F-ddN).

Prodrug ADA Substrate Properties

In this study, a primary characteristic that a successful prodrug must possess is an ability to be a substrate for ADA, the enzyme that hydrolyzes the 6-substituent to a hydroxyl group forming the anti-HIV-active compound, F-ddI (Scheme 4.1). There is large amount of ADA substrate data available for conventional (nonfluorinated) nucleosides that are used as DNA and RNA precursors. These data suggest that certain 6-substituents in the fluoroprodrug series might be more useful than others with regard to both hydrolysis and an increase in lipophilicity. A major problem from the beginning of this effort is that ADA rates of human whole-body deamination for standard compounds such as adenosine are not known. What is known is that the human body contains copious amounts of ADA. For this reason, it was necessary to prepare a number of prodrugs (Tables 4.1 and 4.2) having a range of deamination rates, in the hope that at least a few might provide practical drug candidates. As a consequence of fluorine substitution, the prototype F-ddI prodrug, F-ddA (**1b**), was shown to be deaminated only about 5% as rapidly as ddA (**1a**) (Fig. 4.3).

In the nonfluorinated purine nucleoside series, the 6-halo (F, Cl, Br, I) compounds are well studied, especially the 6-chloro analog.[17,18] All of these conventional nucleosides serve as substrates for ADA, and therefore also serve as guides for the design of our new prodrugs. Among the 6-amino analogs, the N^6-methylamino compound has received the most attention.[15,19,20] This compound is an ADA substrate with a better affinity (K_m), but a slower deamination rate (V_{max}) than adenosine. N^6-ethylamino adenosine also is deaminated, but more slowly than the methyl analog. It was hoped that increasing alkyl size in our series would increase prodrug lipophilic character and still provide prodrugs with practical in vivo ADA hydrolysis rates. The nonfluorinated N^6-hydroxylamino, methoxylamino, and hydrazino purine nucleoside analogs are all substrates with varying hydrolysis rates, while the N^6-dimethylamino analog was shown to be an inhibitor of the

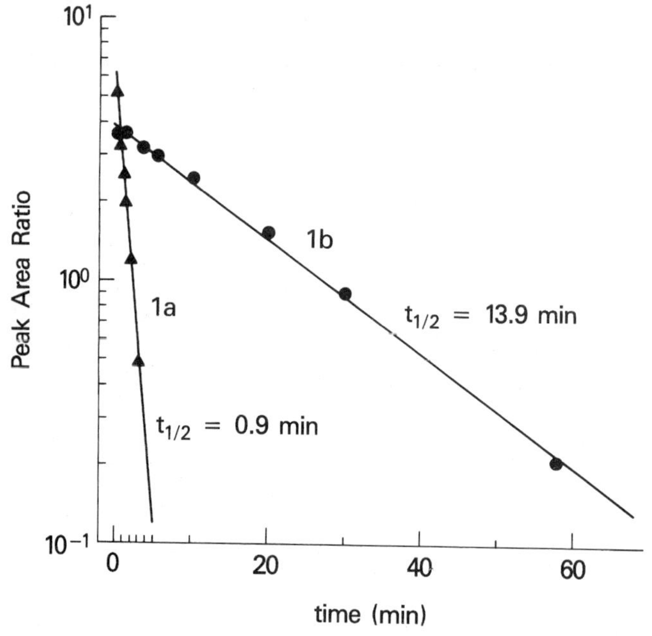

FIGURE 4.3 The effect of fluorine on the rate of ADA-catalyzed hydrolysis (0.3 U/mL) of the 2′,3′-dideoxyadenosine compounds **1a** (9.2 μg/mL) and **1b** (34.5 μg/mL) to their corresponding 2′,3′-dideoxyinosine analogs, **2a** and **2b** at pH 7.5 and 25 °C.

enzyme. As in the N^6-alkylamino series, increasing alkyl group size in the 6-alkoxy series reduced the hydrolysis rate.[17] Substitution in the purine aglycon appears to affect the hydrolysis rates more than sugar substitution. Other workers also have shown that 2′-F sugar substitution in purine nucleosides provides compounds that are ADA substrates, which are usually hydrolyzed more slowly than their nonfluorinated counterparts.[21] This preexisting literature on the hydrolysis rates for nonfluorinated nucleosides provided a guide for the selection of prodrug candidates.

Prodrug Lipophilic Properties

In addition to being a point for ADA-catalyzed hydrolysis, the 6-substituent contributes to the lipophilic properties of the prodrug in a characteristic manner. Early work by Hansch et al. determined substituent constants (π values) for many functional groups, thus allowing the calculation of $\log P$ values for unknown compounds.[22] Based on a CNS structure–activity study on hypnotics by the Hansch group,[9,12] and the $\log P$ values of known anti-HIV agents,[23] we wanted to design molecules that were 10–100 times more lipophilic than AZT, the most lipophilic of the clinically useful RT inhibitors. When this calculation was carried out prior to synthesis for the ADA-hydrolyzable groups mentioned above, a spectrum of potential $\log P$ values was obtained. This variation was desirable since there was no way, a priori, to predict an optimum CNS $\log P$ value in this series prior to in vivo testing. The synthesis and evaluation of a number of compounds was required in order to determine this optimum (Tables 4.1 and 4.2). If a compound is too lipophilic, it will be rapidly sequestered in body fats and be nonbioavailable. If it is too hydrophilic, the prodrug will not cross the BBB well, and will be rapidly excreted.

Prodrug Synthesis

The syntheses of the parent purine molecules (**1a**,**b** and 2**a**,**b**) have been described.[24,25] The simple methyl purine analogs, compunds **3–6** (listed in Table 4.1) were prepared by conventional coupling methods or by methylation of purine nucleosides.[24] Compound **5** was prepared by both methods. The 2′-fluoro nebularine analog (compound **15**, listed in Table 4.2) which cannot be hydrolyzed, was prepared as a reference point for the calculation of the partition coefficients for 6-substituted analogs. The synthetic methods used to prepare the prodrug starting materials (5′-protected analogs of **1b** and **16**) were those described previously, and simplified schemes for the preparation of the prodrugs are shown in Scheme 4.2. Details of this work have been reported.[25,26]

Diazotization followed by photochemically induced radical generation from halocarbon solvents provided the 6-chloro (**16**), bromo (**17**), and iodo (**18**) analogs (Scheme 4.2). The corresponding fluoro compound (**19**) was prepared from the 6-chloro compound (**16**) via an unisolated 6-trimethylammonium compound, followed

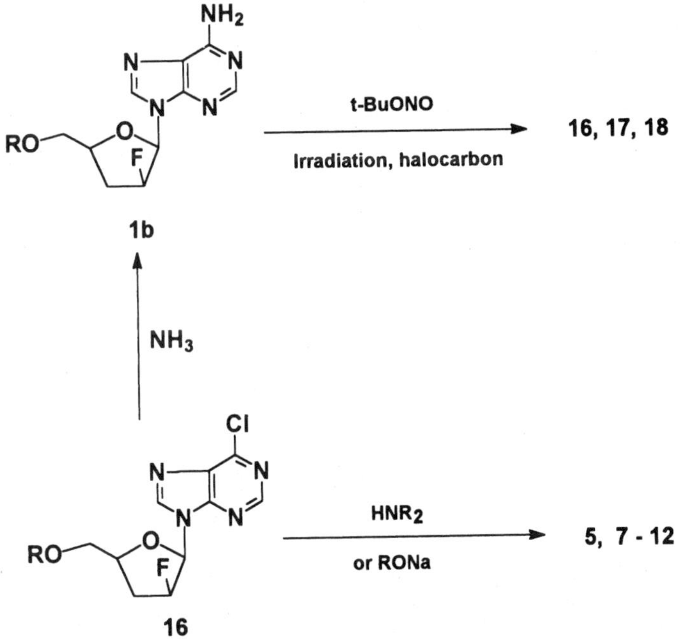

SCHEME 4.2

by fluoride ion displacement. The 6-amino analogs, **5** and **7–12** (Table 4.2), were prepared from **16** by simple chlorine displacement reactions, as were the two alkoxy compounds, **20** and **21**. Compound **14** was prepared from **1b** by benzoylation. The 6-nitro analog, **13**, was isolated as a side product from the photochemical generation of **16** in carbon tetrachloride/t-butyl nitrite. This compound was convertible to **1b** by catalytic hydrogenation.

RESULTS

Since stomach acid plays such a critical role in the development of a convenient oral formulation for an anti-HIV drug, this property was evaluated for most of the prodrugs in pH 1.0 buffer solution at 37 °C. The expected stability of the 2′-fluoro analogs toward cleavage of the glycosidic bond, which is characteristic of nonfluorinated dideoxypurine nucleosides, was observed in all instances. Occasionally, some chemical (nonenzymatic) hydrolysis of an electronegative group in the 6-position occurred with hydroxyl group replacement to produce **2b**. This was observed with the 6-fluoro (**19**) and 6-nitro (**13**) analogs.

Once the prodrugs had been synthesized, $\log P$ values were determined experimentally in octanol/pH 7 buffer using a convenient microscale method developed in the author's laboratory.[23] A comparison of these experimental values with those calculated prior to synthesis showed good general agreement for compounds with nonpolar groups, such as halogen and alkyl substituents. Less quantitative accuracy, however, was found for prodrugs containing polar substituents (e.g., hydroxyl). The experimental $\log P$ values are shown in Table 4.3. Our goal was to produce prodrugs more lipophilic than AZT ($\log P$ 0.05). With the exception of compounds **9**, **10**, **13**, and **15**, all the prodrugs synthesized realized this goal. The benzyloxyamino analog (**11**) was the most lipophilic prodrug prepared—18 times greater than AZT. The 6-alkylamino analogs **5–8** ranged between 2 and 8 times greater, and in the halogen series, the 6-iodo (**18**) and 6-chloro (**16**) compounds were 4 and 2 times more lipophilic than AZT, respectively.

The prodrugs in this study were designed with the possibility for CNS ADA-catalyzed hydrolysis to the anti-HIV agent **2b**. Evaluation of in vitro hydrolysis rates was made with commercially available, purified ADA (EC 3.5.4.4) from bovine intestinal mucosa. This enzyme was used at 1 U/mL (pH 7.0, 37 °C) with a prodrug concentration of 50 μM. The results of this study are shown in Table 4.4. The ubiquitous nature of ADA in the human body suggests that an optimum prodrug might require a relatively slow hydrolysis rate to survive systemic ADA prior to penetrating the BBB. The utility of any prodrug partially converted to **2b** in the plasma would not be lost, however, since **2b** would be active against systemic HIV. The relative rates for the new prodrugs (normalized to F-ddA; **1b**) are seen to vary by a factor of $> 20{,}000$ (fluorine vs. ethoxy). The addition of a 2′-fluorine atom to the sugar (**1b**) slows the deamination rate by a factor of 20 (Fig. 4.3) relative to the nonfluorinated analog, ddA (**1a**). The addition of one methyl group (**5**) to the 6-amino function of **1b** lowers the hydrolysis rate by > 100-fold (Table 4.4). With the

TABLE 4.3 Experimental Log *P* Values for 2′,3′-Dideoxynucleosides[a]

Compound	Log *P*
1a	−0.29
1b	−0.18
2a	−1.24
2b	−1.21
3	0.10
4	0.12
5	0.27
6	0.64
7	0.80
8	0.78
9	−0.98
10	−0.41
11	1.26
13	−0.40
14	0.73
15	−0.40
16	0.32
17	0.44
18	0.63
19	0.06
20	0.24
21	0.80

[a] Octanol/pH 7.0 system; AZT log $P = 0.05$.

addition of two methyl groups (**7**) or an ethyl moiety (**8**), the rates are so slow that no deamination can be detected during the 24-h timeframe of the experiment. The rate of the 6-methoxy analog (**20**) is comparable to that of 6-methylamino (**5**), with a similar rate decrease when methyl is extended to ethyl (**21**). These data show that the concept of increasing carbon alkyl length to increase lipophilicity is probably not a practical approach in this series since the increased lipophilicity is usually accompanied by a dramatic decrease in the ADA hydrolysis rate. The halogens analogs (**16–18**) are hydrolyzed at rates about 1–2% that of the corresponding amino compound, **1b**, and these might be appropriate rates for an in vivo situation. The 6-fluoro analog (**19**) is the only prodrug in the 2′-fluoro-2′,3′-dideoxy series that has a more rapid ADA hydrolysis rate than **1b**. This is in keeping with its chemical hydrolytic instability (hydrolysis at the 6-position).

The HIV-infected, phytohemagglutinin-stimulated peripheral blood mononuclear cell (PHA/PBM) test system[27] was used to measure prodrug activity against the cytopathogenic effects of HIV-1. This system quantitates a decrease in HIV-1 p24 *gag* protein expression relative to an untreated control as a measure of compound activity and potency. Prodrug anti-HIV-1 data are shown in Table 4.4. The activities of the parent compounds, **1a**,**b** and **2a**,**b**, were determined in earlier studies.[8,25] Among the simple methylated derivatives initially prepared (**3–6**), only the N^6-

TABLE 4.4 F-ddI Prodrug Data

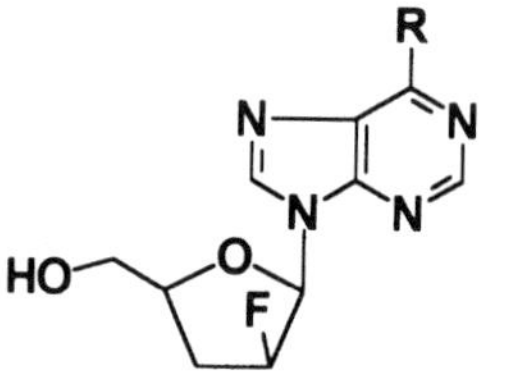

Compound	R	Relative ADA Rate[a]	Anti-HIV Activity[b] (IC_{50}, μM)
1a (ddA)	NA[c]	2080	< 5
1b (F-ddA)	NH_2	100	3
2a (ddI)	NA	NA	< 5
2b (F-ddI)	OH	NA	< 5
5	$NHCH_3$	0.9	13
7	$N(CH_3)_2$	NR[d]	18
8	$NHCH_2CH_3$	NR	11
10	$NHOCH_3$	0.02	> 80
13	NO_2	58	10
16	Cl	2.0	6
17	Br	1.7	7
18	I	0.3	10
19	F	202	< 5
20	OCH_3	0.6	20
21	OCH_2CH_3	0.01	14

[a] Normalized to F-ddA.
[b] PHA-activated PBM cells.
[c] Not applicable.
[d] No reaction.

monomethyl analog, **5**, was active (Table 4.4) with an IC_{50} value of 13 μM. The addition of a methyl group to the aglycon 2-position (**4**, **6**) or to the 8-position (**3**) abolished activity. Substitution at the 2-position of adenosine analogs with any group (except amino) normally abolishes or greatly reduces ADA-catalyzed deamination.[28] This effect may account for the lack of activity of **4** and **6**. An attempt to increase the lipophilicity of **5** by lengthening the N^6 carbon chain from methyl (**5**) to ethyl (**8**) or dimethylamino (**7**) generated active compounds, but ADA hydrolysis was abolished. When hydroxylamino (**9**) was present in the 6-position, activity was observed (IC_{50} 7 μM), but the compound was too unstable to obtain a reproducible ADA hydrolysis rate. The introduction of a methoxyamino (**10**) or a benzyloxyamino (**11**) group produced compounds with increased lipophilicity (Table 4.3), but anti-HIV activity was abolished. Although a reproducible log *P* value could not be determined for the 6-hydrazino analog (**12**), this prodrug did possess activity with moderate potency (IC_{50} 36 μM). The 6-nitro prodrug (**13**) possessed a relatively rapid ADA hydrolysis rate as well fairly good anti-HIV potency (10 μM).

All the halogen analogs (**16–19**) possessed anti-HIV activity with potencies between 5 and 10 μM. ADA hydrolysis rates were in the range of 1–2% of that for F-ddA (**1b**) with the exception of the 6-fluoro compound (**19**) with a rate twice that of F-ddA (**1b**). This very electonegative moiety appears to undergo chemical, as well as enzymatic, hydrolysis to produce **2b**. This is in keeping with the rapid rate for the nitro compound (**13**), which also possesses a strongly electonegative group. The 6-methoxy analog (**20**) had an ADA rate similar to that of most of the halogens and possessed anti-HIV activity. The 6-ethoxy prodrug (**21**) had a relatively high lipophilicity and a very low ADA rate, but also showed activity.

Even though the correlation between ADA hydrolysis rates and anti-HIV activity is imperfect, a qualitative trend can be seen (Table 4.4). In one of our earlier studies it was shown that anti-HIV activity could be increased by the addition of ADA to the prodrug test system and conversely, activity could be abolished by the addition of the powerful ADA inhibitor, 2′-deoxycoformycin, to the system.[24] It also is possible that some of the non-hydrolyzable prodrugs (e.g., **7**, **8**) may have intrinsic activity of their own and do not require hydrolysis. This requires further study.

2′-Fluoro-2′,3′-dideoxyadenosine (F-ddA; **1b**) is currently in phase I/II clinical trial as a general anti-HIV drug.[29] If **1b** is considered to be a prodrug of **2b**, and its clinical ADA deamination pharmacokinetics are examined, it is clear that the human rate of **1b** deamination is very rapid.[30] This indicates that a compound with a significantly slower hydrolysis rate than **1b** might be an appropriate choice as a CNS prodrug. On the basis of these observations, the 6-chloro prodrug (**16**) was chosen for an initial pharmacokinetic study in uninfected rats. This compound was administered as a 100-mg/kg, 60-min infusion. The plasma and cerebrospinal fluid (CSF) concentrations of **16**, and the **2b** formed from it, were determined by high

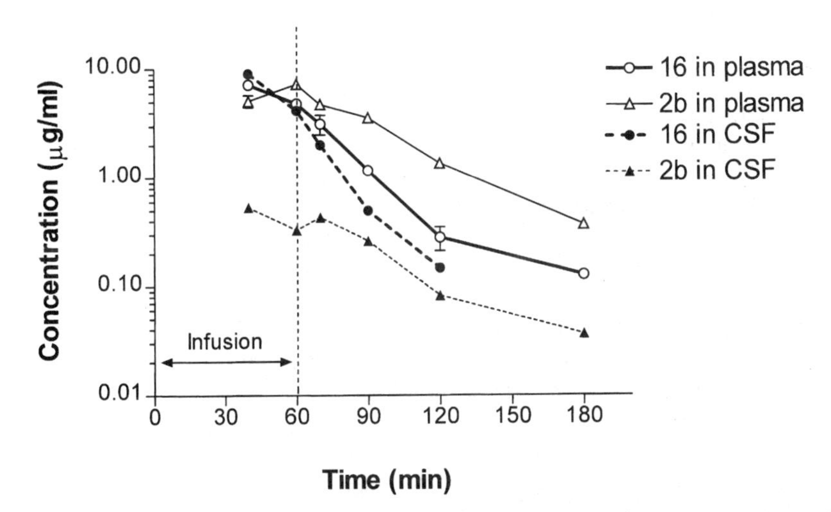

FIGURE 4.4 Concentration vs time profiles for 2′-fluoro-2′,3′-dideoxy-6-chloropurine riboside (**16**) and its in vivo–generated metabolite, F-ddI (**2b**), following a 100 mg/kg, 60-min intravenous infusion of **16** into rats.

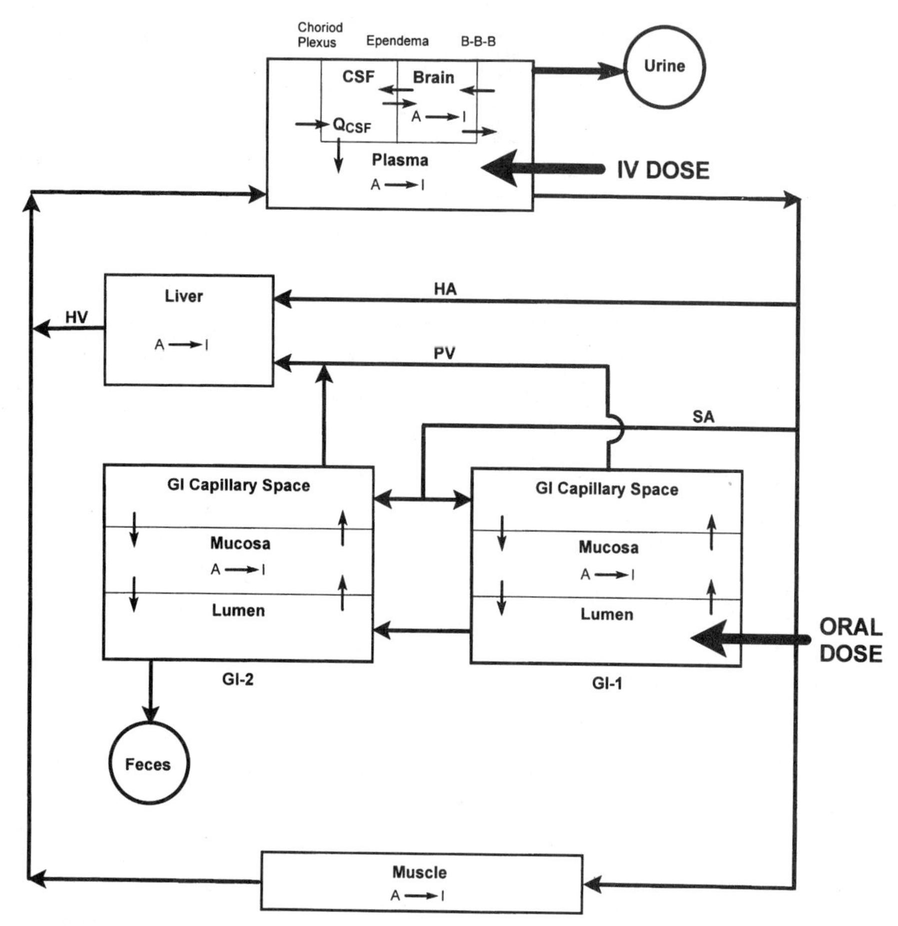

FIGURE 4.5 Schematic diagram for a whole-body physiological pharmacokinetic model for a prodrug (F-ddA) where dose administered is either oral or intravenous. [*Abbreviations*: GI-1, gastrointestinal tract (segment 1); GI-2, gastrointestinal tract (segment 2); HA, hepatic artery; HV, hepatic vein; PV, portal vein; Q_{CSF}, flow rate of CSF in the CNS; SA, splenic artery. A → I indicates compartments where ADA metabolism (**1b** to **2b**) is allowed.]

performance liquid chromatography (HPLC). Figure 4.4 shows that the prodrug **16** was taken up into rat CSF, and formed detectable amounts of F-ddI (**2b**).

Currently, a physiologically based pharmacokinetic model is being developed to predict the in vivo disposition and rate of activation of F-ddI prodrugs.[31] This model incorporates compartments for plasma, liver, muscle, brain, and gastrointestinal tissues (Fig. 4.5) and will utilize data from either oral or intravenous administration of F-ddA (**1b**), F-ddI (**2b**), and the 6-chloro analog (**16**). The model includes tissue volumes, plasma flow rates, metabolic rate constants, and drug permeability in individual tissues, as well as urinary and fecal clearances. There is reason to believe that this model, once finalized, will be scalable from rats to humans, allowing the selection of an optimal ADA-activated prodrug for treatment of HIV in the human CNS.

ACKNOWLEDGMENT

The following National Institutes of Health scientists played indispensable roles in the experimental work described above: James A. Kelley, Victor E. Marquez, Harry Ford, Jr., Hiroaki Mitsuya, Maqbool A. Siddiqui, Jeri S. Roth, Robert J. Lutz, Robert L. Dedrick, Joseph J. Barchi, Takuma Shirasaka, Shizuko Aoki, and Masatoshi Tanaka.

REFERENCES

1. Mitsuya, H.; Broder, S., *Proc. Natl. Acad. Sci.* (USA) 1986, **83**, 1911–1915.
2. Geleziunas, R.; Schipper, H. M.; Wainberg, M. A., *AIDS* 1992, **6**, 1411–1426.
3. Yarchoan, R.; Brouwers, P.; Spitzer, A. R.; Grafman, J.; Safai, B.; Perno, C. F.; Larson, S. M.; Berg, G.; Fischl, M. A.; Wichman, A.; Thomas, R. V.; Brunetti, A.; Schmidt, P. J.; Myers, C. E.; Broder, S., *Lancet* 1987, 132–135.
4. Atwood, W. J.; Berger, J. R.; Kaderman, R.; Tornatore, C. S.; Major, E. O., *Clin. Microbiol. Rev.* 1993, **6**, 339–366.
5. Rall, D. P.; Zubrod, C. G., *Annu. Rev. Pharmacol.* 1962, **2**, 109–128.
6. Marquez, V. E.; Tseng, C. K-H.; Mitsuya, H.; Aoki, S.; Kelley, J. A.; Ford, H.; Roth, J. S.; Broder, S.; Johns, D. G.; Driscoll, J. S., *J. Med. Chem.* 1990, **33**, 978–985.
7. York, J. L., *J. Org. Chem.* 1981, **46**, 2171–2173.
8. Marquez, V. E.; Tseng, C. K-H.; Kelley, J. A.; Mitsuya, H.; Broder, S.; Roth, J. S.; Driscoll, J. S., *Biochem. Pharmacol.* 1987, **36**, 2719–2722.
9. Hansch, C.; Steward, A. R.; Anderson, S. M.; Bentley, D., *J. Med. Chem.* 1967, **11**, 1–11.
10. Craig, P. N., *J. Med. Chem.* 1971, **14**, 680–684.
11. Collins, J. M.; Klecker, R. W.; Kelley, J. A.; Roth, J. S.; McCully, C. L.; Balis, F. M.; Poplack, D. G., *J. Pharmacol. Exp. Therap.* 1988, **245**, 466–470.
12. Hansch, C.; Clayton, J. M., *J. Pharm. Sci.* 1973, **62**, 1–21.
13. Ho, D. H. W.; Pincus, C.; Carter, C. J.; Benjamin, R. S.; Freireich, E. J.; Bodey, G. P., *Cancer Treat. Rept.* 1980 , **64**, 629–633.
14. Raiteri, R.; Marietti, G.; Scolfaro, C.; Sinicco, A., *Med. Sci. Res.* 1989, **17**, 187–188.
15. Corey, J. G.; Suhadolnik, R. J., *Biochemistry* 1966, **4**, 1729–1732, 1733–1735.
16. Baer, H. P.; Drummond, G. I.; Gillis, J., *Arch. Biochem. Biophys.* 1968, **123**, 172–178.
17. Chassy, B. M.; Suhadolnik, R. J., *J. Biol. Chem.* 1967, **242**, 3655–3658.
18. Shirasaka, T.; Murakami, K.; Ford, H.; Kelley, J. A.; Yoshioka, H.; Kojima, E.; Aoki, S.; Broder, S.; Mitsuya, H., *Proc. Natl. Acad. Sci.* (USA) 1990, **87**, 9426–9430.
19. Frederiksen, S., *Arch. Biochem. Biophys.* 1966, **113**, 383–388.
20. Chu, C. K.; Ullas, G. V.; Jeong, K. S.; Ahn, S. K.; Doboszeqski, B.; Lin, Z. X.; Beach, J. W.; Schinazi, R. F., *J. Med. Chem.* 1990, **33**, 1553–1561.

21. Stoeckler, J. D.; Bell, C. A.; Parks, R. E.; Chu, C. K.; Fox, J. J., *Biochem. Pharmacol.* 1982, **31**, 1723–1728.

22. Leo, A., Hansch, C., Elkins, D., *Chem. Rev.* 1971, **71**, 525–616.

23. Ford, H.; Merski, C. L.; Kelley, J. A., *J. Liq. Chromatog.* 1991, **14**, 3365–3386.

24. Barchi, J. J.; Marquez, V. E.; Driscoll, J. S.; Ford, H.; Mitsuya, H.; Shirasaka, T.; Aoki, S.; Kelley, J. A., *J. Med. Chem.* 1991, *34*, 1647–1655.

25. Ford, H.; Siddiqui, M. A.; Driscoll, J. S.; Marquez, V. E.; Kelley, J. A.; Mitsuya, H.; Shirasaka, T., *J. Med. Chem.* 1995, **38**, 1189–1195.

26. Driscoll, J. S.; Siddiqui, M. A.; Ford, H.; Kelley, J. A.; Roth, J. S.; Mitsuya, H.; Tanaka, M.; Marquez V. E., *J. Med. Chem.* 1996, **39**, 1619–1625.

27. Gao, W-Y.; Agbaria, R.; Driscoll, J. S.; Mitsuya, H., *J. Biol. Chem.* 1994, **269**, 12633–12638.

28. Maguire, M. H.; Sim, M. K., *Eur. J. Biochem.* 1971, **23**, 22–29.

29. Little, R. F.; Lietzau, J. A.; Welles, L.; Pluda, J. M.; Kelley, J. A.; Mitsuya, H.; Yarchoan, R., 12th World AIDS Conf. 1998, Geneva, June 28–July 3.

30. Kelley, J. A.; Ford, H.; Roth, J. S.; Welles, L.; Little, R. F.; Malinowski, N. M.; Leitzau, J. A.; Gillim, L. A.; Davignon, J. P.; Driscoll, J. S.; Yarchoan, R., 12th World AIDS Conf. 1998, Geneva, June 28–July 3.

31. Lutz, R. J.; Kelley, J. A.; Driscoll, J. S.; Stephenson, E. C.; Dedrick, R. L., "A Physiological Pharmacokinetic Model for an Anti-HIV Drug," in *Proc. 10th Internat. Conf. Mechanisms in Medicine and Biology*, J. A. Ashton-Miller, ed., Pacific Centre of Thermal Fluids Engineering, Honolulu, 1998, pp. 359–362.

CHAPTER 5

Anti-HIV Phosphotriester Pronucleotides: Basis for the Rational Design of Biolabile Phosphate Protecting Groups

CHRISTIAN PÉRIGAUD, GILLES GOSSELIN, and JEAN-LOUIS IMBACH

Laboratoire de Chimie Bioorganique, Université Montpellier II,
Sciences et Techniques du Languedoc

NUCLEOSIDE ANALOGS AND HIV INFECTION: GENERAL CONSIDERATIONS

During the 1990s, significant progress was made in the treatment of human immunodeficiency virus (HIV) infection. This progress resulted from a better understanding of viral pathogenesis, the development of new techniques for sensitive and accurate quantification of plasma virus levels, as well as the clinical use of an increasing number of antiviral agents in combination.[1,2] Among the different steps in the HIV replicative cycle (Fig. 5.1), the reverse transcription and the viral maturation events involving two essential virus-specific enzymes, namely, reverse transcriptase (RT) and protease, are the targets of three classes of drugs presently available in chemotherapy of the HIV disease.[3,4]

Used initially in monotherapy, nucleoside analogs constitute the first class of anti-HIV agents approved by the U.S. Food and Drug Administration (FDA). Six nucleoside analogs (Fig. 5.2) are currently licensed for the treatment of HIV infection. This number should increase in a near future; for instance, another nucleosidic derivative, adefovir dipivoxil [bis(POM)PMEA, Preveon, Gilead Sciences], is now available under the expanded access program.[5]

All these compounds are 2′,3′-dideoxynucleoside analogs (ddNs) due to substitution or removal of the 3′-hydroxyl group of the natural carbohydrate moiety, the 2′-deoxyribofuranose. The drugs of the 2′,3′-dideoxynucleoside group

Biomedical Chemistry: Applying Chemical Principles to the Understanding and Treatment of Disease, Edited by Paul F. Torrence
ISBN 0-471-32633-x

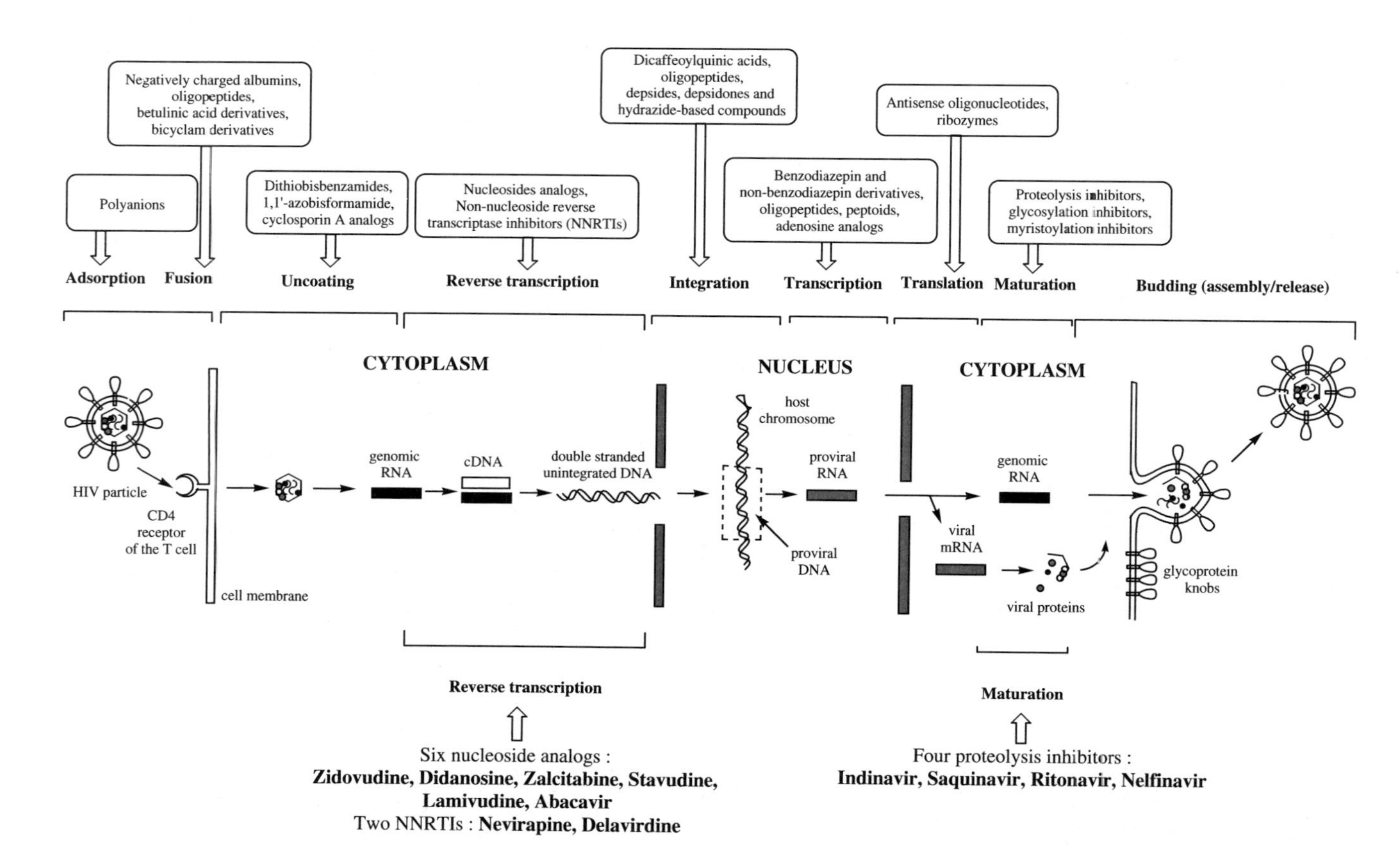

FIGURE 5.1 Replicative cycle of HIV, showing sites of action of potential and currently available antiretroviral drugs.

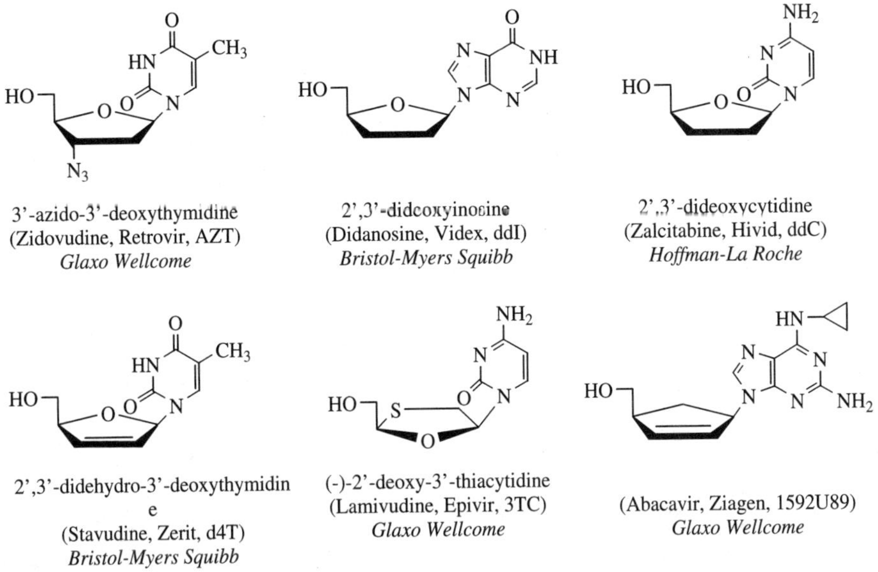

3'-azido-3'-deoxythymidine
(Zidovudine, Retrovir, AZT)
Glaxo Wellcome

2',3'-dideoxyinosine
(Didanosine, Videx, ddI)
Bristol-Myers Squibb

2',3'-dideoxycytidine
(Zalcitabine, Hivid, ddC)
Hoffman-La Roche

2',3'-didehydro-3'-deoxythymidin
e
(Stavudine, Zerit, d4T)
Bristol-Myers Squibb

(-)-2'-deoxy-3'-thiacytidine
(Lamivudine, Epivir, 3TC)
Glaxo Wellcome

(Abacavir, Ziagen, 1592U89)
Glaxo Wellcome

FIGURE 5.2 Chemical structures of the six nucleoside analogs approved for the treatment of HIV infection.

are characterised by a common feature (Fig. 5.3).[6] They have no intrinsic antiviral activity and must be metabolised to their respective 5′-triphosphates (ddNTPs) by nucleotidases, kinases, and/or other activating enzymes present naturally in cells.[7] At the RT level, ddNTPs act either as competitive inhibitors preventing the incorporation of the natural substrates (dNTPs) or as alternate substrates incorporated in the growing DNA chain, leading to termination of the newly synthesized viral nucleic acid.

The factors influencing the ddNTP formation may be of equal or even greater importance than the relative abilities of ddNTPs to inhibit the RT (which are in a narrow range).[8] Of the three anabolic steps of ddNs, the monophosphorylation step is, in many cases, considered as being the most restrictive. Furthermore, the presence and the activity of intracellular enzymes involved in the monophosphorylation of ddNs are often subordinate to the host species, the cell type, and the phase in the cell cycle.[9,10] The dependence on phosphorylation for activation of nucleoside analogs may therefore be problematic in cells where phosphorylating enzyme activity is known to be low or even lacking. The use of phosphorylated derivatives such as ddNMPs could be envisaged, but due to their polar nature under physiological conditions, those compounds are not able to cross the cell membrane efficiently.[11–13] Moreover, they are readily dephosphorylated in extracellular fluids and on cell surfaces by nonspecific phosphohydrolases.[14–16] Consequently, several strategies have been developed in order to improve the therapeutic potential of nucleoside analogues by the use of mononucleotide prodrugs (pronucleotides) *that would be expected to revert back to the corresponding 5′-mononucleotides inside the target cells* (Fig. 5.3).[17–22]

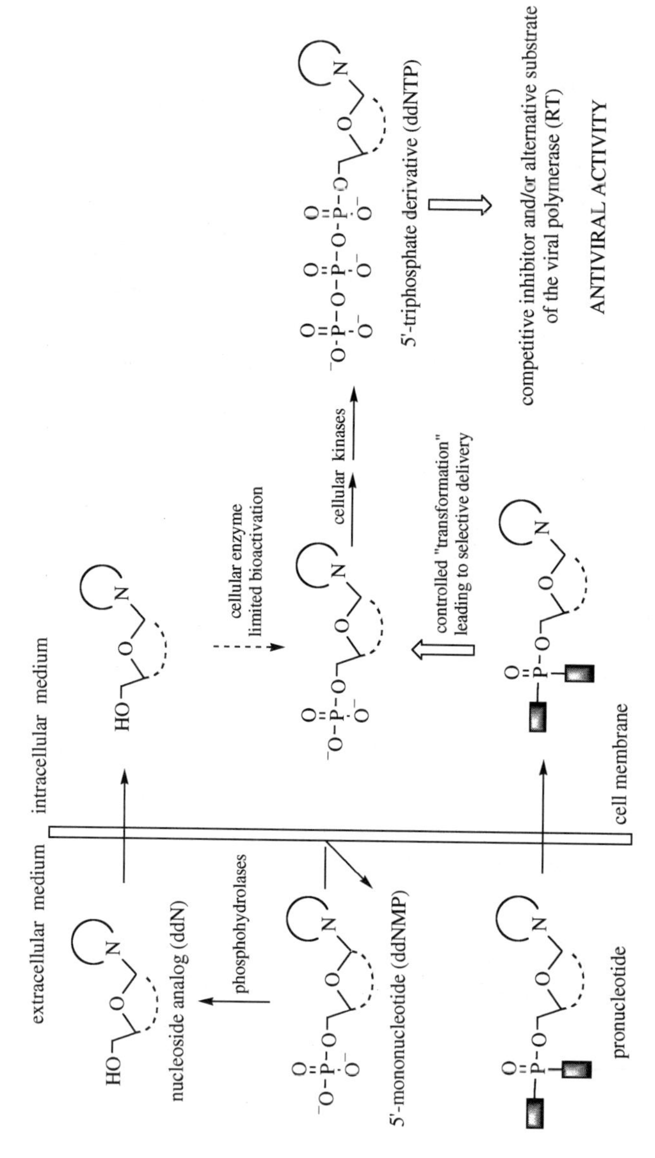

FIGURE 5.3 Mechanism of action of anti-HIV nucleoside analogs and general principle of approach involving mononucleotide prodrugs (pronucleotides).

THE CHEMIST'S APPROACH TO PRONUCLEOTIDE DESIGN

The difficulty of a pronucleotide approach lies in the fact that the transformations involved in the delivery of the 5′-mononucleotide from its corresponding prodrug should be performed selectively inside the cells (Fig. 5.3). Briefly, two strategies have been envisaged requiring either structural modifications or introduction of transient groups in order to reduce or to mask the phosphate negative charges (Fig. 5.4).

Regarding the first strategy, a specific enzymatic system is needed to perform the bioconversion of the structurally modified phosphorylated precursor. This strategy could be illustrated by the design of nucleoside fluorophosphate[18,23] or phosphoramidate[24–27] derivatives (Fig. 5.4). In the second strategy, the hydrolysis of the phosphate-masking group bond requires a difference between the hydrolytic rates in the extra- and intracellular media, and involves either a specific chemical or an enzyme-mediated processes. The need to mask the two charged phosphate oxygens of the 5′-mononucleotide to obtain neutral and lipophilic prodrugs has led to the development of various mononucleoside phosphotriesters. Since no phosphotriesterase activity has been reported in mammalian cells, the hydrolysis of a simple dialkyl phosphotriester involves initially a chemical mechanism. Phosphotriesters (Fig. 5.5) are characterized by phosphoryl (P=O) group consisting of σ bond and $d\pi - p\pi$ bonding formed by overlapping of the phosphorus $3d_{xz}$ orbital and the p orbital of the neighboring oxygen atom. A $d\pi - p\pi$ interaction exists partially in the single bonds expressed between the phosphorus and the other oxygen ligands.[28]

The reactivity of this structure is basically due to the partial positive charge on the phosphorus atom. However, according to the nature of nucleophiles (Fig. 5.6), substitutions may occur not only on the phosphorus atom but also on the α-tetrahedral carbon atom of the ester alkyl group (Fig. 5.6).

The behavior of nucleophiles agrees with the *hard/soft–acid base* (HSAB) concept,[29,30] meaning that "hard" bases such as alkoxide ions will react preferentially with the phosphoryl P, "hard" acid, while the "soft" bases such as mercaptide ion will prefer the tetrahedral C, "soft" acid. Mechanistic studies of the hydrolysis of phosphotriester derivatives have usually been carried out at elevated temperatures, often at extremes of pH, and the data then extrapolated to physiological conditions.[31] In basic media, the P–O bond cleavage is usually observed. In general, associative S_N2 mechanisms [S_N2(P), Fig. 5.7] are observed which involve the formation of a trigonal bipyramidal transition state, in which the entering and leaving groups enter and leave from axial positions.

According to the lifetime of the five-coordinate intermediate and the associated degree of fluxional character, an intramolecular ligand exchange called *pseudorotation* can occur (Fig. 5.8).[32–34] In this process, the leaving ability and the anion stabilization of the departing axial group appears as critical factors in the course of the reaction. The conformational preferences (apicophilicity) between the different phosphorus substituents could lead to an unselective delivery of 5′-mononucleotide from mononucleoside dialkylphosphotriester.

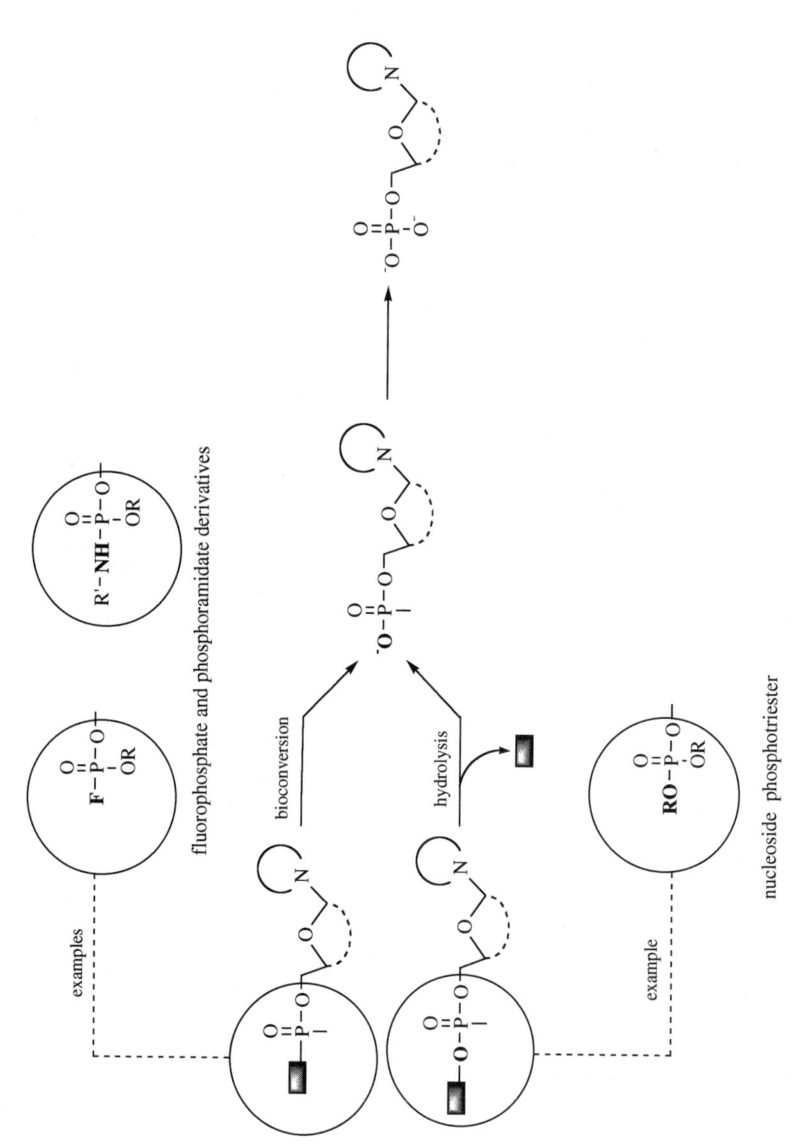

FIGURE 5.4 Concept of prodrugs bearing structural modifications or masking group.

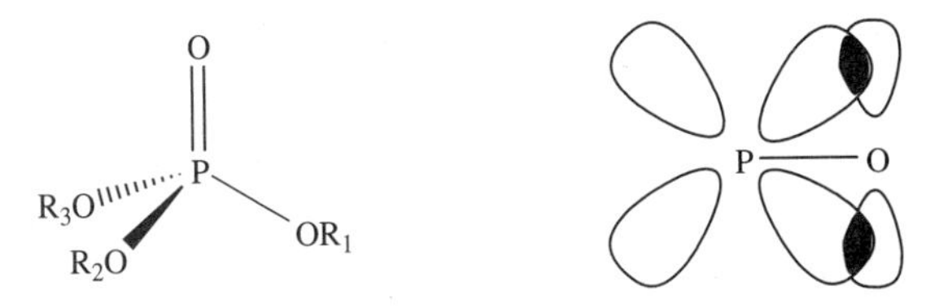

FIGURE 5.5 Tetrahedral structure of phosphotriesters and P–O $d\pi$-$p\pi$ bonding.

In neutral or acidic conditions, the hydrolysis of phosphotriester derivatives proceeds from C–O bond cleavage following a typical S_N2 process, and the dealkylation rate decreases with the increase of the size and the bulk of the alkyl group.[35]

In physiological conditions, the hydrolysis of a dialkyl phosphotriester, which can occur through nucleophilic attacks on phosphorus or α-carbon atoms, is particularly slow. Moreover, the resulting phosphodiester (Fig. 5.6) is extremely stable with respect to further chemical hydrolysis. The presence of a negative charge on the phosphate prevents a second nucleophilic attack on the phosphorus atom, and the hydrolysis proceeds slowly from a predominant C–O bond cleavage.[31] Of course, this greater stability against chemical hydrolysis of the phosphodiester linkage is related to the biological role of DNA.[36]

As a result of pseudorotation phenomenon and chemical stability, approaches based on chemical hydrolysis of dialkyl phosphotriesters were unable to selectively deliver a 5′-mononucleotide.

In an effort to achieve selective hydrolysis, many authors have envisaged the use of aryl groups as transient phosphate protections.[20] Indeed, aryl residues (with pK_a values between 7 and 10) have better leaving-group ability compared to the nucleoside moiety (the pK_a value for the 5′-hydroxyl group is about 13), and diaryl phosphotriesters are known to be more reactive than their alkyl counterparts under nucleophilic conditions. Moreover, some literature data suggest that aryl phosphodiesters would be good substrates for phosphodiesterase activity in contrast to the corresponding aliphatic derivatives.[37,38] Unfortunately, the biological evaluations of simple aryl or diaryl phosphotriester derivatives of bioactive nucleosides, and in some cases the stability studies, have shown that those compounds are not able to deliver the corresponding 5′-mononucleotide intracellularly, and that they act as nucleoside depot forms. In fact, the development of nucleotide prodrugs designed to undergo chemical decomposition processes should take into consideration two limiting parameters: the selectivity of the hydrolysis and the possible toxicity problems associated with nucleophilic substitutions on the phosphorus atom. Many assay methods using various T cells or macrophage-derived cell lines have been developed to determine the in vitro activity of potential anti-HIV agents. Cells were propagated in a culture medium, generally constituted from a nutrient medium (RPMI) complemented by fetal calf serum (10%) partially inactivated by heating (30 min at 56°C). According to its relative chemical stability in the culture medium (pH close to 7.3), the displacement of a lipophilic and neutral

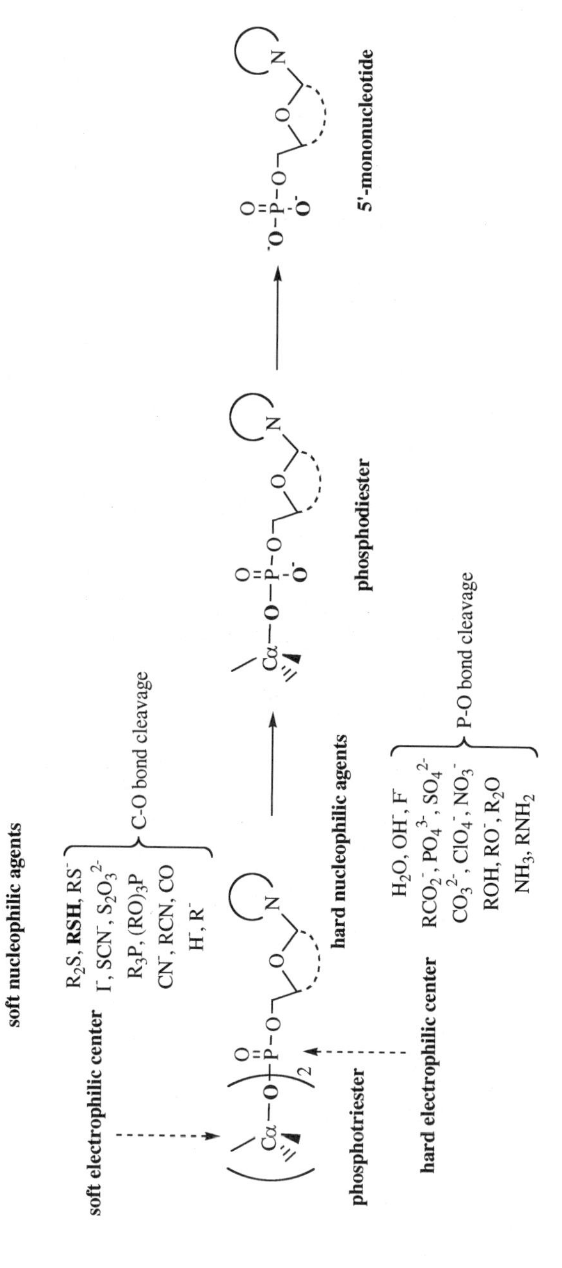

FIGURE 5.6 Mechanisms involved in the chemical hydrolysis of dialkyl phosphotriesters.

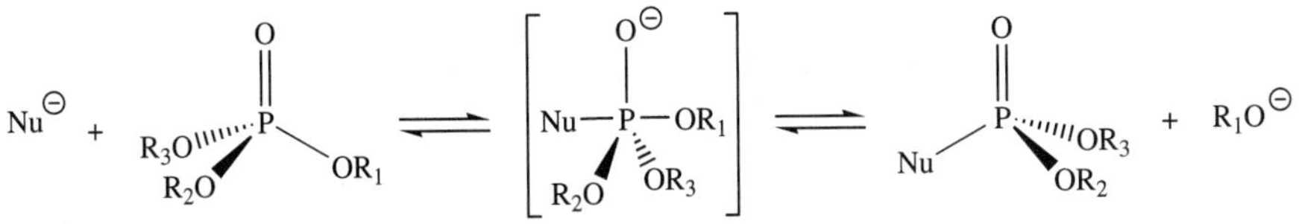

FIGURE 5.7 $S_N2(P)$ mechanism and trigonal bypyramidal-type intermediate.

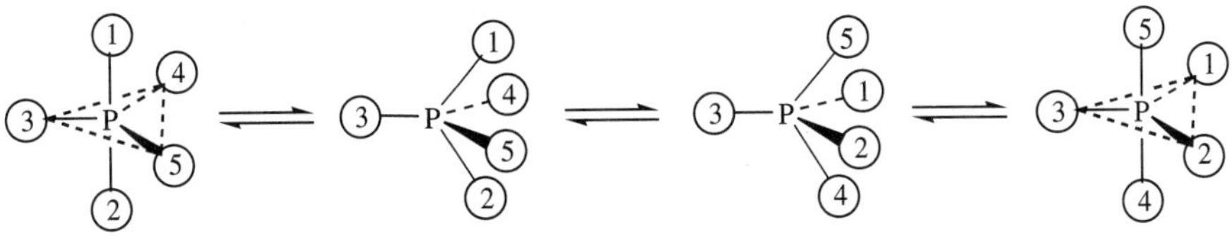

FIGURE 5.8 Pseudorotation phenomenon. In this example, the substituent in position 3 of the trigonal–bypyramidal molecule remains fixed, while the vertical (apical) substituents are pushed backward to produce, with the horizontal (equatorial) substituents, a tetragonal pyramid where the fixed substituent is at the apex. By rotation about the bond from the fixed substituent (the "pivot") to the phosphorus atom, continuation of the process leads to a second trigonal bipyramid that appears to have been rotated by 90° about one of the interatomic bonds.

phosphotriester across the cell membrane by passive diffusion is governed by a driving force (concentration gradient) giving rise to a rapid equilibrium between extra- and intracellular concentrations. Consequently, nonspecific hydrolysis (or microenvironmental pH changes) may be responsible for conversion of the prodrug into the corresponding 5′-mononucleotide inside the cell. In practice, it is important to remember that in vivo administration is a dynamic process and that extended pH variations will be encountered within the body. For example, the stomach pH varies from approximately 1 to 4, the intestinal pH from 5 to 8, and the blood pH is around 7.4. Generally, one would expect the rate of chemical hydrolysis to be higher in acid media than at neutral pH, leading to partial degradation of the orally administrated prodrug before reaching the systemic circulation. Finally, although phosphotriestases have not yet been identified in the mammalian system, data related to the hydrolysis rate of some aryl phosphotriesters in human serum are too rapid to be explained only by chemical hydrolysis.[19,20] In addition to a diminished prodrug accessibility toward infected cells or organs, this observation may not be without toxicological consequences. Indeed, neutral phosphorylated derivatives with good leaving groups attached to the phosphorus atom (Fig. 5.9) are known to interact with hydrolytic enzymes such as esterases possessing both esteratic and proteolytic activities.[39,40] Enzymes hydrolyzing phosphorus compounds but not inhibited by them are called *A-esterases*; those not hydrolyzing phosphorus compounds but inhibited by them are known as *B-esterases*.

In mammals, signs of poisoning are all consistent with the fact that the transmission of nerve impulses is affected. This interference results mainly from the

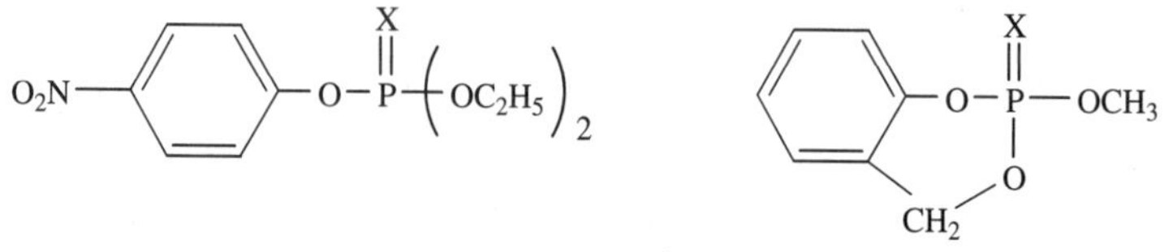

X = O (paraoxon)
X = S (parathion)

X = O (salioxon)
X = S (salithion)

FIGURE 5.9 Example of phosphorus derivatives bearing aryl groups and displaying a toxicity profile due to their interaction with esterases. Paraoxon and its derivatives are potent inhibitors of AChE, while neuropathy target esterase (NTE) has been proposed as the target for initiation of delayed neurotoxicity observed with some saligenin cyclic phosphates such as salioxon or salithion.

inhibition of acetylcholinesterase (AChE), a B-esterase that is widely distributed within the body, in particular in erythrocytes, many nerve cells, ganglia, and brain, and at myoneural (muscle–nerve) junctions. Some aryl phosphotriester derivatives are known to be inhibitors of AChE activity. The proposed mechanisms for this inhibition involve, at the binding site, the nucleophilic attack of a serine hydroxyl group on the phosphorus atom from the opposite side of the aryl group. The selectivity of the enzyme inhibition is influenced by the pK_a of the leaving group in relation to the hydrolysis rate of the P–O bond.[41] In order to circumvent possible toxicity problems, chemical mechanisms involving a C–O bond cleavage rather than P–O bond should be favored in the design of neutral mononucleotide prodrugs. In this respect, the presence of good leaving groups (e.g., aryl residues) attached to the phosphorus should be carefully considered.

THE SATE (*S*-ACYL-2-THIOETHYL) GROUPS AS BIOLABILE PHOSPHATE PROTECTION

Rational

In addition to the attack of nucleophiles described previously (see also Fig. 5.5), intramolecular nucleophilic substitutions (S_Ni) could lead to the fission of the C–O bond of phosphate esters. Indeed, studies of the pH rate–product profile for the aqueous hydrolysis of *O,S*-ethylene phosphorothioates (Fig. 5.10) demonstrated that *O*-phosphorylated derivatives of 2-mercaptoethanol (resulting to exclusive P–S bond cleavage) could eliminate ethylene sulfide in neutral conditions.[42–44]

Pseudorotation is required to bring the sulfur atom into an apical leaving position. The greater leaving ability of the alkylthio ligand and the weakness of the P–S bond (due to less efficient $d\pi$–$p\pi$ bonding) compared to the oxy analogs were cited as factors involved in the fast and selective cleavage of *O,S*-phosphorothioate esters. It is noteworthy that the selectivity of this process has been used as a chemical procedure for sequencing oligonucleotide phosphorothioates.[45]

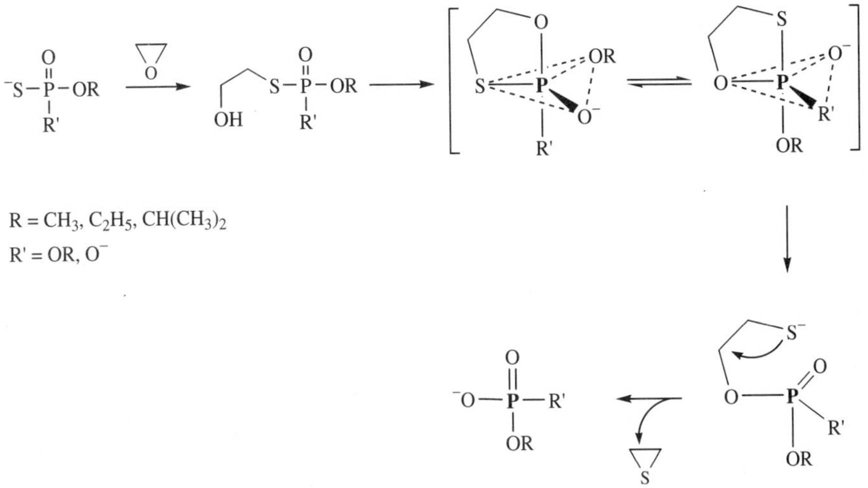

$R = CH_3, C_2H_5, CH(CH_3)_2$

$R' = OR, O^-$

FIGURE 5.10 Proposed mechanisms for the aqueous hydrolysis of *O,S*-ethylene phosphorothioate derivatives.

We decided to use the selectivity of this nucleophilic process to design new phosphate protections incorporating a thioethyl chain, where the sulfur atom of the thiol function is involved in an enzyme-labile bond (Fig. 5.11). This kind of protection is formally constituted by two-component masking groups, and belongs to the double-prodrug concept (*pro*-prodrugs).[46–49] The first hydrolysis step of these masking groups is the cleavage of sulfur bond via an enzymatic activity (preferentially intracellular) resulting in the formation of an unstable 2-mercapto-ethyl ester, giving rise in a second step to a spontaneous release of the corresponding phosphate function (Fig. 5.11).

Our work on this topic started with study of phosphate protections incorporating a thioethyl chain where the thiol is masked by either a disulfide (DTE group)[50,51] or a thioester (SATE group)[50,52]. The resulting bis[*S*-(2-hydroxyethylsulfidyl)-2-thioethyl] and bis(*S*-acyl-2-thioethyl) mononucleoside phosphotriesters (Fig. 5.12) were designed for release through reductase- or carboxyesterase-dependent activation processes, as unstable *O*-2-mercaptoethylphosphotriesters that decompose spontaneously via intramolecular nucleophilic displacement into the corresponding phosphodiesters and ethylene sulfide. Removal of the remaining phosphate protecting group from the phosphodiester derivatives gives rise—by either a similar mechanism and/or a phosphodiesterase activity—to the delivery of the parent 5′-mononucleotides into cells.

It is not the intention of the authors to give an entire overview of all the results obtained with the use of SATE (or DTE) pronucleotides in order to overcome mononucleotide delivery drawbacks. These data can be gleaned in the recent literature.[18–20,53] The current review is restricted to some unpublished chemical aspects and comments about 5′-mononucleotide intracellular delivery by the use of SATE pronucleotide approach.

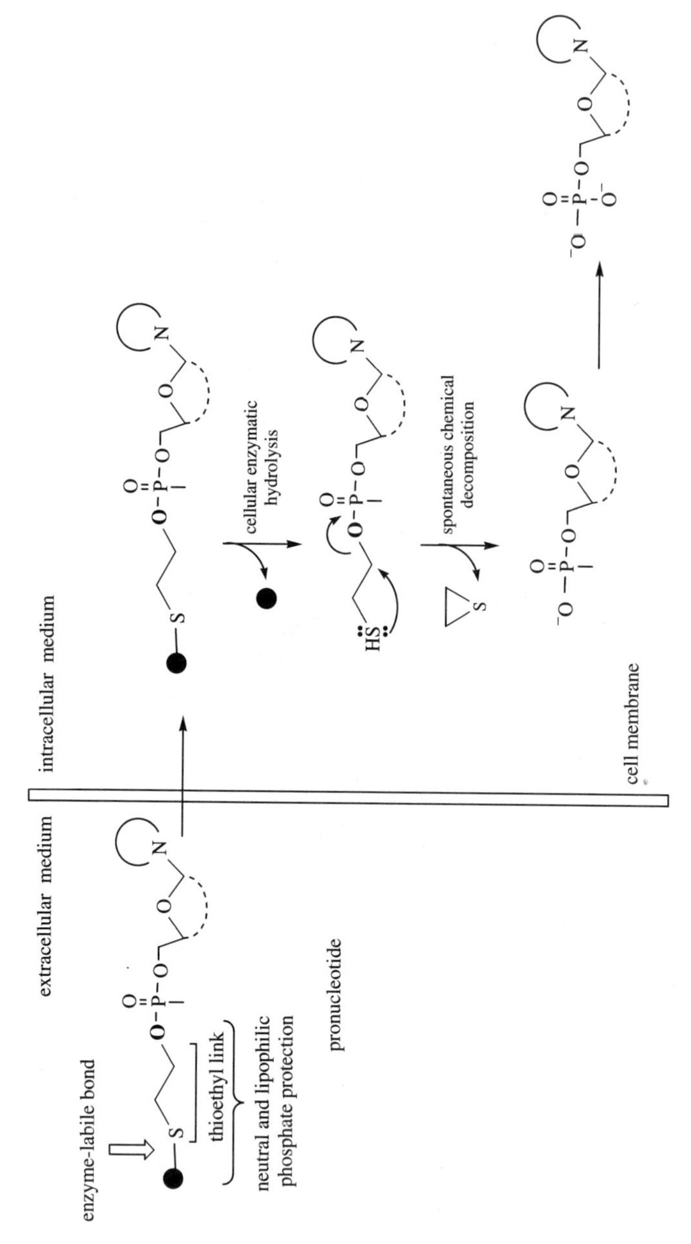

FIGURE 5.11 General principle of the double-prodrug concept using a thioethyl link.

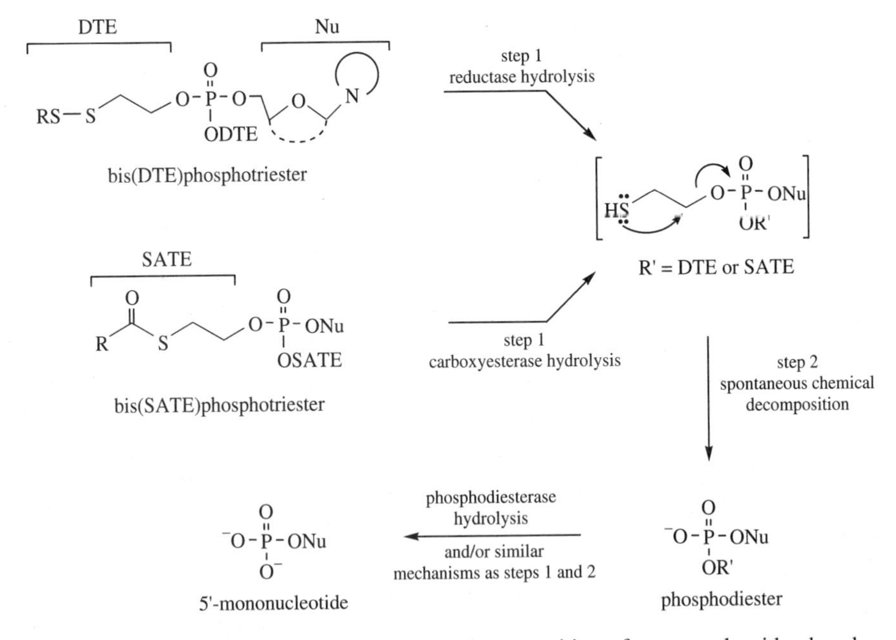

FIGURE 5.12 Proposed mechanisms for the decomposition of mononucleoside phosphotriester derivatives bearing SATE and DTE biolabile phosphate protections.

Chemical Aspects

Mononucleoside phosphotriester derivatives can be synthesized following two common strategies using P^{III} (hydrogenphosphonate, phosphoramidite, etc.) or P^{V} (phosphodiester, phosphomonoester, etc.) intermediates.[54,55] In the case of SATE pronucleotides, the preliminary preparation of the thioester precursors (Fig. 5.13) is required regardless of the synthetic method selected. These precursors could be

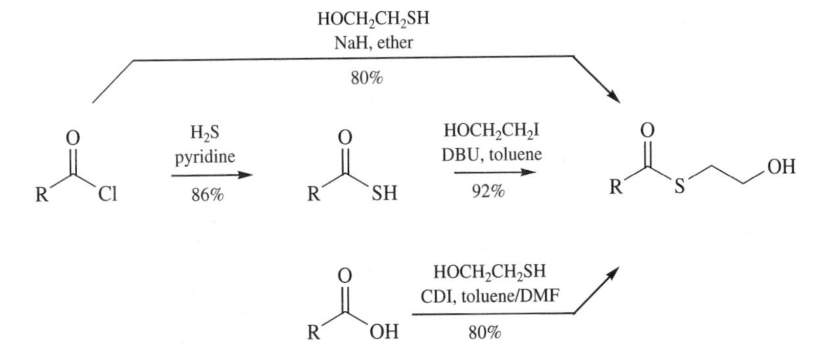

FIGURE 5.13 Synthesis of thioester precursors. The yields given (obtained after distillation) correspond to R = $(CH_3)_3C$. [DBU = 1,8-diazabicyclo(5.4.0)undec-7-ene (1.5-*S*); CDI = 1,1′-carbonyldiimidazole.]

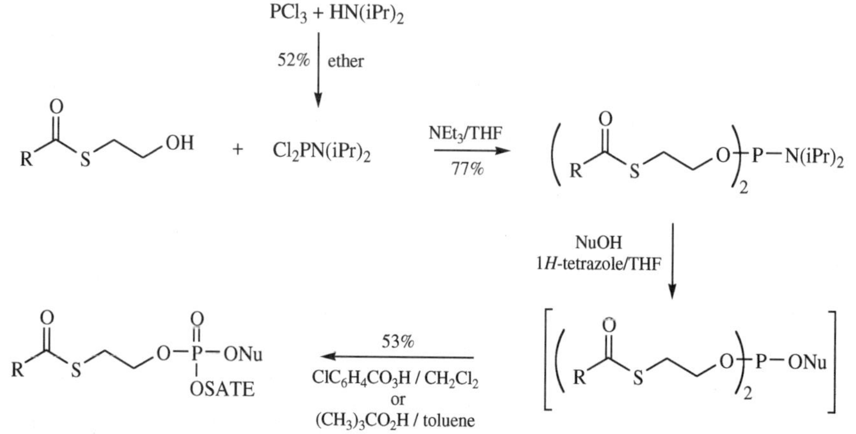

FIGURE 5.14 The phosphoramidite approach applied to the synthesis of SATE pronucleotides. The yields given (after purification) correspond to $R = (CH_3)_3C$, Nu = 3′-azido-3′-thymidin-5-yl.

obtained from the corresponding carboxylic acids, thioacids, or acyl halides according to their commercial availability or their chemical accessibility.

Initially developed using hydrogenphosphonate chemistry,[50] the synthesis of a wide range of mononucleoside bis(SATE)phosphotriesters has been then carried out by coupling the nucleoside analog with an appropriate phosphoramidite reagent, followed by in situ oxidation (Fig. 5.14).[52] In comparison to other nucleoside analogs, the yield of the coupling reaction of AZT (given in Fig. 5.14) with phosphoramidite derivatives is lower owing to the concomitant formation of by-products due to the reactivity of the AZT azido function.[56]

As the phosphoramidite approach necessitates strictly anhydrous conditions it does not seem to be the more appropriate way for the synthesis of large quantities of SATE pronucleotides required for in vivo studies. Consequently, an alternative P^V strategy (Fig. 5.15) can be also used where the corresponding 5′-monucleotide, previously activated, is coupled with the thioester precursor.

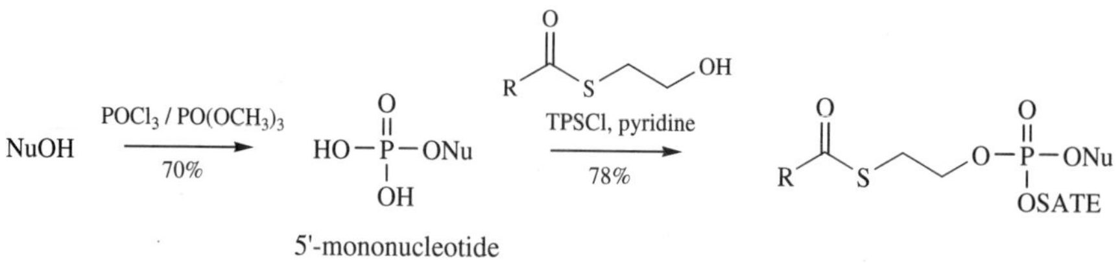

FIGURE 5.15 The phosphomonoester approach applied to the synthesis of SATE pronucleotides. The yields given (after purification) correspond to $R = (CH_3)_3C$, Nu = 3′-azido-3′-thymidin-5-yl. [TPSCl = 2,4,6-triisopropylbenzenesulfonyl chloride.]

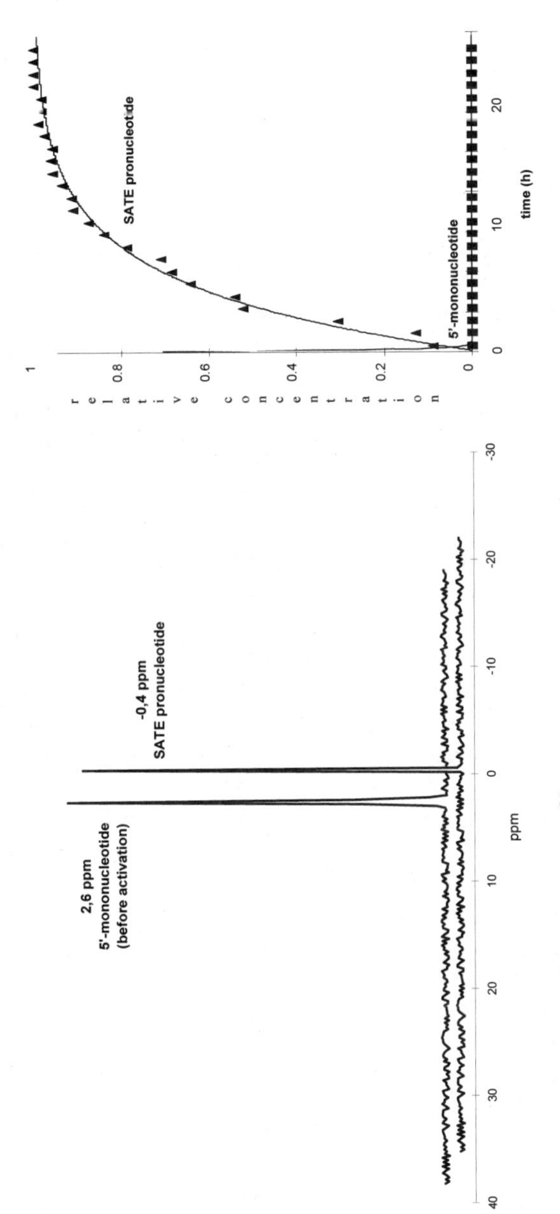

FIGURE 5.16 Examples of ^{31}P NMR spectra profiles and kinetic data obtained during the reaction of AZT 5′-mononucleotide with *S*-2-hydroxyethyl thiopivaloate [R = $(CH_3)_3C$; Fig. 5.15].

The kinetic study followed by ^{31}P NMR (Fig. 5.16) shows the rapid disappearance of the starting material (δ 2.6 ppm) to give the desired bis(SATE)phosphotriester (δ −0.4 ppm) as the single product.

Prerequisites for an In Vitro Evaluation

Pronucleotides have to give rise to the delivery of the 5′-mononucleotides preferentially inside the cells. Direct proof of the decomposition mechanism and its selectivity could be obtained using radiolabeled materials. But the coast of such in vitro assays is often expensive for academic laboratories, and could not be envisaged for all kinds of prodrugs. Automated cell culture experiments may constitute an alternative and attractive method, and two approaches have been proposed to demonstrate the intracellular release of 5′-mononucleotides from pronucleotides.[57] In the first approach, one must use an inactive nucleoside analog as model, of which the corresponding 5′-monosphosphate is further anabolized to the 5′-triphosphate, the latter interfering selectively with the viral polymerase. In the second approach, the pronucleotide must be evaluated for its in vitro inhibitory effects on the replication of the virus in cell lines deficient in the nucleoside kinase responsible for the monophosphorylation of the corresponding nucleoside. In this case, when a biological response is detected in cell cultures, one can assume that it is due only to the release of the 5′-mononucleotide inside the cells and does not result from a possible extracellular decomposition of the prodrug into its parent nucleoside.

The first approach can be illustrated by attemps to deliver phosphorylated forms of 2′,3′-dideoxyuridine (ddU).[50,58,59] Indeed, ddU appears as an appropriate model since this nucleoside analog is inactive against HIV in cell cultures, whereas the corresponding 5′-triphosphate (ddUTP) is a potent HIV reverse transcriptase inhibitor.[60] Furthermore, it has been shown in cell cultures that the delivery of its 5′-mononucleotide (ddUMP) by liposomes results in an anti-HIV activity.[53] The second approach can be exemplified by the biological evaluation of mononucleotide prodrugs in HIV-1-infected nucleoside kinase-deficient cells, as for instance, in CEM/TK$^-$ cells. This cell line should be considered as an ideal in vitro system to investigate the efficiency of nucleotide derivatives for which the metabolism of parent nucleosides is known to be strictly dependent on cellular thymidine kinase–mediated activation. Previous studies on the metabolism of AZT in both uninfected and HIV-infected cells have shown that this nucleoside analog, after permeation of the cell membrane by nonfacilitated diffusion (passive transport), is efficiently metabolised to AZTMP by cytosolic thymidine kinase (TK_1).[61,62] At the opposite to its anti-HIV activity observed in wild-type thymidine kinase positive (TK^+) cell lines, this nucleoside analog is inactive in CEM/TK$^-$ cells (Table 5.1). Consequently, the appearance of an anti-HIV activity in CEM/TK$^-$ cell line for phosphorylated derivatives of AZT [illustrated in Table 5.1 by the evaluation of some bis(SATE)phosphotriester derivatives] will support the hypothesis that these compounds exert their biological effects *via* the intracellular delivery of the corresponding 5′-mononucleotide and so can be considered as pronucleotides.

TABLE 5.1 Antiviral Activity of Some Bis(SATE)phosphotriester Derivatives of AZT Compared to Their Parent Nucleoside in Two Human Lymphoblastoid Cell Lines Infected with HIV-1[52]

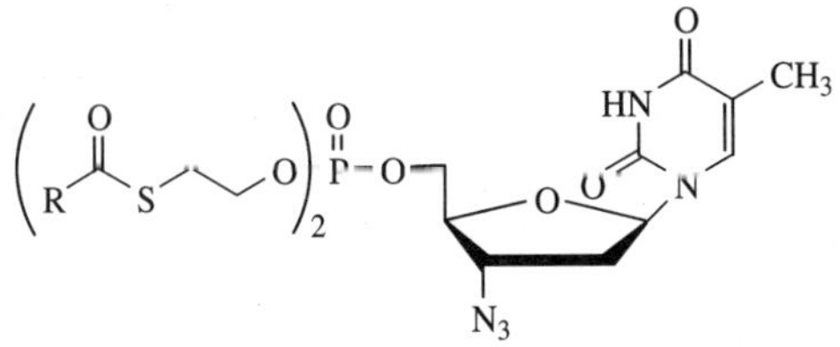

R = CH_3, bis(MeSATE)AZTMP
R = $(CH_3)_2CH$, bis(iPrSATE)AZTMP
R = $(CH_3)_3C$, bis(tBuSATE)AZTMP

	CEM-SS		CEM/TK⁻	
	EC_{50},[a] μM	CC_{50},[b] μM	EC_{50}, μM	CC_{50}, μM
Bis(MeSATE)AZTMP	0.022	93	0.049	> 100
Bis(iPrSATE)AZTMP	0.046	> 10	0.52	> 10
Bis(tBuSATE)AZTMP	0.015	> 10	0.45	> 10
AZT	0.006	> 100	> 100	> 100

[a] EC_{50}, 50% effective concentration or concentration required to inhibit the replication of HIV-1 by 50%.
[b] CC_{50}, 50% cytotoxic concentration or concentration required to reduce the viability of uninfected cells by 50%.

Lipophilicity and Enzymatic Stability in Relation to the In Vitro Biological Activity

It is obvious that in vitro antiviral activities of pronucleotides are related to factors such as their lipophilicity and kinetics of decomposition (in culture medium and inside the cell). Penetration through biological membranes is one of the requisites for a pronucleotide approach. The flux of a compound across the cell membrane due to passive diffusion is directly related to its partition coefficient (P) between the aqueous media (i.e., culture medium in cell culture experiments) and lipid membrane. For a qualitative estimation of the membrane diffusion properties, the partition coefficient (P) of a prodrug is usually determined in an 1-octanol/water mixture. This a priori completely empirical method is considered to be predictive of the tendency of compounds, especially simple ones, to associate with a lipid membrane. Typically, partition coefficients on the order of 100–1000 ($\log P = 2–3$) are required for efficient passive diffusion process.[63] However, the presence in lipophilic molecules of intramolecular bonds, inductive effects, chain branchings, ionizable functions at physiological pH, or possible hydrogen bonds with the aqueous media limit the predictive ability of partition coefficients.

Relative stability in culture medium associated with a rapid intracellular decomposition are key factors for compounds, such as SATE pronucleotides, designed to release selectively inside the cells the parent mononucleotides through enzymatic activation. In fact, lipophilicity, aqueous solubility, and substrate properties for cellular enzymes are indissociable.

To validate the proposed mechanisms involved in the decomposition pathways of SATE pronucleotides, we studied their stability in several media using an "online internal surface reverse phase (ISRP) cleaning" HPLC method developed in our laboratory.[52,64] This technique allows the direct analysis of biological samples without pretreatment. The stability of prodrugs has been evaluated notably in culture medium, and in total CEM-SS cell extracts as a mimic for the intracellular medium. Briefly, during pronucleotide incubation in the investigated media, aliquots were taken out and analyzed by HPLC. Several signals can be observed in the resultant chromatograms (Fig. 5.17). Coinjection with authentic samples (retention time, UV spectra) as well as mass spectra coupling allowed the identification of the decomposition products and their quantification.

The computation of kinetic data (Fig. 5.18) was consistent with consecutive-competitive pseudo-first-order mechanisms, and rate constants of each step were optimized using mono- or polyexponential regressions according to integrated equations.

As illustrated with several SATE pronucleotides (Fig. 5.19), such experiments can serve as useful tools in determining the decomposition mechanisms and explaining the anti-HIV effect observed in CEM/TK$^-$ cell line for mononucleotide prodrugs of AZT.[52]

In all our studies, the bis(SATE)phosphotriester derivatives of AZT decomposed more rapidly in cell extracts than in culture medium, and the decomposition mechanisms differed according to the medium. In cell extracts, the SATE phosphodiester intermediates, resulting from the hydrolysis of the corresponding

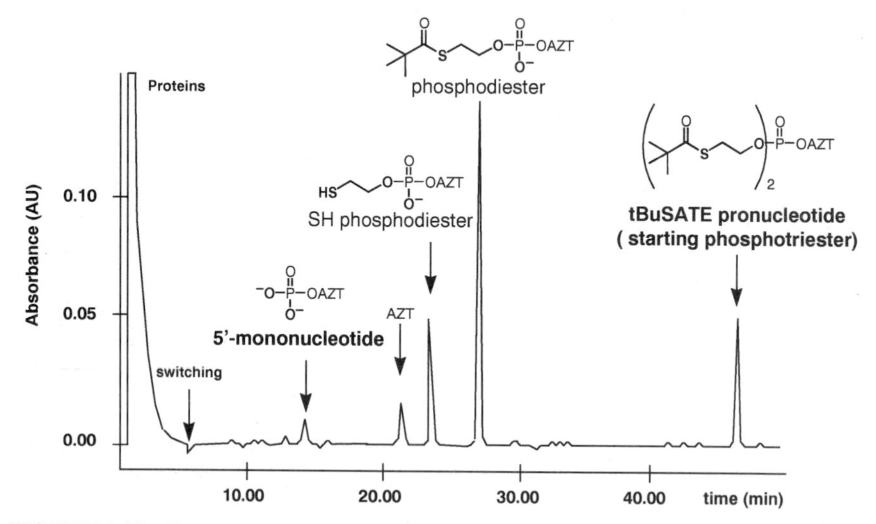

FIGURE 5.17 Example of HPLC chromatogram (UV absorbance 266 nm) after incubation of the bis(tBuSATE)AZTMP (Table 5.1) in total cell extracts from CEM-SS cells.

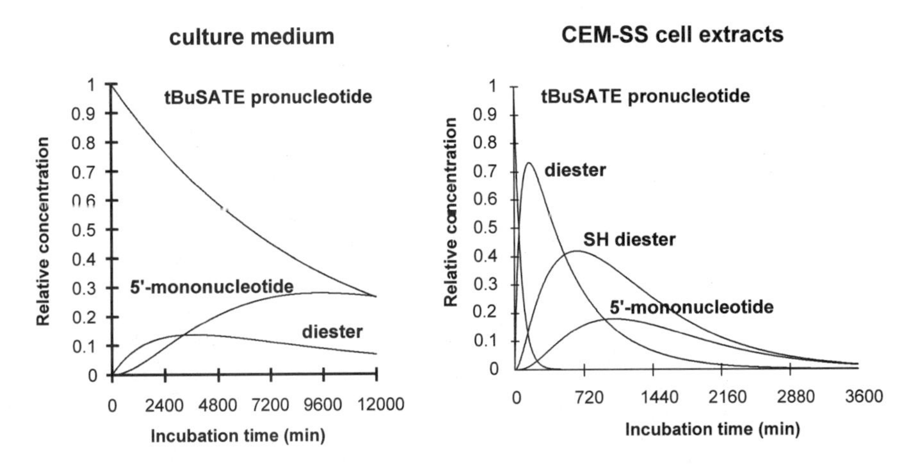

FIGURE 5.18 Example of the decomposition kinetics of the bis(tBuSATE)AZTMP (Table 5.1) in two biological media.[52] The different terms (diester, SH diester) are related to the chemical structures depicted in Fig. 5.17.

phosphotriesters after esterase activation, gave rise to the formation of 5′-mononucleotide mainly through a similar process. In culture medium, this hydrolytic reaction resulted from phosphodiesterase activity present in the fetal calf serum. Furthermore, a sterically hindered environment around the carboxyl group of the SATE moiety increased the enzymatic stability of the corresponding pronucleotides.[52] Thus, by varying the acyl portion of the SATE group, it was possible to modulate the formation rate of 5′-mononucleotide. The validity of these stability studies has been corroborated by cellular pharmacology studies involving radiolabelled SATE phosphotriesters of other bioactive nucleosides analogs.[65,66] Similar metabolites were found in intact cells, and, as expected, the kinetics of each step involved in the decomposition process were observed to be faster than in cell extracts. This point illustrates the potential limitations that can be encountered using the cell extracts model, limitations related to its intrinsic nature. During the preparation of the medium, a decrease or a loss of the enzymatic activity used to trigger the 5′-mononucleotide delivery may be observed as previously reported in cell homogenates.[67] Moreover, the enzymatic content of the CEM-SS cell line studied does not necessarily reflect the enzymatic variability in other cells and tissues. In fact, for the author's experience, decomposition pathways and related kinetics observed in total cell extracts for mononucleotide prodrugs are predictive of their antiviral activity in cell culture experiments. On the other hand, the absence of decomposition for a prodrug or the accumulation of one of its intermediate observed in total cell extracts is not necessary reliable, due to a lack of in vitro biological activity.

In the case of mononucleotide prodrugs of AZT, variation of the lipophilicity does not seem to be a critical factor for in vitro anti-HIV activity in wild-type (thymidine kinase positive, TK^+) cells. In contrast, comparative evaluations in

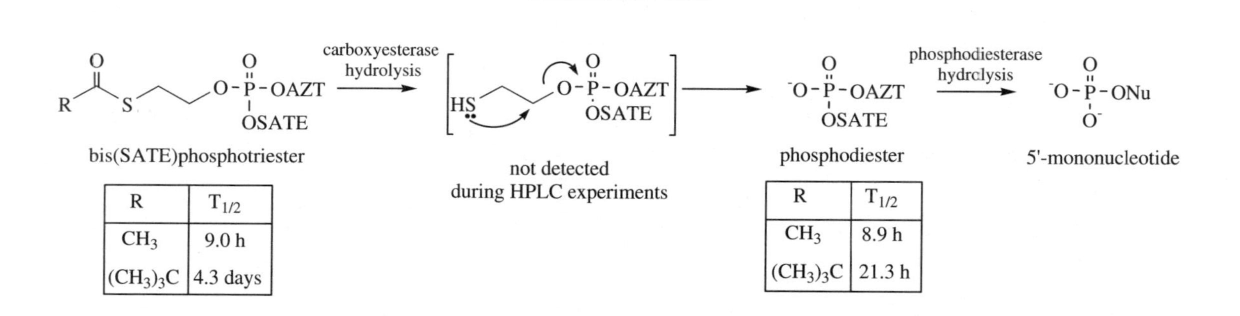

R	$T_{1/2}$
CH_3	9.0 h
$(CH_3)_3C$	4.3 days

R	$T_{1/2}$
CH_3	8.9 h
$(CH_3)_3C$	21.3 h

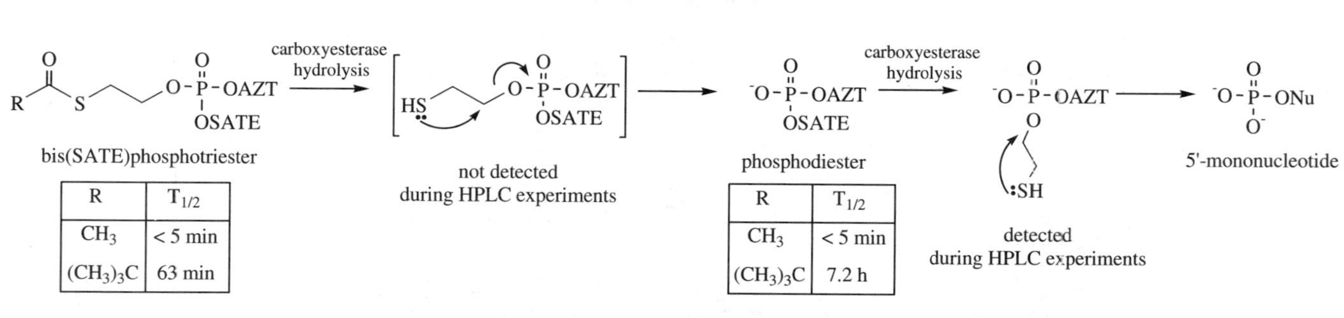

R	$T_{1/2}$
CH_3	< 5 min
$(CH_3)_3C$	63 min

R	$T_{1/2}$
CH_3	< 5 min
$(CH_3)_3C$	7.2 h

FIGURE 5.19 Decomposition mechanisms and calculated half-lives of two bis(SATE)-phosphotriester derivatives of AZT in culture medium and CEM-SS cell extracts.

CEM-TK$^-$ cell line clearly demonstrate that an increase of the enzymatic stability in CEM cell extracts of a pronucleotide is correlated with a decrease in its antiviral activity.[68] Consequently, using AZT as nucleosidic model, a fast intracellular release of the 5′-mononucleotide would be better in order to observe a potent anti-HIV effect in CEM-TK$^-$ cells. This fact could be tentatively explained with regard to the specific metabolism of AZT and the pharmacokinetic parameters of each corresponding mononucleoside phosphotriesters. The second phosphorylation step of AZT catalyzed by cellular thymidylate kinase constitutes the rate-limiting step in the anabolism of AZT to its 5′-triphosphate (AZTTP), and the accumulation of AZTMP causes inhibition of thymidylate kinase, resulting in a blockage of its further phosphorylation to AZT 5′-diphosphate (AZTDP). Thus, the intracellular concentrations of AZTMP are 2–3 orders of magnitude higher than that of AZTTP in both uninfected and HIV-infected cells. In addition, while cells membrane are considered impermeable to nucleotides, several reports have shown that AZTMP is excreted from cells treated with AZT,[69,70] suggesting that cell membrane can be partly permeable to the efflux of anionic compounds. In this context, it would be conceivable that, if the global decomposition kinetic of the phosphotriester is relatively slow inside the cells, the intermediate metabolites (phosphodiesters) could diffuse out of the cells into the culture medium, leading to a decrease of the intracellular concentration of AZTMP. This hypothesis could explain the presence of SATE phosphodiesters, observed during cellular pharmacology studies, in a free enzymatic culture medium.[71]

It is noteworthy that a correlation between a fast decomposition of pronucleotides and a potent anti-HIV activity has been observed only in the case of AZT. Indeed, the validity of a pronucleotide approach using SATE groups as transient phosphate protections has also been demonstrated in vitro using other anti-HIV and anti-HBV nucleoside analogs.[72,73] Applied to nucleoside analogs that are hampered at the first phosphorylation step by a dependence on kinase-mediated phosphorylation or by a rate-limiting step in the anabolism pathway, the use of the corresponding bis(SATE)phosphotriester derivatives leads to markedly enhanced antiviral activity. For example, the SATE pronucleotides of 2′,3′-dideoxyadenosine (ddA), a well known anti-HIV agent,[74] exhibited a very potent anti-HIV effect (illustrated in Table 5.2 through their evaluation in monocyte-derived macrophages) and proved to be superior to their parent nucleoside, with a decrease of between 3 and 4 logs in the 50% effective concentration (EC_{50}) values.[75,76]

Comparative evaluation of pronucleotides bearing different SATE groups showed that phosphotriesters incorporating the more lipophilic and enzymatically resistant *S*-pivaloyl-2-thioethyl group (tBuSATE group, Tables 5.1 and 5.2) generally exhibited similar or superior antiviral effects compared to the most biolabile *S*-acetyl-2-thioethyl phosphate protection (MeSATE).

Furthermore, as illustrated in Table 5.2, an increase in the cytotoxicity could be observed in cell culture experiments for pronucleotides as compared to the parent nucleosides. This may be related to the intracellular accumulation of the phosphorylated forms that possibly interact with host cellular enzymes. This result does not reduce the potential interest of a pronucleotide approach, but indicates the

TABLE 5.2 Antiviral Activity of Some Bis(SATE)phosphotriester Derivatives of ddA Compared to Their Parent Nucleoside in Primary Monocyte-Derived Macrophages Infected with HIV-1[76]

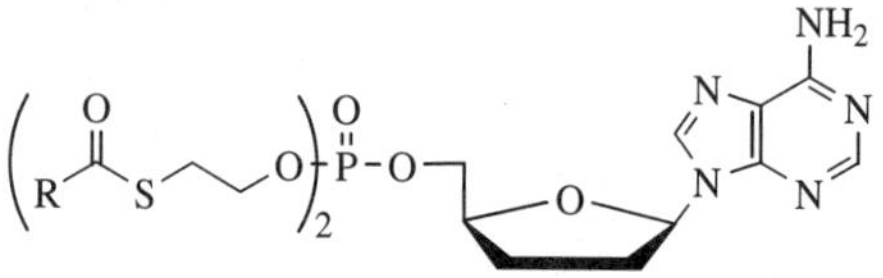

R = CH_3, bis(MeSATE)ddAMP
R = $(CH_3)_2CH$, bis(iPrSATE)ddAMP
R = $(CH_3)_3C$, bis(tBuSATE)ddAMP

	EC_{50},[a] nM	CC_{50},[b] nM
Bis(MeSATE)ddAMP	0.01 ± 0.009	2,000
Bis(iPrSATE)ddAMP	0.04 ± 0.007	1,000
Bis(tBuSATE)ddAMP	0.015 ± 0.008	1,000
ddA	100	100,000

[a,b] See the corresponding footnotes in Table 5.1.

importance of the selectivity of the corresponding 5′-triphosphate derivatives for the viral polymerase.[77]

THE END OF THE BEGINNING

With the awareness, in the early 1980s, that AIDS was caused by a retrovirus, the hunt was on for effective antiviral agents to tackle this devastating disease. Perhaps not surprisingly, the first success was AZT, a nucleoside analog synthesized almost 20 years earlier as a potential anticancer drug. At this time, the mode of action of nucleoside analogs and their interference with cellular or viral replication events have been well established. In the late 1980s, the discovery of acyclovir (an acyclonucleoside) as a potent and selective inhibitor of herpes simplex virus replication completely changed the thinking in the antiviral chemotherapy area. However, the advent of HIV infection has hastened the synthesis and screening of many new series of nucleoside analogs, and combination strategies using these drugs currently represent the major chemotherapeutic approach to the treatment and prevention of AIDS. These unprecedented multidisciplinary investigations have also demonstrated the therapeutic limitations of this class of antiviral agents, limitations due in a large part to the necessity to be metabolized in order to exert their antiviral effect. As a result, the idea has emerged that the need for intracellular enzymatic activation of nucleoside analogs would be partially overcome by the use of mononucleotide prodrugs (pronucleotides). The first bases of an approach using biolabile phosphate protections were established in 1983,[78] and since that time both academic and industrial laboratories have provided insight into the nature of chemical and biological requirements associated with the in vitro selective delivery of 5′-mononucleotide inside cells from neutral precursors. Thus, various mononucleoside phosphotriester structures (Fig. 5.20) have demonstrated their ability to act in vitro as pronucleotides, marking the end of the beginning of a rational development.

To date, the primary issue concerns the in vivo therapeutic potential of such a pronucleotide approach, which may not be without problems, according to the inherent pharmacological objectives, specifically, the (selective) delivery of a biological 5′-mononucleotide into infected cells or tissues. A carboxyesterase-dependent activation process is involved in the decomposition mechanism of the most efficient biolabile phosphate protections. Esterases are widely distributed in organs, tissues, and body fluids of mammalian species.[79,80] Consequently, ester prodrugs, such as (acyloxy)alkyl esters (illustrated by the POM ester derivatives; Fig. 5.20), are rapidly hydrolysed in human after oral absorption, since the gastrointestinal lumen, mucosal cells, and the liver are rich in these enzymes. The use of this kind of transient phosphate protections can be useful for increasing the bioavailability of anti-HIV agents, as demonstrated by the advanced clinical trials of bis(POM)PMEA (Fig. 5.21), an oral prodrug form of the acyclic nucleoside phosphonate PMEA.

Despite this unquestionable therapeutic interest, a presystemic metabolism, preventing the delivery of the prodrugs in infected cells or tissues, may constitute the major limitation in the in vivo development of anti-HIV pronucleotides designed to promote a site-specific delivery by an esterase-mediated activation process. Further research in the field of pronucleotide approaches will take up this challenge, and several strategies can be envisaged. The first one consists in increasing the enzymatic stability of the esterase-labile phosphate protections by the introduction of polar or ionisable (at physiological pH) functions in the immediate or near vicinity of the ester functionality.[68] Another possibility is the modulation of the

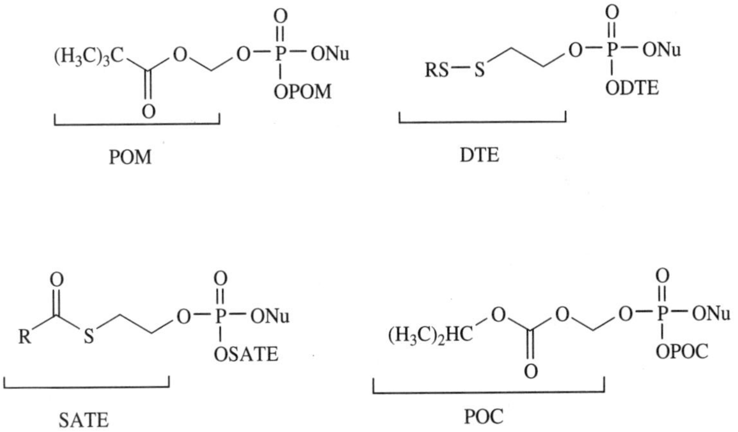

POM

DTE

SATE

POC

FIGURE 5.20 Examples of mononucleoside phosphotriesters bearing biolabile phosphate protecting groups (Nu = nucleoside analog; POM = pivaloyloxymethyl; POC = isopropyloxycarbonyloxymethyl).

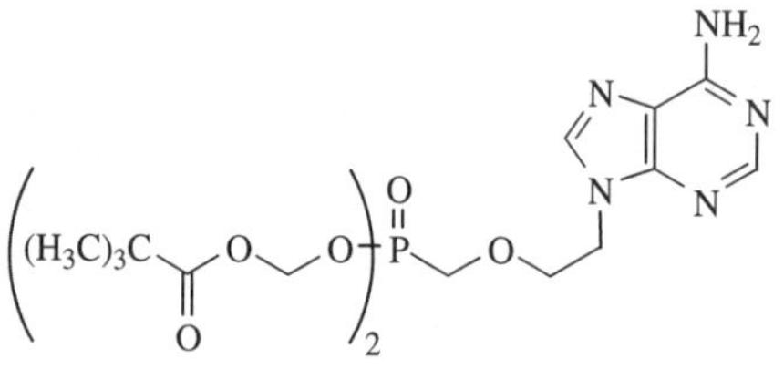

bis(POM)PMEA
(Adefovir dipivoxil, Preveon)
Gilead Sciences

FIGURE 5.21 Chemical structure of bis(POM)PMEA.

kinetic parameters involved in the chemical mechanism leading to the nucleophilic attack on the α-carbon atom. Examples of this strategy could be found in the increase of thioalkyl chain length of SATE phosphate protecting groups.[81,82] Finally, more specific enzymatic activation systems can be searched in order to target cells or tissues according to preferential transport and distribution (site-directed strategy) or to obtain selective bioactivation (site-activated strategy).[83–87] Work in this topic is currently in progress in our group[88] but is not without obstacles due to the limitations of in vitro biological evaluation, unsuitable animal models for routine confirmation of anti-HIV activity, and enzymatic variability between animal species.[89] Many works will be necessary in order to obtain efficient in vivo biolabile phosphate protections, and in the future this challenge is equal to the therapeutic potentialities of pronucleotide strategies.

ACKNOWLEDGMENTS

The original investigations of the authors were supported by grants from the Agence Nationale de Recherches sur le SIDA (ANRS, France). The authors are grateful to S. Peyrottes for a critical reading of the manuscript, and would like to thank all the collaborators whose names are mentioned in the reference list.

REFERENCES

1. Barry, M.; Mulcahy, F.; Back, D. J., *Br. J. Clin. Pharmacol.* 1998, **45**, 221–228.
2. Moyle, G. J.; Gazzard, B. G.; Cooper, D. A.; Gatell, J., *Drugs* 1998, **55**, 383–404.
3. De Clercq, E., *Collect. Czech. Chem. Commun.* 1998, **63**, 449–479.
4. De Clercq, E., *Pure Appl. Chem.* 1998, **70**, 567–577.
5. Vandamme, A. M.; Vanvaerenbergh, K.; Declercq, E., *Antiviral Chem. Chemother.* 1998, **9**, 187–203.
6. Sommadossi, J.-P., *Clin. Infect. Dis.* 1993, **16** (Suppl. 1), S7–S15.
7. Peter, K.; Gambertoglio, J. G., *Pharm. Res.* 1998, **15**, 819–825.

8. Hao, Z.; Cooney, D. A.; Hartman, N. R.; Perno, C. F.; Fridland, A.; DeVico, A. L.; Sarngadharan, M. G.; Broder, S.; Johns, D. G., *Molec. Pharmacol.* 1988, **34**, 431–435.
9. Gao, W.-Y.; Shirasaka, T.; Johns, D. G.; Broder, S.; Mitsuya, H., *J. Clin. Invest.* 1993, **91**, 2326–2333.
10. Gao, W.-Y.; Agbaria, R.; Driscoll, J. S.; Mitsuya, H., *J. Biol. Chem.* 1994, **269**, 12633–12638.
11. Leibman, K. C.; Helldelberg, C., *J. Biol. Chem.* 1955, **216**, 823–830.
12. Roll, P. M.; Weinfeld, H.; Caroll, E.; Brown, G. B., *J. Biol. Chem.* 1956, **220**, 439–454.
13. Lichtenstein, J.; Barner, H. D.; Cohen, S. S., *J. Biol. Chem.* 1960, **235**, 457–465.
14. Schrecker, A. W.; Goldin, A., *Cancer Res.* 1968, **28**, 802–803.
15. Ho, D. H. W., *Biochem. Pharmacol.* 1971, **20**, 3538–3539.
16. Lepage, G. A.; Naik, S. R.; Katakkar, S. B.; Khaliq, A., *Cancer Res.* 1975, **35**, 3036–3040.
17. Meier, C., *Synlett* 1998, 233–242.
18. Arzumanov, A. A.; Dyatkina, N. B., *Russ. J. Bioorg. Chem.* 1996, **22**, 777–794.
19. Krise, J. P.; Stella, V. J., *Adv. Drug Deliv. Rev.* 1996, **19**, 287–310.
20. Périgaud, C.; Girardet, J.-L.; Gosselin, G.; Imbach, J.-L., in *Antiviral Drug Design*; E. De Clercq, ed., JAI Press, London, 1996, Vol. 2, pp. 147–172.
21. Jones, R. J.; Bischofberger, N., *Antiviral Res.* 1995, **27**, 1–17.
22. Alexander, P.; Holy, A., *Collect. Czech. Chem. Commun.* 1994, **59**, 2127–2165.
23. Egron, D.; Arzumanov, A. A.; Dyatkina, N. B.; Krayevsky, A.; Imbach, J.-L.; Aubertin, A.-M.; Gosselin, G.; Périgaud, C., *Nucleosides, Nucleotides* 1999, **18**, 983–984.
24. McIntee, E. J.; Remmel, R. P.; Schinazi, R. F.; Abraham, T. W.; Wagner, C. R., *J. Med. Chem.* 1997, **40**, 3323–3331.
25. Winter, H.; Maeda, Y.; Uchida, H.; Mitsuya, H.; Zemlicka, J., *J. Med. Chem.* 1997, **40**, 2191–2195.
26. Balzarini, J.; Karlsson, A.; Aquaro, S.; Perno, C.-F.; Cahard, D.; Naesens, L.; De Clercq, E.; McGuigan, C., *Proc. Natl. Acad. Sci.* (USA) 1996, **93**, 7295–7299.
27. Valette, G.; Pompon, A.; Girardet, J.-L.; Cappellacci, L.; Franchetti, P.; Grifantini, M.; La Colla, P.; Loi, A. G.; Périgaud, C.; Gosselin, G.; Imbach, J.-L., *J. Med. Chem.* 1996, **39**, 1981–1990.
28. *Nucleic Acids in Chemistry and Biology*; 2nd ed., G. M. Blackburn and M. J. Gait, eds., Oxford Univ. Press, 1996; and references cited therein.
29. Ho, T.-L., *Chem. Rev.* 1975, **75**, 1–20.
30. Pearson, R. G., *Hard and Soft Acids and Bases*, Dowden, Hutchinson and Ross, Stroudsburg, PA, 1973.
31. Shabarova, Z.; Bogdanov, A., in *Advanced Organic Chemistry of Nucleic Acids*, Z. Shabarova, and A. Bogdanov, eds., VCH, Weinheim, 1994, pp. 93–180; and references cited therein.
32. Westheimer, F. H., *Acc. Chem. Res.* 1968, **1**, 70–78.
33. Emsley, J.; Hall, D., *The Chemistry of Phosphorus*; Harper and Row, London, 1976.
34. Holmes, R. R., *Pentacoordinated Phosphorus*, American Chemical Society: Washington, DC, 1980, Vol. 176.
35. Cox, J. R.; Ramsay, O. B., *Chem. Rev.* 1964, **64**, 317–352.

36. Westheimer, F. H., *Science* 1987, **235**, 1173–1178.
37. Landt, M.; Everard, R. A.; Butler, L. G., *Biochemistry* 1980, **19**, 138–143.
38. Schlienger, N.; Beltran, T.; Perigaud, C.; Lefebvre, I.; Pompon, A.; Aubertin, A. M.; Gosselin, G.; Imbach, J. L., *Bioorg. Med. Chem. Lett.* 1998, **8**, 3003–3006.
39. Aldridge, W. N.; Reiner, E., *Enzyme Inhibitors as Substrates*; North-Holland, Amsterdam, 1972.
40. Eto, M., *Biosci. Biotech. Biochem.* 1997, **61**, 1–11; and references cited therein.
41. Ashani, Y.; Snyder, S. L.; Wilson, I. B., *J. Med. Chem.* 1973, **16**, 446–450.
42. Hamer, N. K., *J. Chem. Soc. Chem. Commun.* 1968, 1399.
43. Gay, D. C.; Hamer, N. K., *J. Chem. Soc.* (B) 1970, 1123–1127.
44. Gay, D. C.; Hamer, N. K., *J. Chem. Soc. Perkin Trans. 2* 1972, 929–932.
45. Eckstein, F.; Gish, G., *TIBS* 1989, **14**, 97–100.
46. Stella, V., in *Pro-drugs as Novel Drug Delivery Systems*, T. Higuchi and V. Stella, eds., American Chemical Society, Washington, DC, 1975; Vol. 14.
47. Sinkula, A. A.; Yalkowsky, S. H., *J. Pharm. Sci.* 1975, **64**, 181–210.
48. Bundgaard, H., *Adv. Drug Deliv. Rev.* 1989, **3**, 39–65.
49. Bundgaard, H., *Drugs Fut.* 1991, **16**, 443–458.
50. Périgaud, C.; Gosselin, G.; Lefebvre, I.; Girardet, J.-L.; Benzaria, S.; Barber, I.; Imbach, J.-L., *Bioorg. Med. Chem. Lett.* 1993, **3**, 2521–2526.
51. Puech, F.; Gosselin, G.; Lefebvre, I.; Pompon, A.; Aubertin, A.-M.; Kirn, A.; Imbach, J.-L., *Antiviral Res.* 1993, **22**, 155–174.
52. Lefebvre, I.; Périgaud, C.; Pompon, A.; Aubertin, A.-M.; Girardet, J.-L.; Kirn, A.; Gosselin, G.; Imbach, J.-L., *J. Med. Chem.* 1995, **38**, 3941–3950.
53. Périgaud, C.; Gosselin, G.; Imbach, J.-L., in *Current Topics in Medicinal Chemistry*, J. C. Alexander, ed., Blackwell Science, Oxford, 1997; Vol. 2, pp. 15–29.
54. Slotin, L. A., *Synthesis* 1977, 737–752.
55. Reese, C. B., *Tetrahedron* 1978, **34**, 3143–3179.
56. Lannuzel, M.; Egron, D.; Imbach, J.-L.; Gosselin, G.; Périgaud, C., *Nucleosides, Nucleotides* 1999, **18**, 1001–1002.
57. Gosselin, G.; Imbach, J.-L., *Internatl. Antivir. News* 1993, **1**, 100–102.
58. Sastry, J. K.; Nehete, P. N.; Khan, S.; Nowak, B. J.; Plunkett, W.; Arlingaus, R. B.; Farquhar, D., *Molec. Pharmacol.* 1992, **41**, 441–445.
59. McGuigan, C.; Bellevergue, P.; Sheeka, H.; Mahmood, N.; Hay, A. J., *FEBS Lett.* 1994, **351**, 11–14.
60. Hao, Z.; Cooney, D. A.; Farquhar, D.; Perno, C. F.; Zhang, K.; Masood, R.; Wilson, Y.; Hartman, N. R.; Balzarini, J.; Johns, D. G., *Molec. Pharmacol.* 1990, **37**, 157–163.
61. Furman, P. A.; Barry, D. W., *Am. J. Med.* 1988, **85**, 176–181.
62. Balzarini, J., *Pharm. World Sci.* 1994, **16**, 113–126.
63. Taylor, M. D., *Adv. Drug Deliv. Rev.* 1996, **19**, 131–148.
64. Lefebvre, I.; Pompon, A.; Valette, G.; Périgaud, C.; Gosselin, G.; Imbach, J.-L., *Pharm. Technol. Eur.* 1998, **10**, 4–9.
65. Martin, L. T.; Faraj, A.; Schinazi, R. F.; Imbach, J.-L.; Gosselin, G.; McClure, H. M.; Sommadossi, J.-P., *Antiviral Res.* 1998, **37**, A64.

66. Faraj, A.; Placidi, L.; Périgaud, C.; Cretton-Scott, E.; Gosselin, G.; Martin, L. T.; Pierra, C.; Schinazi, R. F.; Imbach, J.-L.; Sommadossi, J.-P., *Nucleosides Nucleotides* 1999, **18**, 987–988.

67. Daniels, L. B.; Glew, R. H., in *Methods of Enzymatic Analysis. Enzymes 2: Esterases, Glycosidases, Lyases, Ligases*; 3rd ed., H. U. Bergmeyer, J. Bergmeyer, and M. Grassl, eds., Verlag Chemie, Weinheim, 1974, Vol. 4, pp. 217–226.

68. Imbach, J.-L.; Périgaud, C.; Gosselin, G., results presented during 1st Internatl. Conf. Discovery and Clinical Development of Antiretroviral Therapies, St. Thomas, West Indies, U.S. Virgin Islands, Dec. 13–17, 1998.

69. Frick, L. W.; Nelson, D. J.; St Clair, M. H.; Furman, P. A.; Krenitsky, T. A., *Biochem. Biophys. Res. Commun.* 1988, **154**, 124–129.

70. Fridland, A.; Connely, M. C.; Ashmun, R., *Molec. Pharmacol.* 1990, **37**, 665–670.

71. Sommadossi, J.-P., personal communication.

72. Gosselin, G.; Girardet, J.-L.; Périgaud, C.; Benzaria, S.; Lefebvre, I.; Schlienger, N.; Pompon, A.; Imbach, J.-L., *Acta Biochim. Pol.* 1996, **43**, 195–208.

73. Périgaud, C.; Gosselin, G.; Imbach, J.-L., *Antiviral Ther.* 1996, **1** (Suppl. 4), 39–46.

74. Faulds, D.; Brogden, R. N., *Drugs* 1992, **44**, 94–116.

75. Périgaud, C.; Aubertin, A.-M.; Benzaria, S.; Pelicano, H.; Girardet, J.-L.; Maury, G.; Gosselin, G.; Kirn, A.; Imbach, J.-L., *Biochem. Pharmacol.* 1994, **48**, 11–14.

76. Thumann-Schweitzer, C.; Gosselin, G.; Périgaud, C.; Benzaria, S.; Girardet, J.-L.; Lefebvre, I.; Imbach, J.-L.; Kirn, A.; Aubertin, A.-M., *Res. Virol.* 1996, **147**, 155–163.

77. Périgaud, C.; Girardet, J.-L.; Lefebvre, I.; Xie, M.-Y.; Aubertin, A.-M.; Kirn, A.; Gosselin, G.; Imbach, J.-L.; Sommadossi, J.-P., *Antiviral Chem. Chemother.* 1996, **7**, 338–345.

78. Farquhar, D.; Srivastva, D. N.; Kuttech, N. J.; Saunders, P., *J. Pharm. Sci.* 1983, **72**, 324-325.

79. Heymann, E., in *Enzymatic Basis of Detoxification*, W. B. Jacoby, and R. R. Bend, eds., Academic Press: New York, 1980; Vol. 2, pp. 291–323.

80. Satoh, T., *Rev. Biochem. Toxicol.* 1987, **8**, 155–181.

81. Egron, D.; Périgaud, C.; Gosselin, G.; Aubertin, A.-M.; Imbach, J.-L., *Bull. Soc. Chim. Belg.* 1997, **106**, 461–466.

82. Egron, D.; Perigaud, C.; Aubertin, A. M.; Imbach, J. L.; Gosselin, G., *Nucleosides Nucleotides* 1998, **17**, 1725–1729.

83. Stella, V. J.; Himmelstein, K. J., in *Design of Prodrugs*, H. Bundgaard, ed., Elsevier Science Publishers, Amsterdam, 1985, pp. 177–198.

84. Tomlinson, E., *Adv. Drug Deliv. Rev.* 1987, **1**, 87–198.

85. Kearney, A. S., *Adv. Drug Deliv. Rev.* 1996, **19**, 225–239.

86. Meijer, D. K. F.; Jansen, R. W.; Molema, G., *Antiviral Res.* 1992, **18**, 215–258.

87. Vamvakas, S.; Anders, M. W., *Adv. Pharmacol.* 1994, **27**, 479–499.

88. Schlienger, N.; Périgaud, C.; Gosselin, G.; Imbach, J.-L., *J. Org. Chem.* 1997, **62**, 7216–7221.

89. Valette, G.; Girardet, J.-L.; Pompon, A.; Périgaud, C.; Gosselin, G.; Aubertin, A.-M.; Kirn, A.; Imbach, J.-L., *Nucleosides Nucleotides* 1997, **16**, 1331–1335.

PART II

EXPLORATION OF FUNDAMENTAL CHEMICAL PRINCIPLES IN DRUG DESIGN AND DISCOVERY

In Chapter 6, George Kenyon continues the vital anti-HIV drug theme by attempting to synthesize proximal inhibitors of a key enzyme of HIV replication, namely, the *reverse transcritase.*

According to the March 15, 1998 issue of the journal *Cancer*, cancer incidence and death rates from cancer declined from 1990 to 1995. Today many people are living with the desease or have been cured. Nonetheless, cancer is very much on people's minds since half of all males and a third of all females will develop cancer. Three contributions in this part of the book deal specifically with efforts to discover additional therapeutic agents for this multifaceted group of diseases characterized by out-of-control growth and spread.

In Chapter 7, Ludeman writes of a "swords into plowshares" approach in which a chemical warfare agent has been converted into a valuable anticancer agent. In Chapter 8, Burke, Gao, and Yao describe their efforts to mimic phosphorylated molecules that are key to transmission of cellular signals during normal cellular function. Aberrations in this communication process can lead to several cancers. Therefore, the discovery and development of signaling inhibitors is an approach to head off malignant growth. Magnetism as a means to deliver drugs to tumors is explored in Chapter 9 by Pulfer and Gallo.

In a wholly different strategy, in Chapter 10 Jacquesy and Fahy describe their use of "superacids" chemistry to find a new promising anticancer drug, *vinflunine*, now in phase I clinical trial. They relied on the pioneering research of George Olah on "superacid" chemistry. Olah's chemistry involved the use of such unusual reagents as *antimony pentafluoride in fluorosulfuric acid.* No one could have predicted that such seemingly esoteric conditions would lead to the discovery of a potential therapeutic for cancer.

Vinflunine contains *fluorine*, and in Chapter 11 Kirk continues this elemental theme by detailing how medicinal chemists can modulate the bioactivity of molecules for substitution with this most electronegative element in the periodic table. Fluorine-substituted molecules have had a widespread impact on diverse areas of medicine, including cancer chemotherapy (5-fluorouracil), antidepressants (Prozac or fluoxetine), antifungals (flucytosine or 5-fluorocytosine), and others as listed by Kirk.

CHAPTER 6

AIDS: Potential Inhibitors of Human Immunodeficiency Virus Replication

GEORGE L. KENYON
College of Pharmacy, University of Michigan

HIV-1 REVERSE TRANSCRIPTASE AS A DRUG TARGET FOR AIDS

Human immunodeficiency virus (HIV) is widely recognized as the etiologic agent responsible for acquired immunodeficiency syndrome (AIDS).[1] HIV-1, the most commonly found form of this class of retrovirus, contains a vital enzyme that is not found in the human host, namely, the HIV-1 reverse transcriptase (HIV-1 RT).[2] To pinpoint its role, it is important to note that the normal flow of genetic information is described by the statement (called the *central dogma* of molecular biology) that deoxyribonucleic acid (DNA) forms ribonucleic acid (RNA), which encodes for protein (see Scheme 6.1).

After the discovery of reverse transcriptases, the dotted line was added. Reverse transcriptases, which are RNA-dependent DNA polymerases, are essential for the life cycles of HIV-1 and other similar RNA viruses in that their core genetic information is stored in RNA rather than in DNA as is the case in the vast majority of living organisms. HIV-1 RT was early on considered to be a highly promising target enzyme for the development of drugs to combat HIV infection, a prediction that has been borne out in practice. Today, the RT remains the most prominent of such HIV drug targets.

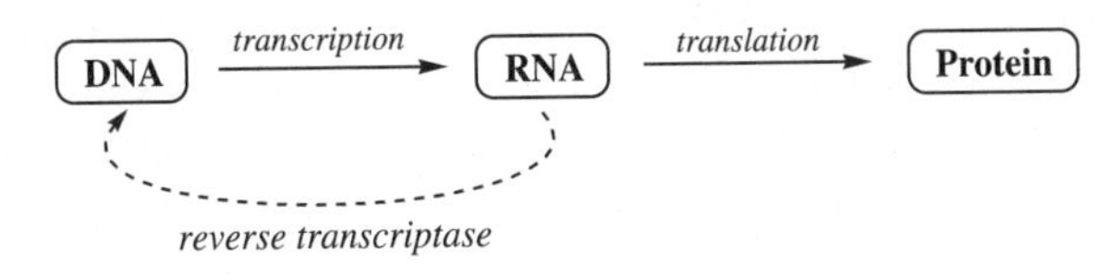

SCHEME 6.1

Biomedical Chemistry: Applying Chemical Principles to the Understanding and Treatment of Disease, Edited by Paul F. Torrence
ISBN 0-471-32633-x

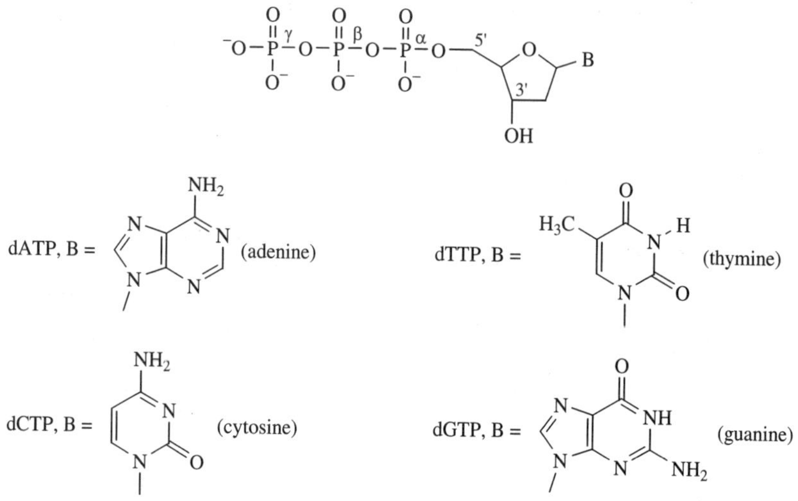

SCHEME 6.2

HIV-1 RT requires a template primer, such as poly (rA)-oligo (dT), and, like all DNA polymerases, uses the four naturally occurring deoxyribonucleoside 5′-triphosphates—dATP, dCTP, dGTP, and TTP—as the substrate "building blocks" for the growing DNA chain. These four structures are represented in the structures shown in Scheme 6.2.

In the course of the DNA polymerization, the α,β-P–O–P linkages of these four nucleoside 5′-triphosphates are cleaved and new phosphate linkages are made between the 3′-OH group with the α-phosphate group at the 5′-position of the next nucleotide that becomes attached as shown in Scheme 6.3.

A common strategy for developing anti-HIV drugs based on inhibiting the HIV-RT is to use nucleosides such as those shown in Scheme 6.4.

These are all dideoxyribonucleoside (ddN) analogs and include AZT (azidothymidine),[3] DCC (2′,3′-dideoxcytidine),[4] and DDI (2′,3′-dideoxyinosine).[5] It should be noted that there are a number of other similar drugs, and these are shown simply as representative examples.

These ddN analogs are all prodrugs and become monophosphorylated, diphosphorylated, and then triphosphorylated in a sequential fashion by intracellular enzyme-catalyzed reactions to generate the corresponding 5′-triphosphate derivatives. These triphosphates, in turn, become substrates for the RT, which accepts them just as it would normal substrates and incorporates these unnatural nucleotides into the growing chains. Since none of these ddN's have a 3′-OH group, however, they cannot add the next nucleotide group. Thus, they are *chain terminators* and, as such, are very effective at stopping the replication of the virus by depriving it of its full-length DNA.[6] Unfortunately, the more these drugs are used, the more they become subject to resistance by the HIV-1 strains that they are attempting to thwart; that is, the HIV strains undergo mutations in the HIV-1 RT gene itself that tend to render the

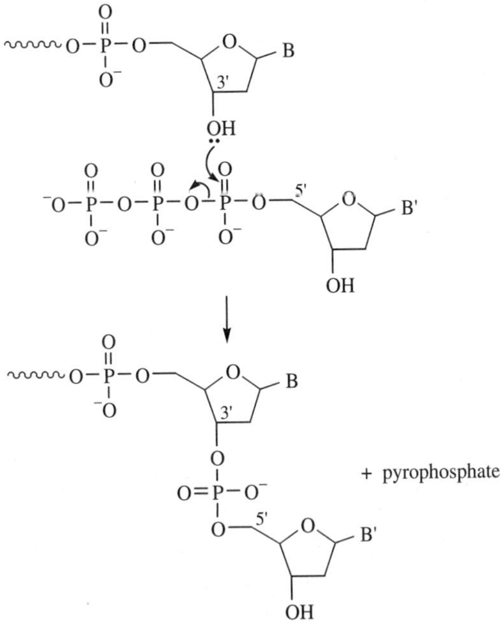

SCHEME 6.3

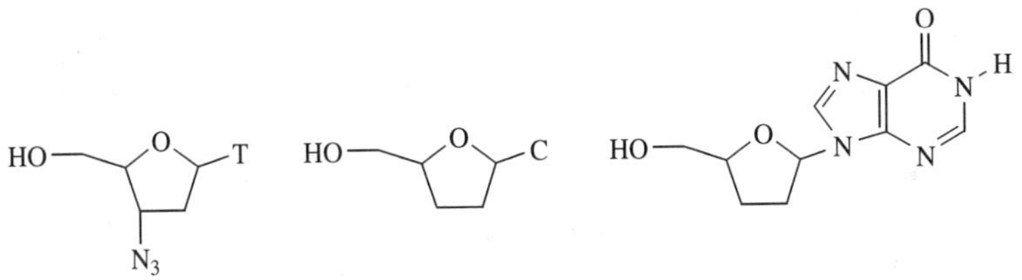

SCHEME 6.4

drug less effective.[7] This is especially well documented in the case of AZT, where it is now known that mutations even somewhat remote from the nucleotide (substrate) binding site can give rise to such resistance.[8] This drug resistance, which is quite common in the case of many drug regimens in the treatment of a variety of diseases, is especially problematic in the treatment of HIV infection because the virus is notorious for having an unusually high mutational rate. (This same high mutational rate has also thwarted numerous attempts to develop a vaccine against HIV.) In light of these drug resistance problems, the author decided to use a different strategy to inhibit the HIV-1 RT. If the α,β-P–O–P linkage could be replaced by a *noncleavable* bond, while still retaining all the other structural features of the normal deoxyribonucleotide 5′-triphosphate substrates, then inhibition could potentially occur with minimal structural perturbations that would lead to resistance.

THE CHEMISTRY OF NONCLEAVABLE (NONHYDROLYZABLE) NUCLEOTIDE ANALOGS

The first reports of noncleavable analogs of nucleotide 5′-triphosphates (NTPs) were made in the 1960s and included both the α,β- and β,γ-methylene analogs, namely, adenosine 5′-[α,β-methylene] triphosphate (AMPCPP) and adenosine 5′-[β,γ-methylene] triphosphate (AMPPCP) (see Scheme 6.5).[9,10] The carbon–phosphorus bonds in these compounds are highly resistant to hydrolysis even under strongly acidic conditions, and these analogs have been utilized as inhibitors, with mixed success, in a variety of biochemical studies over the years.

AMPPCP has been used more often, especially in cases such as ATPases, where it was the goal either to inhibit the hydrolysis of the β,γ-P–O–P linkage or to examine the effects of binding the AMPPCP without the usual β,γ-P–O–P cleavage taking place.

In 1970, we published the first synthesis of adenosine 5′-bis(dihydroxyphosphinylmethyl) phosphinate (AMPCPCP) (see Scheme 6.6), the α,β-bismethylene analog of ATP.[11]

AMPCPCP still remains the most hydrolytically stable of all of the known ATP analogs. It was synthesized by the route shown in Scheme 6.7, starting with bis(dihydroxyphosphinylmethyl) phosphinic acid (PCPCP).[12]

The product was purified using a Dowex 50 (H^+) column that was sufficiently acidic to protonate the adenine base of the AMPCPCP and transform it into a cation, thereby causing it to be held up on the column. At the same time, this acidic treatment also promoted the hydrolysis of the isopropylidene protective group.

Unfortunately, as stable as they are toward hydrolysis, there are some drawbacks to these methylene analogs of the NTPs. Their P–OH groups have pK_a values about 1–2 units above those of their P–O–P counterparts, they have tetrahedral geometries about the methylene groups, and they lack lone pairs that are normally on the bridging oxygen atom. These factors often render them less than optimal as

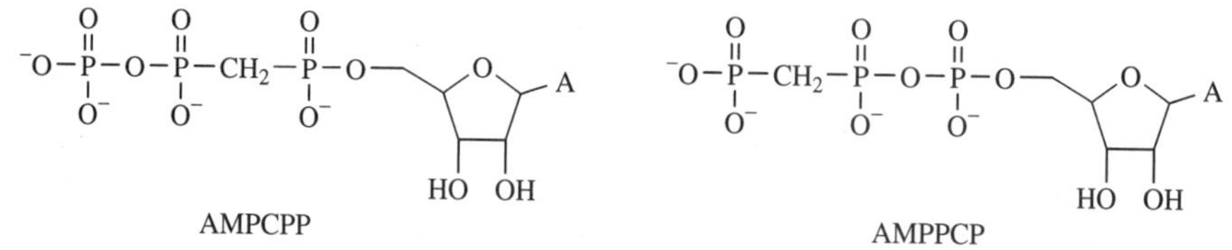

AMPCPP

AMPPCP

SCHEME 6.5

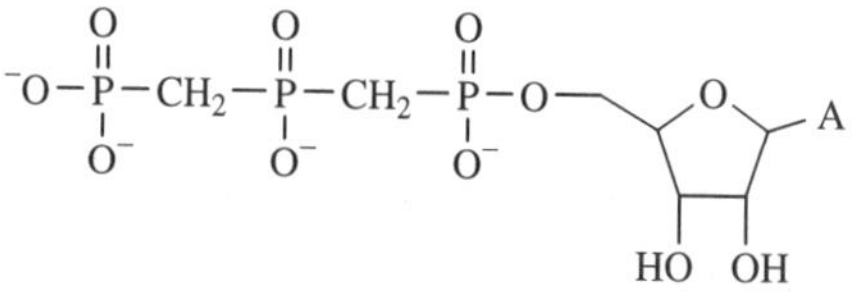

AMPCPCP

SCHEME 6.6

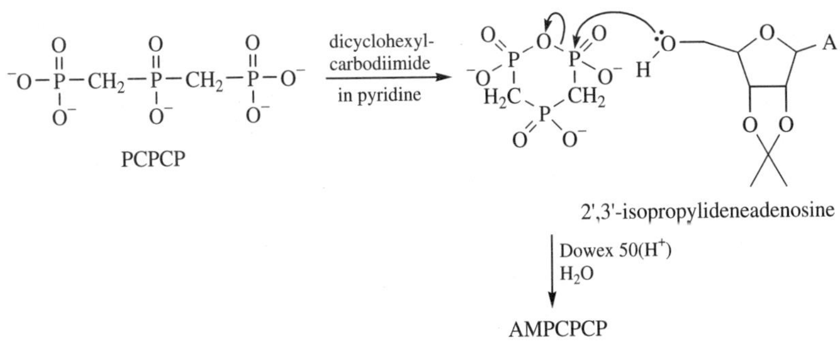

PCPCP

dicyclohexyl-carbodiimide in pyridine

2',3'-isopropylideneadenosine

Dowex 50(H^+) H_2O

AMPCPCP

SCHEME 6.7

inhibitors of NTP-requiring enzymes. In the early 1970s, the search was on for something better, and the corresponding imido N–H (imido) bridging analogs proved to be the answer.

In 1971, Yount and co-workers reported the first synthesis of adenosine 5′-(β,γ-imido) triphosphate (AMPPNP) and showed that it successfully inhibited enzymes that normally cleave the β,γ-P–O–P bond of ATP, especially ATPases and kinases.[13] Since that time, AMPPNP has been very widely used in enzymological studies. Not only is the pK_a value of the γ-P–OH bond similar to that of ATP itself, but also (1) the bridging –N–H still has a lone pair of electrons and therefore can either act as a hydrogen bonding acceptor or coordinate to an electrophilic metal ion and (2) X-ray crystal structures of imidodiphosphate (PNP) and pyrophosphate (see Scheme 6.8) show that the bond lengths and bond angles of the P–N–P and P–O–P linkages are nearly identical.[14]

In the mid 1970s, Tran-Dinh and co-workers published ^{31}P NMR studies on AMPPNP as a function of both pH and divalent metal ion concentration (e.g., Mg^{2+}) that showed anomalous behavior relative to the corresponding behavior of ATP itself.[15,16] Specifically, the β-phosphorus experienced a greater change in chemical shift on titration in the pH 4–10 range than did the γ-phosphorus, whereas with ATP the reverse is true. In light of these somewhat disturbing results, the author developed the hypothesis that the anomalous ^{31}P NMR pH-titration behavior could be due to the tautomeric structures shown in Scheme 6.9.[17] If the pK_a of the N–H proton in the bridge were sufficiently low, the proton could reside, at least part of the time, on oxygen rather than nitrogen.

To examine this hypothesis, the author and a graduate student, Dr. Qi-feng Ma, synthesized AMPPNP that had a specific ^{15}N label in the β,γ-nitrogen position for

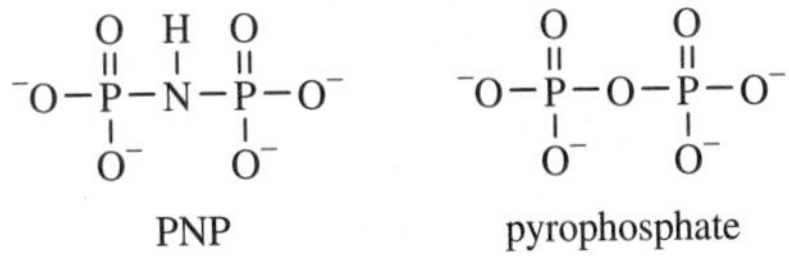

PNP

pyrophosphate

SCHEME 6.8

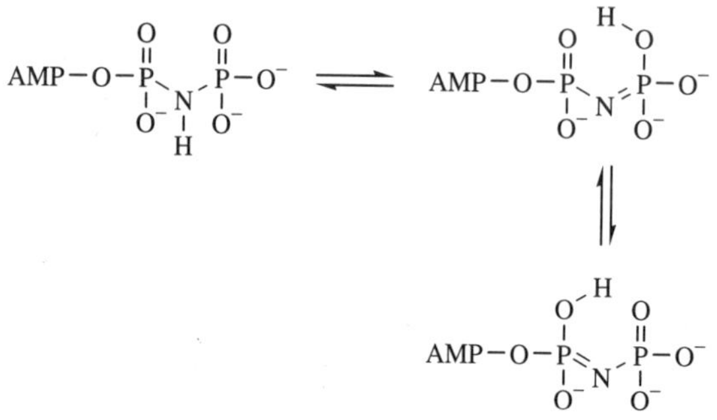

$$AMP{-}O{-}P(=O)(O^-){-}N(H){-}P(=O)(O^-){-}O^- \rightleftharpoons AMP{-}O{-}P(=O)(O^-){-}N{=}P(OH)(O^-){-}O^-$$

$$\rightleftharpoons AMP{-}O{-}P(OH)(O^-){=}N{-}P(=O)(O^-){-}O^-$$

SCHEME 6.9

use in ^{15}N NMR pH-titration studies.[18] The AMPPNP, which contained 99% ^{15}N in the β,γ-bridging position, was first prepared as shown in Scheme 6.10[18] (where AMP is the adenosine-5′-monophosphate moiety).

Later, the author and Dr. Ma found a remarkable paper by Emsley and Moore,[19] who carried out the conversion shown in Scheme 6.11.

The author recognized that this inorganic product, trichloro(dichlorophosphoryl)imino phosphorane, could be hydrolyzed in high yield in basic aqueous solution to form PNP[17] (see Scheme 6.12).

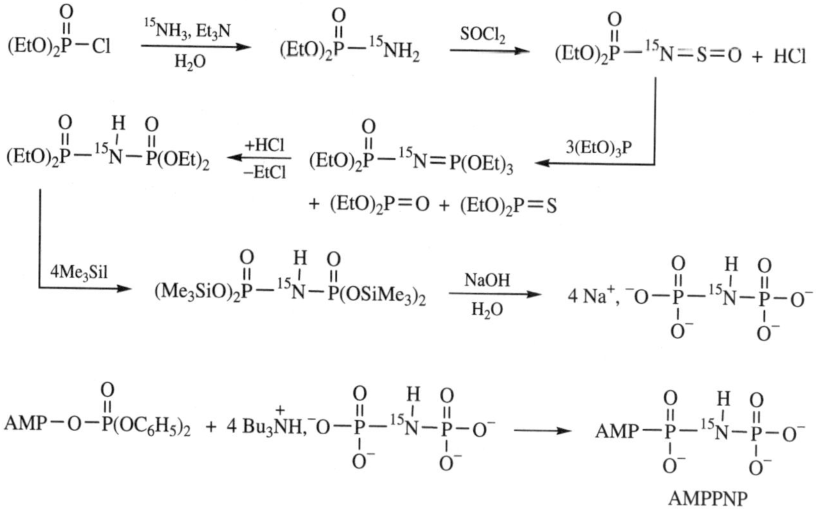

$$(EtO)_2P(O){-}Cl \xrightarrow[H_2O]{^{15}NH_3,\ Et_3N} (EtO)_2P(O){-}{}^{15}NH_2 \xrightarrow{SOCl_2} (EtO)_2P(O){-}{}^{15}N{=}S{=}O + HCl$$

$$\xrightarrow{3(EtO)_3P} (EtO)_2P(O){-}{}^{15}N{=}P(OEt)_3 + (EtO)_2P{=}O + (EtO)_2P{=}S$$

$$\xrightarrow[-EtCl]{+HCl} (EtO)_2P(O){-}{}^{15}NH{-}P(O)(OEt)_2$$

$$\xrightarrow{4Me_3SiI} (Me_3SiO)_2P(O){-}{}^{15}NH{-}P(O)(OSiMe_3)_2 \xrightarrow[H_2O]{NaOH} 4\,Na^+,\ {}^-O{-}P(O)(O^-){-}{}^{15}NH{-}P(O)(O^-){-}O^-$$

$$AMP{-}O{-}P(O)(OC_6H_5)_2 + 4\,Bu_3\overset{+}{N}H,\ {}^-O{-}P(O)(O^-){-}{}^{15}NH{-}P(O)(O^-){-}O^- \longrightarrow AMP{-}P(O)(O^-){-}{}^{15}NH{-}P(O)(O^-){-}O^-$$

AMPPNP

SCHEME 6.10

$$4\,PCl_5 + (NH_4)_2SO_4 \xrightarrow[146\,°C,\ 1\,h]{s\text{-tetrachloroethane}} 2\,Cl_3P{=}N{=}P(O)Cl_2 + 8\,HCl$$

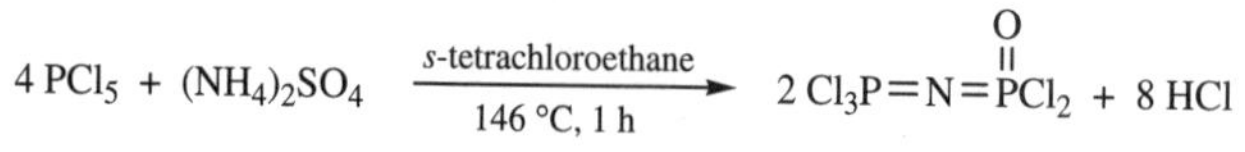

SCHEME 6.11

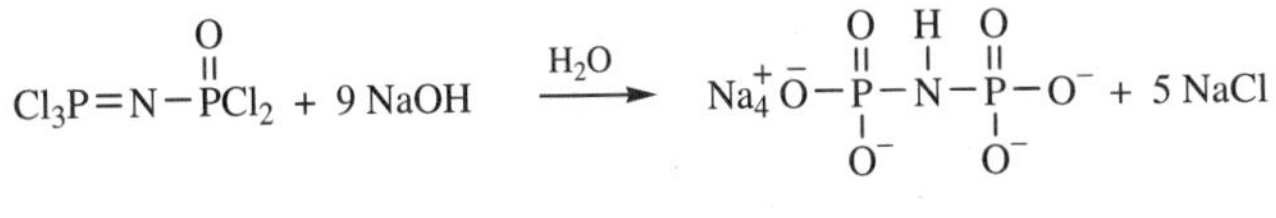

$$Cl_3P{=}N{-}\overset{\overset{\displaystyle O}{\|}}{P}Cl_2 + 9\,NaOH \xrightarrow{H_2O} Na_4^+\,\bar{O}{-}\overset{\overset{\displaystyle O}{\|}}{\underset{\underset{\displaystyle O^-}{|}}{P}}{-}\overset{\overset{\displaystyle H}{|}}{N}{-}\overset{\overset{\displaystyle O}{\|}}{\underset{\underset{\displaystyle O^-}{|}}{P}}{-}O^- + 5\,NaCl$$

SCHEME 6.12

Using ^{15}N-labeled ammonium sulfate, the author and Mr. Ma now had access to a streamlined synthesis of ^{15}N-labeled PNP. The overall synthesis was also readily adapted to the synthesis of ^{17}O-labeled PNP, the use of which is discussed below.

Both the ^{15}N-labeled PNP and the ^{15}N-labeled AMPPNP were examined by ^{15}N NMR spectroscopy at both pH $\sim$ 11 and pH $\sim$ 7. Proton decoupling was employed to simplify the spectra. Only small chemical-shift differences between these pH extremes were observed in both cases, and the spectra were consistent with the expected coupling to both of the flanking ^{31}P nuclei. Then, at pH 11, proton-*coupled* spectra were recorded for both the ^{15}N-labeled PNP and the ^{15}N-labeled AMPPNP. Both showed clean doubling of the signals, indicating that, even at the high pH, the N–H bond was intact. Also, when the ^{15}N-labeled AMPPNP was titrated with Mg^{2+} ion at pH 9.4, the chemical shift increased by only a relatively small amount ($\sim$ 1 ppm) between 0 and 1 Mg^{2+} ion per AMPPNP molecule. And, again, the proton-coupled ^{15}N NMR spectrum of AMPPNP in the presence of excess Mg^{2+} ion showed clear presence of the N–H coupling.

Finally, ^{17}O-labeled PNP was used, in turn, to, prepare AMPPNP ^{17}O-labeled in the β and γ-phosphorus oxygen positions,[17] as shown in Scheme 6.13.

When the ^{17}O NMR pH-titration curve was determined, it showed ^{17}O NMR chemical shifts in good agreement with those determined for a variety of phosphates and phosphate analogs, confirming that protonation does indeed occur on the oxygens. Moreover, the pK_a values obtained from this titration curve, 10.37 ± 0.12 and 7.41 ± 0.10, were in good agreement with those found by ^{31}P NMR. Therefore, it was the ^{31}P NMR pH titration of AMPPNP that was considered to be anomalous. This work provided another example of the unreliability of ^{31}P NMR to quantitate charge neutralization of the phosphoryl oxygens in phosphates.[20] Thus, the overall conclusion is that the widely reported N–H tautomeric structures for both PNP and AMPPNP are correct, and, if the tautomers shown above exist, they must be very minor contributors.

Secure in the knowledge that AMPPNP is truly isosteric with ATP, and armed with the newfound facile synthesis of large quantities of PNP, the author and Dr. Ma set out to meet one of the unfulfilled challenges of the field of noncleavable NTP analogs, namely, the synthesis of the α,β-imino analog of ATP (AMPNPP). Fortunately, a key paper in the field of nucleotide synthesis by Davisson, Poulter, and co-workers appeared in 1987[21] (the year before our synthesis was reported), which

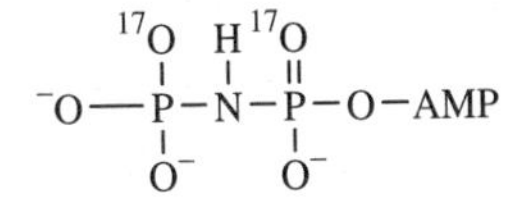

$$^{-}O{-}\overset{\overset{\displaystyle ^{17}O}{|}}{\underset{\underset{\displaystyle O^-}{|}}{P}}{-}\overset{\overset{\displaystyle H}{|}}{N}{-}\overset{\overset{\displaystyle ^{17}O}{\|}}{\underset{\underset{\displaystyle O^-}{|}}{P}}{-}O{-}AMP$$

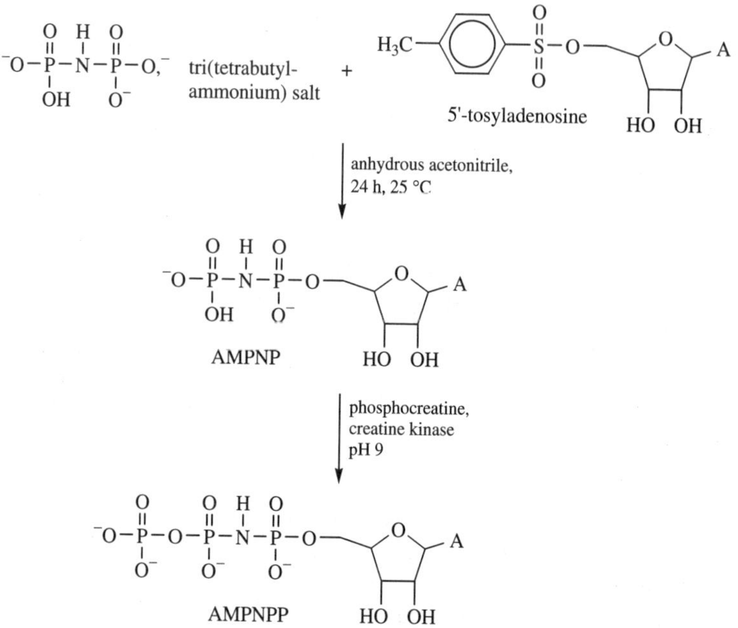

SCHEME 6.14

helped pave the way; and the author's synthesis, shown in Scheme 6.14, was patterned after their procedures.[22]

In the same timeframe, an abstract appeared by Tomasz[23] et al., who showed that the ADP analog, AMPNP, could be prepared by direct addition of adenosine to trichloro[(dichlorophosphory)imino] phosphorane followed by basic hydrolysis.

Both AMPNP and AMPNPP proved to be good substrates for rabbit muscle creatine kinase, with $V_{\max}$ values and $V_{\max}/K_{\mathrm{m}}$ values within approximately an order of magnitude of those for ADP and ATP, respectively.[22] These results are striking in that the corresponding α,β-methylene analogs, AMPCP and AMPCPP, had been reported to be rather poor substrates for the same enzymes with relative rate values of 10^{-5} those of ADP and ATP, respectively.[24] Thus, AMPNP and AMPNPP are at least 10^5-fold more reactive in the creatine kinase reaction than their corresponding methylene analogs. Further work showed that AMPNP was a good substrate for pyruvate kinase and that AMPNPP was a good substrate for hexokinase as well.[22]

AMPNPP should inhibit a broad range of enzymes that ordinarily catalyze the cleavage of the α,β-P–O–P linkage of ATP, including, for example, adenylate cyclase, acetyl CoA synthase, glucose 1-phosphate adenyltransferase, the aminoacyl tRNA synthases, the glutamine synthase adenylylating enzyme, and adenylate kinase. To date, AMPNPP has been examined in only a very few enzyme systems.

Next, we set out to prepare the last remaining member of the noncleavable imino analogs of ATP, adenosine 5′-(α,β;β,γ-diimido) triphosphate (AMPNPNP, the diimido analog of ATP),[25] as shown in Scheme 6.15.

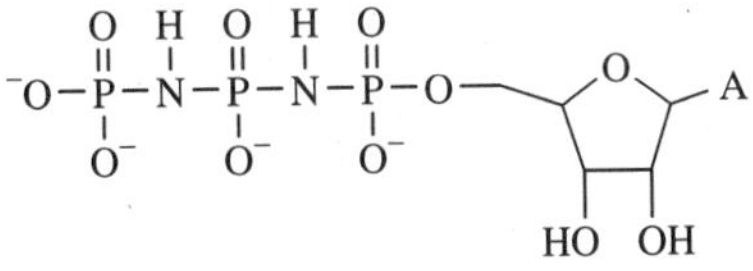

AMPNPNP

SCHEME 6.15

AMPNPNP had never been synthesized before, and the author believed that it could be prepared from the pentasodium salt of diimidotriphosphate (PNPNP). The synthesis of PNPNP, we believed, could be achieved by the hydrolysis of the known compound, [*P,P*-dichloro-*N*-(dichlorophosphinyl) phosphinimyl] phosphorimidic trichloride,[26] which, in turn, could be prepared from the treatment of trichloro[(dichlorophosphoryl)imino] phosphorane with hexamethyldisilazane and PCl_5 in *s*-tetrachloroethane (see Scheme 6.16).

When the tetrabutylammonium salt of PNPNP was treated with 5′-tosyladenosine in CH_2CN, it yielded two products, the expected one, AMPNPNP, and, surprisingly, the isomeric $AMP(NP)_2$,[25] as shown in Scheme 6.17.

An examination of the literature revealed that there was a single enzyme that could, in principle, at least, accept all the possible mono- and disubstituted methylene and imido analogs of ATP as potential substrates. The enzyme is *S*-adenosylmethionine (Adomet) synthase, which catalyzes the reaction shown in Scheme 6.18.[27]

In collaboration with Professor G. D. Markham, the author and Dr. Qi-feng Ma examined all of these analogs as potential substrates in this reaction, including AMPCPP, AMPPCP, AMPCPCP, AMPNPP, AMPPNP, AMPNPNP, and $AMP(NP)_2$.[25] In each case, the imido mono- and disubstituted analogs proved to

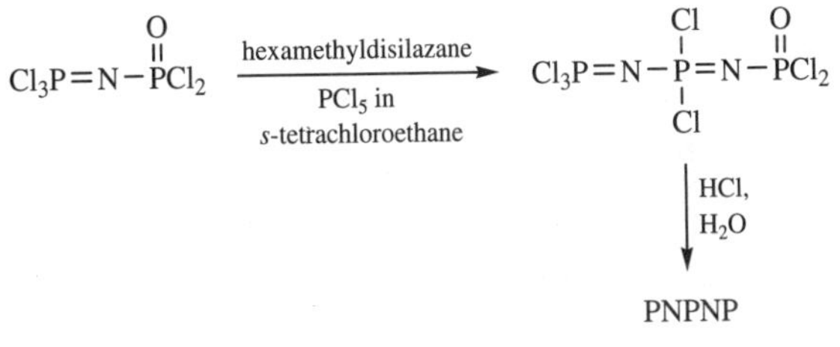

$Cl_3P{=}N{-}P(O)Cl_2$ → (hexamethyldisilazane, PCl_5 in *s*-tetrachloroethane) $Cl_3P{=}N{-}PCl_2{=}N{-}P(O)Cl_2$ → (HCl, H_2O) PNPNP

SCHEME 6.16

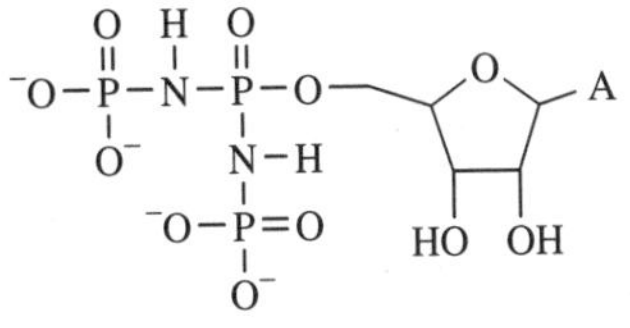

$AMP(NP)_2$

SCHEME 6.17

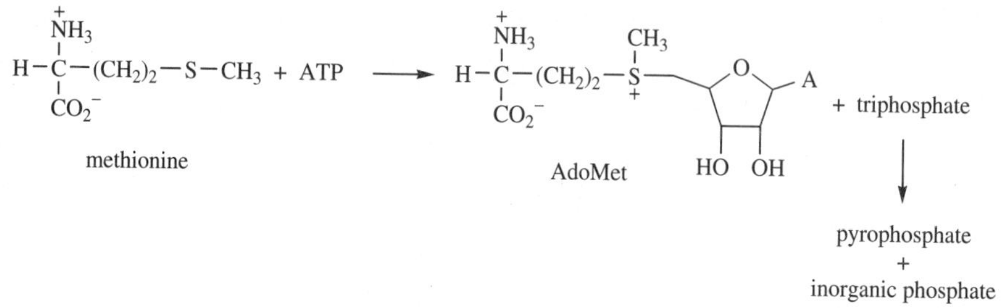

$$H{-}\overset{\overset{+}{NH_3}}{\underset{CO_2^-}{C}}{-}(CH_2)_2{-}S{-}CH_3 + ATP \longrightarrow H{-}\overset{\overset{+}{NH_3}}{\underset{CO_2^-}{C}}{-}(CH_2)_2{-}\overset{CH_3}{\underset{+}{S}}{-}\text{(ribose, HO OH)}{-}A + \text{triphosphate}$$

methionine — AdoMet

triphosphate → pyrophosphate + inorganic phosphate

SCHEME 6.18

be better substrates than their corresponding methylene analogs. AMPCPCP showed no detectable activity as a substrate ($< 10^{-5}\ V_{max}$ of ATP). In contrast, AMPNPNP reacted rapidly to give a single turnover of product per active site, followed by turnovers that are at least 1000-fold slower. This means that AMPNPNP can be used to quantify active-site concentrations. The branched $AMP(NP)_2$ was also not a substrate, but instead was a competitive inhibitor of the enzyme, greater than 100-fold more potent than ADP. AMPNPP was found to be a surprisingly potent inhibitor (with an inhibition constant ~60-fold lower than the K_m for ATP). It was also a substrate. AMPNPP was found to bind about 1000-fold more tightly than AMPCPP.

In another extension of this same kind of imido chemistry shown above, the *N*-methylated derivative of PNP, *N*-methylimidodiphosphate[PN(Me)P], was prepared for the first time as shown in Scheme 6.19.[28]

The first two steps, containing an unusual O → N methyl rearrangement, had been reported earlier by Riesel et al.[29] On basic hydrolysis, the salt of PN (Me)P was obtained in 80% yield. Synthetic procedures similar to those used for the preparation of the corresponding nonmethylated imido analogs were first used to prepare the α,β-*N*-methylimido analogs, AMPN(Me)P (the ADP analog) and AMPN(Me)PP (the ATP analog). Both were shown to be substrates in the forward and reverse directions of the creatine kinase reactions, respectively. AMPN(Me)P was shown to have a V_{max} value about 40% that of ADP but had a K_m value approximately 100 times greater than that for ADP. The β,γ-*N*-methylimido analog, AMPPN(Me)P, was prepared from PN(Me)P and adenosine 5′-monophosphate (AMP) using the

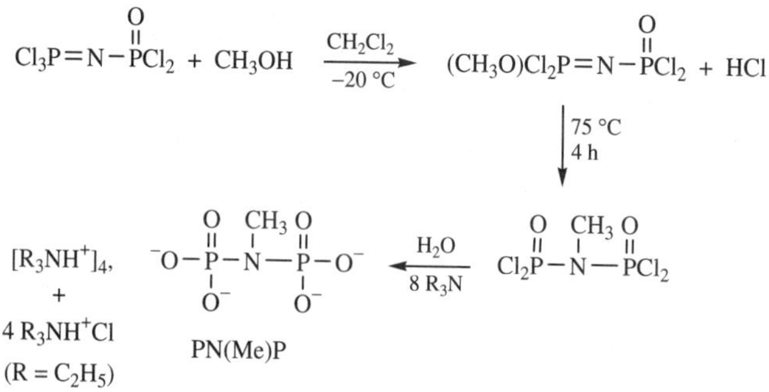

$$Cl_3P{=}N{-}\overset{O}{\overset{\|}{P}}Cl_2 + CH_3OH \xrightarrow[-20\ °C]{CH_2Cl_2} (CH_3O)Cl_2P{=}N{-}\overset{O}{\overset{\|}{P}}Cl_2 + HCl$$

$$\downarrow 75\ °C,\ 4\ h$$

$$[R_3NH^+]_4,\ {}^-O{-}\overset{O}{\overset{\|}{\underset{O^-}{P}}}{-}\overset{CH_3}{N}{-}\overset{O}{\overset{\|}{\underset{O^-}{P}}}{-}O^- \xleftarrow[8\ R_3N]{H_2O} Cl_2\overset{O}{\overset{\|}{P}}{-}\overset{CH_3}{N}{-}\overset{O}{\overset{\|}{P}}Cl_2$$

$+\ 4\ R_3NH^+Cl$ $(R = C_2H_5)$ — PN(Me)P

SCHEME 6.19

Michelson synthesis. AMPPN(Me)P appears to be rejected by creatine kinase since the author and Dr. Ma could detect activity neither as a substrate nor as an inhibitor. It is hoped that eventually longer and more complex N–R substituents (other than methyl) can be introduced synthetically so that, for example, potential alkylating agents can be built in to create novel affinity labels. But, so far at least, the author's and Dr. Ma's attempts to introduce alkyl groups larger than methyl have been unsuccessful.[28]

At this stage, the author and Dr. Ma were ready to tackle the synthesis of the α,β-imido analogs of the 2′-deoxyribonucleotides. The first set were all thymidine 5′-triphosphate (TTP) analogs, the syntheses of which are shown in Scheme 6.20.[30]

At the same time, the author and Dr. Ma prepared the α,β: β,γ-diimido analog of TTP (TMPNPNP) as well by the method shown in Scheme 6.21.[30]

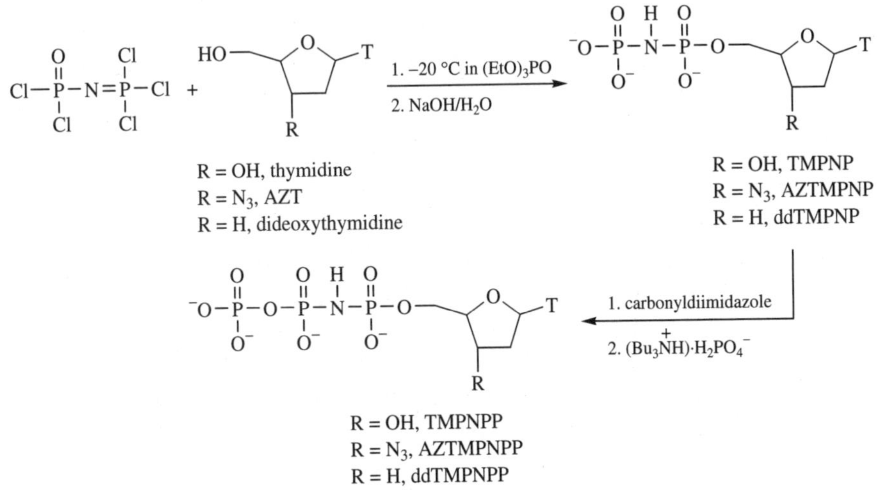

1. −20 °C in $(EtO)_3PO$
2. $NaOH/H_2O$

R = OH, thymidine
R = N_3, AZT
R = H, dideoxythymidine

R = OH, TMPNP
R = N_3, AZTMPNP
R = H, ddTMPNP

1. carbonyldiimidazole
2. $(Bu_3\overset{+}{N}H)\cdot H_2PO_4^-$

R = OH, TMPNPP
R = N_3, AZTMPNPP
R = H, ddTMPNPP

SCHEME 6.20

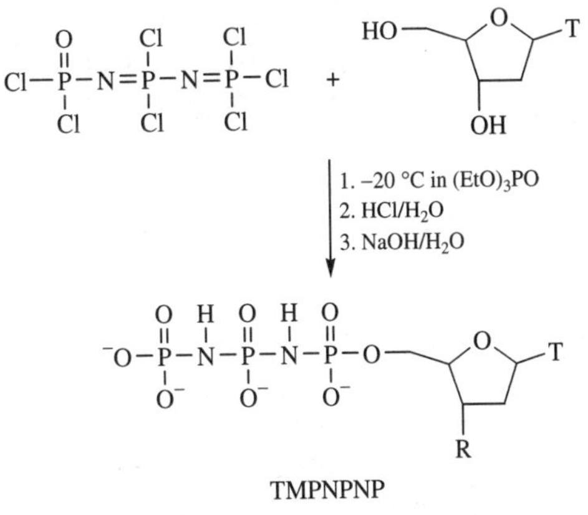

1. −20 °C in $(EtO)_3PO$
2. HCl/H_2O
3. $NaOH/H_2O$

TMPNPNP

SCHEME 6.21

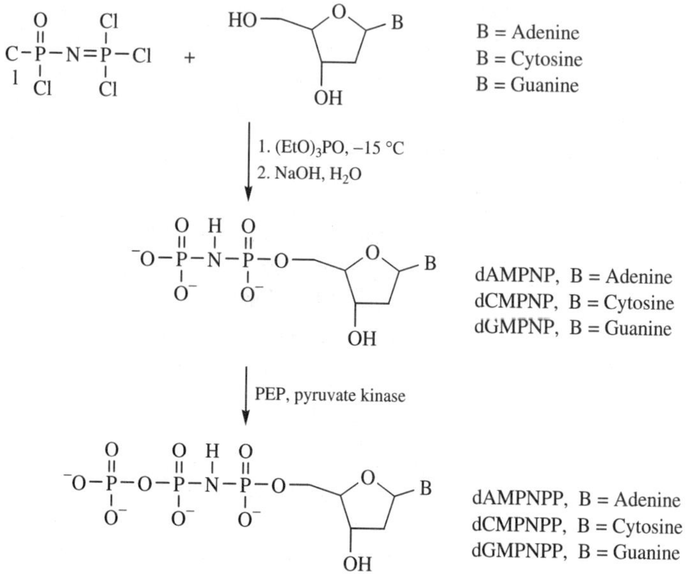

B = Adenine
B = Cytosine
B = Guanine

1. $(EtO)_3PO$, −15 °C
2. NaOH, H_2O

dAMPNP, B = Adenine
dCMPNP, B = Cytosine
dGMPNP, B = Guanine

PEP, pyruvate kinase

dAMPNPP, B = Adenine
dCMPNPP, B = Cytosine
dGMPNPP, B = Guanine

SCHEME 6.22

This synthesis closely parallels the author's earlier reported synthesis of AMPNPNP. The author and Dr. Ma also prepared 3′-azidothymidine 5′-(β,γ-imido) triphosphate (AZTMPPNP) from imidodiphosphate (AZPMP) using published methods for preparing AZT triphosphate itself.

In recent synthetic work, the author, Dr. Rongshi Li and Dr. Angelika Muscate prepared the remaining deoxyribonucleotide analogs as shown in Scheme 6.22.[31]

Improved conditions for the synthesis of TMPNPP and AZTMPNPP, as well as 5-iodo-2′-deoxyuridine 5′-(α,β-imido) triphosphate (IdUMPNPP), are shown in Scheme 6.23.[31]

INHIBITION OF HIV-1 REVERSE TRANSCRIPTASE BY VARIOUS IMIDO NUCLEOTIDE ANALOGS

Armed with this series of imido analogs of various deoxyribonucleotides, the author set out to examine their abilities to inhibit the HIV-1 reverse transcriptase.[30,31] Initially, only the TTP analogs were tested, and these results are shown in Table 6.1.[30]

The assay mixture (in a total of 50 μL) for the kinetic study contained the following: 50 mM Tris-HCL (pH 8.0), 6 mM $MgCl_2$, 8 mM dithiothreitol, 80 mM KCl, 0.4 μL poly (rA)-oligo (dT) (as template primer), different inhibitors, and variable amounts of (^{3}H) TTP (4–20 μM, 0.2–1 μCi per 20 μL). After incubation at 37 °C for 30 min, the mixtures were quenched using 15 μL of 0.5 M ethylenediaminetetraacetic acid (EDTA), and the template primers were collected on 2-

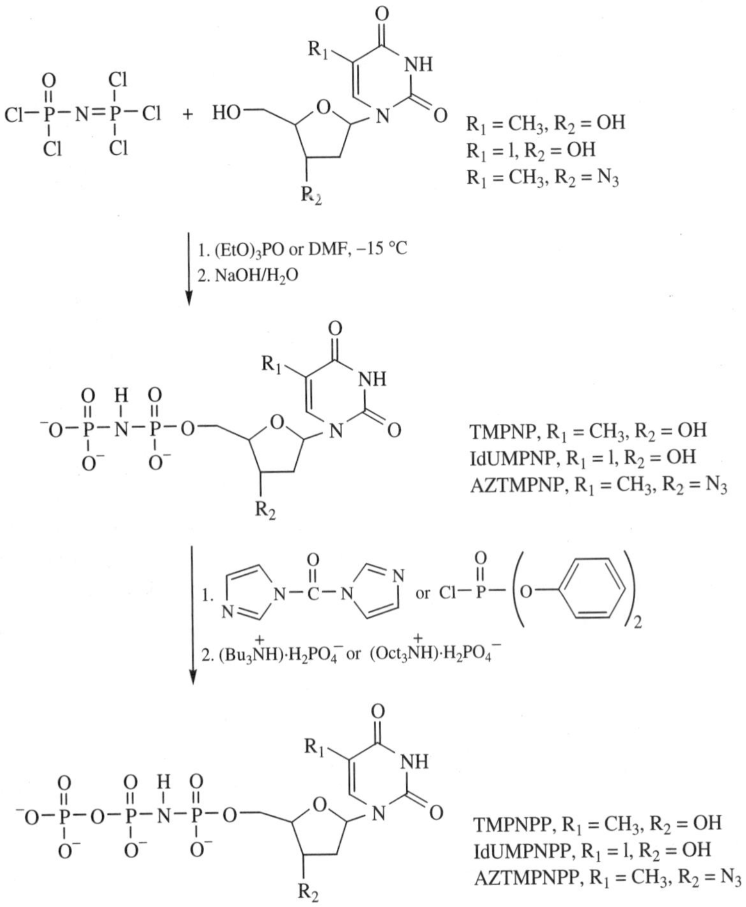

$R_1 = CH_3, R_2 = OH$
$R_1 = I, R_2 = OH$
$R_1 = CH_3, R_2 = N_3$

1. $(EtO)_3PO$ or DMF, −15 °C
2. $NaOH/H_2O$

TMPNP, $R_1 = CH_3, R_2 = OH$
IdUMPNP, $R_1 = I, R_2 = OH$
AZTMPNP, $R_1 = CH_3, R_2 = N_3$

1. [carbonyldiimidazole] or $Cl{-}P(O)(OPh)_2$
2. $(Bu_3\overset{+}{N}H){\cdot}H_2PO_4^-$ or $(Oct_3\overset{+}{N}H){\cdot}H_2PO_4^-$

TMPNPP, $R_1 = CH_3, R_2 = OH$
IdUMPNPP, $R_1 = I, R_2 = OH$
AZTMPNPP, $R_1 = CH_3, R_2 = N_3$

SCHEME 6.23

(diethylamino) ethyl cellulose (DE-81) paper. The incorporation rate of the TMP moieties was measured by counting the tritium on the DE-81 paper, and the K_i values were calculated from double reciprocal plots.

The three monosubstituted α,β-imido analogs, TMPNPP, ddTMPNPP, and AZTMPNPP, as expected, were strict competitive inhibitors relative to the native TTP substrate. Their K_i values ranged between 2.4 and 23 μM. Since these analogs are all noncleavable at the scissile bond, one would expect that their K_i values should be close to their dissociation constants (K_d values). It is notable that TTP itself has a reported K_d value of 10 μM, which is 4 times *higher* than that observed for its closest structural analog, TMPNPP. Since the true substrates for HIV-RT are Mg^+–nucleotide complexes, the apparent enhanced affinity of TMPNPP over TTP may reflect an altered distribution among the various metal–chelate complexes present in solution. Although the azido group of AZTPNPP is not so well tolerated, it still had a K_i value within an order of magnitude the same as that for TMPNPP. The same was

TABLE 6.1 Inhibition of HIV-1 RT by TTP Analogs

Compounds	Inhibition Type	K_i, μM
TMPNPP	Competitive	2.4 ± 0.1
DdTMPNPP	Competitive	15 ± 0.9
AZTMPNPP	Competitive	22 ± 1.3
AZTMPPNP	Competitive	0.087 ± 0.005
TMPNPNP	Competitive	19.1 ± 1.0
AZTTP	Competitive	0.13 ± 0.012[a]
		0.0036 ± 0.0004[a]

[a] The inhibitory potential of AZTTP for HIV-1 RT is template-dependent.[6] With a template of defined sequence, AZTTP gave a K_i value of 0.13 ± 0.012 μm. With poly(rA)-oligo(dT) as a template primer, its K_i value was found to be 0.0036 ± 0.0004.

true for both ddTMPNPP and the diimido analog, TMPNPNP. The fact that TMPNPNP was apparently more poorly bound than the TMPNPP indicates that the extra NH in the β,γ-bridging position is disadvantageous. Monosubstituted analogs were consistently better inhibitors than their disubstituted counterparts in all enzymes so far tested. Interestingly, all of these TMP analogs were shown to be strictly competitive inhibitors with respect to TTP.

Note that both AZTTP and AZTMPPNP appear to be superior to all the other inhibitors. This comparison is deceptive, however, since AZTTP and AZTMPPNP, unlike any of the other analogs shown in Table 6.1, are capable of being *chain terminators*;[6] that is, they contain cleavable α,β-P–O–P linkages and can therefore be turned over (once) by the HIV-1 RT and have their AZTMP moiety incorporated into the growing DNA chain. But then, because they have the 3′-azido group instead of the normal 3′-OH group, they cannot accept the next nucleotide as a substrate. The fact that AZTMPNPP gave rise to an order of magnitude higher K_i value than did TMPNPP actually suggests that AZTTP is likely to be somewhat poorer in forming the initial Michaelis complex with HIV-1 RT than is TTP itself.

The best of these noncleavable α,β-imido thymidine analogs, TMPNPP, was also tested as an inhibitor of DNA polymerase I (Klenow fragment) under conditions closely comparable to those used in the assay described above for the HIV-1 RT. An IC_{50} value (inhibition constant at 50% inhibition) for TMPNPP of 4000 μM was found. Under comparable conditions, an IC_{50} value for TMPNPP for the HIV-RT was found to be 10 μM. Thus, TMPNPP is ~400-fold more effective at inhibiting the HIV-1 RT than at inhibiting the DNA polymerase I (Klenow). Since the reported K_m values for TTP are close to the same for HIV-1 RT and the DNA polymerase ($K_m = 3.9$ μM for DNA polymerase I[32]; $K_m = 5.9$ μM for HIV-1 RT[33]), the selective inhibition of HIV-1 RT over the DNA polymerase I by TMPNPP apparently is due solely to introduction of the imido group. Ideally, one would like to see an X-ray crystal structure of TMPNPP bound to the HIV-1 RT to shed light on this intriguing result.

In a follow-up study, the three other α,β- imidonucleotides—dAMPNPP, dCMPNPP, and dGMPNPP—as well as IdUMPNPP—were all examined with respect to their IC_{50} values with the HIV-1 RT.[31] These data, along with comparable data for TMPNPP, AZTMPNPP, and tetrahydroimidazo[4,5,1-*jk*] [1,4] benzodiozepin-2-(1*H*)-thione (TIBO), are given in Table 6.2. (TIBO is a well-known "non-nucleoside" inhibitor of the HIV-1 RT and is occasionally used as a drug to combat HIV infection.) K_i values for both TMPNPP and AZTMPNPP are included for comparison purposes. K_i values are typically five- to sevenfold lower than IC_{50} values measured under these conditions.

This study allowed, for the first time, an evaluation of the relative binding interactions of all four noncleavable α,β-imido nucleotide analogs of the natural substrates, dATP, dCTP, dGTP, and TTP. Because these imido analogs cannot turn over to give product, only binding to the enzyme–primer template complex in the active site is being measured. These imido analogs are bound in the order TMPNPP > dGMPNPP > dCMPNPP > dAMPNPP, although all were apparently bound within a factor of 3 of one another. As reflected earlier in the determination of their relative K_i values, AZTMPNPP was bound considerably less tightly than TMPNPP. Finally, IdUMPNPP, with a very bulky iodo substituent on the 5 position of the uracil ring, had an IC_{50} value twice as low as that for TMPNPP and quite comparable to that for TIBO, a known HIV drug. Clearly, there is substantial bulk tolerance on the enzyme around the 5 position of the uracil base, a fact that may be exploitable in the design of future generations of inhibitors. Also, the IdUMPNPP would appear to be an ideal candidate for cocrystallization trials with HIV-1 RT because of its iodo group (a heavy atom) as well as its relatively high apparent binding potency. Indeed, a cocrystal structure of the HIV-1 RT in the presence of *any* of the α,β-imido analogs would be very useful since it would permit our first glimpse at the chemical interactions that are involved in binding of the nucleotide 5′-triphosphate substrate to the active complex normally leading to catalytic turnover.

TABLE 6.2 Inhibition Activity of HIV-RT by dNMPNPP

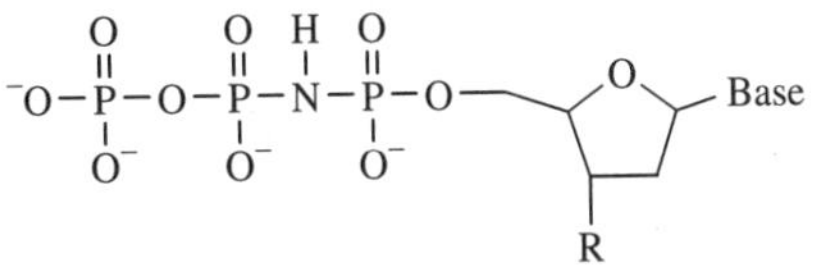

$^{-}O-P(=O)(O^{-})-O-P(=O)(O^{-})-N(H)-P(=O)(O^{-})-O-$ [sugar ring with O, Base, R]

Compounds	R	Base	IC_{50}, μM[a]	K_i, μM[b]
TIBO	—	—	5 ± 0.5	—
dAMPNPP	OH	Adenine	46 ± 3.2	—
dCMPNPP	OH	Cytosine	38 ± 4.5	—
dGMPNPP	OH	Guanine	32 ± 2.8	—
TMPNPP	OH	Thymine	14 ± 1.5	2.2 ± 0.1
IdUMPNPP	OH	5-Iodouracil	7 ± 1.1	—
AZTMPNPP	N_3	Thymine	110 ± 9.9	22 ± 1.3

[a] IC_{50} ± SD (three determinations for each compound).
[b] Data from Ref. 28.

How close are these α,β-imido nucleotide analogs to being drugs that could inhibit HIV replication in humans? Since AZTTP is one of the drugs most commonly used throughout the world to combat HIV-1 infection (at least at the time of this writing), the structures of the α,β-imido nucleotide analogs themselves are extremely similar. The challenge, of course, is one of appropriate in vivo delivery to the HIV-1 RT active sites. In the case of AZTTP, the nucleoside AZT is administered as a prodrug, and intracellular kinases are relied on to generate the actual drug, AZTTP, in situ. The α,β-imido nucleotide analogs, such as TMPNPP, for example, cannot be so generated and delivered. Other delivery systems, such as liposomal delivery, will be needed if these α,β-imido nucleotide inhibitors are to succeed in the clinic. [Actually, it may be that the thymidine 5′*-diphosphate* analogs, e.g., thymidine 5′-(α,β-imido) diphosphate (TMPNP), can be used as prodrugs that could, in principle, be converted intracellularly to their corresponding triphosphate analogs).] Potency is also an issue, and it would obviously be desirable to design α,β-imido nucleotide analogs with K_i values substantially lower than those that we have identified so far. The results with IdUMPNPP give rise to hope that this may be achievable.

ACKNOWLEDGMENTS

This work was generously supported by grants AR 17323 and GM 39552 from the National Institutes of Health. The author wishes to acknowledge all the co-authors for this work cited in the reference list. Special recognition should go to Dr. Qi-feng Ma, who performed most of the work on the imido nucleotide analogs.

REFERENCES

1. Hirsch, M. S., *J. Infect. Dis.* 1990, **161**, 845.
2. DeClercq, E., *AIDS Rese. Hum. Retroviruses* 1992, **8**, 119.
3. Mitsuya, H.; Weinhold, K. J.; Furman, P. A.; St. Clair, M. H.; Lehrman, S. N.; Gallo, R. C.; Bolognesi, D.; Barry, D. W.; Broder, S., *Proc. Natl. Acad. Sci.* (USA) 1985, **82**, 7096.
4. Cooney, D. A.; Dahal, M.; Mitsuya, H.; McMahon, J. B.; Nadkarni, M.; Balzarini, J.; Broder, S.; Johns, D. G., *Biochem. Pharmacol.* 1986, **35**, 2065.
5. Yarchoan, R.; Mitsuya, H.; Thomas, R. V.; Phida, J. M.; Hartman, N. R.; Perno, C. F.; Marczyk, K. S.; Allain, J. P.; Johns, D. G.; Broder, S., *Science* 1989, **245**, 412.
6. Ma, Q.; Bathurst, I.C.; Barr, P. J.; Kenyon, G. L., *Biochemistry* 1992, **31**, 1375.
7. Hecht, F.; Grant, R.; Petropoulos, C.; Dillion, B.; Chesney, M.; Tian, H.; Hellmann, N.; Bandrapalli, N.; Digilio, L.; Branson, B.; Kahn, J., *N. Engl. J. Med.* 1998, **339**, 307.
8. Boyer, P.; Tantillo, C.; Jacobo-Molina, A.; Nanni, R.; Ding, J.; Arnold, E.; Hughes, S., *Proc. Natl. Acad. Sci.* (USA) 1994, **91** 4882.
9. Myers, T. C.; Nakamura, K.; Flesher, J. W., *J. Am. Chem. Soc.* 1963, **85**, 3292.
10. Myers, T. C., Nakamura, K.; Danielzadeh, A. B., *J. Org. Chem.* 1965, **30**, 1517.
11. Trowbridge, D. B.; Kenyon, G. L., *J. Am. Chem. Soc.* 1970, **92**, 2181.

12. Trowbridge, D. B.; Yamamoto, D. M.; Kenyon, G. L., *J. Am. Chem. Soc.* 1972, **94**, 3816.
13. Yount, R. G.; Babcock, D.; Ballantyne, W.; Ohala, D., *Biochemistry* 1971, **10**, 2484.
14. Larsen, M.; Willet, R.; Yount, R. G., *Science* 1969, **166**, 1510.
15. Tran-Dinh, S.; Roux, M.; Ellenberger, M., *Nucleic Acids Res.* 1975, **2**, 1101.
16. Tran-Dinh, S.; Roux, M., *Eur. J. Biochem.* 1977, **76**, 245.
17. Reynolds, M.; Gelt, J. A.; Demou, P. C., Oppenheimer, N. J.; Kenyon, G. L., *J. Am. Chem. Soc.* 1983, **105**, 6475.
18. Reynolds, M. A.; Oppenheimer, N. J.; Kenyon, G. L., *J. Labelled Compd. Radiopharm.* 1981, **18**, 1357.
19. Emsley, J.; Moore, J.; Udy, P. B., *J. Chem. Soc. A* 1971, 2863.
20. Jaffe, E. K.; Cohn, M., *Biochemistry* 1978, **17**, 652.
21. Davisson, V. J.; Davis, D. R.; Dixit, V. M.; Poulter, C. D., *J. Org. Chem.* 1987, **52**, 1794.
22. Ma, Q.; Babbitt, P. C.; Kenyon, G. L., *J. Am. Chem. Soc.* 1988, **110**, 4060; 8267.
23. Tomasz, J.; Willis, R. C.; Kent, S. L.: Robins, R. K.; Vaghefi, M. M., Medicinal Chemistry Section, 3rd Chemical Congress of North America, Toronto, Canada, June 5–10. 1988, Abstract No. 89.
24. Milner-White, E. J.; Rycroft, D. S., *Eur. J. Biochem.* 1983, **133**, 169.
25. Ma, Q.; Kenyon, G. L.; Markham, G. D., *Biochemistry* 1990, **29**, 1412.
26. Riesel, L.; Somieski, R. Z., *Anorg. Allg. Chem.* 1975, 411(2), 148.
27. Tabor, C. W.; Tabor, H., *Adv. Enzymol. Relat. Areas Molec. Biol.* 1984, **56**, 251.
28. Ma, Q.; Reynolds, M. A.; Kenyon, G. L., *Bioorg. Chem.* 1989, **17**, 194.
29. Riesel, L.; Willfahrt, M.; Grosse, W.; Kindscherowsky, P.; Chodak, V. A.; Kabachnik, M. I. Z., *Anorg. Allg. Chem.* 1977, **435**, 61.
30. Ma, Q.; Bathurst, I. C.; Barr, P. J.; Kenyon, G. L., *J. Med. Chem.* 1992, **35**, 1938.
31. Li, R.; Muscate, A.; Kenyon, G. L., *Bioorganic Chemistry* 1996, **24**, 251.
32. Slater, J. P.; Tamir, I.; Loeb, L. A.; Mildivan, A. S., *J. Biol. Chem.* 1972, **247**, 6784.
33. Hizi, A.; Tal, R.; Shaharabany, M.; Loya, S., *J. Biol. Chem.* 1991, **266**, 6230.

CHAPTER 7

From Nerve Agent to Anticancer Drug: The Chemistry of Phosphoramide Mustard

SUSAN M. LUDEMAN

Duke Comprehensive Cancer Center and Department of Medicine,
Duke University Medical Center

The term *cancer* is applied to a collection of about 100 diseases with the common characteristic of abnormal cells that grow rapidly, invade adjacent tissues, and often disseminate to other locations, forming secondary tumors.[1] These secondary growths, or metastases, may occur at great distances from the primary tumor, and in early stages they are often difficult to detect. Localized treatments such as surgery and radiation generally focus on the primary tumor and can miss metastatic sites. Chemotherapy, on the other hand, provides the possibility of destroying cancer cells at any location in the body. Of course, the circulation of therapeutics throughout the body exposes normal cells as well as cancer cells to the drugs, and this often results in debilitating side effects. The challenge, then, in chemotherapy is to design drugs that will act selectively on cancer cells without harming healthy cells.

ALKYLATING AGENTS AND DNA

Much of the success that has been realized within cancer chemotherapy has resulted from strategies in which DNA function is disrupted. Of the drugs that target DNA, alkylating agents are among the most clinically effective.[2,3] Alkylating agents are compounds that have electrophilic alkyl groups in the parent structure or that generate such groups through metabolism. Alkylation occurs when the reactive alkyl moiety is attacked by a nucleophile resulting in the formation of a covalent bond. In DNA, the nitrogenous bases have many electron rich sites. To determine how the

Biomedical Chemistry: Applying Chemical Principles to the Understanding and Treatment of Disease, Edited by Paul F. Torrence
ISBN 0-471-32633-x

alkylation of a base in DNA can result in cell death, one must understand the chemistry of the alkylator and how this affects the biological function of DNA.

NITROGEN MUSTARDS

Alkylating agents are subclassified by chemical structure into a number of groups; of these, the nitrogen mustards are the oldest and the most widely used clinical alkylators.[4] Ironically, nitrogen mustards are historically based in chemical warfare.[5] The original "mustard gas" (sulfur mustard) was used during World War I, and among its many toxic effects was leukopenia (suppression of white blood cells). Nitrogen mustards, dating from World War II, also lowered white blood cell levels and were found to attack rapidly dividing cells (a characteristic of cancer cells). From these observations, it was deduced that such compounds might be useful against leukemia and in 1942, the nitrogen mustard mechlorethamine ($R = CH_3$) was used in a clinical trial.[6] This drug, commonly known as "nitrogen mustard," was the first synthetic compound that elicited an anticancer response in humans and its use marked the beginning of modern cancer chemotherapy.[5]

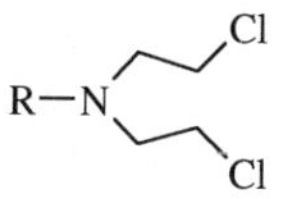

R—N Cl Cl

S Cl Cl

Nitrogen Mustards 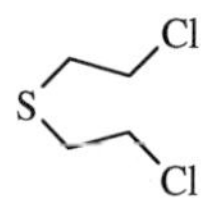 Sulfur Mustard

While the clinical use of mechlorethamine was historical, its success was tempered by the realization that drugs with greater selectivity for cancer cells were needed. When searching for improved drugs, it is common to make systematic structural changes to the parent molecule and then to compare the biological activities of the new compounds with that of the original.[7] Through such methodical variations, the structural facets necessary to a therapeutic response can be determined. At the same time, those moieties that can be altered without detrimental effects or, more importantly, with beneficial effects can also be pinpointed. Applying this method of analog design relative to mechlorethamine, a unifying result among the many compounds synthesized was that both chloroethyl chains were required for the greatest cytotoxicity. This bifunctional requirement becomes understandable when considering the mechanism by which nitrogen mustards alkylate nucleophiles in general and DNA in particular.

Nitrogen mustards alkylate through a two-step mechanism that is initiated by the *intra*molecular nucleophilic attack of nitrogen on the β carbon of the chloroethyl chain with the concomitant loss of chloride. Subsequently, the intermediate aziridinium ion undergoes an *inter*molecular attack by a nucleophile resulting in ring opening and alkylation of the nucleophile [Eq. (7.1)].[4,8] This second step is driven by electronic as well as steric factors (i.e., relief of charge and ring strain). Repetition of this sequence with the second chloroethyl chain and another nucleophile results in bisalkylation [Eq. (7.2)]. In the case of DNA, the two nucleophiles are electron-rich nitrogen centers of nucleotide residues, one on each strand of a double helix. The net result of bisalkylation in DNA is an interstrand crosslinkage that "ties together" the two strands of the helix. Because DNA replication requires the separation of the strands, crosslinking prevents this process and thereby promotes cell death.

R–N(CH₂CH₂Cl)₂ —(–Cl⁻)→ aziridinium ion (R–N⁺) + :Nu⁻ → R–N(CH₂CH₂Nu)(CH₂CH₂Cl) (7.1)

R–N(CH₂CH₂Nu)(CH₂CH₂Cl) —(–Cl⁻)→ aziridinium ion (R–N⁺) + :Nu⁻ → R–N(CH₂CH₂Nu)₂ (7.2)

For the alkylating agents, the ability to crosslink DNA is generally accepted as an important factor in achieving a therapeutic response. Related compounds that are monofunctional and, therefore, alkylate only one strand of DNA ("point mutations") are known to result in mutagenic effects.[9] Such changes in DNA function without an attendant cell death can lead to new occurrences of cancer.

PHOSPHORAMIDIC MUSTARDS

When it was determined that two chloroethyl chains were intricate to crosslinking and the therapeutic activity of mustards, the focus of structure modification became centered on the nitrogen mustards rather than sulfur mustard. Because of the differences in valency between sulfur and nitrogen, the nitrogen mustards offered greater opportunities for improved designs through variations of the third ligand, R [Eq. (7.1)]. This group could be considered the carrier moiety: one that not only would serve to deliver the active group to the desired target but also could control the alkylating activity of the mustard by preventing alkylation until the target site was reached.

It can be argued that chemical reactivity is dependent on one or both of two factors: stereochemistry (including sterics) and electronics. For the nitrogen mustards, it is particularly easy to modulate reactivity through electronics because the first step in the alkylation sequence is directly related to the nucleophilicity of the nitrogen atom. For simple alkyl nitrogen mustards (R = alkyl), the nitrogen is reasonably nucleophilic as reflected by the fact that aziridinium ion formation is generally facile under physiological conditions (pH 7.4, 37 °C).* To decrease the rate of this reaction, then, one need only decrease the nucleophilicity of the nitrogen. A comparison of the chemistry of amines and amides provides one example of how this might be accomplished.

The structural feature underlying the chemistry of nitrogen in amines as well as amides is the ability of this atom to share its lone pair of electrons. In a simple alkyl amine, this electron pair is localized at nitrogen and, therefore, is generally available for sharing with an electron-deficient center. Conversely, the same electrons on an amide nitrogen are delocalized through resonance with the carbonyl group. In fact, the C–N bond order of a simple amide is typically around 1.5, reflecting the significant double-bond character of this linkage. As a result, amide nitrogens are not very basic or nucleophilic relative to amine nitrogens.

Thus, on the basis of the chemical principles outlined above, it could be predicted that incorporating the nitrogen mustard functionality into an amide [e.g., R = RC(*O*)] would diminish if not totally eliminate the alkylating capabilities of the mustard by decreasing the nucleophilicity of the nitrogen atom. But there is another factor to consider: How could the alkylating moiety in such an analog be "activated" in the targeted cancer cell? The goal is to control alkylating activity, not eliminate it!

In 1948, it was reported that phosphamidase, an enzyme thought to catalyze P–N bond hydrolysis in phosphoramides [$R_2P(O)$-NR_2], was present in relatively high concentrations in certain tumor cells.[10,11] It was believed that the prevalence of such an enzyme in cancer cells presented an opportunity for selective drug activation. This belief was based on the knowledge that the chemistry of phosphoramides paralleled that of "regular" amides. By analogy to the electronic nature of amides, therefore, it was predicted that phosphoramidic mustards would be ineffective alkylating agents due to resonance of the mustard nitrogen with the electron withdrawing phosphoryl group [Eq. (7.3)]. However, these compounds would be metabolized preferentially in tumor cells to an active alkylator, nornitrogen mustard (NNM), by the action of phosphamidase. As an amine with increased nucleophilicity at nitrogen, NNM would spontaneously cyclize and begin the alkylation sequence.

$$-P(O^-)=N^+(CH_2CH_2Cl)_2 \longleftrightarrow -P(=O)-\ddot{N}(CH_2CH_2Cl)_2 \xrightarrow{\text{Phosphamidase}} -P(=O)-OH + H\ddot{N}(CH_2CH_2Cl)_2 \quad (7.3)$$

Phosphoramidic Mustards

Nornitrogen Mustard

This design strategy led to the synthesis of hundreds of candidate compounds and cyclophosphamide (CP; clinically known as *Cytoxan*) emerged as the phosphor-

*The electron-withdrawing nature of the chloroethyl groups makes such amines weak bases. For example, the pK_a value for mechlorethamine (R = CH_3) is 6.8 and that for nornitrogen mustard (R = H) is 7.4. Thus, the nitrogen atom in each of these compounds is significantly deprotonated at physiological pH (7.4) and the electron pair is available to participate in the formation of an aziridinyl ring.

amidic mustard with the best combination of cytotoxicity and selectivity.[11,12] As predicted, the nitrogen mustard group in CP was "deactivated" through resonance and the parent drug itself had no alkylating activity. Metabolism was required for a therapeutic response.

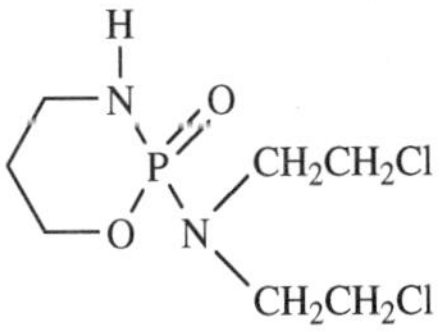

Cyclophosphamide; CP

It has been four decades since CP was first synthesized and during this time, no other phosphoramidic mustard has proved more effective. In fact, CP has become the most widely used of any chemotherapeutic drug in that it is clinically beneficial against the broadest range of human cancers.[3] This is indeed amazing because the phosphamidase premise of activation that contributed to the design of this compound was subsequently found to be invalid.[11]

Metabolism of Cyclophosphamide

Cyclophosphamide (CP) was designed as a prodrug, that is, as a compound that itself would not act as an alkylating agent because of decreased nucleophilicity of the mustard nitrogen. As predicted, CP per se had no alkylating activity, but it did undergo enzyme-mediated activation to a compound with potent cytotoxicity. It was unexpected, however, that this metabolism was not catalyzed by a hydrolytic phosphamidase but rather that it was the result of oxidative hepatic P-450 enzymes.

The oxidation of CP by P-450 triggers a cascade of metabolites, each with many possibilities for undergoing additional chemical reactions both spontaneous and enzymatic.[4,13] Scheme 7.1 is an abbreviated version of this metabolism but one that includes the most important contributors to CP toxicity and selectivity. The metabolism of CP is initiated by an enzyme-mediated oxidation at the C4 position to give 4-hydroxycyclophosphamide (4-HO–CP). The hemiaminal moiety [R–CH(OH)–NR_2] in 4-HO–CP allows for a spontaneous (and reversible) ring-opening reaction to give its tautomer aldophosphamide (AP). This interconversion of 4-HO–CP and AP is strictly analogous to the ring-opening and -closing reactions in glucose that occur through a hemiacetal [R–CH(OH)–OR].

AP is a pivotal metabolite that partitions between pathways of detoxification and cytotoxicity. In the presence of the enzyme aldehyde dehydrogenase (ALDH), AP is oxidized to the nontoxic carboxyphosphamide. In the absence of significant ALDH activity, AP is subject to an α,β-elimination reaction that produces acrolein and phosphoramide mustard (PM), the latter of which is generally believed to be responsible for crosslinking DNA.[4]

Several structural features in AP allow for the facile production of PM at physiological pH: (1) as is true of aldehydes in general, the protons α to the carbonyl

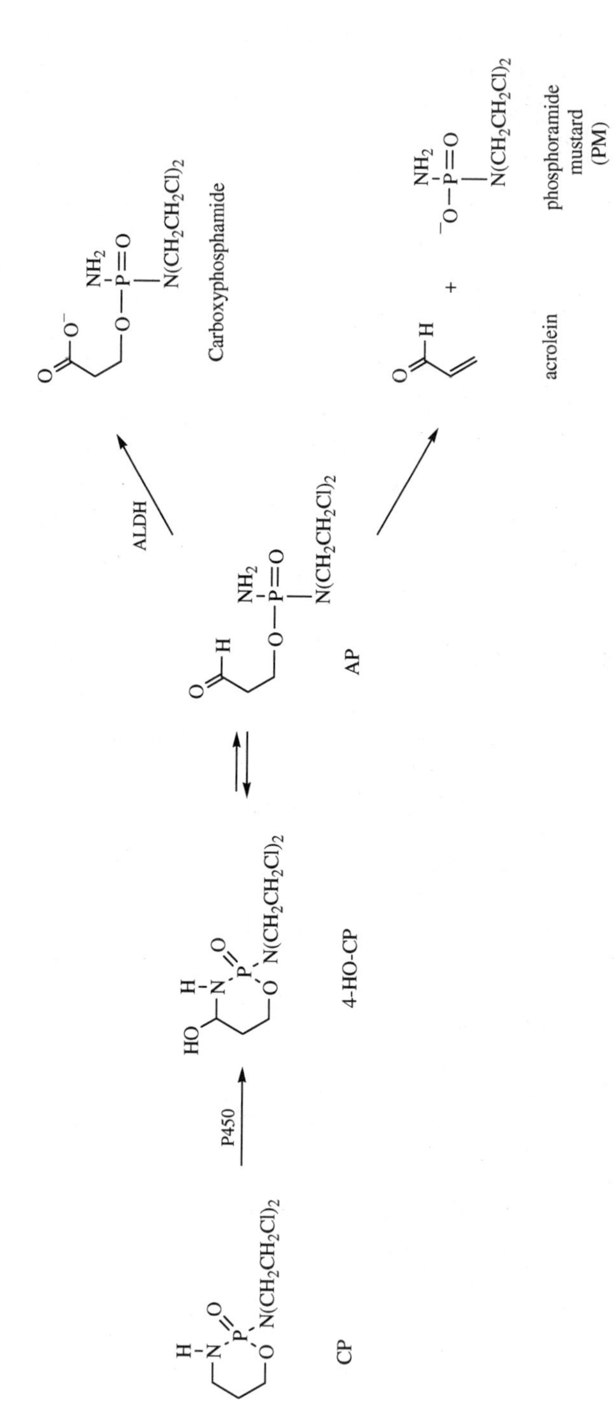
CP
P450
4-HO-CP
AP
ALDH
Carboxyphosphamide
acrolein
phosphoramide mustard (PM)

SCHEME 7.1

in AP are acidic; (2) the elimination reaction provides for a product (acrolein) with conjugation; and (3) PM acts as a good leaving group by virtue of the formation of a resonance stabilized PO_2^- group. An analogous fragmentation does not occur in vivo with carboxyphosphamide because this molecule lacks one of the important structural driving forces, the protons α to a carboxylate ion are relatively nonacidic and cannot easily be removed by the weak bases found in the in vivo environment. To illustrate this point, the half-life* of AP in the elimination reaction is 38 min at pH 7.4, 37 °C, 1 M lutidine buffer;[14] under the same conditions, the half-life for carboxyphosphamide is 7 days (and its mechanism of decomposition may not include elimination).[15]

The detoxification of AP to carboxyphosphamide by ALDH accounts for one mechanism of selectivity associated with CP therapy.[16] In general, normal cells have higher levels of ALDH relative to cancer cells. Thus, the conversion to carboxyphosphamide occurs with greater frequency in healthy cells, thereby sparing them from the production of the cytotoxic PM. On the other hand, the enzyme that protects normal cells can do the same for cancer cells. Those cancers that are resistant to the effects of CP treatment often have high levels of ALDH.

PHOSPHORAMIDE MUSTARD AND ITS MECHANISM OF ALKYLATION

Phosphoramide mustard is the first metabolite generated by CP that is capable of spontaneous alkylation reactions. Its half-life under physiological conditions (pH 7.4, 37 °C) is just 18 min.[17] How is it that the mustard is so reactive in this structure but not in CP? In both PM and CP, the nitrogen atom of the mustard group is part of a phosphoramide and would be expected, at least at first inspection, to be of decreased nucleophilicity relative to an amine nitrogen. Could it be that the mechanism of alkylation for PM is not initiated by the *intra*molecular formation of an aziridinium ion but rather that it is a one-step *inter*molecular S_N2 reaction? The rate of the latter would *not* be independent of the electronic character of the mustard nitrogen, but nucleophilicity of this atom would not play as pivotal a role as it would in the cyclization mechanism. Another consideration is that the aziridinium ion has never been directly detected by any chromatographic or spectroscopic technique. On the other hand, the inability to observe a species does not mean that it does not exist. Its lifetime may be transient or it may be formed in concentrations too low to be observed by current methodologies. Knowing the mechanism of alkylation is important to understanding the total drug chemistry and the nature of the interaction with DNA. For PM, the question of mechanism was readily answered through a labeling experiment using PM synthesized with deuterium in the β positions of the chloroethyl chains.[18]

*Because AP interconverts with 4-HO–CP (and other metabolites), its lifetime is characterized by an "apparent" half-life. The actual half-life, in the absence of any reversible interconversions, would be shorter.[14]

The consequences of the direct substitution versus aziridinium ion mechanism on label positioning are shown in Scheme 7.2. In a one-step substitution reaction, the label is retained in the β position of the ethylene groups in the product. In the cyclization reaction, the two carbons in the chloroethyl chain become equivalent in an unlabeled aziridinium ion intermediate. As a result, each carbon atom can be attacked by a nucleophile with equal probability. In the case of a specifically labeled starting material, minimal secondary kinetic isotope effects would be expected,[19,20] but the key result is label scrambling (as shown in Scheme 7.2).

Reaction of labeled PM with a strong nucleophile in an irreversible reaction provided a product mixture that considered as a whole, revealed an essentially equal distribution of deuterium between the α and β positions of each ethylene moiety.[18] This experiment established that the mechanism of alkylation by PM occurs through aziridinium ion intermediates.

Theoretical Studies of Phosphoramide Mustard Structure

Having established experimentally that alkylation by PM occurs through an aziridinyl intermediate, the question remains as to why this mustard nitrogen is nucleophilic while the corresponding nitrogen in CP (or any intervening metabolite) is not. The answer must be based in electronics and a comparison of bond lengths gives some clue as to the electronic differences between PM and CP. The data shown in the structures below are derived from the crystal structure of CP (as the hydrate)[21,22] and ab initio calculations for gas-phase structures of PM.[23] While these bond length values [in angstroms (Å)] were determined by very different techniques, comparisons can be based on the finding that theoretical calculations of CP structure are consistent with the experimental crystallograpy results.[23]

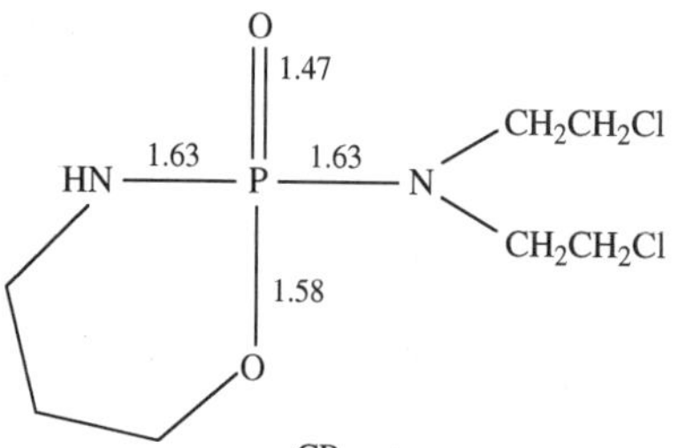

O
1.47
HN 1.63 P 1.63 N CH_2CH_2Cl CH_2CH_2Cl
1.58
O

CP

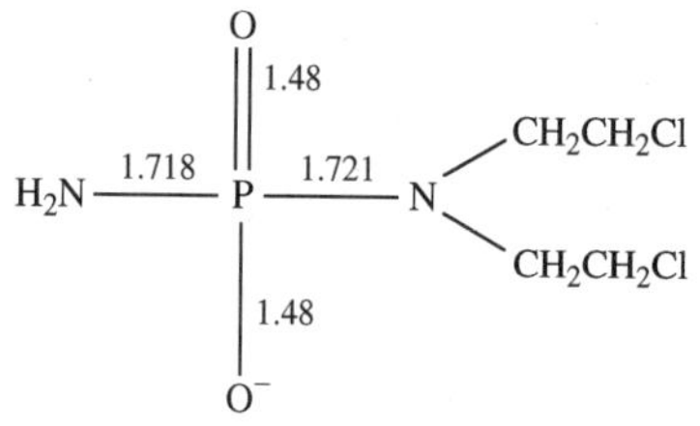

O
1.48
H_2N 1.718 P 1.721 N CH_2CH_2Cl CH_2CH_2Cl
1.48
O^-

PM

The bond between the phosphorus atom and the nitrogen of the mustard group is shorter in CP than in the anionic PM (the structure that is relevant under physiological conditions). This suggests that there is less double-bond character to this bond in PM and, therefore, that the lone pair of electrons on the mustard nitrogen is more localized relative to that in CP. The decrease in a resonance contribution from the mustard nitrogen in PM is "offset" by an increase in the participation of an oxygen atom. The bond representing P–O^- in PM is shorter relative to its counterpart (P–OR) in CP.

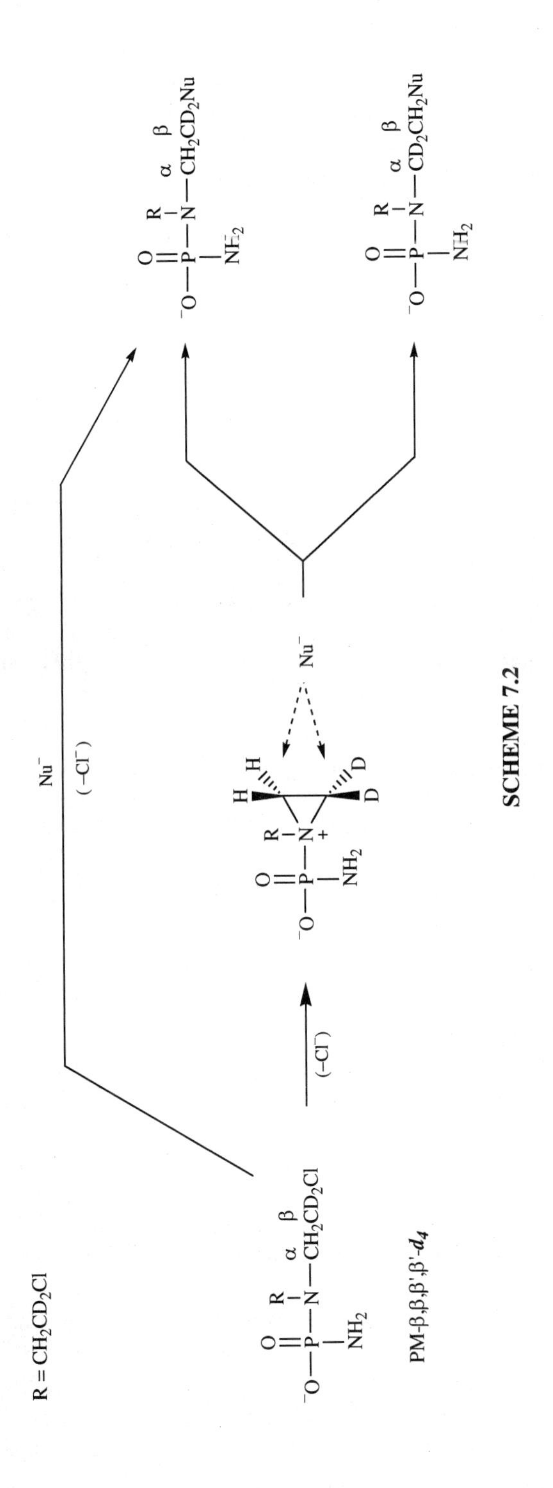

SCHEME 7.2

That these electronic characteristics reflect an increased reactivity of the mustard nitrogen in PM can be seen in theoretical calculations of charge density.[23] The atomic charge for the nitrogen mustard in PM under aqueous conditions has been calculated to be −0.56. Similar data are currently unavailable for CP, but the neutral structure of PM, as shown in the following diagram with a P–OH group, reveals great similarity with CP in terms of bond lengths:

O
1.459
CH_2CH_2Cl
H_2N 1.656 P 1.654 N
CH_2CH_2Cl
1.601
OH

The calculated atomic charge for the mustard nitrogen in the neutral structure is −0.38. This may be used as an indicator of the charge on the corresponding atom in CP.

While caution must be exercised when trying to extrapolate gas-phase data to the aqueous phase and also when using models, these data suggest that the anionic nature of PM under physiological conditions contributes to the increased reactivity of the alkylating moiety relative to that in CP.

Theoretical Studies of Aziridinium Ion Structure

To understand the total chemistry of PM, it is of interest to examine the structure of the aziridinium ion formed in the first step of the alkylation mechanism. Using ab initio calculations, gas-phase, equilibrium structures were determined for ethyl analogs of PM and its aziridinium ion. The calculated bond lengths (Å) are:

O
1.482
CH_2CH_3
H_2N 1.727 P 1.705 N
CH_2CH_3
1.484
O^-

O
1.485
H_2N 1.668 P 1.887 $^+$N
CH_2CH_3
1.462
O^-

At the time of this analysis, the ethyl analogs were used to reduce the complexity of the calculations. Reported after this analysis were similar calculations for PM, as shown in the preceding section.[23] The good agreement between the results for PM and its diethyl congener support the reliability of the ethyl analogs as models.

As stated earlier, alkylation occurs because of electronic and ring strains associated with the three-membered ring. An examination of the bond lengths in the model aziridinium ion reveals another structural factor that profoundly influences the stability of this intermediate. Formation of an aziridinium ion increases the bond length between phosphorus and the aziridinyl nitrogen from 1.703 to 1.887 Å. By way of comparison, the anion of phosphoramidic diacid $[(H_3N^+)PO_3^{2-}]$ has been cited as having a "special place in P–N chemistry" because this ion is considered to have an authentic P–N single bond free of π contributions (bond length = 1.77 Å).[24] The P–N bond in the aziridinium model, therefore, would be predictably unstable because of its unusual length as well as its decreased double-bond character. Notably, increased susceptibility to hydrolytic bond cleavage would be pH-independent. This is in contrast to any P–N bond hydrolysis that might occur in PM itself, prior to aziridinium ion formation.[25]

The bond angles calculated for the ethyl aziridinium ion model in its gas-phase, equilibrium geometry are shown in the diagrams following this paragraph. An examination of these angles reveals that the phosphorus atom, the two oxygens and the nitrogen of the NH_2 group are nearly coplanar. As shown in the illustration on the left, the sum of the three bond angles made by these atoms is 350.3°. This value approaches the sum (360°) of the bond angles given by a similar array of atoms in a trigonal plane. The calculated bond angles shown on the right indicate that the aziridinyl moiety is nearly perpendicular to the other three ligands; this geometry is very vulnerable to a hydrolysis mechanism in which the aziridinyl moiety leaves from an apical position through a trigonal bipyramide transition state:*

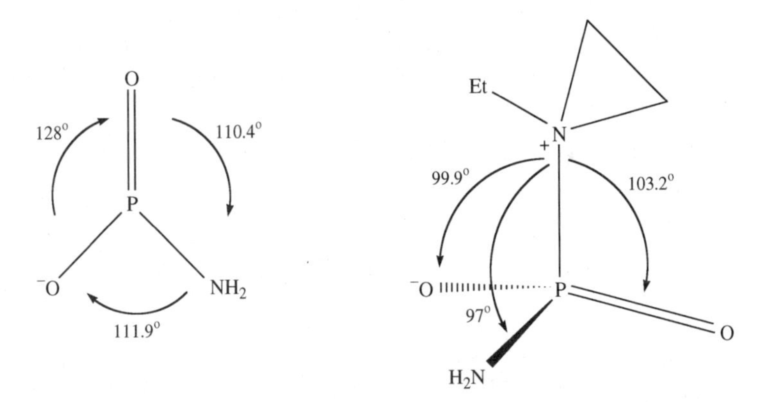

Again it must be stated that caution must be used in applying gas-phase data to the aqueous condition. However, calculations of the type discussed here often give similar results for species in gas and solution phases. Given this caveat, these structural data reveal that the chemistry of PM and its aziridinium ions must include considerations not only of alkylation but also of P–N bond hydrolysis. In the first aziridinium ion formed from PM, cleavage of the $P–N^+$ bond would give chloroethylaziridine (CEZ), a compound that has been reported as a cytotoxic CP metabolite and which is a bisalkylating agent itself.[26,27]

*In our original presentation of this data, two of the bond angles in the aziridinium ion model were inadvertently reversed (97° and 99.9° on the right); the angles are correct as shown here.[34]

PHOSPHORAMIDE MUSTARD AND P–N BOND HYDROLYSIS

Throughout the long written record of CP chemistry, metabolic schemes that have included a P–N bond hydrolysis pathway for PM have shown nornitrogen mustard (NNM) as the product. The intramolecular cyclization of NNM (see the first footnote in this chapter, in section on phosphoramidic mustards) might then account for the formation of the known metabolite, chloroethylaziridine (CEZ).[26,27] While NNM does, in fact, produce CEZ at physiological pH, the theoretical studies discussed in the preceding section provide substantive cause to question whether the formation of CEZ from PM requires the intermediacy of NNM. As a framework for experimental studies, three pathways for CEZ formation can be considered, as shown in Scheme 7.3. Path **A** represents the traditional view. Path **B** is a variation on that and incorporates experimental and theoretical data showing that, under aqueous conditions, the nitrogen of the NH_2 moiety is more likely to become protonated than is the nitrogen of the mustard group (presumably a condition required for hydrolysis).[28] Path **C** reflects the theoretical calculations that predict P–N bond scission of the aziridinium ion (**2**) formed from PM rather than bond cleavage in PM (**1**) itself.

A common feature of paths **A** and **B** is that they both require the formation of NNM. If NNM can be excluded as a product of P–N bond hydrolysis in PM, then both these pathways can be ruled out. A sample of PM at neutral pH was monitored by ^{1}H NMR spectroscopy over timc. NNM could not be detected in any spectrum even though resonances for CEZ were clearly visible. When the NMR sample was

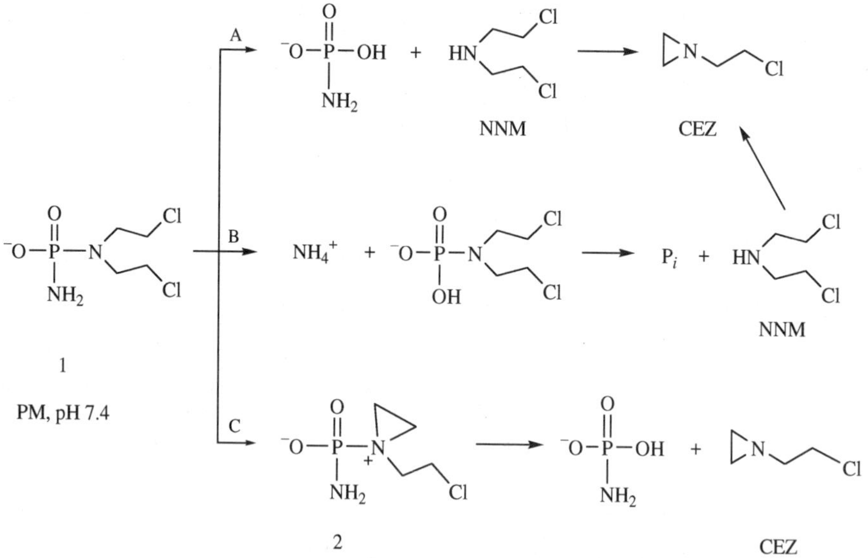

SCHEME 7.3 Pathways for CEZ formation (where P_i = inorganic phosphate). (Reprinted with permission from *J. Med. Chem.* 1998, **41**, 515–529. Copyright 1998 American Chemical Society.)

spiked with authentic NNM, well-resolved signals for NNM appeared where none had been detected previously.

The absence of detectable amounts of NNM was not, however, sufficient to declare paths **A** and **B** invalid. If the the kinetics of NNM formation and disappearance were similar, it was possible that this compound could not be detected in NMR samples of PM because the concentration of transient NNM would be too low to be detected by ^{1}H NMR. To probe this, a mathematical simulation was carried out to determine what the concentration of NNM might be if it were an intermediate by any pathway. For the simulation, it was assumed that the rate of formation of NNM could not be greater than the rate of PM disappearance (half-life $= 18$ min, $k = 0.038\ \mathrm{min}^{-1}$ at pH 7.4, 37°C).[17,29] The rate of disappearance of NNM was determined using authentic material and ^{1}H NMR spectroscopy (half-life $= 17$ min, $k = 0.0411\ \mathrm{min}^{-1}$, pH 7.4, 37°C). Simulations[29] were run assuming a conservative estimate of 10% conversion of PM to NNM. The results of the simulation showed that maximal NNM concentration would occur $\sim$25 min after PM began to react and that this maximal concentration would be $\sim$35% that of the final concentration of CEZ. Since CEZ was readily observed in the NMR spectra, these data suggested that if NNM were the precursor to CEZ, its formation in solution should have been detectable in an unambiguous manner. There were no resonances of the appropriate intensity (or any intensity) for NNM in the ^{1}H NMR spectrum representing $t = 30$ min.

These data provided arguments against the formation of NNM and, therefore, against significant contributions from paths **A** and **B** at neutral pH (Scheme 7.3). Thus, formation of CEZ via path **C** was considered most likely, especially in view of the theoretical data discussed in the preceding section. A simple consideration of the electronic structure of aziridinium ion **2** clearly indicated that the P–N$^+$ bond would be susceptible to cleavage. But perhaps the most compelling argument for the formation of CEZ (at pH 7.4) from aziridinium ion **2** rather than from NNM was derived from an analysis of the ^{31}P NMR experiments discussed in a following section. These experiments demonstrated that the incidence of P–N bond scission in PM was dependent on the strength and concentration of added nucleophiles. Direct hydrolysis of PM to NNM would be a process uninfluenced by the presence of nucleophiles other than water.

THE PARTITIONING OF PHOSPHORAMIDE MUSTARD AMONG ALKYLATION AND P–N BOND HYDROLYSIS REACTIONS

In the studies described above, experimental (NMR) as well as theoretical (quantum-chemical calculations) data were obtained which demonstrate that P–N bond scission in PM occurs under neutral conditions. Most notably, the evidence also supports the conclusion that this P–N bond hydrolysis does *not* occur in the parent molecule to produce NNM. Rather, it is the aziridinium ion (**2**) formed from PM that hydrolyzes to give chloroethylaziridine (CEZ). These data raise important questions with respect to the mechanism of drug action. Does P–N bond scission in **2** represent a loss of therapeutic activity by siphoning drug away from the alkylation pathway? Or does it constitute the production of an additional DNA alkylating agent, CEZ? Furthermore, which pathway is predominant, and how might variations in conditions affect the incidence of hydrolysis versus alkylation? Would nucleophile strength and concentration impact the product distribution? At first inspection, one might expect that the aziridinium ion would be more likely to undergo an alkylation reaction in the presence of a strong nucleophile. But many nucleophiles in vivo are relatively weak and, of course, water itself can act as a nucleophile. How much do these compounds participate in alkylation? How much therapeutic alkylation actually takes place? These issues were probed through NMR investigations that quantified the partitioning of PM (**1**) among spontaneous alkylation and P–N bond hydrolysis reactions as a function of nucleophile strength and concentration.

The complex set of spontaneous reactions available to PM (**1**) will be broadly characterized as alkylation and P–N bond hydrolysis reactions as illustrated in Scheme 7.4, where an aziridinium ion is the branch point for these competing pathways. Alkylation reactions will include the reaction of any nucleophile with an aziridinium ion (**2, 4, 9**) to give an acyclic product. In Scheme 7.4, these alkylation reactions are illustrated for the tripeptide glutathione (GSH), the thiol group of which is a strong nucleophile, and for water, which is ubiquitous and representative of a weak nucleophile. Alkylation of GSH by PM represents an irreversible detoxification and another mechanism of resistance in cells with elevated levels of GSH.[16,30] At one time, alkylation reactions between GSH and PM were reported not to occur in the absence of enzymatic assistance;[31] subsequently, it was shown that such reactions are spontaneous.[32] The alkylation sequences are **2** → **3** → **5**; **2** → **8** → **10**; and **2** → **3/8** → **6**. P–N bond hydrolysis reactions are those involving hydrolytic cleavage of the $P–N^+$ bond in an aziridinium ion to produce CEZ (or an analog thereof) and phosphoramidic diacid (**2/4/9** → **7**). As concluded in the previous section, direct P–N bond scission of PM giving NNM is not significant at neutral pH and, therefore, is not included in the scheme for PM chemistry.

^{31}P NMR Studies of PM Reactivity

Using ^{31}P NMR, the reactivity of PM (11 mM) was studied in the presence of zero, 11, 22, 55, and 110 mM GSH at 37 °C, pH 7.4, and using a nonnucleophilic buffer (0.22 M BisTris). Signal identifications were based on a number of considerations, including (1) use of authentic materials, (2) timeframes for signal disappearance and/or growth, and (3) similar work in the literature.[17,18,28,29,32,33] Because of the sensitivity of ^{31}P NMR chemical shifts to pH, temperature, concentration, and solvent, literature values of chemical shifts for a given compound may vary somewhat.

With 110 mM GSH, the resonance for PM (δ 13.20) decreased with the concomitant growth of a major, but transient, resonance at δ 13.29 (Fig. 7.1). The disappearance of this intermediate was linked to the formation of a reasonably stable product at δ 13.33. This sequence, well documented for thionylations of PM, represented the conversion of PM to monoglutathionylated **3** and, ultimately,

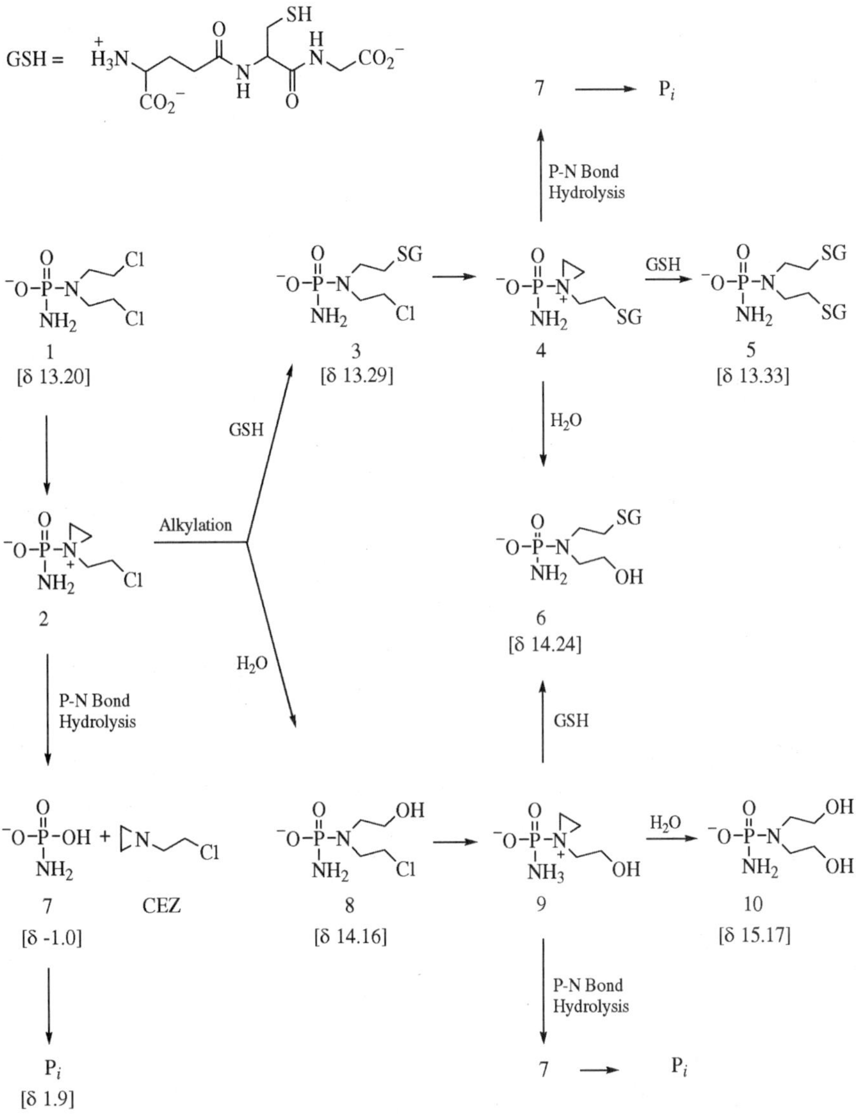

SCHEME 7.4 Alkylation and P–N bond hydrolysis reactions (where P_i = inorganic phosphate; ^{31}P chemical shifts shown in brackets). (Reprinted with permission from *J. Med. Chem.* 1998, **41**, 515–529. Copyright 1998 American Chemical Society.[34])

bisglutathionylated **5**.[17,18,29,32] An additional signal at δ 14.24 appeared during the course of the reaction, and the persistence of this resonance indicated that it represented a stable end product (half-life much greater than that of any precursor). Even after 2 h of reaction time, however, this signal accounted for only 4% of the total phosphorus intensity, while that for bisglutathionylated **5** was 72% (Fig. 7.1). The signal did not appear in spectra of solutions without GSH. These data were

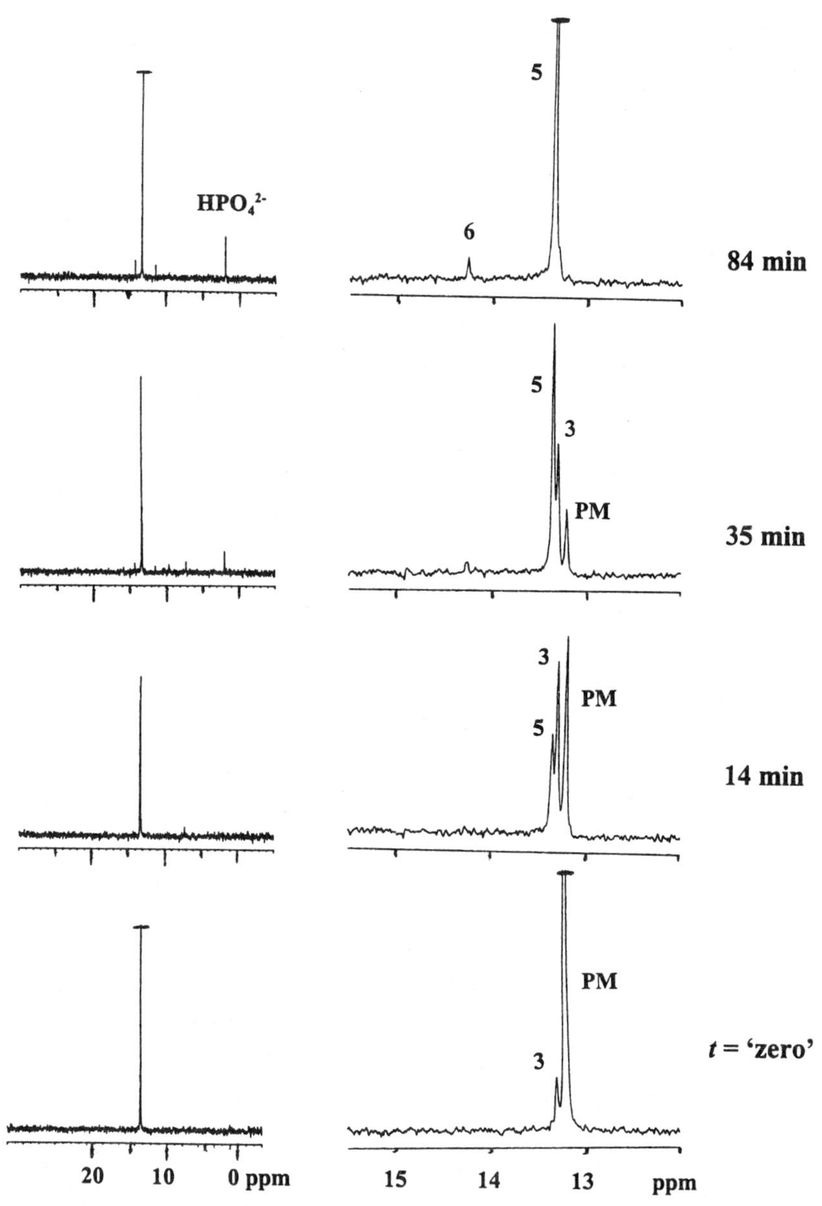

FIGURE 7.1 ^{31}P NMR (202.5 MHz) time-averaged spectra recorded as a function of time for the reactions of 11 mM PM with 110 mM GSH in BisTris (0.22 M, pH 7.4) at 37 °C. The designated times refer to the start of spectral acquisition; time "zero" indicates several minutes after placement of the sample in the NMR probe (and 10–15 min after sample preparation). Left-hand column: full display. Right-hand column: partial display, expanded to show individual signals. Chemical shifts (δ): PM, 13.20; **3**, 13.29; **5**, 13.33; **6**, 14.24; inorganic phosphate (P_i), 1.9. Signals shown "cropped" at times zero and 84 min were off scale on these displays. (Reprinted with permission from *J. Med. Chem.* 1998, **41**, 515–529. Copyright 1998 American Chemical Society.[34])

consistent with assignment of the resonance to monoglutathionylated, monohydroxylated **6**, the product formed from two sequential alkylation reactions involving different nucleophiles (i.e., GSH and water).

In the absence of any GSH, one NMR experiment was initiated quickly enough such that the only signal observed in the "time zero" ^{31}P spectrum was that for PM (Fig. 7.2). Two additional signals appeared simultaneously in the spectrum taken at $t = 7$ min. These resonances were attributed to monohydroxylated intermediate **8** (δ 14.16) and P–N bond hydrolysis product **7** ($\delta - 1.0$), as supported through considerations of kinetic characteristics, literature reports[17,29,32,33] and/or use of authentic material (**7**).[28] In subsequent spectra, signals appeared that were consistent with bishydroxylated **10** (δ 15.17) and inorganic phosphate (δ 1.9).

Spectra obtained from solutions of PM with 11 mM GSH contained features of those obtained with 0 and 110 mM GSH. Increasing concentrations of GSH (22 and 55 mM) provided spectra that were more similar to those obtained with 110 mM GSH. This trend resulted from a decreased influence of water as a nucleophile and an increase in the product distribution derived from reactions in which GSH was the nucleophile (Fig. 7.3).

Kinetic Analysis of NMR Data for Phosphoramide Mustard Reactivity

A kinetic model[34] was constructed consisting of a system of rate equations that mathematically described the precursor-product pathways shown in Scheme 7.4. The analysis utilized ^{31}P NMR peak height data as a measure of component concentration.

Table 7.1 ("A" columns) shows the distribution of reaction products as predicted by kinetic modeling of the NMR data. The distribution estimates reflect the fact that, except at the highest GSH concentration, there was appreciable reduction in the GSH concentration in the reaction mixtures over time as a result of consumption of GSH in the alkylation reactions. In the presence of only a weak nucleophile, namely, water (55 M), it was predicted that just 10% of the PM completed the bisalkylation pathway giving bishydroxylated **10**. The remainder of the PM ultimately gave rise to products of P–N bond hydrolysis (69% CEZ and 21% CEZ analogs). Increasing concentrations of a stronger nucleophile, specifically, GSH, resulted in a decrease in the generation of P–N bond hydrolysis products and an increase in the formation of bisalkylation products and, in particular, those products derived from alkylation of the stronger nucleophile [i.e., **6** and, especially, **5**].

Table 7.1 ("B" columns) also shows the calculated distribution of reaction products *given constancy of the GSH concentrations* (as would obtain were the PM concentration low relative to the initial GSH concentration). Comparing the data in columns A and B, the results are virtually the same under conditions where GSH consumption is negligible (i.e., [GSH] $\gg$ [PM]; 55 and 110 mM GSH). When [GSH] $\approx$ [PM], there is more variance. This is a reflection of the fact that the data in column B are based on a constant GSH concentration while those in column A are not.

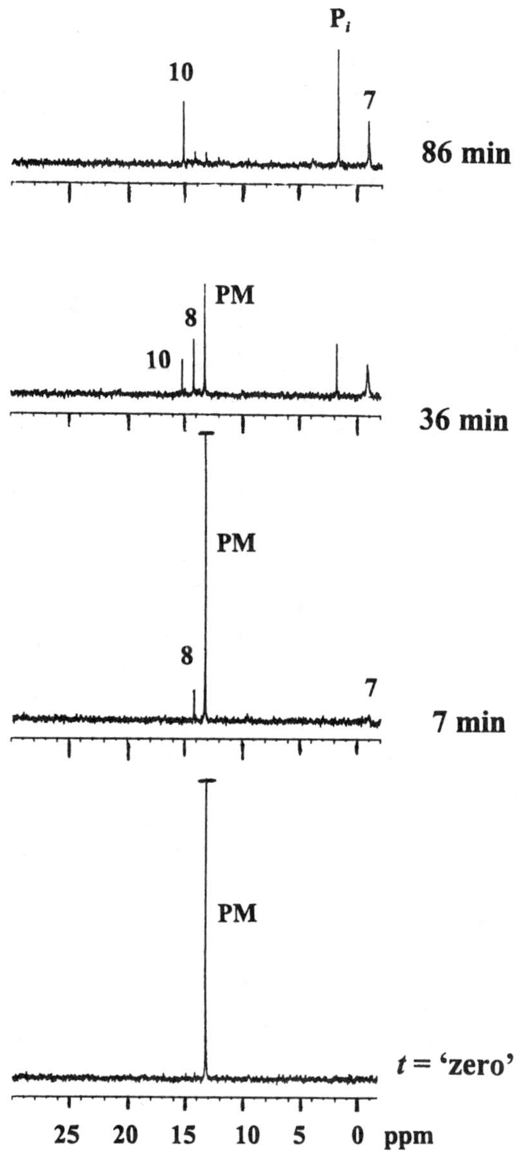

FIGURE 7.2 ^{31}P NMR (202.5 MHz) time-averaged spectra recorded as a function of time for the reactions of 11 mM PM in BisTris (0.22 M, pH 7.4) at 37 °C. The designated times refer to the start of spectral acquisition; time "zero" indicates several minutes after placement of the sample in the NMR probe (and 10–15 min after sample preparation). Chemical shifts (δ): PM, 13.20; **8**, 14.16; **10**, 15.17; **7**, – 1.0; inorganic phosphate (P_i), 1.9. Signals for PM at times zero and 7 min were off-scale on these displays. (Reprinted with permission from *J. Med. Chem.* 1998, **41**, 515–529. Copyright 1998 American Chemical Society.[34])

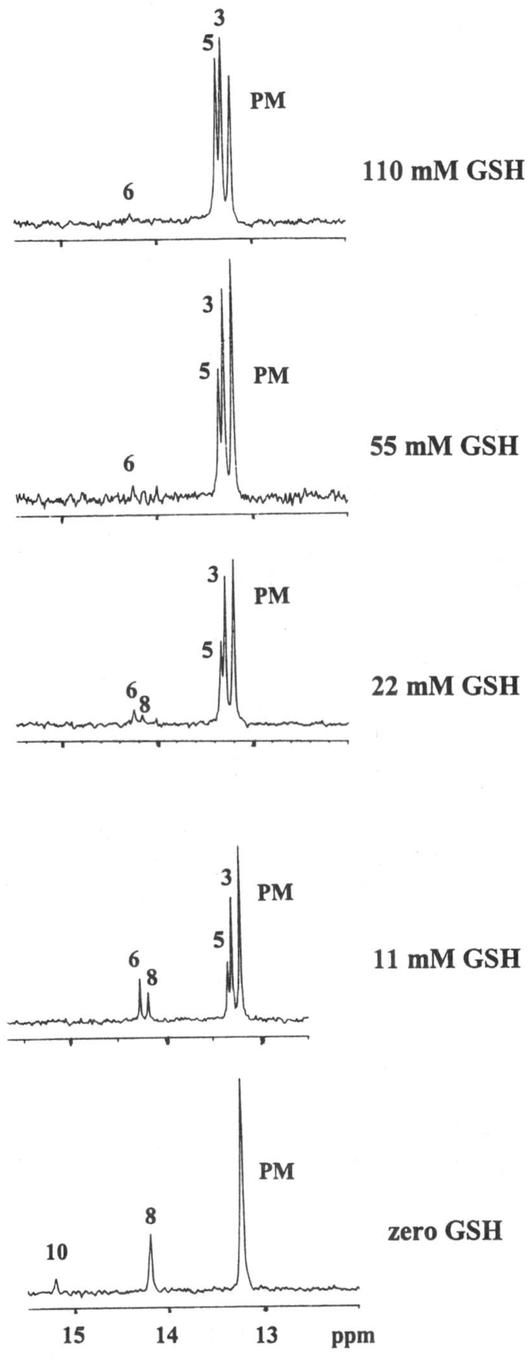

FIGURE 7.3 ^{31}P NMR (202.5 MHz) time-averaged spectra recorded for the reactions of 11 mM PM with varying initial concentrations of GSH in BisTris (0.22 M, pH 7.4) at 37 °C. Each spectrum shown was recorded at $t = \sim 21$–22 min (slightly greater than one half-life for PM). (Reprinted with permission from *J. Med. Chem.* 1998, **41**, 515–529. Copyright 1998 American Chemical Society.[34])

TABLE 7.1 Distribution (%) of the Alkylation End Products Predicted for Reactions of 11 mM PM in 0.22 M BisTris at pH 7.4, 37 °C

mM GSH	Compound **5**		Compound **6**		Compound **10**		P–N Bond Hydrolysis	
	A[a]	B[b]	A	B	A	B	A	B (CEZ)[c]
0	NA[d]	NA	NA	NA	10	10	90	90(69)
11	22	41	19	14	4	1	55	44(25)
22	45	61	11	11	< 1	< 1	44	28(15)
55	55	59	NM[e]	11	NM	< 1	45	30(16)
110	75	76	NM	7	NM	< 1	25	17(9)

[a] For the data in the columns labeled "A", modeling was done assuming that the GSH concentration decreased over time as GSH was consumed in the alkylation reactions.

[b] For the data in the columns labeled "B", modeling was done assuming constancy of GSH concentration.

[c] P–N bond hydrolysis can occur in any aziridinium ion in Scheme 7.4. The first number in column B reflects the total amount of PM that eventually undergoes P–N bond hydrolysis. The number in parentheses is specifically the amount of CEZ formed.

[d] Not applicable.

[e] Not modeled.

Based on the reaction partitioning given in Table 7.1, then, the general predictions and implications for reactions of PM (11 mM) at pH 7.4 are as follows:

1. In the presence of a high, unchanging concentration of a weak nucleophile (55 M H_2O), only 10% of the original PM is predicted to successfully undergo bisalkylation of the weak nucleophile (water). A large percentage (90%) of the original PM undergoes P–N bond hydrolyis through aziridinium ion intermediates, 69% of which is converted to the bisalkylator CEZ (and with 21% accounting for CEZ analogs that are not bisalkylating agents).
2. In the presence of a high, unchanging concentration of a strong nucleophile (110 mM GSH), 83% of the original PM is predicted to undergo bisalkylation while 17% of the parent drug goes on to provide for products of P–N bond hydrolysis (9% of which is CEZ).

The distribution of the alkylation products is heavily weighted toward bisalkylation of the strong nucleophile (76%). Concentrations of GSH between the two extremes give intermediate values of alkylation and P–N bond hydrolysis. Under conditions that relate to intracellular concentrations of GSH (3–5 mM), the extent of P–N bond hydrolysis would be significant.

While this study used a very strong (GSH) and a very weak (H_2O) nucleophile, there is literature to support the generality of the data and their implications. Lu and Chan have reported a 32% conversion to CEZ for 0.14 mM PM in 67 mM phosphate at pH 7.4.[27] At that low concentration of PM, the concentration of phosphate would, for all practical purposes, be constant. Comparisons can be made, therefore, between the results for phosphate (a nucleophile of moderate strength) and those given in Table 7.1 for GSH under similar conditions (55 mM, column B). With the stronger

TABLE 7.2 Mean Value of the Rate Constant for Aziridinium Ion Formation (pH 7.4, 37 °C)

Reaction	k, min^{-1}
PM → **2**	0.0406
3 → **4**	0.0359
8 → **9**	0.0486

nucleophile GSH, less PM is converted to CEZ (16%). These results are consistent with the conclusion that product distribution is linked to nucleophile strength; in other words, the stronger the nucleophile, the greater the yield of alkylation end products and, necessarily, the lower the yield of P–N bond hydrolysis products (e.g., CEZ).

SUBSTITUENT EFFECTS ON THE RATE OF ALKYLATION

Through monoalkylation, the nucleophile becomes a substituent and, as such, must have an effect on the rate of the second alkylation reaction. Quantification of this influence on PM reactivity is important to an understanding of how monoconjugation may affect subsequent crosslinking and inactivation reactions. In the NMR study described above, the relative effects of Cl, SG, and OH on the second alkylation reaction were determined by calculating the rate constants for the rate-limiting step in each alkylation reaction, that is, the formation of each aziridinium ion **2**, **4**, and **9** (Table 7.2). Relative to Cl as a substituent, SG was found to retard aziridinium ion formation while OH accelerated it. Although this is a trend opposite to what would have been predicted based on electronegativities and their impact on the nucleophilicity of the reacting nitrogen, the same rate retardation effect has been reported for analogous reactions of GSH and nitrogen mustards.[35] In any case, these data illustrate that the rates of the first and second alkylation reactions of DNA could be quite different. If the actual crosslink is formed slowly, this might provide more opportunity for DNA repair (enzymatic removal of the alkyl group).[5]

CEZ REACTIVITY

Relevant to the therapeutic significance of CEZ is its ability to act as an alkylating agent. The reported half-life for CEZ is 20 h in 0.067 M phosphate, pH 7.4, 37 °C.[27] The pK_a of the nitrogen atom in CEZ is 6.62, and the stability of the aziridinyl ring in this molecule, relative to that in aziridinium ion **2**, for example, can be attributed to its being neutral at pH 7.4. In the presence of the strong nucleophile GSH, the stability of CEZ drops significantly; in 0.1 M phosphate in D_2O, pD ~ 7, 37 °C, the half-life of CEZ (1 mM) with GSH (10 mM) is approximately an hour. As would be expected for a bimolecular substitution where attack by a nucleophile is rate-limiting, the alkylating activity of CEZ will be dependent on the strength and concentration of reactant nucleophiles.

SUMMARY

For any anticancer agent, there is generally one pathway that leads to a therapeutic response. Deviations from this pathway are often important contributors to other effects such as detoxification and drug resistance. Factors that govern the mechanistic fate of a drug are complex, but they generally can be explained, at least in part, through considerations of molecular structure and electronics. Such is the case for cyclophosphamide and its metabolites.

Consistent with the chemical basis for its design, CP is a prodrug. Contrary to the biological basis for its design, the active metabolite that is generally believed to be responsible for crosslinking DNA is phosphoramide mustard not nornitrogen mustard. An examination of the chemistry of PM, as described in this chapter, reveals that another metabolite, chloroethylaziridine, must also be considered as a potential crosslinking agent. CEZ was first reported[26] as a cytotoxic CP metabolite in 1968, but it has received scant attention in the literature since then.[27,34,36,37]

The cumulative experimental and theoretical data cited above support the hypothesis that, at neutral pH, P–N bond hydrolysis occurs in aziridinium ion **2** to give CEZ (path **C**, Scheme 7.3). It is noteworthy that hydrolytic P–N bond scission in **2** would be a pH-independent process. This is in contrast to any P–N bond hydrolysis that might occur in PM itself, prior to aziridinium ion formation. Although path **B** (Scheme 7.3) may not contribute to the chemistry of PM at neutral pH, it remains the likely mechanism by which PM hydrolyzes under acidic conditions (e.g., $pH \leq 5$). This is based on the conclusion that the pK_a value of 4.9 that is associated with PM relates specifically to the nitrogen of the NH_2 moiety.[28] The pK_a value for the nitrogen mustard (as well as the acid) is lower. Thus, the NH_2 group will protonate first thereby allowing for the reaction depicted in path **B**.

In view of these findings regarding the production, or lack thereof, of NNM from PM at neutral pH, numerous literature reports must be reconsidered. References to the "well-known formation of NNM from PM" are often based on earlier works that actually are speculative in nature. This problem is compounded when the conditions of pH are not included. In a given analysis, other sources of NNM must also be considered. For example, it has been shown that the CP metabolite carboxyphosphamide hydrolyzes to NNM under acidic conditions; coupled with sample handling, this could account, at least in part, for the NNM found by some (but not others) in patient urine and plasma.[15,27,33,38]

Reactions between GSH and PM are spontaneous and irreversible and, therefore, represent a mode of detoxification. Nucleophile strength and concentration impact on the distribution of alkylation and hydrolysis products given by PM. In short, the stronger and more concentrated the nucleophile, the greater the yield of alkylation end products and, necessarily, the lower the yield of P–N bond hydrolysis products (e.g., CEZ).

The implications of the kinetic data to the in vivo situation are many. Given the multitude of endogenous nucleophiles available, one can imagine the impact on the distribution of possible products in an expanded Scheme 7.4. Cellular phosphate, alone, is expected to participate in a significant way, as has been shown for the alkylation reactions of the nitrogen mustard melphalan.[39] Among all the nucleophiles in vivo, it is hard to predict the relative nucleophilicity of the N7 position of a guanine residue in an intact DNA helix (the site of alkylation by PM[5]—and/or CEZ?). On the other hand, alkylation and P–N bond hydrolysis need not be mutually exclusive pathways when considering the mechanism of DNA cross-linking. Alkylation of a guanine residue by aziridinium ion **2** could provide the initial link; subsequent formation of the second aziridinium ion could lead to loss of the phosphoramidate group. The actual crosslink would be completed through a CEZ-like intermediate with second order kinetics.

Given the reactivity of PM, it is almost a wonder that any of this species survives to crosslink DNA. This is not meant to suggest that PM is not the ultimate alkylator derived from CP but rather that other contributors must also be considered. Because a significant portion of aziridinium ion **2** partitions to hydrolysis, the relevance, if any, of CEZ in the therapeutic efficacy of CP treatment must be investigated.

ACKNOWLEDGMENTS

This work was supported in part by Public Health Service Grant CA16783 from the National Cancer Institute (Department of Health and Human Services). Some material that appears here has been published previously (literature citation 34), and I would like to thank my co-authors on that publication for their permission to excerpt portions for this chapter.* I also thank Dr. Paul F. Torrence, Dr. Michael P. Gamcsik, and Ms. Erica Aldrich-Reck for their suggestions and review. I am particularly grateful to Dr. Sally M. Winston for her organizational advice. This chapter was completed for Shannon.

REFERENCES

1. Prescott, D. M.; Flexer, A. S., *Cancer: The Misguided Cell*, 2nd ed.; Sinauer Associates, Sunderland, MA, 1986, **49**, pp. 36–37.
2. Cooper, M. R.; Cooper, M. R., in *American Cancer Society Textbook of Clinical Oncology*, A. I. Holleb, D. J. Fink, and G. P. Murphy, eds., The American Cancer Society, Atlanta, 1991, pp. 47, 57–58.
3. Colvin, M., in *Cancer Medicine*, 3rd ed., J. F. Holland, E. Frei, D. W. Kufe, D. L. Morton, and R. R. Weichselbaum, eds., Lea and Febiger, Malvern, PA, 1991, pp. 733–754.

*Beginning with the section titled "Theoretical Studies of Aziridinium Ion Structure," portions of each section were excerpted with permission from *J. Med. Chem.* 1998, **41**, 515–529. Copyright 1998 American Chemical Society.[34]

4. Colvin, M.; Chabner, B. A., in *Cancer Chemotherapy: Principles and Practice*; B. A. Chabner, and J. M. Collins, eds., Lippincott, Philadelphia, 1991, pp. 276–313.
5. Pratt, W. B.; Ruddon, R. W.; Ensminger, W. D.; Maybaum, J., *The Anticancer Drugs*; Oxford Univ. Press, New York, 1994, pp. 17–18, 108–112.
6. Gilman, A.; Phillips, F. S., *Science* 1946, **103**, 409.
7. Kier, L. B.; Tute, M. S., in *Principles of Medicinal Chemistry*, 2nd ed., W. O. Foye, ed., Lea & Febiger: Philadelphia, 1986, pp. 39–47.
8. Gamcsik, M. P.; Hamill, T. G.; Colvin, M., *J. Med. Chem.* 1990, **33**, 1009–1014.
9. Saffhill, R.; Margison, G. P.; O'Connor, P. J., *Biochem. Biophys. Acta* 1985, **823**, 111–145.
10. Gomori, G., *Proc. Soc. Exp. Biol. Med.* 1948, **69**, 407–409.
11. Hill, D. L., *A Review of Cyclophosphamide*, Charles C. Thomas, Springfield, IL, 1975, pp. 12–15, 25–26.
12. Arnold, H.; Bourseaux, F., *Angew. Chem.* 1958, **70**, 539–544.
13. Ludeman, S. M.; Zon, G.; Colvin, O. M.; Gamcsik, M. P.; Shulman-Roskes, E. M., *Curr. Topics Med. Chem.* 1993, **1**, 155–171.
14. Zon, G.; Ludeman, S. M.; Brandt, J. A.; Boyd, V. L.; Özkan, G.; Egan, W.; Shao, K.-L., *J. Med. Chem.* 1984, **27**, 466–485.
15. Ludeman, S. M.; Ho, C.-K.; Boal, J. H.; Sweet, E. M.; Chang, Y. H., *Drug Metab. Disp.* 1992, **20**, 337–338.
16. Colvin, O. M., in *Anticancer Drug Resistance: Advances in Molecular and Clinical Research*, L. J. Goldstein, and R. R. Ozols, eds., Kluwer Norwell, MA, 1994, pp. 249–262.
17. Engle, T. W.; Zon, G.; Egan, W., *J. Med. Chem.* 1982, **25**, 1347–1357.
18. Colvin, M.; Brundrett, R. B.; Kan, M.-N. N.; Jardine, I.; Fenselau, C., *Cancer Res.* 1976, **36**, 1121–1126.
19. Springer, J. B.; Colvin, M. E.; Colvin, O. M.; Ludeman, S. M., *J. Org. Chem.* 1998, **63**, 7218–7222.
20. March, J., in *Advanced Organic Chemistry*, 2nd ed., McGraw-Hill, New York, 1977, pp. 204–207.
21. Karle, I. L.; Karle, J. M.; Egan, W.; Zon, G.; Brandt, J. A., *J. Am. Chem. Soc.* 1977, **99**, 4803–4807.
22. Garcia-Blanco, S.; Perales, A., *Acta Crystallogr.* 1972, **B28**, 2647–2652.
23. Millis, K. K.; Colvin, M. E.; Shulman-Roskes, E. M.; Ludeman, S. M.; Colvin, O. M.; Gamcsik, M. P., *J. Med. Chem.* 1995, **38**, 2166–2175.
24. Emsley, J.; Hall, D., in *The Chemistry of Phosphorus*, J. Wiley, New York, 1976, p. 382.
25. Engle, T. W.; Zon, G.; Egan, W., *J. Med. Chem.* 1979, **22**, 897–899.
26. Rauen, H. M.; Norpoth, K., *Klin. Wochenschr.* 1968, **46**, 272–275.
27. Lu, H.; Chan, K. K., *J. Chromatogr. B: Biomed. Appl.* 1996, **678**, 219–225.
28. Gamcsik, M. P.; Ludeman, S. M.; Shulman-Roskes, E. M.; McLennan, I. J.; Colvin, M. E.; Colvin, O. M., *J. Med. Chem.* 1993, **36**, 3636–3645.
29. Boal, J. H.; Williamson, M.; Boyd, V. L.; Ludeman, S. M.; Egan, W., *J. Med. Chem.* 1989, **32**, 1768–1773.

30. Colvin, O.; Friedman, H.; Gamcsik, M.; Fenselau, C.; Hilton, J., *Advan. Enzyme Regul.* 1993, **33**, 19–26.
31. Yuan, Z.-M.; Smith, P. B.; Brundrett, R. B.; Colvin, M.; Fenselau, C., *Drug Metab. Disp.* 1991, **19**, 625–629.
32. Dirven, H. A. A. M.; Venekamp, J. C.; van Ommen, B.; van Bladeren, P. J., *Chem.-Biol. Interact.* 1994, **93**, 185–196.
33. Watson, E.; Dea, P.; Chan, K. K., *J. Pharm. Sci.* 1985, **74**, 1283–1292.
34. Shulman-Roskes, E. M.; Noe, D. A.; Gamcsik, M. P.; Marlow, A. L.; Hilton, J.; Hausheer, F. H.; Colvin, O. M.; Ludeman, S. M., *J. Med. Chem.* 1998, **41**, 515–529.
35. Gamcsik, M. P.; Millis, K. K.; Hamill, T. G., *Chem.-Biol. Interact.* 1997, **105**, 35–52.
36. Chan, K. K.; Zheng, J. J.; Wang, J. J.; Dea, P.; Muggia, F. M., *Proc. Am. Assoc. Cancer Res.* 1994, **35** (1785), 300.
37. Shulman-Roskes, E. M.; Noe, D.; Marlow, A.; Colvin, O. M.; Gamcsik, M. P.; Ludeman, S. M., *Proc. Am. Assoc. Cancer Res.* 1996, **37** (2590), 379.
38. Jardine, I.; Fenselau, C.; Appler, M.; Kan, M.-N.; Brundrett, R. B.; Colvin, M., *Cancer Res.* 1978, **38**, 408–415.
39. Bolton, M. G.; Hilton, J.; Robertson, K. D.; Streeper, R. T.; Colvin, O. M.; Noe, D. A., *Drug Metab. Disp.* 1993, **21**, 986–996.

CHAPTER 8

Phosphoryltyrosyl Mimetics as Signaling Modulators and Potential Antitumor Agents

TERRENCE R. BURKE, JR., YANG GAO, and ZHU-JUN YAO

Laboratory of Medicinal Chemistry, Division of Basic Sciences, National Cancer Institute, National Institutes of Health

BACKGROUND

The Role of PTKs in Communicating Extracellular Information to the Nucleus

Normal cellular function requires the transfer of information from the extracellular environment to the nucleus, where control of cellular response can be regulated by gene transcription. Conceptually, such a process must involve an initial conversion of extracellular stimuli into forms appropriate for intracellular transmission. Such communication is frequently accomplished by means of *signal transduction* pathways, in which stimuli at the cell surface (induced by either environmental factors or specific intercellular messengers) are sequentially recognized at the cell surface, transmitted through the cell membrane, and then carried through the cytoplasm to the cell nucleus.[1]

Eukaryotic cells have developed a remarkable approach toward this problem of information transfer, in which extracellular binding of ligands to cell surface receptors is translated into intracellular chemical signals through the phosphorylation of hydroxyl-containing amino acid residues. These phosphorylations are carried out by *kinases*, which transfer the γ-phosphate of ATP to target hydroxyl residues. Of particular importance for signaling by growth factors and cytokines, are protein–tyrosine kinases (PTKs), which phosphorylate tyrosyl residues.[2] A typical PTK-dependent signaling pathway is shown in Figure 8.1, where binding of a growth factor to its extracellular receptor results in activation of the intracellular PTK,

Biomedical Chemistry: Applying Chemical Principles to the Understanding and Treatment of Disease, Edited by Paul F. Torrence
ISBN 0-471-32633-x

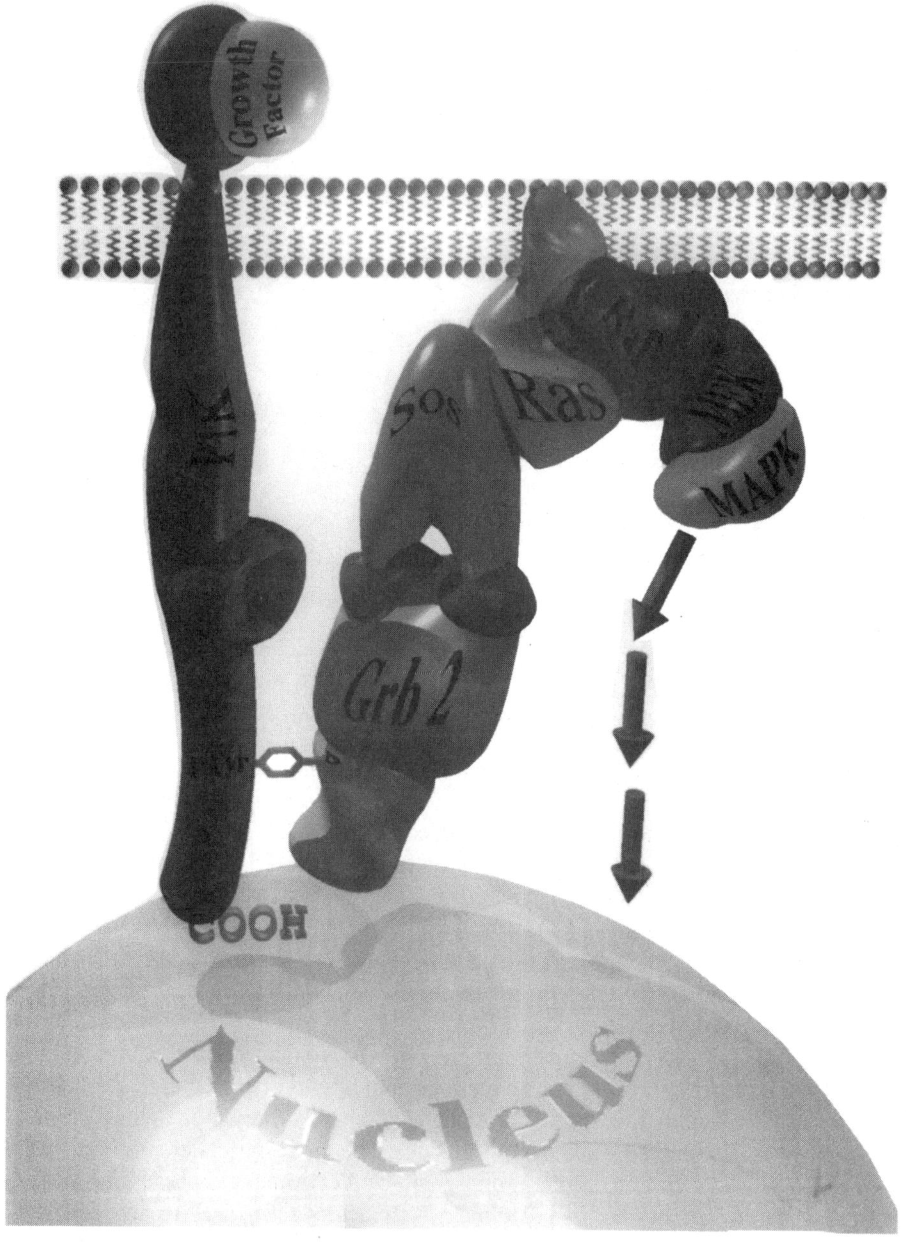

FIGURE 8.1 Cartoon depiction of prototypical growth factor receptor-mediated signal transduction through the pTyr–Grb2–Ras pathway, highlighting the key role pTyr residues.

which constitutes the cytoplasmic component of the receptor. The result of this PTK activation is the phosphorylation of one or more of its own tyrosyl residues (termed *autophosphorylation*); the net effect is to convert the initial extracellular interaction of a cell with a growth factor into an intracellularly situated pTyr residue (**1**). By analogy to mechanical signaling devices, this may be viewed as "flipping a switch" or changing a binary 0 to a 1.

Functional Consequences Induced by Generation of the pTyr Pharmacophore

As seen above, initiation of cytoplasmic signaling is achieved through the generation of pTyr residues. Subsequent transmission of these signals can be achieved by several different means, most commonly the recognition and binding to the pTyr residues by other signaling proteins. This phenomena of pTyr recognition and binding is achieved by modular protein units such as Src homology 2 (SH2)[3] and phosphotyrosine binding (PTB)[4] domains, which acquire high affinity for protein ligands once a pTyr pharmacophore has been introduced into the ligand. There are several families of homologous SH2 (and PTB) domains, which recognized different pTyr residues depending on the amino acid sequences surrounding the target pTyr.[5] In this manner, when the Tyr residue is in its unphosphorylated state (the switch is in the OFF position) there is no binding affinity, while in the phosphorylated, pTyr state (the switch is flipped ON), high sequence-dependent affinity is induced. The consequences of these pTyr-dependent bindings can be threefold:[6] (1) the enzymatic activity of the binding protein can be modulated;[7,8] (2) in the case of PTKs, substrate can be delivered specifically to a PTK for phosphorylation; and (3) the binding protein can be translocated to a specific cytoplasmic location for the construction of oligomeric signaling complexes.[9] As shown in Figure 8.1, by combining these mechanisms, signaling cascades can be achieved that carry a message from the outer cell membrane to the nucleus where cellular activation can occur.

Creation and Destruction of the pTyr Residue: PTKs versus PTPs

The analogy raised above, that generation of pTyr motifs through the actions of PTKs can be viewed as flipping a switch, or in digital terms, "changing a 0 to 1," implies that there must also be an OFF mechanism, or a way of converting a 1 back to a 0. In cellular terms, this function is served by *protein–tyrosine phosphatases* (PTPs), which remove the pTyr phosphate group and return the Tyr residue to its unphophosphorylated state. While the most frequent effect of PTPs is therefore to down-regulate PTK-dependent signaling, a growing number of instances are being found where the overall net effect of a PTP may be to enhance PTK signaling. Examples of this include the PTP CD45, which is an activator of T-cell signaling,[10] and the SH-PTP2, which functions in insulin and epidermal growth factor (EGF) signaling.[11] One manner in which such activation may occur is by phosphorolysis of pTyr residues which function in inhibitory manners.[12,13]

The Role of pTyr Residues in Disease Processes and the Potential for Therapeutic Intervention

In light of the important role played by signal transduction in normal cell growth and function, it is not surprising that aberrations in these signaling pathways can contribute to a variety of diseases, including several cancers.[14] For this reason, development of pTyr-dependent signaling inhibitors has become a significant research objective.[15–18] Since PTK-dependent signaling consists of three conceptually distinct branches, all of which involve the pTyr pharmacophore (its generation by PTKs, its recognition and binding by protein modules such as SH2 domains, and its destruction by PTPs), each leg of this "signaling triad" offers potential for therapeutic intervention through the development of inhibitors. In parallel with this functional triad subdivision of pTyr-dependent signalling, development of inhibitors can therefore be categorized into three broad families consisting of (1) PTK catalytic site-directed inhibitors,[19,20] (2) SH2/PTB domain-directed inhibitors,[21,22] and (3) PTP-directed inhibitors.[23]

The pTyr Motif as a Starting Point for Inhibitor Design

Because pTyr residues play critical roles in all branches of the PTK signaling triad, the structure of phosphotyrosine provides a useful framework for inhibitor design.[24] Since the roles of pTyr residues vary for each signaling branch, theoretically these differences can be utilized at the molecular level to design pTyr motif-based inhibitors directed at individual legs. Central to these efforts is the application of synthetic organic chemistry for specific modification of the basic pTyr motif. It is therefore instructive to delineate important structural aspects of pTyr residues and the manners in which they interact with proteins of the respective signaling pathways and then to point out examples where synthetic organic chemistry has allowed the preparation of inhibitors predicated on these interactions. Figure 8.2 depicts the components of a pTyr residue that may be relevant to inhibitor design. Because pTyr residues exist as parts of larger protein scaffolds, features of both the pTyr residue as well as surrounding protein determine ultimate biological

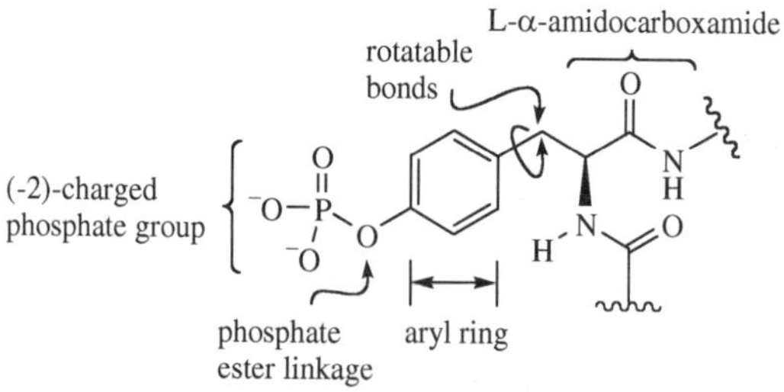

FIGURE 8.2 Structural features of a pTyr residue of potential importance for inhibitor development.

interactions. Therefore the design of pTyr-based inhibitors is dependent on features of the pTyr residue itself as well as additional structural components intended to mimic aspects of the surrounding protein. In the following paragraphs, examples illustrating the application of this approach to each signaling triad leg are provided. Rather than attempting an exhaustive review of the field, this chapter draws primarily from efforts of the authors'laboratory as exemplary of the types of approaches being put forth in this area.

FUNDAMENTALS OF INHIBITOR DESIGN

PTK Catalytic Site-Directed Inhibitors

Uncharged Tyrosyl Analogs PTKs function by simultaneously binding ATP and tyrosyl-containing substrate, with subsequent direct transfer of the γ-phosphate to the 4′-hydroxyl of the tyrosyl residue (Fig. 8.3). On the basis of this model, depending on inhibitor structure, design of PTK catalytic site-directed agents predicated on the pTyr motif could conceptually be designed to compete at the tyrosyl binding site or simultaneously at both the tyrosyl and ATP binding sites ("bisubstrate" inhibitors). Early PTK inhibitors, such as the fermentation product erbstatin (**2**),[25] were polyhydroxylated vinylbenzene-based structures in which the hydroxylated aryl ring and heteroatom-containing vinyl sidechain were envisioned to function as mimetics of the tyrosyl ring and sidechain, respectively. Polyhydroxylated styryls such as **3**[26] subsequently became models on which to design inhibitors directed at PTK tyrosyl binding sites.[27] Bicyclic compounds such as **4** and **5** were also designed as conformationally constrained styryl analogs, in which the rotationally mobile phenyl–vinyl bond has been constrained to planarity.[28,29] The fundamental characteristic of this class of pTyr mimetics is the omission of phosphate-mimicking functionality.

Charged Tyrosyl Analogs In contrast to the polyhydroxylated styryls and bicyclics that conceptually represent mimetics of the uncharged tyrosyl portion of the ATP·tyrosyl complex (Fig. 8.3), analogs have also been prepared that contain additional charged moieties intended to extend the tyrosyl structure in order to include elements of the ATP molecule. Examples of such "bisubstrate" inhibitors include **6** and **7** by Traxler et al.[30,31] and phosphonate-containing styryls **8**[32] and the bicyclic **9**.[33]

A characteristic of both uncharged and charged PTK catalytic site-directed inhibitors has been that the "pTyr structure" has been most effectively mimicked as either the "styryl" nucleus, or other similar planar moiety. Additionally, while a limited number of bisubstrate inhibitors employing charged moieties have been reported, the majority of tyrosyl-based catalytic site-directed inhibitors are uncharged species, perhaps reflecting the fact that uncharged tyrosyl species are initially recognized and bound by PTKs. As discussed below, this is in stark contrast to tyrosyl motif-based analogs directed against SH2 domain and PTP branches of the

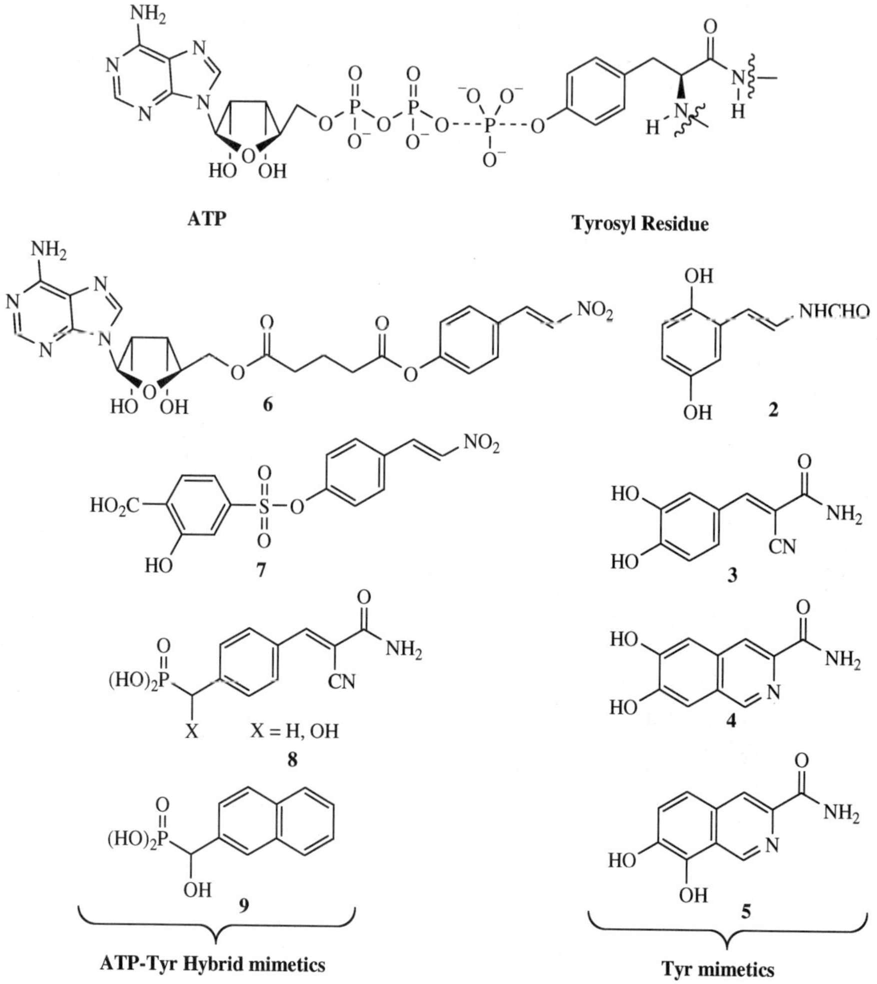

FIGURE 8.3 Depiction of the PTK-mediated transfer of an ATP γ-phosphate to the 4′-hydroxyl of a Tyr residue (top) with examples of PTK catalytic site-directed inhibitors shown below.

PTK signaling triad, where the phosphate group provides a key component of pTyr recognition, and therefore, where highly charged functional groups serve as critical recognition components of inhibitors.

SH2 Domain-Directed Inhibitors

General Considerations SH2 domains are modular structures of approximately 100 amino acid residues found in several signaling molecules, which specifically recognize and bind pTyr-containing sequences. Several families of SH2 domains are known, each of which recognizes different pTyr-containing sequences depending on the *C*-proximal amino acids adjacent to the target pTyr residue. It had long been known that short (containing 4–5 amino acids) pTyr-containing peptides modeled on native sequences of high affinity SH2 domain binding sites, could effectively compete with and inhibit the binding of cognant full-length proteins to SH2 domains. The basis for this highly localized (occurring within a few residues of the pTyr) and sequence-specific binding was made evident when X-ray crystal lographic structures became available for high-affinity SH2 domain·ligand complexes.[34] Here it was shown that SH2 domains share a common general structure in which critical binding of pTyr residues occurs in a well-formed pocket, with secondary interactions of neighboring specificity-determining *C*-proximal residues occurring in fashions unique to each subfamily of SH2 domain. Because of this, binding of pTyr-containing ligands to SH2 domains is highly dependent on both a pTyr-like moiety, which can bind within the pTyr-binding pocket, and secondary recognition features, which provide interactions outside this pocket (Fig. 8.4). Design of SH2 domain-directed inhibitors must account for pTyr-binding pocket interactions as well as specificity determinants outside the pTyr-binding pocket. Because of the common defining roles played by pTyr residues in ligand binding to all SH2 domains, considerable effort has been expended on development of pTyr mimetics that can serve as central components of SH2 domain-directed inhibitors. Therefore, in contrast to PTK catalytic site-directed agents, where binding species are Tyr residues in their unphosphorylated states (with inhibitors being uncharged), for SH2 domains (and PTPs) it is phosphotyrosyl residues that are recognized, with the "phosphoryl component" of the pTyr residues providing critical determinants of binding affinity. For this reason, highly charged phosphate-mimicking moieties are frequently essential components of pTyr motif-derived inhibitors directed at SH2 domain or PTP legs of the PTK signalling triad.

Structural Features of pTyr Residues Amenable to Mimetic Modification Development of pTyr mimetics for SH2 domain-directed inhibitors is dominated by the fact that the phosphoryl portion of the pTyr residue is critical for high-affinity binding of peptide ligands to SH2 domains. This is one principal reason why the phosphate moiety serves as the foundation of PTK-dependent signaling: The addition of a single phosphate group by a PTK induces downstream signaling events that are missing in the absence of the phosphoryl moiety (similar to flipping a switch or changing a 0 to a 1). However, even though in physiological contexts pTyr residues are critical components of high-affinity SH2 domain binding interactions, pTyr residues themselves are not ideally suited for inhibitor development. This is due primarily to the lability of phosphate esters to hydrolysis by cellular phosphatases. A consideration of secondary importance is also the potentially poor cell membrane penetration of inhibitors bearing highly charged phosphate species. Therefore, in developing SH2 domain-directed inhibitors, a major task facing the chemist is to devise pTyr mimetics that are stable toward phosphatases, yet retain high SH2 domain binding affinity. In order to achieve this objective, it is important to understand the manner in which pTyr residues are recognized by SH2 domains.

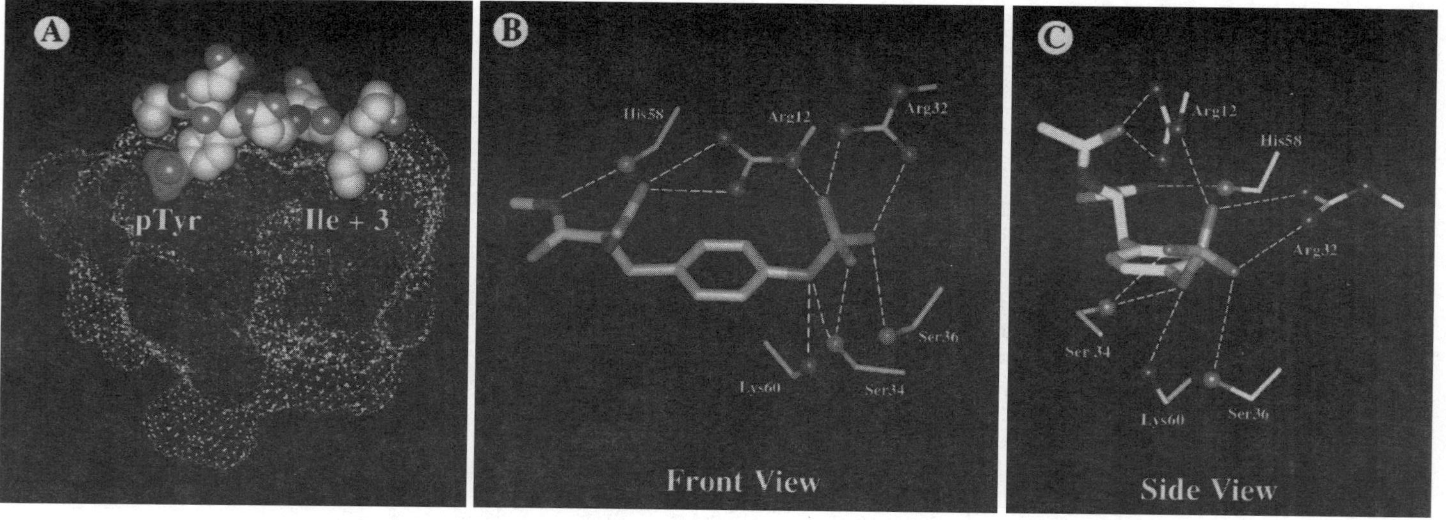

FIGURE 8.4 Binding of a pTyr-containing peptide to the p56^{lck} SH2 domain:[35] (*a*) the pTyr-containing peptide (rendered as a space-filling model) bound to the SH2 domain (depicted as a cutaway surface view) showing the manner in which the pTyr and pTyr + 3 (Ile) residues bind in well-formed pockets; (*b*) a closeup of the pTyr binding pocket highlighting important binding interactions; (*c*) a side view of panel *b* showing binding interactions from another angle.

Interactions of pTyr Residues with SH2 Domains The molecular basis for the binding of pTyr residues to SH2 domains is evident from numerous X-ray structures of SH2 domains complexed to pTyr-containing ligands. With few exceptions, bidentate chelation of the doubly charged phosphate anion is achieved by αA and βB Arg residues (nomenclature as defined by Eck et al.[35]) with a secondary role being played by hydrogen bonding to the ester bridge oxygen linking the tyrosyl ring to the phosphate group (Fig. 8.4). These binding interactions highlight the importance for phosphate recognition of both the phosphoryl (−2) charge and the ester oxygen that joins the phosphorus to the phenyl ring. Other potentially important components of pTyr residues, are sidechain and phenyl ring torsion angles (see Figs. 8.2 and 8.4). It should also be noted that two additional critical pTyr hydrogen bonding interactions common to many SH2 domains are formed between the tyrosyl carboxamido nitrogen and the backbone carbonyl of a conserved SH2 domain βD His residue as well as between the αA Arg and the tyrosyl α-amido carbonyl (Fig. 8.4). It can therefore be seen that, although the phosphate group provides a dominant theme in the development of pTyr mimetics, it is one of several factors which must be considered in the design of pTyr mimetics.

Specifics of Inhibitor Design The finding that small pTyr-containing peptides can effectively inhibit the binding of SH2 domains to larger peptides and proteins in a sequence specific manner[36] provides a starting point for development of peptide-based inhibitors. Because of the hydrolytic lability of the pTyr phosphate ester in the presence of cellular phosphatases, a significant effort has been expended on the design and synthesis of phosphatase-resistant pTyr mimetics. Initially much of this research was directed at phosphorus-containing analogs.

Phosphonate-Based Inhibitors Prior to the first demonstration that short pTyr-containing peptides can serve as effective SH2 domain binding antagonists, the pTyr mimetic, phosphonomethyl phenylalanine (Pmp, **10**) had already been reported, in which the phosphate ester linkage is replaced with a methylene unit (Fig. 8.5).[37] This analog was subsequently prepared in a protected form suitable for facile peptide synthesis,[38] and incorporated into SH2 domain-directed peptides.[39] It was found, however, that binding potency against a PI3 kinase SH2 domain was reduced fivefold relative to parent pTyr-containing peptide.[40] This loss of potency was attributed potentially to an elevation of the Pmp phosphonate pK_a values such that while at physiological pH approximately 50% of Pmp residues would exist in a monoionized state, the parent pTyr residue would exist essentially in the completely diionized form. Presumably the preferred ionization state for binding is the diionized form. A second potential reason for reduced binding of Pmp was that hydrogen bonding interactions, originally present between the SH2 domain and the pTyr phosphate ester oxygen (see Fig. 8.4), would be missing in the Pmp residue, where this oxygen is replaced by a methylene unit.

To compensate for these factors, a series of Pmp analogs were prepared in which a hydroxyl (HPmp, **11**), one fluorine (FPmp, **12**), or two fluorines (F_2Pmp, **13**) were appended at the critical phosphonate methylene (Fig. 8.5).[41] It was shown that each

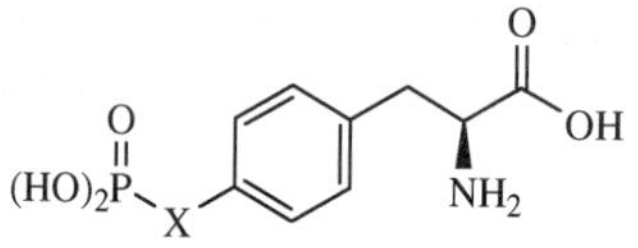

1 pTyr X = O
10 Pmp X = CH_2
11 HPmp X = CHOH
12 FPmp X = CHF
13 F_2Pmp X = CF_2

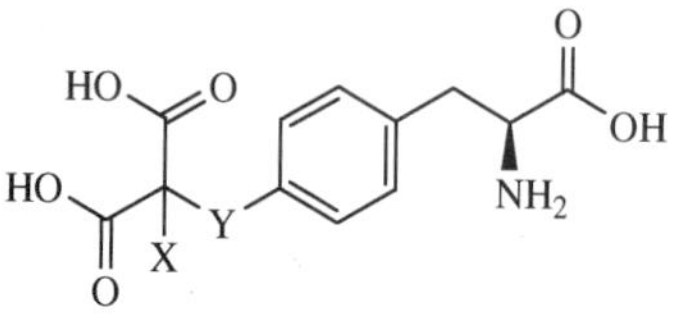

14 OMT Y = O X = H
15 FOMT Y = O X = F
16 Y = CH_2 X = H

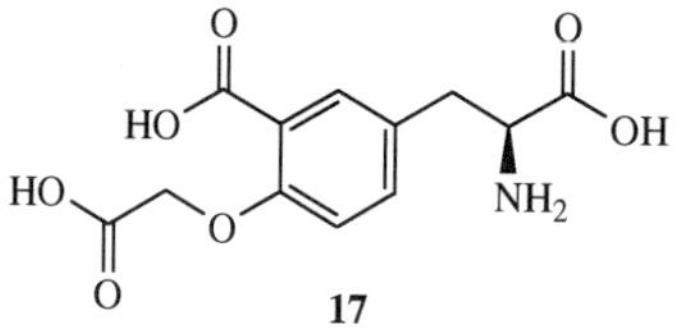

17

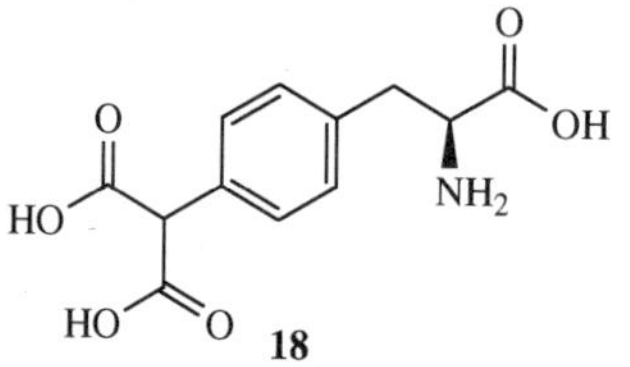

18

FIGURE 8.5 Structures of several pTyr mimetics that have been used in SH2 domain and PTP-directed inhibitors. Of note is the different manners in which the pTyr phosphate group has been mimicked in each case using either phosphonate or biscarboxylic-containing moieties.

fluorine decreased the phosphonate pK_a value by approximately 0.5 pK_a units,[42] such that F_2Pmp exhibited an ionization state equivalent to that of the parent pTyr residue. Moreover, when these new pTyr mimetics were examined in a PI3 kinase SH2 domain binding assay, it was found that the F_2Pmp-containing peptide displayed the full inhibitory potency of the parent pTyr-containing peptide, yet was stable to phosphatases. Further binding studies using F_2Pmp-containing peptides directed against other SH2 domains showed that the potency of F_2Pmp residues relative to pTyr were dependent on the SH2 domain examined, such that F_2Pmp was more potent than pTyr against Src SH2 domains, and less potent against Grb2 SH2 domains. This finding was an early indication that the selectivity of inhibitors for different SH2 domains could be affected by interactions within the pTyr binding pocket as well as by traditional interactions in peptide-specificty regions outside the pocket.

As shown in Fig. 8.5, a number of phosphorus-containing analogs have been prepared as pTyr mimetics for use in SH2 domain binding systems. The binding of each of these is presumably dependent on interactions between their phosphonate oxygens and key αA and βB arginine residues which normally bind to the pTyr phosphate oxygens. As pointed out above, additional binding interactions can also occur between the αA arginine and the tyrosyl α-amido carbonyl (Fig. 8.4). Such auxiliary binding originating from functionality appended off the tyrosyl α-position (see Fig. 8.5), has been exploited in a number of inhibitors, resulting in net binding affinities higher than parent pTyr-containing peptides.

The underlying rationale for using phosphonate-based pTyr mimetics has been to obtain SH2 domain inhibitors that are stable to phosphatases in cellular contexts yet retain high binding potency. However, high SH2 domain binding potency afforded by phosphonate-based species must be balanced by the limitations of cell membrane transport presented by such highly charged molecules. Cell-based studies with F_2Pmp-containing peptides have therefore relied on laborious techniques such as microinjection or membrane permeabilization to overcome problems of cell membrane transport.[11,43] Limited applications of bioreversible phosphonate protection have also been examined.[44] More recently, high-affinity Grb2 SH2 domain binding has been observed in whole cells by direct application to cell media without the need either for prodrug derivatization or artificial introduction techniques, using an *N*-oxalyl Pmp-based pTyr mimetic.[45] In these studies, however, effective intracellular binding was significantly reduced relative to extracellular potency, highlighting a need for nonphosphonate-based pTyr mimetics that may potentially afford greater cell membrane penetration.

Non-Phosphorus-Containing Biscarboxylic-Based Inhibitors In order to derive pTyr mimetics, which may afford alternate avenues for delivery through cell membranes, a number of non-phosphonate-containing analogs have been examined. Common to all these mimetics is the use of carboxy functionality to maintain key binding interactions normally afforded by the pTyr phosphate group. On the basis of precedence as to where α,α- biscarboxyl-containing moieties have been incorporated as phosphate mimetics in agents directed at unrelated enzymes, the *O*-malonyl tyrosyl (OMT, **14**)[46] and fluoro-OMT (FOMT, **15**)[47] compounds as well as the related analog **16** having a methylene bridge rather than ether attachment to the aryl ring[48] were prepared as pTyr mimetics. OMT was shown to exhibit moderate inhibitory potencies in several SH2 domain systems (approximately 25-fold loss of potency relative to corresponding pTyr-containing peptides.[49] In more recent work, OMT variants have been prepared that either spatially separate the original malonyl α,α-biscarboxyls (compound **17**)[45] or remove the ether oxygen linking the malonyl group to the aryl ring (compound **18**).[50] Mimetics **17** and **18** have exhibited good inhibitory potency against Grb2 SH2 domain systems, with **18** approaching the potency of phosphorus-containing Pmp-based inhibitors.

Non-Phosphorus-Containing Monocarboxylic-Based Inhibitors Since biscarboxylic-containing pTyr mimetics maintain the (−2) charge of phosphonate-based mimetics, they potentially afford similar cell membrane penetration problems. Therefore, efforts have focused on developing pTyr mimetics bearing less formal charge (Fig. 8.6). Early work by Glaxo showed that removing one carboxyl from malonyl analog **18**, to give carboxymethyl phenylalanine (cmF, **19**) resulted in a 450-fold loss of potency relative to pTyr (Fig. 8.6).[48] In spite of this dramatic

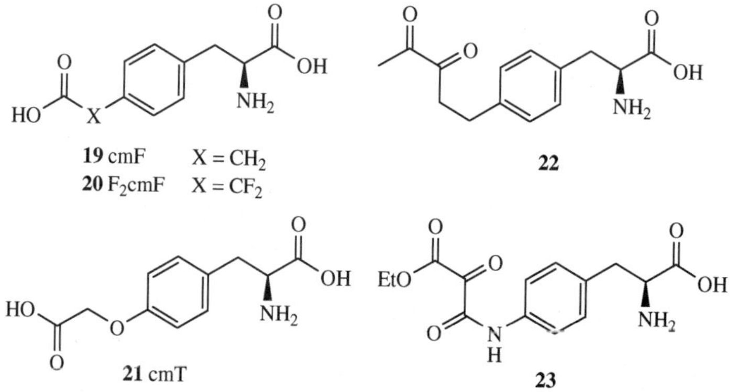

19 cmF X = CH_2
20 F_2cmF X = CF_2
22
21 cmT
23

FIGURE 8.6 Structures of several monocarboxylic and uncharged pTyr mimetics used in SH2 domain-directed inhibitors.

reduction in affinity, the X-ray structure of a cmF-bearing peptide complexed to the lck SH2 domain revealed good correspondence between the carboxymethyl oxygens and two of the parent pTyr phosphate oxygens.[51] However, as was seen above with Pmp analogs, an obvious potential limitation of the cmF residue is its lack of functionality at the methylene bridge between the carboxylate and the aryl ring, which could afford binding interactions similar to that observed with the pTyr phosphate ester oxygen (see Fig. 8.4). Reminiscent of modifications in the Pmp series that resulted in F_2Pmp, replacement of this methylene with a diflouromethylene (compound **20**, Fig. 8.6) or by insertion of an ether oxygen (analog **21**) resulted in loss of binding potency.[52] Nonetheless, combining an Nα- oxalyl moiety with the cmF reside **19** in a high affinity ß-bend mimicking patform, provided low-micromolar to submicromolar affinity Grb2 SH2 domain binding potency (an approximate 10–20-fold loss of potency relative to the corresponding Nα-acetyl pTyr-containing ligand).[52]

Uncharged pTyr Mimetics An essential component of both the phosphonate and carboxylate-based pTyr mimetics is their use of charged moieties to mimic the ionic interaction of a phosphate group with positively charged Arg residues within SH2 domain pTyr-binding pockets (see Fig. 8.4). In an alternate approach, uncharged α-dicarbonyls such as **22** and vicinal tricarbonyls such as **23** (Fig. 8.6) have been employed as affinity labels intended to covalently bond to these Arg residues. While their lack of charge gives such agents potential advantages in cell membrane penetration, it also removes a critical component of phosphate recognition needed for high affinity binding, and such analogs have exhibited only weak potency.[53]

Conformationally Constrained Inhibitors As shown in Figure 8.2, pTyr residues contain a number of structural features of potential importance for interaction with biological macromolecules. Among these are multiple single bonds within the sidechain that allow for a range of conformations in aqueous solution. In the process of binding to SH2 domains, entropy penalties must be paid for the selection of proper binding conformations. As a general principle, conformationally constrained analogs have often been used to enhance binding affinity by reducing these entropy penalties. In the case of SH2-domain-directed pTyr mimetics, tricyclic analogs such as **24** have been designed that contain within their skeletons tyrosyl residues having sidechain torsion angles constrained to angles approximating those observed in a pTyr reside ligated to the p56lck SH2 domain (Fig. 8.7).[54,55] Benazocine analog **24** is remarkable in simultaneously constraining three such torsion angles. Importantly, when **24** is docked into the SH2 domain, the appended constraining framework is situated away from the protein, minimizing stearic interference. Although biological evaluation is forthcoming, initial model studies with simplified analogs potentially support the design principle.[54]

PTP-Directed Inhibitors

Along with PTKs and SH2 domains, PTPs provide a third leg of the pTyr-dependent signaling triad. Inhibitors of PTPs may afford valuable tools for studying the functions of these enzymes in cellular processes, and may potentially provide new

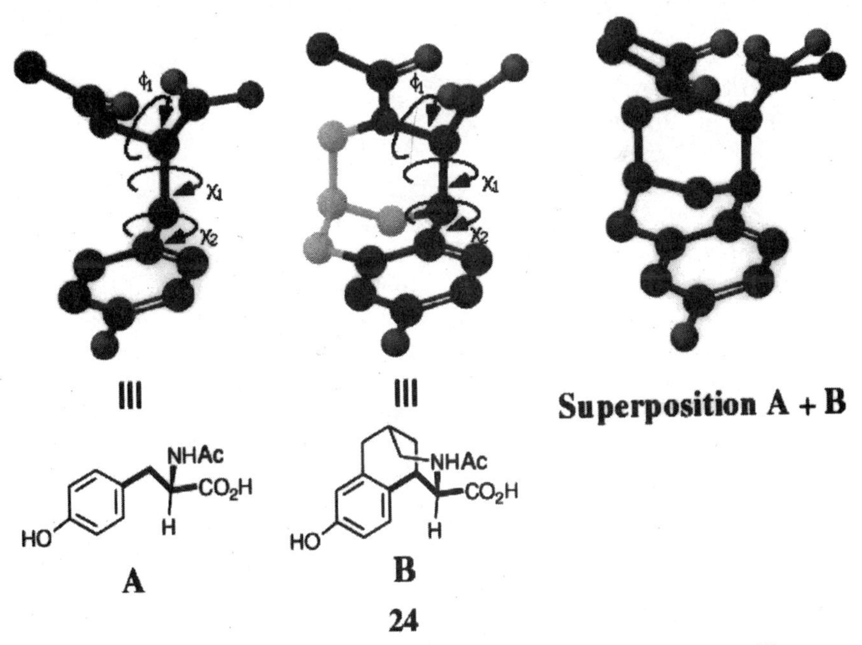

FIGURE 8.7 (*a*) A pTyr residue in its p56lck SH2 domain-bound conformation[35] (top, ball-and-stick rendering; bottom, symbolic depiction); (*b*) tricyclic Tyr analog **24** (in light shade) with the Tyr residue contained within it shown in dark (the three constrained Tyr sidechain bonds are explicitly shown); (*c*) superposition of *a* and *b* showing the very close match of Tyr sidechain torsion angles.

therapeutic approaches toward a variety of diseases, including diabetes and certain cancers. Similar to SH2 domains, pTyr residues provide critical components of substrate recognition by PTPs. For this reason, pTyr mimetics have become valuable motifs for structure-based PTP inhibitor design.[23,24] In order to understand the rational foundation behind the use of pTyr mimetics in PTP inhibitor development, it is important to examine the manner in which pTyr-containing substrates are recognized and bound by PTPs.

Important pTyr Binding Interactions For PTPs, a critical catalytic component is the active-site signature motif "Cys–X_5–Arg" (where "X" is any residue). This motif occurs in PTPs as well as in two other families of phosphatases that hydrolyze pTyr phosphate residues: low-molecular-weight phosphatase and cdc25 phosphatases.[56] Using PTP1B as a prototypical model, it has been shown that the active site consists of an approximately 9-Å-deep cleft bounded on top and bottom by the aryl rings of Tyr46 and Phe182, with the signature Cys215–X_5–Arg221 residues forming a semicircle at the base of the cleft (Fig. 8.8).[57] The function of the signature residues is to coordinate and stabilize the pTyr phosphate moiety for SN2 attack by the Cys215 thiolate anion, followed by release of tyrosyl product and formation of a thiophosphate intermediate.[58] The resulting thiophosphate ester is hydrolyzed by attack of a water molecule under the general base catalysis of Asp181. The Asp residue is brought into correct proximity by its neighbor Phe182, which along with the Tyr46, sandwiches the Tyr aryl ring during the substrate binding process. The key Arg221 residue provides direct ionic chelation of the phosphate group throughout the entire catalytic cycle. In addition to these interactions, residues situated outside, yet proximal, to the catalytic cleft serve important roles in substrate binding. Among these is Arg47, which facilitates the selection of pTyr peptide substrates bearing acidic acid residues.[59] Important aspects of the pTyr residue as they relate to PTP substrate recognition can therefore be categorized as (1) a phosphate-mimicking group that can interact with the positively charged Arg221 moiety and that can be coordinated within the loop formed by the Cys–X_5–Arg signature residues, (2) an aryl ring system that is compatible for binding between the two aromatic sidechains (Tyr46 and Phe182) that form the boundaries of the catalytic cleft, and (3) additional functionality that can interact with residues positioned on the periphery of the catalytic cleft, such as the Arg47 residue (Fig. 8.8).

Comparison of pTyr Binding to PTPs with SH2 Domains In comparing binding of pTyr residues by PTPs to those of SH2 domains, significant similarities and differences are found that can be exploited for inhibitor design. As shown in Figures 8.4 and 8.8, both the manner in which the pTyr phosphate moiety and aryl rings are bound, and the sidechain torsion angles are different for these two ligand–protein interactions. One additional factor not obvious from Figures 8.4 and 8.8, is that, while SH2 domains bind the pTyr residue such that one side is facing the protein while the opposite side is facing outward into solvent, for PTPs the pTyr

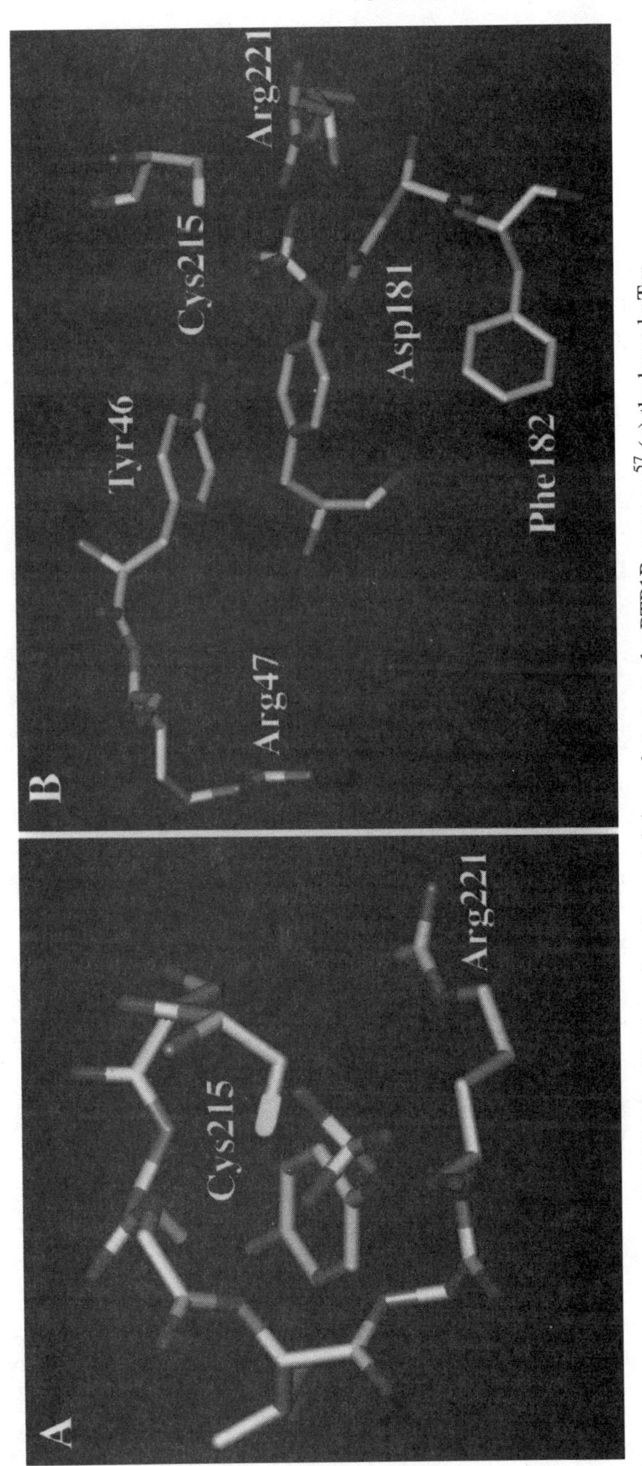

FIGURE 8.8 Binding of a pTyr-containing substrate to the PTP1B enzyme.[57] (*a*) the bound pTyr group showing its orientation to the signature motif "Cys215–X_5–Arg221" loop; (*b*) an alternate view showing relative orientations of key enzyme residues as the relate to the pTyr residue.

residue is buried and surrounded by protein. These differences potentially allow discrimination of pTyr mimetic-based inhibitors for either PTPs or SH2 domains.

Historical Progression of Inhibitor Development

Phosphonate-Containing Peptide-Based Inhibitors Since the K_m value for free pTyr is 3–4 orders of magnitude higher than the best protein/peptide substrates, it is evident that important aspects of substrate recognition occur outside the pTyr binding cleft. To take advantage of these more extended binding interactions, short high- affinity peptide sequences have been used as display platforms to examine the PTP affinity of various pTyr mimetics. One such approach has used thiophosphorylated analogs, in which the pTyr phosphate group has been replaced by a thiophosphate group that is less readily hydrolyzed by PTPs.[60] Another approach has used Pmp residues (**10**), described above for use in SH2 domain-directed inhibitors. Pmp-bearing peptides have exhibited competitive inhibition, with K_i values in the range of 10–30 μM, which is significantly higher than K_m values of pTyr-containing peptides.[59,61] Note was therefore taken of Pmp-based SH2 domain-directed inhibitors, where replacement of the Pmp phosphonate methylene a difluoromethylene (F_2Pmp **12**) in some cases resulted in enhanced binding. When this was done with Pmp-containing PTP inhibitors modeled on the high-affinity substrate sequence Ac–Asp–Ala–Asp–Glu–Xxx–Leu–amide, where Xxx = pTyr or pTyr mimetic, it was shown that for Xxx = F_2Pmp an IC_{50} value of 200 nM was obtained against PTP1B. This represented a remarkable 1000-fold enhancement in potency relative to the corresponding Pmp-containing peptide (IC_{50} = 200 μM).[62] In later studies kinetic evidence indicated that the enhancing effect of the fluorines was due to specific interactions within the catalytic site, and not to effects on phosphonate pK_a values.[63] These studies were the first indication of the extremely high PTP affinity of aryldifluoromethylphosphonates, a fact that would subsequently be applied to the design of new non-peptide-based PTP inhibitors.

Non-Phosphonate-Containing Peptide-Based Inhibitors In parallel with SH2 domain-directed inhibitor development, significant efforts have been devoted to finding non-phosphonate-containing alternatives to pTyr mimetics. Among these are several carboxylic-acid-based analogs already presented above for use against SH2 domains. When incorporated into the Ac–Asp–Ala–Asp–Glu–Xxx–Leu–amide sequence, OMT (**14**) has shown an IC_{50} value of 10 μM, which is 10-fold more potent than for Xxx = Pmp.[64] Addition of a fluorine atom to give FOMT (**15**) resulted in yet another 10-fold enhancement in potency (IC_{50} = 1 μM),[47] with cyclization of the FOMT-containing peptide using a thioether linkage providing even higher affinity (K_i = 170 nM).[65] A recent X-ray structure of the PTP1B enzyme ligated to this cyclic peptide demonstrated the manner in which the FOMT residue is bound within the catalytic site.[66] Several dicarboxylic- and monocarboxylic-containing pTyr mimetics previously discussed as SH2 domain inhibitors, have also been examined for their PTP affinity using the Ac–Asp–Ala–Asp–Glu–Xxx–Leu–amide sequence. Notable among these is the high affinity of bis carboxylic analog **17** (K_i = 3.6 μM)[67] and the extremely poor affinity of monocarboxylic-based analogs **19–21**.[68]

Phosphonate-Containing Non-Peptide-Based Inhibitors While PTP inhibitor development using peptides affords a convenient and uniform platform for examining affinity of pTyr mimetics, one important aim of this type of research is to provide information for the development of nonpeptide, small molecule inhibitors. In this light, initial findings of the good potency of F_2Pmp-containing peptides indicated the potential high PTP affinity of aryldifluoromethylphosphonates. In a study designed to isolate and examine the PTP affinity of aryldifluoromethylphosphonate moieties devoid of peptide components, it was found that while a simple phenyl difluoromehtylphosphonic acid (**25**) showed extremely low affinity, extending the aryl system to a bicylcic naphthyl system (compound **26**) resulted in reasonable affinity (IC_{50} = 180 μM) (Fig. 8.9).[69] The X-ray structure of this naphthyl difluoromethylphosphonic acid ligated to the PTP1B enzyme was subsequently solved,[70] providing a basis for further structure-based elaboration of the difluoromethylphosphonic acid nucleus. Examples of this work (Fig. 8.9) include the addition of a hydroxyl at the naphthyl 4 position to replace a water molecule observed in the crystal structure (compound **27**, which displayed a doubling of potency, K_i = 90 μM)[70] and the appending of a glutamic acid residue to allow potential interactions with an Arg47 residue situated just exterior to the pTyr binding pocket (compound **28**, K_i = 12 μM).[71] The X-ray structures of dipeptide mimetic **28** as well as the simpler monomeric pTyr mimetic **29** (K_i = 22 μM) ligated to PTP1B have recently (at the time of writing) been reported.[66] In related work, aryl bis(difluoromethylphosphonic acids) such as **30** (K_i = 4.5 μM) have also recently been reported.[72]

Non-Phosphonate-Containing Non-Peptide-Based Inhibitors Although the difluoromethyl moiety serves as a high-affinity phosphate mimetic in PTP systems, difficulties of cell membrane transport are potentially a serious limitation. Non-

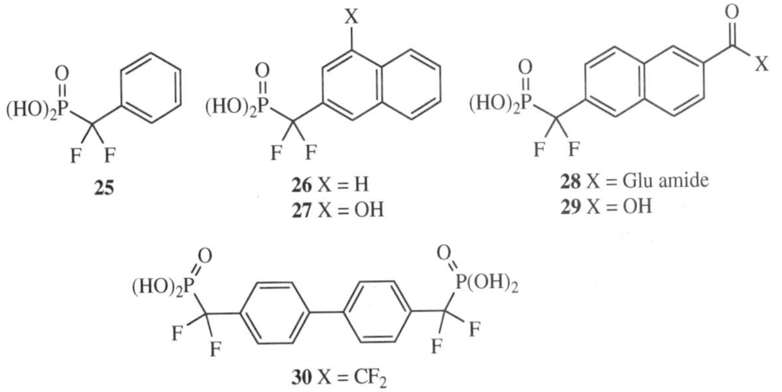

25

26 X = H
27 X = OH

28 X = Glu amide
29 X = OH

30 X = CF_2

FIGURE 8.9 Small molecule difluorophosphonate-containing PTP inhibitors.

phosphonate-based small molecule inhibitors are therefore being examined as mimetics of peptide inhibitors. Using 6-(phosphonodifluoromethyl)naphthalene-2-carboxylic acid (**29**) as a model, recent work has examined non-phosphonate-containing alternatives. In one approach predicated on the high binding affinity of Ac–Asp–Ala–Asp–Glu–Xxx–Leu–amide, where Xxx = pTyr mimetic **17** ($K_i = 3.6\ \mu M$),[67] it was noted that the phenyl ring of **17** can bind within the PTP catalytic site in two possible rotational conformers (A and B, Fig. 8.10). The isomeric naphthyl compounds **32** and **33** were therefore prepared as conformationally constrained mimetics of these two possible conformers. Although the affinity of these analogs proved to be much lower than corresponding difluoromethyl-based compound **29**, isomer **32** ($K_i = 90\ \mu M$) was significantly more potent than isomer **31**

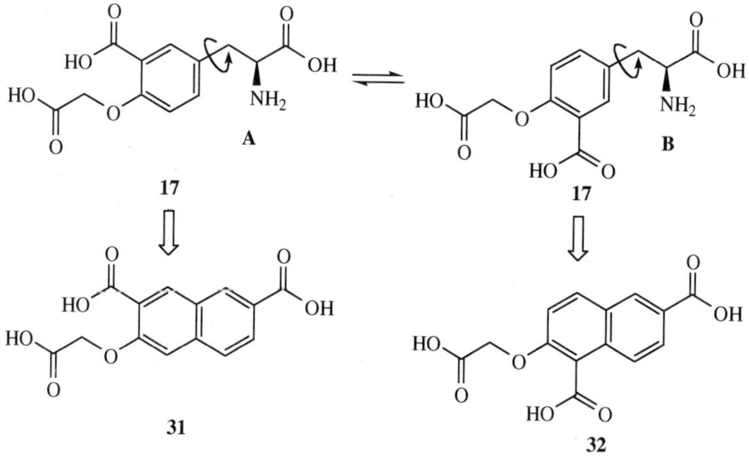

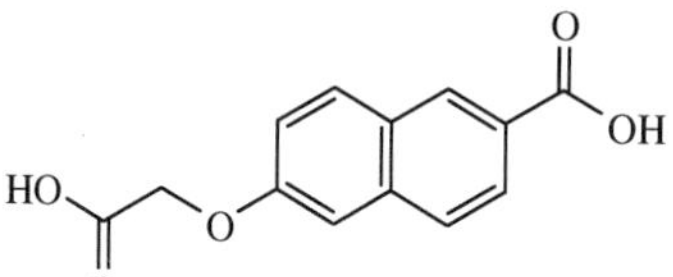

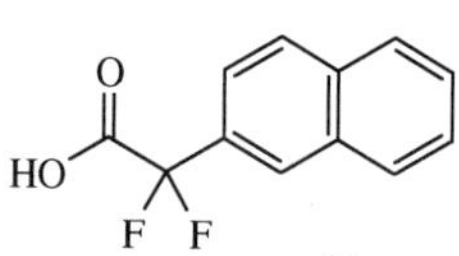

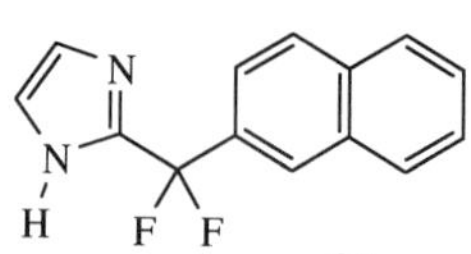

FIGURE 8.10 Non-phosphonate-containing PTP-directed agents. Structures **32** and **33** were designed as conformationally constrained small molecule mimetics of rotational isomers A and B respectively, of potent pTyr mimetic **17**.

($K_i = 250\,\mu M$), consistent with modeling of the flexible parent **17** within the PTP catalytic site.[68] The importance of the 5-carboxyl group was indicated by compound **33**, which is lacking this carboxyl and that displays a 45-fold loss of potency ($K_i = 9.4\,mM$).[68] Recent work comparing naphthyl-based compound **34** ($IC_{50} = 230\,\mu M$) with **35** ($IC_{50} = 640\,\mu M$) also identified the tetrazolyldifluoromethyl moiety as a potentially valuable replacement for the carboxydifluoromethyl group as a PTP-directed phosphate.[73]

CONCLUSIONS

This chapter has attempted to demonstrate the utility of phosphotyrosyl-based structures for the development of PTK-dependent signaling inhibitors. An emergent theme has been that pTyr motifs represent a conceptual Roseta stone for the design of signal transduction inhibitors. Implicit in this, is the essential role of synthetic organic chemistry for the translation of design concepts into actual compounds that can be subjected to biological evaluation. In the special case of pTyr mimetics, particular problems present themselves related to phosphate mimicry within the specific contexts of SH2 domain and PTP-binding interactions. Here, issues concerning inhibitory potency, and bioavailability are being addressed in the design and synthesis process. In these efforts, a major consideration has been the search for non- phosphate-containing agents which maintain binding interactions displayed by diionic arylphosphate species of parent pTyr residues. The examples presented herein show clear examples where synthetic organic chemistry has allowed a structure-based progression from initial peptides bearing enzymatically labile phosphate groups to hydrolytically stable phosphonate-based mimetics and finally to non-phosphorus-containing small molecule inhibitors. Much remains yet to be done, and this area promises to be a fertile field of medicinal chemistry research for the foreseeable future.

REFERENCES

1. Edwards, D. R., *Trends Pharmacol. Sci.* 1994, **15**, 239.
2. Karin, M.; Hunter, T., *Curr. Biol.* 1995, **5**, 747.
3. Pawson, T., *Adv. Cancer Res.* 1994, **64**, 87.
4. Eck, M. J., *Structure* 1995, **3**, 421.
5. Zhou, S. Y.; Cantley, L. C., *Trends Biochem. Sci.* 1995, **20**, 470.
6. Panayotou, G.; Waterfield, M. D., *Bioessays* 1993, **15**, 171.
7. Hubbard, S. R.; Wei, L.; Elis, L.; Hendrickson, W. A., *Nature* 1994, **372**, 746.
8. Ohnishi, H.; Kubota, R.; Ohtake, A.; Sato, K.; Sano, S. I., *J. Biol. Chem.* 1996, **271**, 25569.
9. Pawson, T., *Nature* 1995, **373**, 573.
10. Shaw, A.; Thomas, M. L., *Curr. Opin. Cell. Biol.* 1991, **3**, 862.

11. Xiao, S.; Rose, D. W.; Sasaoka, T.; Maegawa, H.; Burke, T. R., Jr.; Roller, P. P.; Shoelson, S. E.; Olefsky, J. M., *J. Biol. Chem.* 1994, **269**, 21244.
12. Zheng, X. M.; Wang, Y.; Pallen, C. J., *Nature* 1992, **359**, 336.
13. Gautier, J.; Solomon, M. J.; Booher, R. N.; Bazan, J. F.; Kirschner, M. W., *Cell* 1991, **67**, 197.
14. Bridges, A., *Chemtracts Org. Chem.* 1995, **8**, 73.
15. Rosen, J.; Day, A.; Jones, T. K.; Turner-Jones, E. T.; Nadzan, A. M.; Stein, R. B., *J. Med. Chem.* 1995, **38**, 4855.
16. Levitzki, A., *Curr. Opin. Cell. Biol.* 1996, **8**, 239.
17. Saltiel, A. R.; Sawyer, T. K., *Chem. Biol.* 1996, 3, 887.
18. Persidis, A., *Nat. Biotechnol.* 1998, **16**, 1082.
19. Fry, D. W., *Exp. Opin. Invest. Drugs* 1994, **3**, 577.
20. Lawrence, D. S.; Niu, J. K., *Pharmacol. Ther.* 1998, **77**, 81.
21. Beattie, J., *Cell. Signal.* 1996, **8**, 75.
22. Shoelson, S. E., *Curr. Opin. Chem. Biol.* 1997, **1**, 227.
23. Burke, T. R., Jr.; Zhang, Z. Y., *Biopolymers* 1998, **47**, 225.
24. Burke, T. R., Jr.; Yao, Z.-J.; Smyth, M. S.; Ye, B., *Curr. Pharm. Design* 1997, **3**, 291.
25. Nakamura, H.; Iitaka, Y.; Imoto, M.; Isshiki, K.; Naganawa, H.; Takeuchi, T.; Umezawa, H., *J. Antibiot.* (Tokyo) 1986, **39**, 314.
26. Gazit, A.; Yaish, P.; Gilon, C.; Levitzki, A., *J. Med. Chem.* 1989, **32**, 2344.
27. Burke, T. R., Jr., *Drugs Fut.* 1992, **17**, 119.
28. Burke, T. R., Jr.; Ford, H.; Osherov, N.; Levitzki, A.; Stefanova, I.; Horak, I. D.; Marquez, V. E., *Bioorg. Med. Chem. Lett.* 1992, **2**, 1771.
29. Burke, T. R., Jr.; Lim, B.; Marquez, V. E.; Li, Z.-H.; Bolen, J. B.; Stefanova, I.; Horak, I., *J. Med. Chem.* 1993, **36**, 425.
30. Traxler, P. M.; Wacker, O.; Bach, H. L.; Geissler, J. F.; Kump, W.; Meyer, T.; Regenass, U.; Roesel, J. L.; Lydon, N., *J. Med. Chem.* 1991, **34**, 2328.
31. Peterli, S.; Stumpf, R.; Schweizer, M.; Sequin, U.; Mett, H.; Traxler, P., *Helv. Chim. Acta* 1992, **75**, 696.
32. Burke, T. R., Jr.; Li, Z. H.; Bolen, J. B.; Marquez, V. E., *J. Med. Chem.* 1991, **34**, 1577.
33. Saperstein, R.; Vicario, P. P.; Strout, H. V.; Brady, E.; Slater, E. E.; Greenlee, W. J.; Ondeyka, D. L.; Patchett, A. A.; Hangauer, D. G., *Biochemistry* 1989, **28**, 5694.
34. Birge, R. B.; Hanafusa, H., *Science* 1993, **262**, 1522.
35. Eck, M. J.; Shoelson, S. E.; Harrison, S. C., *Nature* 1993, **362**, 87.
36. Songyang, Z.; Shoelson, S. E.; Chaudhuri, M.; Gish, G.; Pawson, T.; Haser, W. G.; King, F.; Roberts, T.; Ratnofsky, S.; Lechleider, R. J.; Neel, B. G.; Birge, R. B.; Fajardo, J. E.; Chou, M. M.; Hanafusa, H.; Schaffhausen, B.; Cantley, L. C., *Cell* 1993, **72**, 767.
37. Marseigne, I.; Roques, B. P., *J. Org. Chem.* 1988, **53**, 3621.
38. Burke, T. R., Jr.; Russ, P.; Lim, B., *Synthesis* 1991, **11**, 1019.
39. Shoelson, S. E.; Chatterjee, S.; Chaudhuri, M.; Burke, T. R., Jr., *Tetrahedron Lett.* 1991, **32**, 6061.
40. Domchek, S. M.; Auger, K. R.; Chatterjee, S.; Burke, T. R., Jr.; Shoelson, S. E., *Biochemistry* 1992, **31**, 9865.

41. Burke, T. R., Jr.; Smyth, M.; Nomizu, M.; Otaka, A.; Roller, P. P., *J. Org. Chem.* 1993, **58**, 1336.
42. Smyth, M. S.; Ford, H., Jr.; Burke, T. R., Jr., *Tetrahedron Lett.* 1992, **33**, 4137.
43. Wange, R. L.; Isakov, N.; Burke, T. R., Jr.; Otaka, A.; Roller, P. P.; Watts, J. D.; Aebersold, R.; Samelson, L. W., *J. Biol. Chem.* 1995, **270**, 944.
44. Stankovic, C. J.; Surendran, N.; Lunney, E. A.; Plummer, M. S.; Para, K. S.; Shahripour, A.; Fergus, J. H.; Marks, J. S.; Herrera, R.; Hubbell, S. E.; Humblet, C.; Saltiel, A. R.; Stewart, B. H.; Sawyer, T. K., *Bioorg. Med. Chem. Lett.* 1997, **7**, 1909.
45. Yao, Z.-J.; King, C. R.; Cao, T.; Kelley, J.; Milne, G. W. A.; Voigt, J. H.; Burke, T. R., Jr., *J. Med. Chem.* 1999, **42**, 25.
46. Ye, B.; Burke, T. R., Jr., *Tetrahedron Lett.* 1995, **36**, 4733.
47. Burke, T. R., Jr.; Ye, B.; Akamatsu, M.; Ford, H.; Yan, X. J.; Kole, H. K.; Wolf, G.; Shoelson, S. E.; Roller, P. P., *J. Med. Chem.* 1996, **39**, 1021.
48. Gilmer, T.; Rodriquez, M.; Jordan, S.; Crosby, R.; Alligood, K.; Green, M.; Kimery, M.; Wagner, C.; Kinder, D.; Charifson, P.; Hassell, A. M.; Willard, D.; Luther, M.; Rusnak, D.; Sternbach, D. D.; Mehrotra, M.; Peel, M.; Shampine, L.; Davis, R.; Robbins, J.; Patel, I. R.; Kassel, D.; Burkhart, W.; Moyer, M.; Bradshaw, T.; Berman, J., *J. Biol. Chem.* 1994, **269**, 31711.
49. Ye, B.; Akamatsu, M.; Shoelson, S. E.; Wolf, G.; Giorgetti-Peraldi, S.; Yan, X. J.; Roller, P. P.; Burke, T. R., Jr., *J. Med. Chem.* 1995, **38**, 4270.
50. Gao, Y.; Burke, T. R., Jr., *SYNLE* (in review).
51. Beaulieu, L.; Cameron, D. R.; Ferland, J. M.; Gauthier, J.; Ghiro, E.; Gillard, J.; Gorys, V.; Poirier, M.; Rancourt, J.; Wernic, D.; Linas-Brunet, M.; Betageri, R.; Kirrane, T.; Sharma, R.; Hickley, G.; Patel, U.; Proudfoot, J.; Moss, N.; Cardozo, M.; Jakes, S.; Lukas, S.; Kabcenell, A.; Tong, L.; Ingraham, R., 216th National American Chemical Society Meeting, Boston, Aug. 23–27, 1998. MEDI 263 1998.
52. Burke, T. R., Jr.; Luo, J.; Yao, Z.-J.; Gao, Y.; Zhao, H.; Milne, G. W. A.; Guo, R.; Voigt, J. H.; King, C. R.; Yang, D., *Bioorg. Med. Chem. Lett.* 1999.
53. Mehrotra, M. M.; Sternbach, D. D.; Rodriguez, M.; Charifson, P.; Berman, J., *Bioorg. Med. Chem. Lett.* 1996, **6**, 1941.
54. Burke, T. R., Jr.; Barchi, J. J., Jr.; George, C.; Wolf, G.; Shoelson, S. E.; Yan, X., *J. Med. Chem.* 1995, **38**, 1386.
55. Ye, B.; Yao, Z. J.; Burke, T. R., Jr., *J. Org. Chem.* 1997, **62**, 5428.
56. Denu, J. M.; Dixon, J. E., *Curr. Opin. Chem. Biol.* 1998, **2**, 633.
57. Jia, Z. C.; Barford, D.; Flint, A. J.; Tonks, N. K., *Science* 1995, **268**, 1754.
58. Pannifer, A. D. B.; Flint, A. J.; Tonks, N. K.; Barford, D., *J. Biol. Chem.* 1998, **273**, 10454.
59. Zhang, Z. Y.; Maclean, D.; Mcnamara, D. J.; Sawyer, T. K.; Dixon, J. E., *Biochemistry* 1994, **33**, 2285.
60. Kole, H. K.; Garant, M. J.; Kole, S.; Bernier, M., *J. Biol. Chem.* 1996, **271**, 14302.
61. Chatterjee, S.; Goldstein, B. J.; Csermely, P.; Shoelson, S. E., "Phosphopeptide Substrates and Phosphonopeptide Inhibitors of Protein-Tyrosine Phosphatases," in *Peptides: Chemistry and Biology*, J. E. Rivier and J. A. Smith, eds., Escom Science Publishers, Leiden, Netherlands, 1992, pp. 553–555.
62. Burke, T. R., Jr.; Kole, H. K.; Roller, P. P., *Biochem. Biophys. Res. Commun.* 1994, **204**, 129.

63. Chen, L.; Wu, L.; Otaka, A.; Smyth, M. S.; Roller, P. P.; Burke, T. R., Jr.; Denhertog, J.; Zhang, Z. Y., *Biochem. Biophys. Res. Commun.* 1995, **216**, 976.
64. Kole, H. K.; Ye, B.; Akamatsu, M.; Yan, X.; Barford, D.; Roller, P. P.; Burke, T. R., Jr., *Biochem. Biophys. Res. Commun.* 1995, **209**, 817.
65. Roller, P. P.; Wu, L.; Zhang, Z. Y.; Burke, T. R., Jr., *Bioorg. Med. Chem. Lett.* 1998, **8**, 2149.
66. Groves, M. R.; Yao, Z.-J.; Roller, P. P.; Burke, T. R., Jr.; Barford, D., *Biochemistry* 1998, **37**, 17773.
67. Burke, T. R., Jr.; Yao, Z. J.; Zhao, H.; Milne, G. W. A.; Wu, L.; Zhang, Z. Y.; Voigt, J. H., *Tetrahedron* 1998, **54**, 9981.
68. Gao, Y.; Burke, T. R., Jr., manuscript in preparation.
69. Kole, H. K.; Smyth, M. S.; Russ, P. L.; Burke, T. R., Jr., *Biochemical J.* 1995, **311**, 1025.
70. Burke, T. R., Jr.; Ye, B.; Yan, X. J.; Wang, S. M.; Jia, Z. C.; Chen, L.; Zhang, Z. Y.; Barford, D., *Biochemistry* 1996, **35**, 15989.
71. Yao, Z. J.; Ye, B.; Wu, X. W.; Wang, S. M.; Wu, L.; Zhang, Z. Y.; Burke, T. R., Jr., *Bioorg. Med. Chem.* 1998, **6**, 1799.
72. Taylor, S. D.; Kotoris, C. C.; Dinaut, A. N.; Wang, Q. P.; Ramachandran, C.; Huang, Z., *Bioorg. Med. Chem.* 1998, **6**, 1457.
73. Kotoris, C. C.; Chen, M.-J.; Taylor, S. D., *Bioorg. Med. Chem. Lett.* 1998, **8**, 3275.

CHAPTER 9

Targeting Tumors Using Magnetic Drug Delivery

SHARON K. PULFER and JAMES M. GALLO

Department of Pharmacology, Fox Chase Cancer Center

INTRODUCTION

In the process of drug development, the question of how to deliver the compound of interest to the target organ or disease site must always be answered. Conventional drug administration routes, including oral, intravenous, and intraarterial modes, result in substantial distribution of the compound to normal, healthy tissue as well as to the disease site. Therefore, higher doses must be administered in order to reach the necessary concentration at the site of interest. For drugs with a low therapeutic index, the drug concentration that is required to see an effect may result in toxic side effects from the activity of the drug on normal tissue. To limit these systemic toxicities, regional or targeted drug delivery methods have emerged to increase local drug concentrations while decreasing the systemic exposure of normal tissues. To this end, magnetically responsive drug delivery systems have been developed that have shown increased local drug concentrations for targeted delivery of anticancer drugs. This chapter outlines the basic principles involved in targeting tumors using magnetic drug delivery systems, reviews the different types of systems that have been developed, and examines the future prospects for magnetic drug therapy.

MAGNETIC DRUG DELIVERY

The idea of applying the basic concept of magnetism to the area of drug delivery was first proposed in the 1960s.[1] These systems were developed as a means of targeting a specific site to increase local drug concentrations and reduce the toxic side effects associated with the systemic administration of these compounds. Antineoplastic

Biomedical Chemistry: Applying Chemical Principles to the Understanding and Treatment of Disease, Edited by Paul F. Torrence
ISBN 0-471-32633-x

agents used in systemic chemotherapy generally have a narrow therapeutic index because of their indiscriminate action on normal as well as tumor cells. This action decreases the concentration of drug that actually reaches the tumor site, requiring larger systemic doses of the drug to reach cytotoxic levels at the tumor site and leads to dose-limiting toxicity. By targeting the local tumor site, magnetic drug carriers allow higher tumor drug concentrations to be attained while reducing systemic toxicity.

Principle

The theory behind magnetic drug delivery is based on the attractive force between a magnetic field applied at the target organ site and the magnetic material (typically Fe_3O_4) dispersed within a drug-loaded particle. The particles are injected into the blood supply of the target organ in the presence of an external magnetic field of sufficient strength to retain the particles at the target site. Magnetic drug localization is based on the competitive forces between tumor blood flow and the external magnetic force. For the particles to be retained at the tumor, the external magnetic force must exceed the linear blood flow rates in the arteries and capillaries.[2] This increased retention time allows the particles to deliver the drug locally while avoiding rapid clearance by the mononuclear phagocytes of the reticuloendothelial system (RES), which occurs with other intravascularly administered drug carriers. This magnetic retention is dependent on magnetic field strength and gradient, time of application, and location of the magnet in relation to the target site. While retained at the tumor site, the magnetic field must be applied long enough for the particle to leave the vasculature to deliver the drug to the tumor cells. This is facilitated by the differences in permeability that exist between tumor and normal tissue. Increased endothelial permeability in tumor tissue can accelerate particle extravasation and increase drug delivery to tumor cells. Controlled release of the cytotoxic drug from the magnetic particle to tumor tissue following extravasation limits drug exposure to normal and endothelial tissues, and this enhanced specificity results in an increased therapeutic index of the drug in question.

Benefits and Limitations

Magnetic drug delivery systems are indicated in situations where specificity of the administered drug is necessary for therapeutic benefit. In vivo studies of magnetic drug systems have shown much higher fractions of the administered dose retained in the target area, resulting in a fundamental change in distribution of the administered drug. This may be attributed to more effective drug transport by the magnetic particles, which allows minimal drug release in systemic areas but through magnetic retention provides for controlled release of the drug in the target tissue. This type of delivery system is particularly beneficial in situations where administration of the indicated drug results in toxic systemic effects from nonspecific action of the drug on blood or non-target tissues or the drug in question is unstable, expensive or subject to extensive metabolism. However, magnetic drug delivery is inherently limited to target sites that are easily accessible to the external magnetic field and that also have a good blood supply. There are also formulation issues involved in the fact that the drug must withstand the magnetic carrier loading process.

TYPES OF MAGNETIC DRUG DELIVERY SYSTEMS

Since the early 1970s, a variety of magnetic drug delivery systems have been reported. These systems have shown considerable diversity in both their composition and formulation. In theory, there are two methods of synthesizing a magnetic drug delivery system. The first is to take an existing micrometer- or submicrometer- sized ferromagnetic particle and attach the drug of interest either through covalent attachment (conjugation) via active surface sites on the particle or through adsorption of the drug onto the particle surface. The other method involves the addition of a magnetically active component (usually magnetite, Fe_3O_4) during synthesis of the carrier system itself, whether it is a microsphere-, microcapsule-, nanosphere-, or liposome-based system with the drug attached through chemical conjugation, adsorption, or encapsulation methods. Following is an overview of the various magnetic drug delivery systems that have been reported in the literature, including their general synthetic scheme and in vitro and in vivo targeting results in an effort to show the potential capabilities and therapeutic benefits of magnetic drug delivery.

Ferromagnetic Particles

Magnetic Methotrexate Conjugates One type of magnetic drug carrier involves chemical conjugation of the drug to an existing magnetic particle containing active surface groups. Various magnetic conjugates involving the cytotoxic drug, methotrexate, were reported previously.[3] The most promising formulation involved carbodiimide coupling to attach aminated methotrexate to carboxyl-terminated magnetic particles with drug loading of 6.3% w/w and a particle size of 1 μm. Intraarterial administration of this magnetic conjugate in the presence of a 6000-G magnetic field for only 15 min resulted in 3.5- to 5-fold increased brain concentrations compared to the same dose of methotrexate (3 mg/kg) administered as a solution with tissue concentrations measured at 15, 30, and 45 min postinfusion.[4] Methotrexate levels were also lower in nontarget organs except for the lung. Confocal and light microscopy showed a predominantly extravascular particle distribution in tumor tissue, but mostly vascular distribution for normal brain tissue. However, a dose-dependent toxicity was observed 45 min postinfusion most likely due to particle aggregation and redistribution to the lung. Given the higher local concentrations in tumor tissue for the magnetic methotrexate conjugate and their extravascular distribution, decreasing the administered dose or increasing magnetic field strength or time may alleviate this toxicity and provide targeted drug delivery.

Drug Adsorbed Ferromagnetic Particles

Carminomycine Ferrocarbon Particles Another type of magnetic drug delivery system involves adsorption of the drug onto existing magnetic particles. Micrometer- and submicrometer-sized particles consisting of a ferromagnetic core with a carbon coating to which the drug, carminomycine, was adsorbed were evaluated for their ability to target various carcinosarcomas in rats.[5] Increased survival times compared to controls were reported for all groups with up to 60% of the animals recovering completely in certain tumor types. The biodistribution of carminomycine (8 mg/kg) showed three-fold higher tumor concentrations following intravenous administration of the magnetic adsorbent with a magnetic field for 30 min compared to drug administered in solution form and the particles injected without a magnetic field. A preliminary clinical study in which these particles were administered to patients with stage 3 and 4 disease involving various cancers was also reported.[5] Although little data were given, their preliminary report indicated tumor regression in the majority of patients with 11 of 13 terminal stage patients in one study exhibiting clinical regression and five complete recoveries. A significant amount of particles were also observed in the lymph nodes, which is the most likely site for metastases, indicating that this magnetic delivery system was able to target the actual tumor site as well as potential metastatic sites.

Paclitaxel Ferrocarbon Particles Recently, a magnetic adsorbent particle system has also been reported in which the anticancer drug, paclitaxel was adsorbed onto an alloy of iron and activated carbon.[6] In vitro studies reported drug loading of 7.5% w/w with 35% of the drug released after 24 h, increasing to a total of 62% after 72 h under physiological conditions. Magnetic retention of these particles under simulated arteriole flow rates was also possible and cytotoxicity assays showed IC_{50} values of 5 ng/mL. However, the ability of these particles to target tumors in vivo was not reported.

Epirubicin-Bound Ferrofluid Another magnetic adsorption system, epirubicin-bound ferrofluid, has also been reported in the literature although little data were given as to the synthesis, composition, drug loading, and release kinetics of this particle.[7,8] Animal studies involving human kidney and colon carcinoma xenograft mouse models in which 1 mg/kg of these particles were injected intravenously (I V) with a 15-min magnetic field application time (0.2–0.5 T) showed distinct regression of tumors.[7] More importantly, a human phase I study was completed involving 14 patients with advanced solid tumors of various origins. The particles (5–100 mg/m^2 doses of epirubicin) were administered I V over 15 min with application of a magnetic field (0.5–0.8 T) at the tumor site during infusion, continuing for 45 min postinfusion. In general, administration of the magnetic particles was reported to be well tolerated with tumor localization in about half of the patients without organ toxicity as determined through visual assessment of certain patients with open wounds and the results of magnetic resonance tomography scans.[8] Possible pharmacokinetic (PK) advantages of this magnetic drug delivery system was unable to be determined due to high variability. However, intraindividual pharmacokinetic data in two patients showed lower peak concentrations following magnetic drug targeting compared to systemic drug administration. Drug-related toxicity at doses greater than $50\,mg/m^2$ and difficulty in maintaining a constant magnetic field at the tumor site were also reported.

Polymeric Magnetic Particles

The third type of magnetic drug carrier system involves addition of a magnetically responsive component (Fe_3O_4) added during synthesis of a microsphere-, microcapsule-, nanosphere-, or liposome-based carrier system. The majority of these systems involve encapsulation of the drug within the magnetic particle, although drug conjugation and adsorption strategies are also possible. Drug encapsulation provides a number of potential advantages for drug delivery such as providing controlled drug release, protection against drug inactivation by metabolism, and modulation of drug resistance. Magnetic drug-encapsulated particles may offer an advantage over other types of magnetic drug delivery systems by limiting drug exposure to the external surroundings, thus minimizing systemic drug exposure.

Various materials that have been proposed for the matrix of magnetic encapsulated particles such as phospholipids (liposomes),[9] starch,[10] ethylcellulose,[11] alginate,[12] polylactic acid,[13,14] polyalkylcyanoacrylate,[15] polymethylmethacrylate,[16] polyethyleneimine,[17] chitosan,[18,19] dextran,[20] and albumin.[21] In vivo targeting studies of various magnetic drug-encapsulated particles have reported successful particle localization in target organs with decreased concentrations in nontarget organs of various animal models. Ear vein experiments have also shown increased particle localization of four- to eightfold increases in target concentrations after application of a magnetic field following administration of magnetic starch microspheres.[10] Kidney targeting experiments also resulted in a threefold increase in target organ concentrations in the presence of a magnetic field after injection of magnetic dactinomycin-loaded polyalkylcyanoacrylate nanoparticles.[15] Tail vein experiments have shown a 60-fold increase in target concentrations in the presence of a magnetic field for magnetic indomethacin-loaded polymethylmethacrylate nanoparticles.[16] Tumor targeting experiments have also shown complete remission of rabbit VX2 bladder tumors following administration of mitomycin C-ethylcellulose microcapsules.[11]

However, of all magnetic drug delivery systems reported in the literature, magnetic albumin microspheres have been the most widely reported and well characterized. These microspheres are prepared by heat or chemical stabilization of a water-in-oil emulsion of an aqueous solution of drug, magnetite (Fe_3O_4) particles and albumin dispersed in an oil phase.[21] These particles have been well characterized with controlled in vitro drug release rates reported and numerous examples of target specificity in vivo demonstrated.[22–35] Rat tail targeting studies demonstrated increased particle retention of magnetic albumin microspheres containing radiolabeled adriamycin at the magnet site dependent on magnetic field strength (4000, 6000, or 8000 G) when applied for 30 min.[21] Particle localization was observed 24 h after removal of the magnetic field. Particles were observed in the interstitial space of the endothelium, preventing delocalization of the particles after magnetic field application and serving as an extravascular depot to release drug at the target site.[21] Studies examining the effect of magnetic field application time (5–60 min) showed 7–10-fold increases in target site radioactivity compared to nontarget organs regardless of application time.[36] At 60 min postinjection, these concentrations were twice those obtained when the drug was administered intravenously in solution form.[36] Other studies have reported particle internalization and increased drug concentrations of up to 16-fold at the target tail site in normal rats with the extravascular appearance of microspheres at 2 h.[23,25–27] Tumor targeting studies of subcutaneous Yoshida sarcomas implanted in the tail have also shown transport of these particles across the capillary endothelium with microspheres located extravascularly as early as 10 min after dosing with a corresponding increase in drug concentration at the target organ with 77% of the animals exhibiting total remission after only one regimen of drug therapy.[32–34] Tail vein experiments in normal animals with 5-fluorouracil-loaded as well as camptothecin-loaded magnetic albumin microspheres have also shown enhanced target concentrations compared to free drug administrations.[37,38] Magnetic albumin microspheres have also shown the ability to target other organs other than the tail. Lung and kidney targeting experiments with magnetic albumin microspheres resulted in 2.5-fold increases in target organ concentrations in the presence of a magnetic field.[29,30] Tumor targeting experiments in lung have also reported increased survival times in animals with AH 7974 lung metastases following administration of magnetic adriamycin-loaded albumin microspheres compared to nontreated controls and free-drug solution.[31]

MAGNETIC CATIONIC MICROSPHERES FOR TARGETED DRUG DELIVERY

Given the targeting effectiveness of magnetic carrier systems in vivo as discussed above, it represents a promising approach for the targeted drug delivery of brain tumors. Chemotherapy of brain tumors has remained one of the most challenging aspects of drug delivery because of the impermeable nature of the blood–brain barrier (BBB), the lack of tumor specificity, and the development of drug resistance. The tight intercellular junctions and the small number of pinocytotic vesicles and fenestrations associated with the BBB block the passage of many drugs resulting in subtherapeutic drug concentrations and the growth of resistant cells. Although the distinct morphological changes associated with the BBB of tumors known as the blood–tumor barrier (BTB)[39–43] is generally associated with increased permeability, this has been shown to vary widely among tumors.[44–46]

Magnetic drug delivery provides a means of targeting the brain tumor site to increase local drug tumor concentrations, minimize drug interaction with normal tissues and limit phagocytosis by the reticuloendothelial system (RES). The magnetic field increases the retention time of the particles and their interaction with brain capillary endothelial cells. These cells, as well as tumor cells, are anionic in nature, due to the presence of glycosaminoglycans (GAGs) on the luminal surface.[47–49] Because of this anionic nature, cationic particles have been proposed to bind electrostatically with the anionic glycosaminoglycans (GAGs) on the cell surface and increase the residence time of the charged particles within the tumor, leading to higher intracellular drug concentrations.[50] These electrostatic forces as well as magnetic interactions help to retain the particle at the BBB or BTB and release entrapped drug, maintaining a high concentration gradient from blood to brain. In vitro studies have demonstrated that cationic polymers can bind and become endocytized by brain capillary endothelial cells.[51,52] Nonmagnetic cationic chitosan–methotrexate conjugates were found to enhance brain concentrations following intraarterial injection in normal rats compared to free drug in solution form.[53,54] Magnetic cationic particles allow for both magnetic retention of the particles at the local tumor site as well as electrostatic interaction with the endothelial cells facilitating extravascular uptake of the particles and increasing tumor concentrations. In normal animals, magnetic cationic particles consisting of oxantrazole-loaded magnetic chitosan microspheres were found to retain 4% of the administered dose in the brain following intraarterial injection with a magnetic field (6000 G) applied for 30 min.[19] Oxantrazole concentrations remained constant 120 min postinfusion, indicating retention and extravasation of the particles, while oxantrazole administered in solution form was below the detection limit in brain tissue at the same timepoint.

Magnetic Aminodextran Microspheres (MADMs)

Method and Physicochemical Characterization Recently, the synthesis and in vivo targeting ability of another type of magnetic cationic particle, the novel magnetic aminodextran microsphere (MADM), was introduced for the targeted drug delivery of brain tumors.[20] Briefly, these biodegradable microspheres are synthesized by crosslinking a water-in-oil emulsion of dextran and ferrofluid with cyanogen bromide. Cationic particles are created by covalently binding a diamine (1,6-hexanediamine) using cyanogen bromide after microsphere formation with the cationic nature controlled through amine reaction time. This control over the cationic nature and the biocompatability of MADM are improvements over the previously reported cationic magnetic particles of chitosan. MDM and MADM particles have a mean diameter of 1–2 μm as determined through SEM and laser light-scattering particle analysis with a magnetite (Fe_3O_4) content of 16% w/w. The amine groups of MADM have also been shown to be stable at physiological temperature and pH with the cationic nature of MADM confirmed through competitive dye displacement assays. In vitro, MADM particles were shown to be endocytized to a greater extent by rat glioma-2 (RG-2) cells than their neutral counterpart, magnetic dextran particles (MDM).[20]

In Vivo *Tissue Distribution Studies* Given this preferential uptake of the cationic particles, the ability of MADM to target intracerebral RG-2 brain tumors in

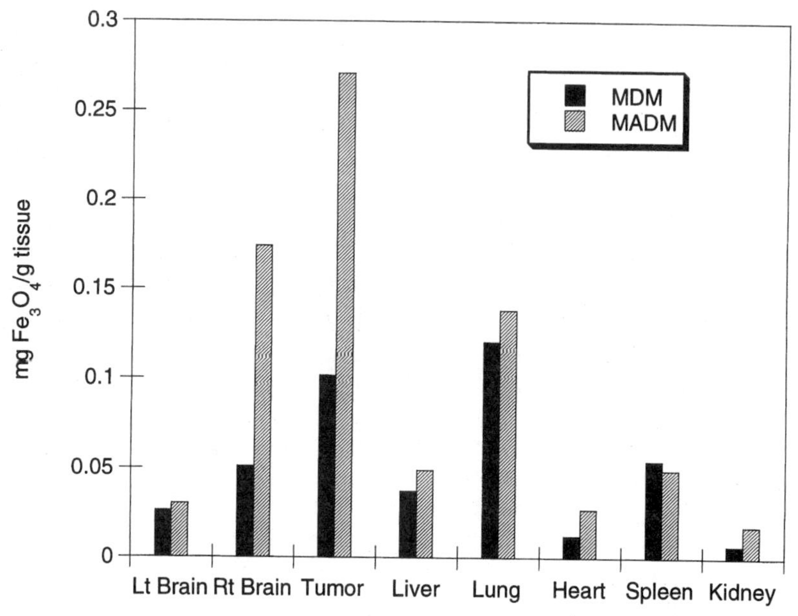

FIGURE 9.1 In vivo tissue distribution of RG-2 tumor bearing animals following intraarterial administration of magnetic cationic aminodextran (MADM) particles (25 mg/kg) and magnetic neutral dextran (MDM) microspheres (25 mg/kg) suspended in 0.1% Tween 80/saline in RG-2 tumor bearing animals in the presence of a 0 or 6000-G magnetic field. Animals were sacrificed at 30 min and 6 h postinjection. Values are reported as the average ($n = 12$) magnetite concentrations for all groups studied normalized to organ weight (mg Fe_3O_4/g tissue).

vivo was examined.[55] MDM and MADM particles (25 mg/kg) were administered intraarterially in the presence and absence of a 6000-G magnetic field for 30 min and evaluated for both tumor and nontarget organ concentrations in an intracerebral RG-2 rat tumor model. Cationic MADM particles were able to target brain tissue compared to nontarget organs for all treatments compared to neutral MDM particles (Fig. 9.1). Total brain tissue concentrations were obtained at levels corresponding to a maximum of 12% of the total dose administered for MDM groups and 32% of MADM for the various groups studied. MADM was also able to specifically target tumor tissue compared to normal brain tissue (Fig. 9.2) at levels two- to four-fold higher than neutral MDM particles and remained constant over time, indicating retention and extravasation of the cationic particles. Transmission electron microscopic (TEM) analysis revealed MADM located in the interstitial space of tumor cells with no direct evidence of endocytosis by tumor cells (Fig. 9.3). Neutral MDM particles were only observed in brain capillaries, providing additional support of electrostatic interactions. Comparison of these cationic and neutral 1 μm particles to smaller magnetic particles of 10–20 nm indicated that higher tumor concentrations were obtained for smaller magnetic particles.[56] However, the high tumor selectivity and concentrations observed over time for MADM and their localization

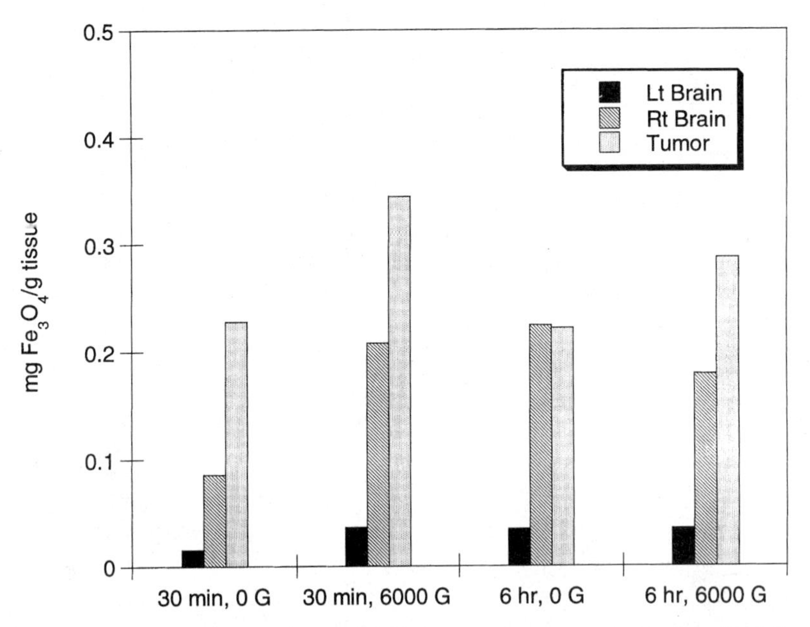

FIGURE 9.2 Brain concentrations of RG-2 tumor-bearing animals ($n = 3$) following intraarterial administration of magnetic cationic aminodextran (MADM) particles (25 mg/kg) suspended in 0.1% Tween 80/saline in the presence of a 0 or 6000-G magnetic field. Animals were sacrificed at 30 min and 6 h postinjection. Values are reported as mean ($n = 3$) magnetite concentrations normalized to organ weight (mg Fe_3O_4/g tissue).

in the interstitial space of tumor cells suggested a basis for further investigation and development of smaller drug-loaded cationic magnetic particles for targeted drug delivery to brain tumors.

Magnetic Cationic Liposomes

The formulation and in vitro targeting studies of small magnetic cationic liposomes have recently been reported in an effort to decrease cationic particle size and provide for lipophilic drug incorporation.[57] Magnetic and nonmagnetic cationic liposomes were approximately 200 nm in diameter and contained the anticancer drug, paclitaxel (76 μM) with drug release within 24 h under physiological conditions. Preliminary in vitro studies with rat glioma-2 (RG-2) cells showed all particles to be equally cytotoxic as free drug in solution with preferential uptake of cationic particles compared to neutral particles in the presence or absence of serum proteins (Fig. 9.4). Cationic particles also provided up to 10-fold higher intracellular drug concentrations compared to neutral particles (Fig. 9.5). Given the preferential uptake and higher accumulation of these magnetic cationic particles along with the in vivo evidence for larger magnetic cationic particles reported above, further development of small, cationic particles may provide improved drug delivery of chemotherapeutic agents for the treatment of brain tumors.

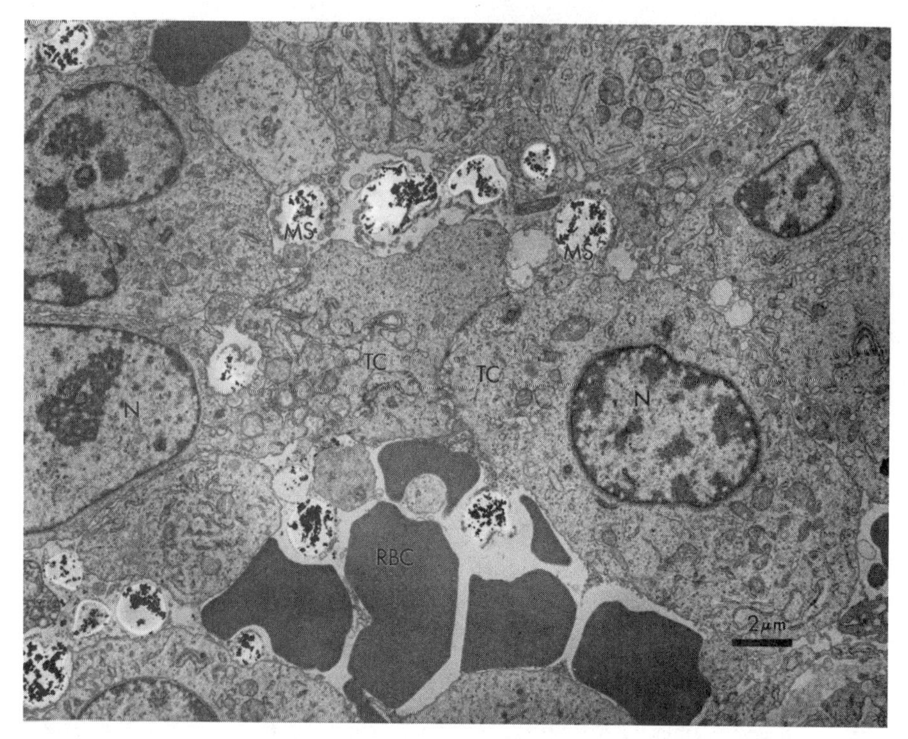

FIGURE 9.3 TEM image of tumor tissue from a rat glioma-2 (RG-2) bearing rat sacrificed 30 min following intraarterial administration of MADM (25 mg/kg) suspended in 0.1% Tween 80/saline in the presence of a 6000-G magnetic field showing MADM (labeled MS) in the interstitial space adjacent to the tumor cell membrane (TC). *N* represents the nucleus of the tumor cell and RBC represents a red blood cell.

CONCLUDING REMARKS AND FUTURE PROSPECTS

Throughout the development of magnetic drug carrier systems that began in the late 1960s and continues today, there have been numerous examples of remarkable particle localization in vivo in various animal models. However, these systems have yet to translate into a viable magnetic targeting strategy for human use. Although the two clinical trials to date involving magnetic drug carriers have alluded to the efficacy of magnetic targeting, they also bring up many questions that must be answered for the future development of magnetic carriers in clinical use. The main obstacle that must be overcome is the need for better-designed magnets that are compact, portable, and capable of producing a high magnetic field and field gradient that can be applied and retained at various parts of the body without difficulty, discomfort, or side effects to the patient. The impact of these magnetic fields on the biological system also needs to be examined as previous studies have shown possible effects on erythrocyte function and blood flow.[58–60]

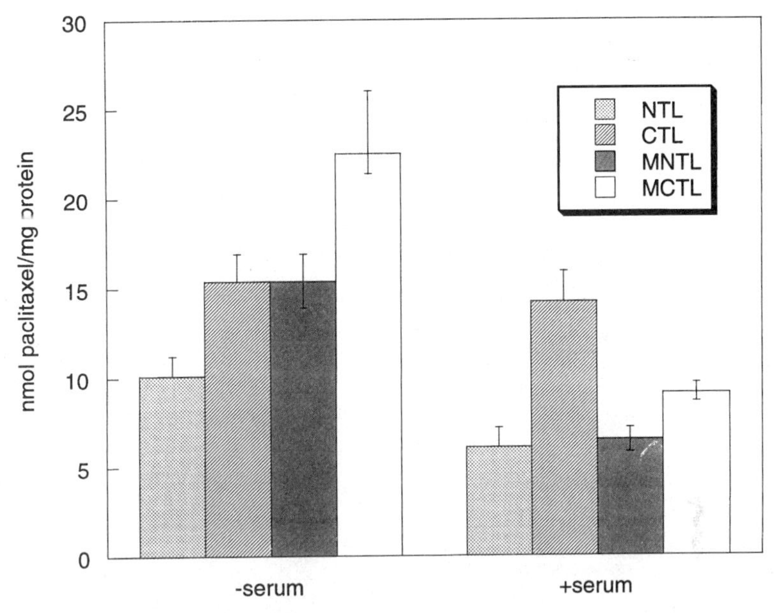

FIGURE 9.4 Cellular uptake of non-magnetic neutral (NTL) and cationic (CTL) paclitaxel-loaded liposomes and magnetic neutral (MNTL) and cationic (MCTL) particles in the presence and absence of 10% FBS in rat glioma-2 (RG-2) cells incubated for 1 h at a drug concentration of 0.88 μM. Values represent the mean ($n = 3$) cell-associated drug concentration normalized to cellular protein ± the standard deviation.

There have also been little data to date on the feasibility of large-scale production and reproducibility of a magnetic drug carrier system as well as the economic aspect of this therapy. Another concern about the use of magnetic targeting systems, is the effect of magnetic particles in living organisms over time. To date, there has been little reported in the literature as to the long-term effect of magnetic particles in animal models. This question of the biological fate of magnetic particles as well as the others mentioned above need to be addressed for regulatory approval of the development of magnetic therapy in humans.

The design of the ideal magnetic drug carrier will differ with regard to its physical and chemical characteristics for each specific application. However, an ideal magnetic drug carrier should be biocompatible and highly responsive in the presence of an external magnetic field with the ability to avoid clearance by the reticuloendothelial (RES) system. The applications of magnetic targeting systems also varies widely, but involve situations with a need for increased local drug concentrations and shielding of nontarget tissues that cannot be achieved in other drug delivery systems. The target area should have an adequate blood supply for the administration of the magnetic particles and accessibility to the external magnetic field. After administration of the particles to the target area, they can release drugs (chemotherapy) for the desired effect or in combination with heat (hyperthermia therapy) or radiation (radiotherapy).

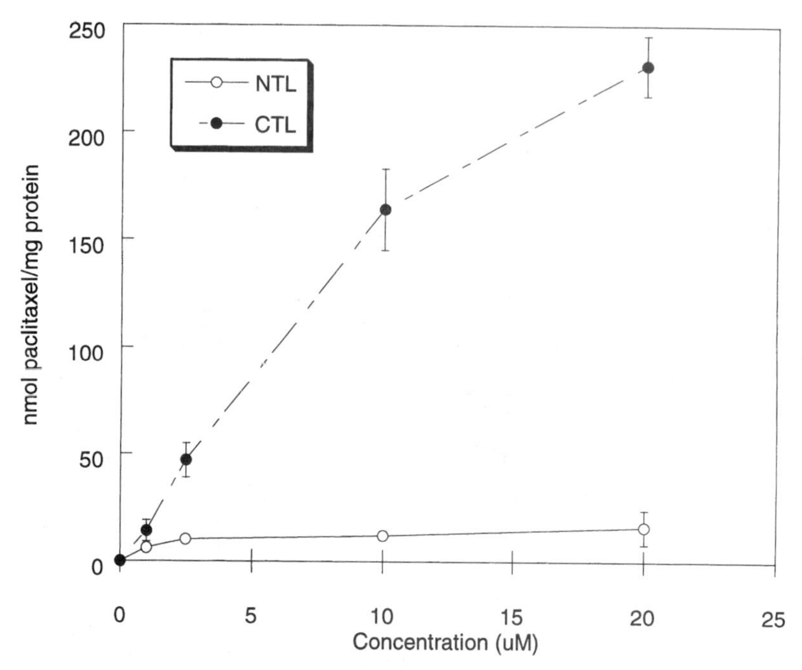

FIGURE 9.5 Cellular uptake of nonmagnetic neutral (NTL) and cationic (CTL) paclitaxel-loaded liposomes in rat glioma-2 (RG-2) cells incubated for 1 h at various drug concentrations in serum-free media. Values represent the mean ($n = 3$) cell-associated drug concentration normalized to cellular protein ± the standard deviation.

The use of magnetic drug carriers in chemotherapy is particularly attractive for tumors such as brain tumors that have limited access through other treatment methods such as surgery, radiation, and systemic chemotherapy. The therapeutic ineffectiveness of chemotherapy for metastatic brain tumors has resulted in poor prognosis rates and represents an attractive model for the use of magnetic drug targeting systems. Animal in vivo models have shown the effectiveness of magnetic drug carriers to increase brain tumor concentrations and overall brain targeting. In combination with cationic particles involving electrostatic interactions with the brain endothelium and tumor cells, these cationic magnetic targeting systems have demonstrated even higher levels of dose retention and tumor targeting in animal models. The location of these particles in the interstitial space of tumors can also provide the extravascular release of a drug, thus promoting tumor cell uptake without the need for the carrier itself to be taken up intracellularly. This ability of magnetic drug delivery systems to target the interstitial space also provides for future research into the design of magnetic carriers containing interstitial fluid targets. These results as well as the investigations mentioned above into the design of magnets suitable for in vivo applications, the biological fate and effect of magnetic particles as well as the feasibility and economical impact of large-scale production may help design magnetic drug carrier systems suitable for use in humans and realize their full potential for targeted drug delivery.

ACKNOWLEDGMENT

Partial funding provided from grant NS 34634 from the NIH.

REFERENCES

1. Meyers, P. H.; Cronic, F.; Nice, C. M., Jr., *Am. J. Roentgenol. Rad. Ther. Nucl. Med.* 1963, **90**, 1068–1077.
2. Senyei, A.; Widder, K.; Czerlinski, G., *J. Appl. Phys.* 1978, **49**, 3578–3583.
3. Devineni, D.; Blanton, C. D.; Gallo, J. M., *Bioconjugate Chem.* 1995, **6**, 203–210.
4. Devineni, D.; Klein-Szanto, A.; Gallo, J. M., *J. Neuro-Onc.* 1995, **24**, 143–152.
5. Kuznetsov, A. A.; Harutyunyan, A. R.; Dobrinsky, E. K.; Filippov, V. I.; Malenkov, A. G.; Vanin, A. F.; Kuznetsov, O. A. in *Scientific and Clinical Applications of Magnetic Carriers*; U. Häfeli, W. Schütt, J. Teller, and M. Zborowski, eds., Plenum Press, New York, 1997, pp. 379–389.
6. Allen, L. M.; Kent, T.; Wolfe, C.; Ficco, C.; Johnson, J. in *Scientific and Clinical Applications of Magnetic Carriers*, U. Häfeli, W. Schütt, J. Teller, and M. Zborowski, eds., Plenum Press, New York, 1997, pp. 481–494.
7. Lübbe, A. S.; Bergemann, C.; Huhnt, W.; Fricke, T.; Riess, H.; Brock, J. W.; Huhn, D., *Cancer Res.* 1996, **56**, 4694–4701.
8. Lübbe, A. S.; Bergemann, C.; Riess, H.; Schriever, F.; Reichardt, P.; Possinger, K.; Matthias, M.; Dorken, B.; Herrmann, F.; Gurtler, R.; Hohenberger, P.; Haas, N.; Sohr, R.; Sander, B.; Lemke, A.J.; Ohlendorf, D.; Huhnt, W.; Huhn, D., *Cancer Res.* 1996, **56**, 4686–4693.
9. Kiwada, H.; Sato, J.; Yamada, S.; Kato, Y., *Chem. Pharm. Bull.* 1986, **34**, 4253–4258.
10. Mosbach, K.; Schroder, U., *FEBS Lett.* 1979, **102**, 112–116.
11. Kato, T.; Nemoto, R.; Mori, H.; Abe, R.; Unno, K.; Goto, A.; Murota, H.; Harada, M.; Homma, M., *Appl. Biochem. Biotechnol.* 1984, **10**, 199–211.
12. Saslawski, O.; Weingarten, C.; Benoit, J. P.; Couvreur, P., *Life Sci.* 1988, **42**, 1521–1528.
13. Häfeli, U. O.; Sweeney, S. M.; Beresford, B. A.; Sim, E. H.; Macklis, R. M., *Jo. Biomed. Mater. Res.* 1994, **28**, 901–908.
14. Häfeli, U. O.; Sweeney, S. M.; Beresford, B. A.; Humm, J. L.; Macklis, R. M., *Nucl. Med. Biol.* 1995, **22**, 147–155.
15. Ibrahim, A.; Couvreur, P.; Roland, M.; Speiser, P., *Jo. Pharma. Pharmacol.* 1983, **35**, 59–61.
16. Vyas, S. P.; Malaiya, A., *Jo. Microencaps.* 1989, **6**, 493–499.
17. Kharkevich, D. A.; Alyautdin, R. N.; Filippov, V. I., *J. Pharm. Pharmacol.* 1989, **41**, 286–288.
18. Hassan, E. E.; Parish, R. C.; Gallo, J. M., *Pharmaceutical Research* 1992, **9**, 390–397.
19. Hassan, E. E.; Gallo, J. M., *J. Drug Target.* 1993, **1**, 7–14.
20. Pulfer, S. K.; Gallo, J. M., in *Scientific and Clinical Applications of Magnetic Carriers*, U. Häfeli, W. Schütt, J. Teller, and M. Zborowski, eds., Plenum Press, New York, 1997, pp. 445–455.

21. Widder, K. J.; Senyei, A. E.; Scarpelli, D. G., *Proc. Soc. Exp. Biol. Med.* 1978, **58**, 141–146.
22. Gallo, J. M.; Hung, C. T.; Gupta, P. K.; Perrier, D. G., *J. Pharmacokin. & Biopharm.* 1989, **17**, 305–326.
23. Gallo, J. M.; Gupta, P. K.; Hung, C. T.; Perrier, D. G., *J. Pharm. Sci.* 1989, **78**, 190–194.
24. Gupta, P. K.; Hung, C. T., *J. Pharm. Sci.* 1989, **78**, 745–748.
25. Gupta, P. K.; Hung, C. T.; Rao, N. S., *J. Pharm. Sci.* 1989, **78**, 290–294.
26. Gupta, P. K.; Hung, C. T., *J. Microencaps.* 1990, **7**, 85–94.
27. Gupta, P. K.; Hung, C. T., *Life Sci.* 1990, **46**, 471–479.
28. McArdle, C. S.; Lewi, H.; Hansel, D.; Kerr, D. J.; McKillop, J. H.; Willmott, N., *Br. J. Surg.* 1988, **75**, 132–134.
29. Morimoto, Y.; Sugibayashi, K.; Okumura, M.; Kato, Y., *J. Pharmacobiodynam.* 1980, **3**, 264–267.
30. Morimoto, Y.; Okumura, M.; Sugibayashi, K.; Kato, Y., *J. Pharmacobiodynam.* 1981, **4**, 624–631.
31. Sugibayashi, K.; Okumura, M.; Morimoto, Y., *Biomaterials* 1982, **3**, 181–186.
32. Widder, K. J.; Morris, R. M.; Poore, G.; Howard, D. P., Jr.; Senyei, A. E., *Proc. Nat. Acad. Sci. USA* 1981, **78**, 579–581.
33. Widder, K. J.; Marino, P. A.; Morris, R. M.; Howard, D. P.; Poore, G. A.; Senyei, A. E., *Eur. J. Cancer Clin. Oncol.* 1983, **19**, 141–147.
34. Widder, K. J.; Morris, R. M.; Poore, G. A.; Howard, D. P.; Senyei, A. E., *Eur. J. Cancer Clin. Oncol.* 1983, **19**, 135–139.
35. Widder, K. J.; Senyei, A. E.; Ranney, D. F., *Cancer Res.* 1980, **40**, 3512–3517.
36. Senyei, A. E.; Reich, S. D.; Gonczy, C.; Widder, K. J., *J. Pharm. Sci.* 1981, **70**, 389–391.
37. Fruitwala, M. A.; Sanghavi, N. M., *Drug Deliv.* 1996, **3**, 5–8.
38. Sonavaria, V. J.; Jambhekar, S.; Maher, T., *Proc. Internatl. Control. Rel. Bioact. Mater.* 1994, **21**, 194–197.
39. Greig, N. H., in *Implications of the Blood-Brain and Its Manipulation*, E. A. Neuwelt, ed., Plenum, New York, 1989; Vol. 1, pp. 311–368.
40. Greig, N. H.; Jones, H. B.; Cavanagh, J. B., *Clin. Expl. Metastasis* 1983, **1**, 229–246.
41. Hirano, A.; Ghatak, N. R.; Becker, N. H.; Zimmerman, H. M., *Acta Neurophathol.* **1974**, *27*, 93–104.
42. Deane, B. R.; Lantos, T. A., *J. Neurol. Sci.* 1981, **49**, 67–77.
43. Nishao, S.; Ohta, M.; Abe, M.; Kitamura, K., *Acta Neuropathol.* 1983, **59**, 1–10.
44. Groothius, D. R.; Fischer, J. M.; Lapin, G.; Bigner, D. D.; Vick, N. A., *J. Neuropathol. Exp. Neurol.* 1982, **41**, 164–185.
45. Warnke, P. C.; Friedman, H. S.; Bigner, D. D.; Groothius, D. R., *Cancer Res.* 1987, **47**, 1687–1690.
46. Levin, V. A.; Freeman-Dove, M.; Landahl, H. D., *Arch. Neurol.* 1975, **32**, 785–791.
47. Wusteman, F. S., in *Biochemical Interactions at the Endothelium*, A. Cryer, ed., Elsevier, Amsterdam, 1983, pp. 79–109.
48. Ausprunk, D. H.; Boudreau, C. L.; Nelson, D. A., *Am. J. Pathol.* 1981, **101**, 353–366.
49. Simionescu, N.; Simionescu, M.; Palade, G. E., *J. Cell Biol.* 1981, **90**, 605–613.

50. Gallo, J. M.; Hassan, E. E., *Pharm. Res.* 1988, **5**, 300–304.
51. Kumagai, A. K.; Eisenberg, J. B.; Pardridge, W. M., *J. Biol. Chem.* 1987, **262**, 15214–15291.
52. Smith, K. R.; Borchardt, R. T., *Pharm. Res.* 1989, **6**, 466–473.
53. Sanzgiri, Y. D.; Blanton, C. D., Jr.; Gallo, J. M., *Pharm. Res.* 1990, **7**, 418–421.
54. Sanzgiri, Y. D.; Blanton, C. D.; Gallo, J. M., *Polym. Adv. Technol.* 1992, **3**, 317–321.
55. Pulfer, S. K.; Gallo, J. M., *J Drug Target.* 1998, **6**, 215–227.
56. Pulfer, S. K.; Ciccotto, S. L.; Gallo, J. M., *J Neuro-Oncol.* 1999, **41**, 99–105.
57. Pulfer, S. K.; Gallo, J. M., *2nd Internat. Conf. Scientific and Clinical Applications of Magnetic Carriers*, Cleveland, OH, May 28–30, 1998, p. 39.
58. Okazaki, M.; Maeda, N.; Shiga, T., *Experientia* 1986, **42**, 842–843.
59. Okazaki, M.; Maeda, N.; Shiga, T., *Eur Biophys J* 1987, **14**, 139–145.
60. Okazaki, M.; Kon, K.; Maeda, N.; Shiga, T., *Physiol. Chem. Phys. Med. NMR* 1988, **20**, 3–14.

CHAPTER 10

Cancer: Superacid Generation of New Antitumor Agents

JEAN-CLAUDE JACQUESY

Laboratoire de Synthèse et Reactivité des Substances Naturelles,
Faculté des Sciences, Université de Poitiers

JACQUES FAHY

Division de Chimie Medicinale V, Centre de Recherche Píerre Fabre

INTRODUCTION

Vinblastine (**1**) and vincristine (**2**) are the lead compounds of the widely recognized antimitotic *Vinca* alkaloids, which represent a chemical class of major interest in cancer chemotherapy. These naturally occurring complex molecules are present in minute quantities in the leaves of the Madagascan periwinkle, *Catharanthus roseus* (L.) G. Don.[1] They were first identified and then isolated in the late 1950s independently by the group of Pr. Noble at the Western University of Ontario, London, Canada, and by researchers at the Eli Lilly Laboratories in Indianapolis, USA.[2] *Vinca* alkaloids are constituted from two building blocks: a cleavamine-type derivative arising from a rearrangement of catharanthine (**3**), and vindoline (**4**).

Vinca alkaloids have been employed in clinical practice for more than thirty years and remain widely used to this day; the clinical chemotherapeutic utility of both vinblastine and vincristine was firmly established by 1965. The binding to tubulin, an ubiquitous protein responsible for the mitotic spindle, and the subsequent arrest of cells in mitosis are generally accepted as the key events in delineating their mechanism of action.[3]

During the 1970s extensive chemical research has been undertaken in order to obtain semisynthetic derivatives of vinblastine (**1**) with the aim of identifying more active and less toxic drugs exhibiting a wider spectrum of anticancer efficacy. In the

Biomedical Chemistry: Applying Chemical Principles to the Understanding and Treatment of Disease, Edited by Paul F. Torrence
ISBN 0-471-32633-x

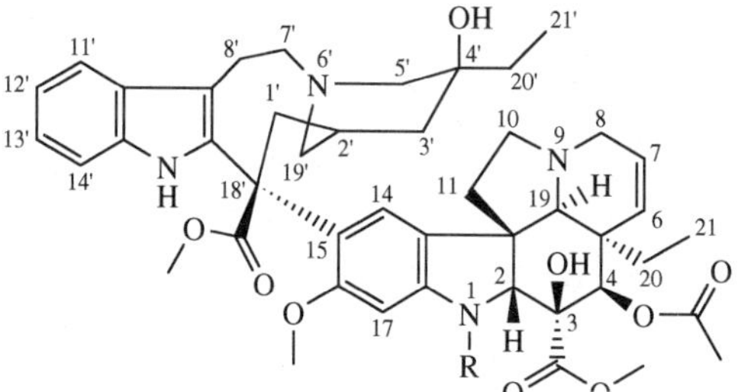

1 : R = CH_3
2 : R = CHO

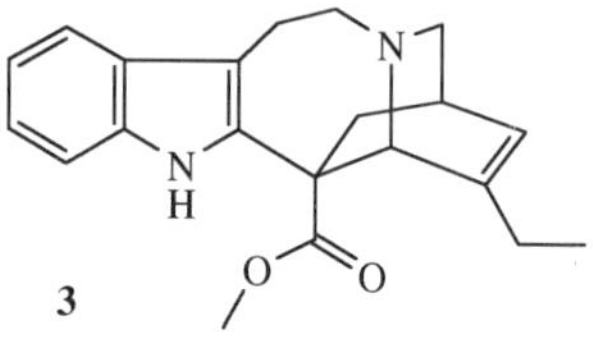

3

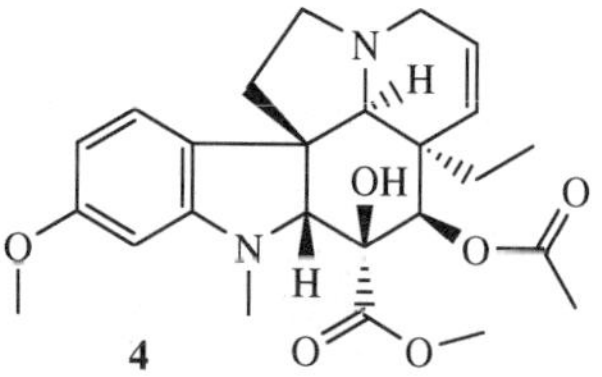

4

meantime, the first synthesis of *Vinca* alkaloids was achieved by Potier's group at the CNRS in Gif-sur-Yvette, France.[4a]

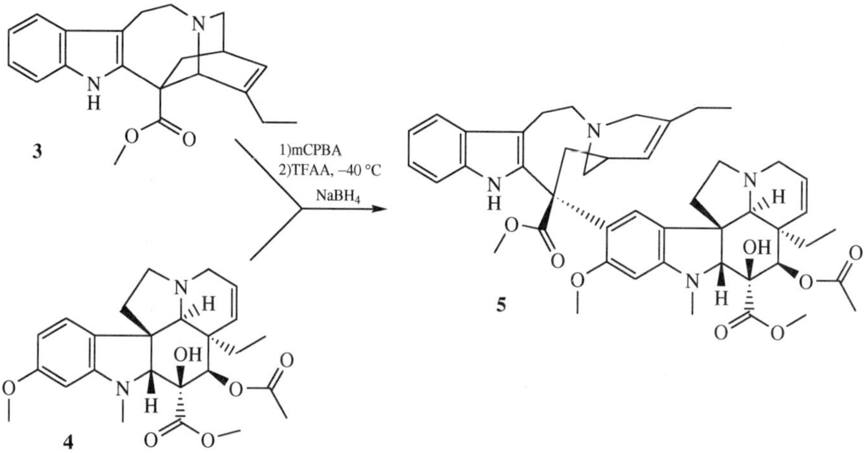

3

1)mCPBA
2)TFAA, –40 °C
$NaBH_4$

5

4

The methodology was based on the biomimetic coupling of catharanthine (**3**) and vindoline (**4**) through a modified Polonovski reaction, leading to the natural anhydrovinblastine (**5**) in high yield. Subsequently, several other groups have investigated different strategies aimed at the partial synthesis of vinblastine (**1**) using vindoline (**4**) as the starting material and elaborating the cleavamine moiety. The main successful contributions have come from the groups of Potier,[4b] Kuehne,[5] and Magnus.[6] The tremendous amount of work up to 1992 has been comprehensively reviewed by Atta-ur-Rahman.[7] Since that time, however very little new chemistry involving the vinblastine family of molecules has been reported. Enantioselectivity of the earlier vinblastine synthesis developed by Magnus[8] and Kuehne[9] has been improved. Cephalosporin-conjugate derivatives have been synthesized with the aim of targeting tumors in vivo.[10] Functionalization at C3′ of anhydrovinblastine (**5**) has been described by Sundberg et al. by reaction of the iminium intermediate with dialkylzinc reagents, but the resulting products have been evaluated only for their tubulin polymerization inhibitory properties.[11] A new methodology has been employed to introduce substituents such as nucleotide derivatives onto the C4 vindoline moiety, but their pharmacological properties have not yet been described.[12]

Vinca Alkaloids in Clinical Development

Since the discovery of the chemotherapeutic efficacy of *Vinca* alkaloids, several hundreds have been synthesized and evaluated for their pharmacological activities. Many of these syntheses used vinblastine (**1**) as the starting substrate, since for a long time this was the only material readily available in sufficient quantity for such studies. For a few of them only, clinical trial data have been reported. Many comprehensive reviews of this subject have been published, the most exhaustive covering the different aspects ranging from their isolation, medicinal chemistry, pharmacology, to their clinical usage.[13]

The majority of the compounds evaluated pharmacologically have been obtained by modifications of the vindoline moiety, bearing several reactive centers:

- An alcohol function at C4
- A substituent at N1
- Modification of the carbomethoxy group at C3.

In this review, only those derivatives for which clinical trial data have been reported are considered.

A large number of such derivatives have been synthesized by researchers at the Eli Lilly Company; the first series was constituted by α-aminoacyl vinblastine analogs modified at C4, leading to vinglycinate (**6**). This compound was the first semisynthetic *Vinca* alkaloid to enter phase I clinical trials in 1967.[14]

Subsequently a large series of amido derivatives at C3 were developed[15] from which vindesine **7** was selected for clinical trials on the basis of its superior efficacy compared to vinblastine or vincristine in pharmacological studies.[16] Vindesine (**7**) was registered in Europe in 1980, and is now available in several countries.

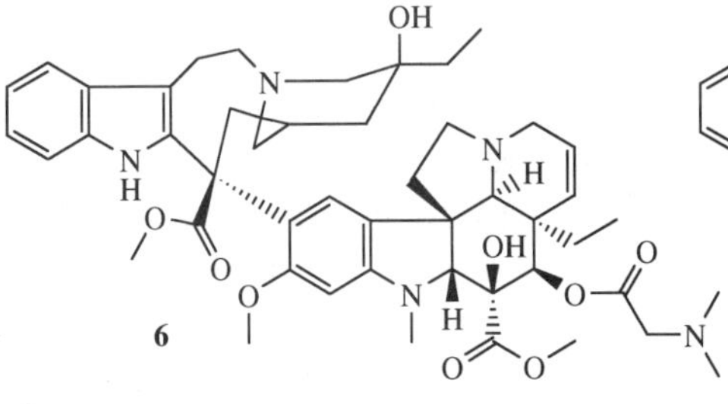

6 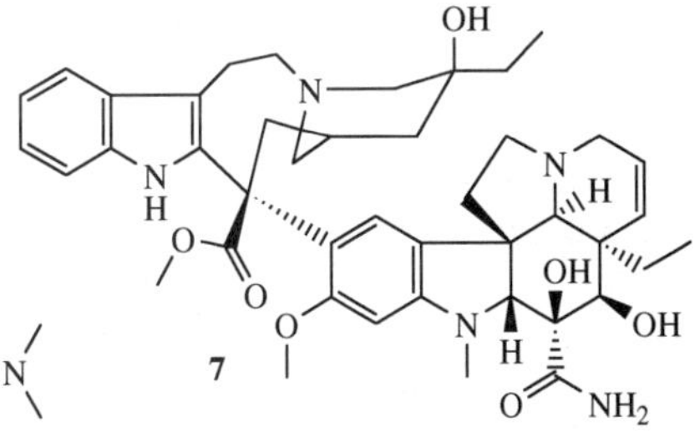 **7**

During the search for original derivatives, Potier and co-workers investigated the functionalization of the newly available anhydrovinblastine (**5**).[4a] Thus C′ ring contraction was first achieved when a second Polonovski reaction was applied to **5**, generating vinorelbine (**8**) as a prototype of new *Vinca* alkaloid derivatives.[17]

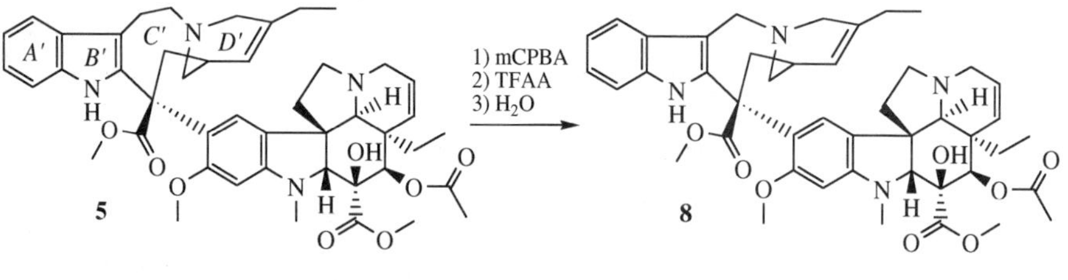

To date, vinorelbine (**8**) is the only product of this class with clinical efficacy to be modified in its cleavamine moiety. It was identified as having major antitumor properties with reduced toxic side effects. Clinical trials started in 1981,[18] and this compound (Navelbine) was first registered in France in 1989 by Pierre Fabre Medicament (France) for the treatment of the lung carcinoma (NSCLC), then for advanced breast cancer in 1992. It is now marketed worldwide, and several further indications are under clinical investigation, including an oral formulation.

Following a strategy similar to that of Eli Lilly, the Belgian pharmaceutical company Omnichem synthesized a family of 3-aminoacyl vinblastinoyl congeners.[19] Among this series, vintriptol (**9**) entered phase I clinical trials in 1983, but subsequently in 1992 its development was discontinued.

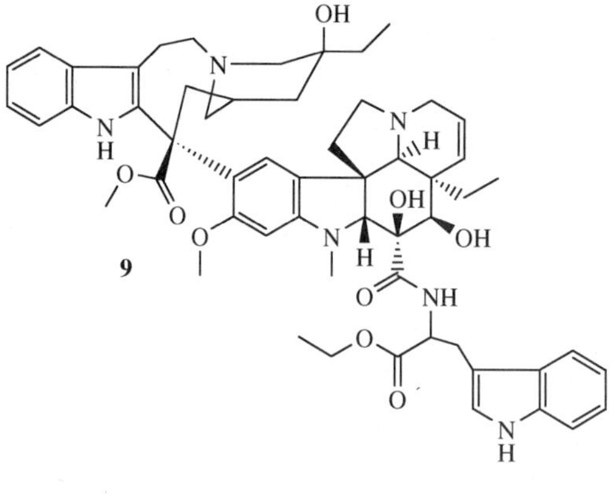

During the same period, the Eli Lilly group put forward two new derivatives for clinical evaluations: vinzolidine (**10a**) (R = CH_2CH_2Cl) and vinepidine (**11**). Two function modifications in the vindoline moiety are involved in the vinzolidine structural originality. As a starting point, reaction of vinblastine **1** with an excess of methyl isocyanate followed by warming led to the methyloxazolidinedione derivative (**10b**) (R = CH_3). Then further chemistry was undertaken and a large series of C3 spiro derivatives were prepared.[20] Marked antitumor efficacy of vinzolidine by the oral route prompted the Eli Lilly group to start clinical trials involving this route of administration, which was unusual with *Vinca* alkaloids. However, significant patient to patient intervariability was observed, and clinical development was discontinued after a further trial by the IV route indicated no significant benefit compared to vinblastine (**1**).

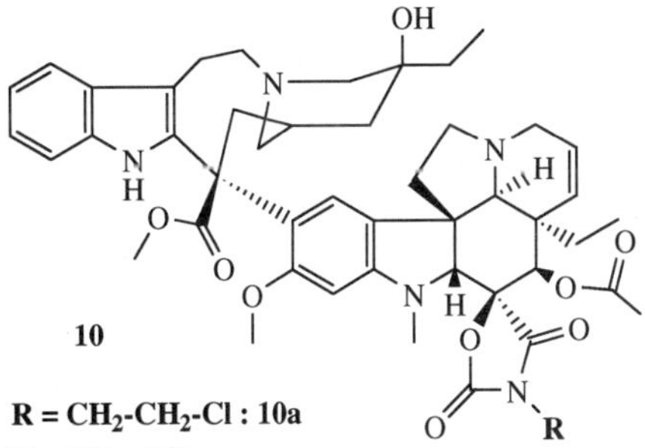

10

11

R = CH_2-CH_2-Cl : 10a
R = CH_3 : 10b

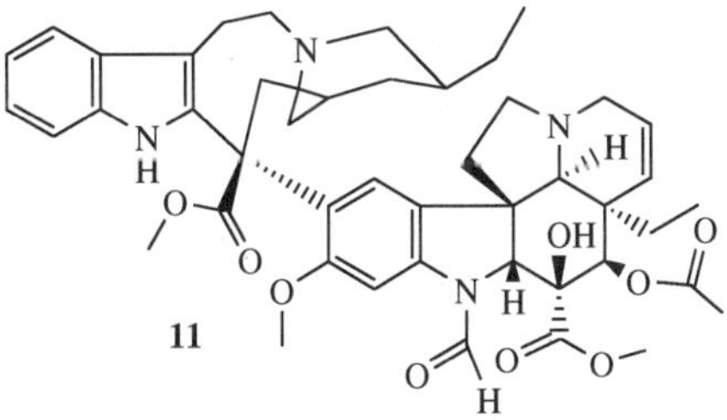

Vinepidine (**11**) is one of the few compounds bearing a modification in the upper part of the molecule for which intensive pharmacological studies have been undertaken.[21] It corresponds to 4′-*epi*-4′-deoxyvincristine and was identified by its increased tubulin binding affinity relative to vinblastine (**1**).[22] Clinical studies started in 1983, but were discontinued on account of neuromuscular toxicity similar to that of vincristine (**2**).[23]

An original strategy based on vinblastine derivatives linked to monoclonal antibodies, such as KS1/4-vinblastine (**12**), has also been elaborated.[24] The aim was to design compounds that specifically targeted the chondroitin sulfate proteoglycan surface antigen common to several carcinomas, with the potential to deliver vinblastine derivatives to malignant tissues and permitting the administration of otherwise lethal doses. Clinical trials started in 1988, but no further development of this interesting approach has been reported.

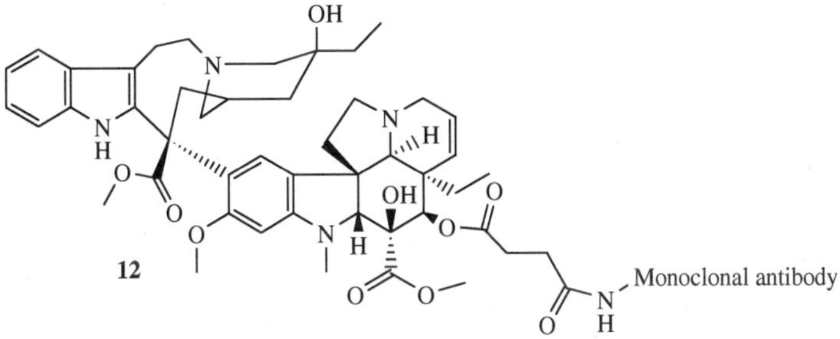

12

Monoclonal antibody

More recently, the French pharmaceutical company Servier has synthesized a series of aminophosphonate derivatives substituted at C3. This chemical approach, similar to that of vinglycinate (**6**) or vintriptol (**9**), was based on the concept that α-amino phosphonic acids can be considered as bioisosters of natural amino acids. The stereochemistry of the phosphonate group proved highly determinant of both in vitro and in vivo activities. Vinfosiltine (**13**) was selected for further development on the basis of its unusual potency compared to the reference compounds, vinblastine (**1**) and vincristine (**2**).

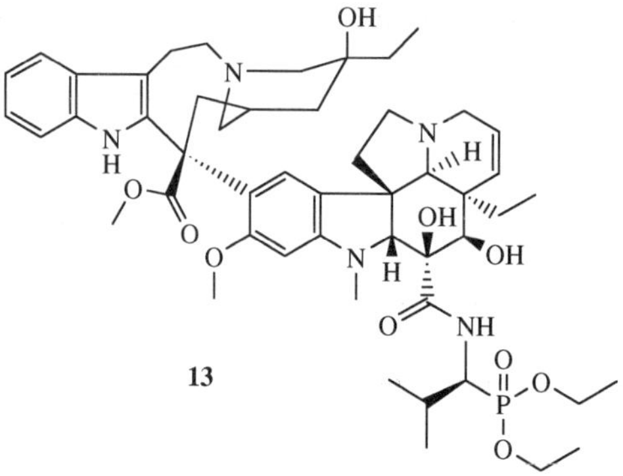

13

Vinfosiltine (**13**) was approximately 10 times as cytotoxic as vinblastine (**1**) when evaluated in vitro for its inhibition of murine leukemia L1210 cell proliferation. Using the murine leukemia P388 (IP/IP) in vivo model, the optimal active dose was 0.15 mg/kg versus 3.0 mg/kg for vinblastine (**1**).[25] However, although overall promising results of preclinical studies were published, no evidence of marked benefit in phase I and phase II clinical trials relative to the other *Vinca* alkaloids has been obtained, and development was discontinued in 1993.[26]

In conclusion, despite these numerous efforts in the fields of chemistry and biology since the early 1970s, only two semisynthetic *Vinca* alkaloid derivatives, vindesine (**7**) and vinorelbine (**8**) achieved the rank of approved anticancer drugs.

REQUIREMENTS FOR DESIGNING NEW *VINCA* ALKALOIDS

The success of vinorelbine (**8**) in human chemotherapy has encouraged the French Pierre Fabre group to continue the search in the field of *Vinca* alkaloids for more active compounds, active against other tumor types and/or with a lower/different spectrum of toxic side effects. The main requirement was to find a way further to exploit the antitumor properties of the *Vinca* alkaloids, knowing that several hundreds of such compounds have been synthesized and quite a large number have been evaluated for their potential anticancer activity. As we have seen, most of them have been obtained using conventional chemistry around the vindoline nucleus. On the other hand, little is known of the molecular interactions between tubulin and *Vinca* alkaloids and how these impact on cytotoxicity. After numerous efforts, the structure (3.8-Å resolution) by electron crystallography of the α,β-tubulin dimer stabilized with paclitaxel has been established only recently.[27] This work is considered a major advance in cytoskeleton research,[28] and further investigations will undoubtedly allow a more precise characterization of the binding sites of the different ligands including *Vinca* alkaloids, taxanes, or even new molecules recently identified. Nowadays, indeed, any rational design of compounds based on the modeling of drug–tubulin interactions still remains outside the realms of possibility.

Identifying an Original Strategy

We were interested in identifying an original chemical approach that conceivably could induce dramatic changes in the skeleton of the *Vinca* molecule. Investigation of the reactivity of these highly functionalized compounds in superacidic media appeared to represent an interesting challenge, since very few such examples have been described in the literature. The term "superacid" appeared for the first time in 1927 in a paper published by Conant and Hall when these authors observed the enhanced acidity of sulfuric acid and perchloric acid in nonaqueous solutions, leading to the formation of salts with weak bases such as carbonyl compounds.[29] It was not through until the 1960s, when Olah demonstrated the importance of these "superacid solutions" in stabilizing carbocations and Gillespie then proposed to define as a superacid any system more acidic than 100% sulfuric acid, namely, $H_0 \leq -12$.[30,31] One of the strongest acidic systems is obtained with hydrogen fluoride (HF), in combination with a fluorinated Lewis acid such as antimony pentafluoride SbF_5. For the 1 : 1 [HF : SbF_5] system, the H_0 acidity scale has been estimated to be of the order of −30, specifically, 10^{18} as acidic as pure sulfuric acid![32]

Since the early 1970s we have studied the reactivity of functionalized organic substrates, especially natural products, in the [HF : SbF_5] system. Under these superacidic conditions these compounds are (poly)protonated and their reactivity is modified dramatically when compared to what is observed in conventional acids. Novel and selective reactions have been discovered first using simple substrates as model compounds, then, when applied to polyfunctional natural products such as steroids, terpenes, and alkaloids: isomerization of saturated or unsaturated ketones, dearomatization of phenolic derivatives, ionic hydrogenation of enones and phenols, oxidation of nonactivated C–H bond, and so on.[33]

Reactivity of *Aspidosperma* Alkaloids

Aspidosperma alkaloids have generated a great deal of interest in organic synthesis because they are recognized as clinically useful derivatives and for the intellectual challenge they represent. Using a completely different approach, we have investigated the reactivity of these highly functionalized substrates in superacids, leading to the modification of a series of new *Vinca* derivatives.

We previously reported on the electrophilic hydroxylation of indolenines and indoles by hydrogen peroxide in [HF : SbF_5]: monohydroxylation of the benzene ring was observed leading to isomeric phenolic derivatives, protonated hydrogen peroxide $H_3O_2^+$ reacting on the protonated substrate.[34] Similar oxidation of the corresponding indolines was more selective, occurring at C4 or C6 (indole numbering).[35]

We anticipated carrying out a similar hydroxylation of readily available tabersonine (**14a**) to obtain its 16-methoxy analog (**14b**), a potential precursor of vindoline **4** that constitutes the lower half of the antineoplastic agents vinblastine (**1**) and vinorelbine (**8**).[14,36]

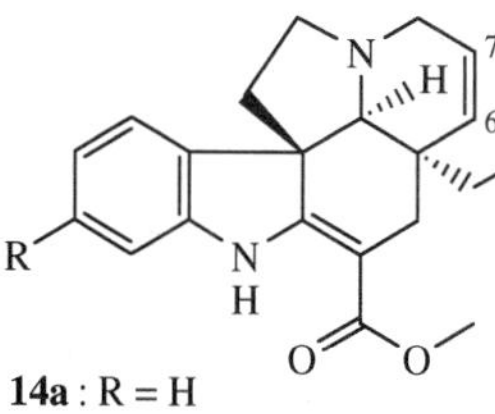

14a : R = H
14b : R = OCH_3

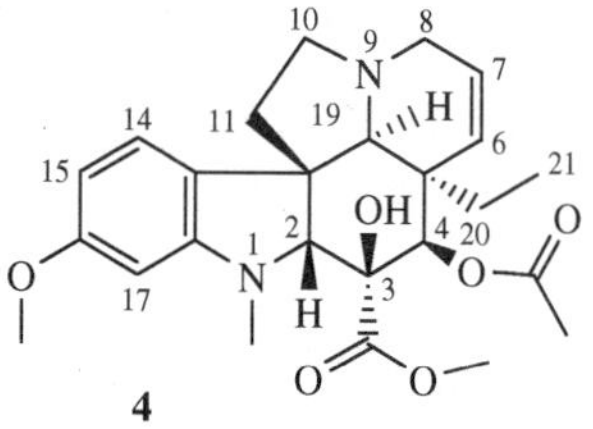

4

Unfortunately, only isomeric fluorhydrins (**15**) were obtained; the C6–C7 double bond was more reactive than the benzene ring.[37]

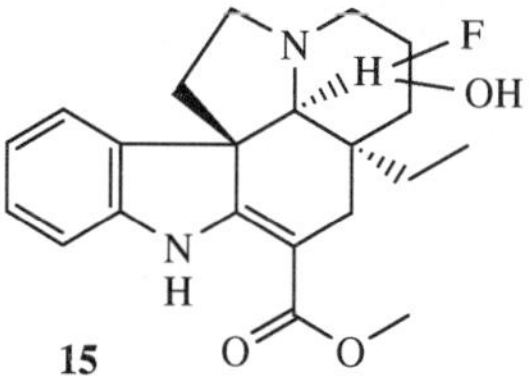

15

This problem could be circumvented starting from vincadifformine and tabersonine derivatives **16a** and **16b**, with hydroxylation leading to the desired hydroxyderivatives **16c** (40%) and **16d** (37%), respectively.[38]

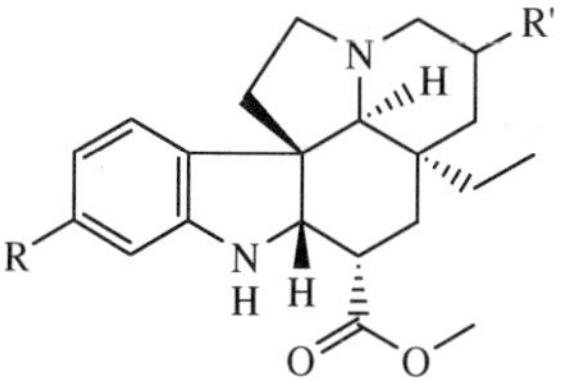

16a : R = R' = H
16b : R = H, R' = OH
16c : R = OH, R' = H
16d : R = R' = OH

The reactivity of vindoline **4** in superacids gave unexpected results.[39] In the presence of NBS, NCS, or H_2O_2 in [HF : SbF_5] (molar ratio 25 : 1), 7β-substituted-20-fluoro-6,7-dihydro vindolines **17b–d** were obtained, respectively; the (20*S*)-stereoisomer was by far the major product (20*S* : 20*R* ratio = 12 to 15). This unusual reaction implies reaction of the electrophile (Br^+, Cl^+, and OH^+ equivalent) on the β face of the C6–C7 double bond followed by a 1,3-hydride shift from C20 to C6 and fluorination at C20. These compounds were incorporated into dimeric derivatives by condensation with catharanthine **3** by known procedures.[4b,40] The resulting compounds displayed lower pharmacological activity than did the parent anhydrovinblastine **5**, confirming the importance of the C6–C7 double bond for antimitotic activity.[41]

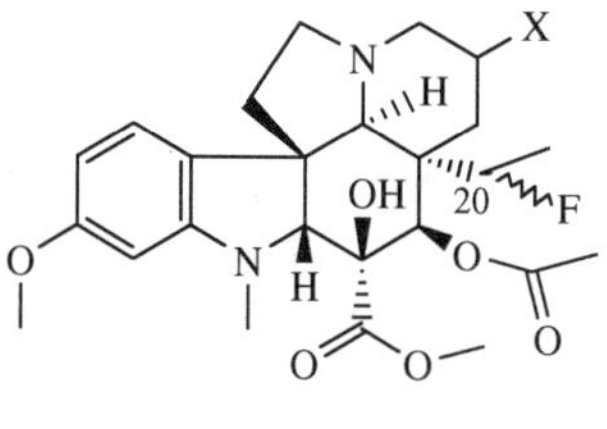

17a : X = H
17b : X = Br (49%)
17c : X = Cl (55%)
17d : X = OH (26%)

In more acidic conditions ([HF : SbF_5] molar ratio 10 : 1), vindoline **4** and its 4-*O*-deacetyl derivative rearrange to the cathovaline-like compounds **18a** and **18b**, respectively, and furanic diastereoisomers (**19**). The ether (**18a**) was obtained almost quantitatively with fluoride elimination when (20*S*) (**17a**) was reacted in [HF : SbF_5], whereas the (20*R*) analog was completely unreactive.[42]

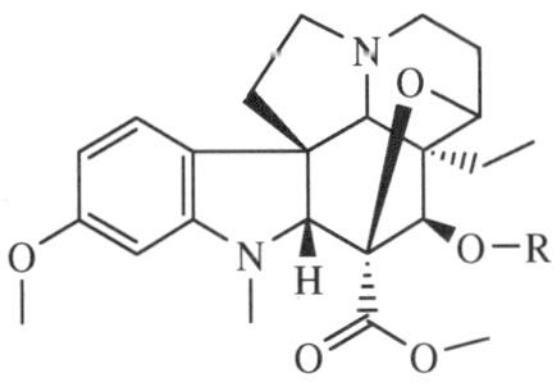

18a : R = OAc
18b : R = H

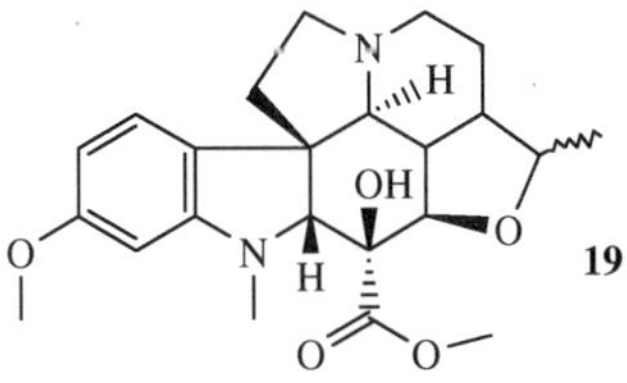

19

Compounds **18a** and **18b** have been previously prepared from vindoline (**4**) using *Streptomyces* cultures.[43]

These preliminary and unprecedented results prompted us to investigate under superacidic conditions the behavior of bis(indole alkaloids) in which vindoline (**4**) constitutes the lower half of the molecule.

Reactivity of Dimeric *Vinca* Alkaloids

As described previously, reactions of vindoline (**4**) in superacids occur specifically at the 6,7-double bond, which is included in a piperidine nucleus. A similar pattern is found in the cleavamine moiety of anhydrovinblastine (**5**) and vinorelbine (**8**), consequently offering another potential reacting center. Reactions of dimeric *Vinca* alkaloids in superacidic media could be expected at both cleavamine and vindoline moieties.

To verify the stability of a dimeric molecule, the first experiments were conducted with vinorelbine (**8**) in the [HF : SbF_5] superacid medium at −35 °C; 45% of the starting material (**8**) was recovered, together with 43% of the two isomeric derivatives (**20**) (Scheme 10.1).[44] Similar results have been reported previously for vinblastine (**1**), which was dehydrated in pure sulfuric acid, leading to a mixture of ethylenic compounds corresponding to 3′,4′-anhydrovinblastine (**5**) and the two isomers of 4′,20′-anhydrovinblastine.[45] These preliminary results were encouraging, with the whole *Vinca* alkaloid molecule remaining intact even under a priori extremely drastic conditions.

Therefore, vinorelbine (**8**) was reacted in [HF : SbF_5] at −40 °C under the conditions described above for vindoline (**4**), in the presence of one equivalent of

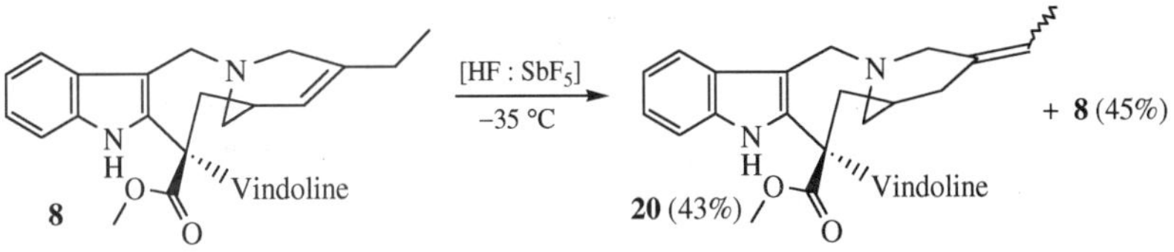

8 → [HF : SbF_5], −35 °C → **20** (43%) + **8** (45%)

SCHEME 10.1 Reactivity of vinorelbine **8** in superacidic medium.

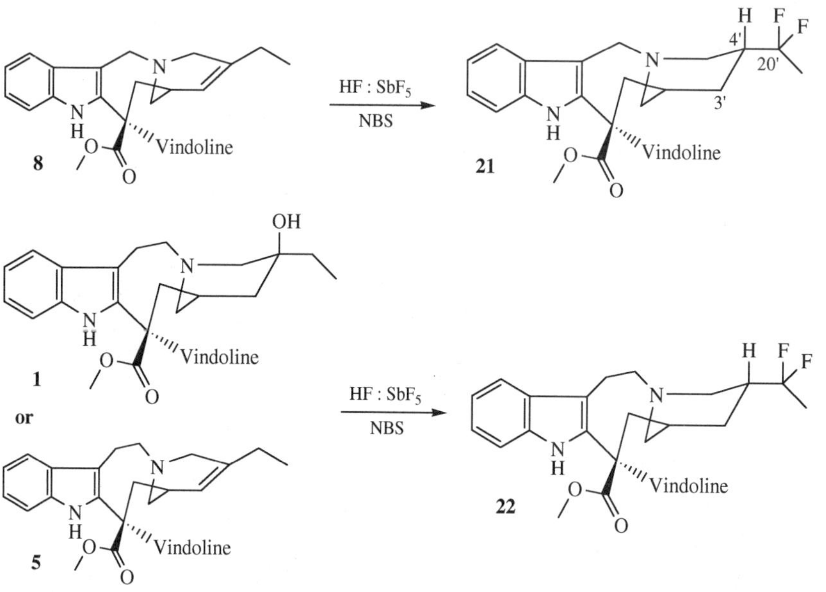

SCHEME 10.2 Reaction of *Vinca* alkaloids in the presence of an electrophilic reagent.

NBS as an electrophilic reagent (Scheme 10.2). Reaction occurred immediately, numerous derivatives being detected in the crude mixture. However one of the peaks, detected by HPLC analysis, was slightly higher than the others, and was observed in each experiment. Careful chromatographic purification was undertaken, leading finally to a compound of molecular weight = 816 ($MH^+ = 817$) containing no bromine atom, as shown by mass spectroscopy. Rapid examination of the ^1H NMR spectrum indicated that the vindoline moiety remained intact. Furthermore, the disappearance of the H3′ signal and a modification of the triplet corresponding to the methyl group at C21′ were observed. A complete NMR study has permitted the structure to be established as 20′,20′-difluoro-3′,4′-dihydrovinorelbine (**21**) (vinflunine). The chemical shift of H21′ was 1.60 ppm (triplet) instead of 1.02 ppm for vinorelbine (**8**), with a characteristic coupling constant $^3J_{H,F} = 18.9$ Hz. In the ^{13}C NMR spectrum, the C20′ signal was found at 125.4 ppm (triplet) with a coupling constant $^1J_{C,F} = 242$ Hz. Reactions involving either vinblastine (**1**) or anhydrovinblastine (**5**) under the same conditions afforded the difluorinated homologue 20′,20′-difluoro-4′-deoxyvinblastine (**22**) (Scheme 10.2).

These intriguing results, corresponding to an oxidative difluorination, were unexpected and prompted us to examine the mechanism of this amazing reaction. With the aim of decreasing the rate of the reaction and of revealing possible intermediates, we have searched for milder electrophilic reagents. Chloromethanes such as the conventional solvents CCl_4, $CHCl_3$, and CH_2Cl_2 were selected since they are known precursors of superelectrophiles in superacidic media. The resulting ionic species CCl_3^+, $CHCl_2^+$, and CH_2Cl^+, respectively possess a strong hydride abstraction power[46] and therefore act as oxidizing agents. They also have been used for the chlorination of alkanes.[47] Then a $CHCl_3$ solution of anhydrovinblastine (**5**) was added to a [HF : SbF_5] mixture at a lower temperature (−50°C). An HPLC kinetic study has permitted the identification of two peaks having the same area appearing rapidly, then disappearing simultaneously with the formation of the difluorinated (**22**); these compounds are clearly intermediates in the fluorination reaction. They correspond to the C20′ diastereoisomers of 20′-chloro-4′-deoxyvinblastine (**23**). The chlorofluorinated derivative (**24**) was also isolated from

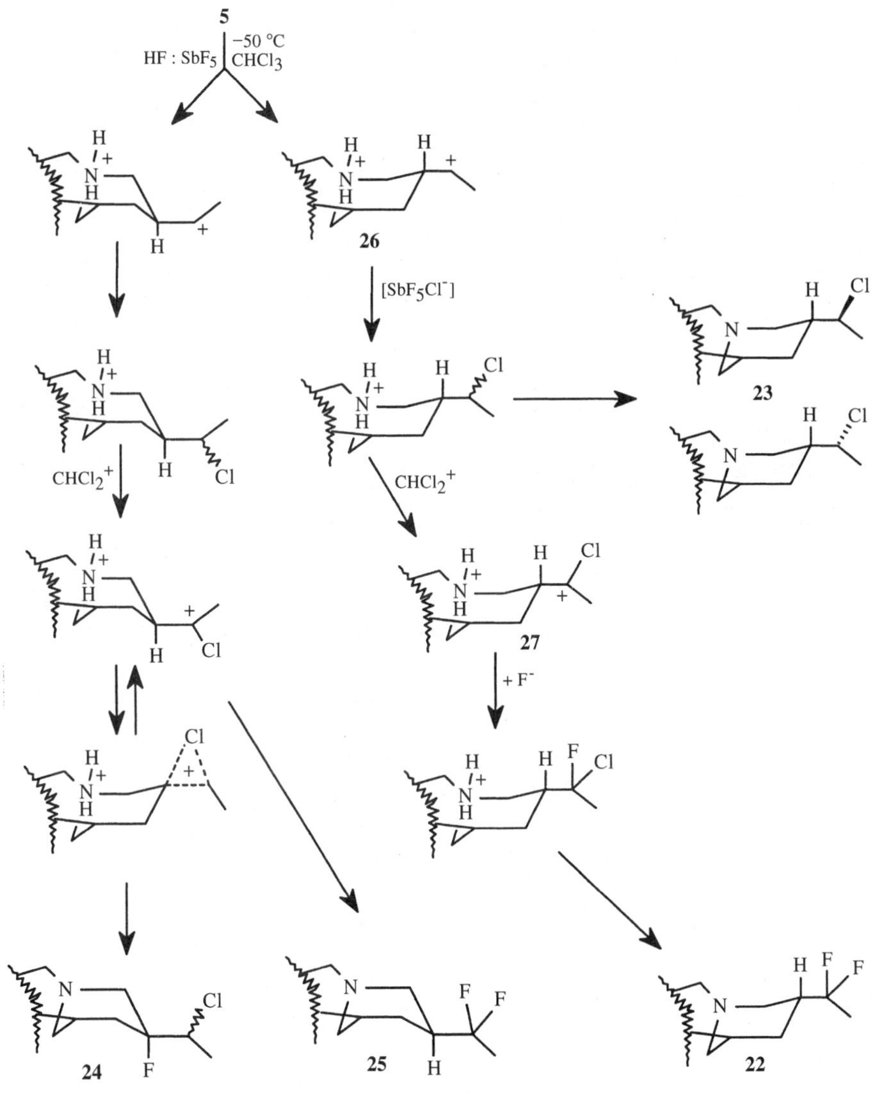

SCHEME 10.3 Postulated reaction mechanism.

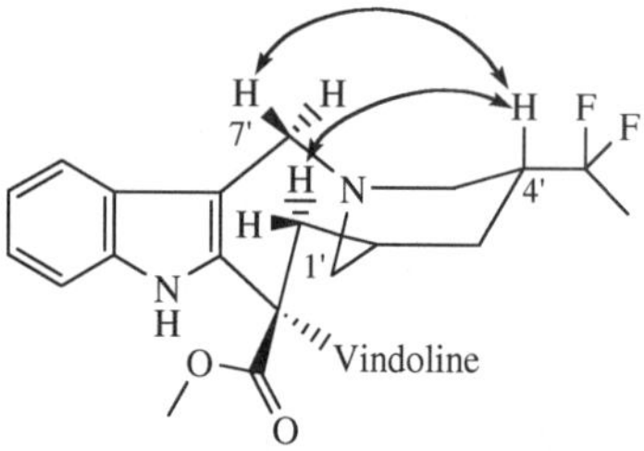

SCHEME 10.4 Determination of C-4′ configuration (NOESY).

the reaction in approximately 6% yield. The main by-product (15%) observed under these conditions was the C4′ epimeric difluorinated compound (**25**). Taking into account all these findings, the postulated mechanism is depicted in Scheme 10.3: formation of the cation **26** at C20′ results most likely from protonation of the 3′,4′-double bond at C3′, followed by a hydride shift from C20′ to C4′. Reaction of **26** with a complex chloride anion such as SbF_5Cl^- leads to the intermediates **23**. Hydride abstraction at C20′ yields the ion **27**, which can trap a fluoride ion and, after halogen exchange, gives the difluorinated compound **22** (> 50% yield).[44]

The (*R*)-C4′ configuration of **21** and **22** was first established by nuclear Overhauser effect spectroscopy (NOESY) experiments; molecular modelling calculations favored the chair conformation of the piperidine ring, with predicted NOEs between H4′ and H7′, and H4′ and H1′ (Scheme 10.4).

These effects were definitely observed. Afterward this result was confirmed by X-ray crystallography of compound **22**.

Under these unusual conditions, we have highlighted new and unexpected reactions occurring at nonactivated bonds according to conventional chemistry. This provides a novel example of *gem*-difluorination of an allylic methylene, involving chlorinated intermediates.[48]

To extend this unprecedented use of superacid chemistry to the synthesis of other semisynthetic derivatives, we have investigated the reactivity of *Vinca* alkaloids in the presence of compounds able to act as reducing agents in superacids. Cycloalkanes such as cyclohexane or methylcyclopentane have been used for such purposes.[49] When vinblastine **1** or anhydrovinblastine **5** were treated with an excess of methylcyclopentane in [HF : SbF_5], the reduced compound **28** was obtained in 65% yield (Scheme 10.5). This ionic hydrogenation is fully stereoselective; no derivative having the 4′*S* configuration has been detected in the reaction mixture. It should be pointed out that catalytic hydrogenation leads to the 4′*S*-reduced compound. Also outstanding is the regioselectivity of the reaction, the 6,7- double bond of the vindoline moiety remaining intact even in the presence of a large excess of reducing agent.[50]

Compound **28** is a natural product, previously named *deoxyvinblastine A*. Refluxing leurosine in ethanol with Raney nickel[51] has been claimed as a preparative method, but its isolation necessitates a difficult separation of the two C4′ epimers. Vinorelbine **8** undergoes similar ionic reduction when placed under the same conditions, leading to the previously unknown reduced derivative **29** in 70% yield,

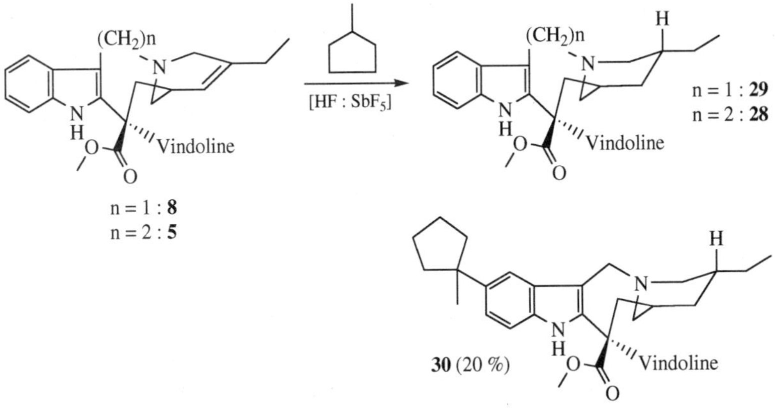

SCHEME 10.5 Ionic hydrogenation in superacidic media.

with compound **30** (~ 20%) as a by-product. The analogous alkylated product has not been detected with vinblastine (**1**) or anhydrovinblastine (**5**).[50]

PHARMACOLOGY

Among the series of new derivatives obtained through superacid chemistry, vinflunine (**21**) has been selected for its outstanding pharmacological properties in comparison with the main *Vinca* alkaloids currently used in the clinic, vinblastine (**1**), vincristine (**2**), and vinorelbine (**8**).

In Vitro Tubulin Interactions

The ability to inhibit tubulin polymerization in vitro was followed according to the method described by Gaskin and Cantor.[52] Activities, reflected by the IC_{50}, corresponding to the concentration of test compound inhibiting 50% of tubulin polymerization, are reported in Table 10.1.[53]

These IC_{50} values were all within a narrow micromolar range, indicating that this tubulin polymerization assay appears not sufficiently discriminatory to permit any clear classification of the evaluated compounds.

TABLE 10.1 Effects of Compounds on Tubulin Assembly

	Inhibition of Tubulin Assembly IC_{50}, μM	
Compound	Microtubular Proteins	Purified Tubulin
Vinblastine (**1**)	2.2	2.7
Vincristine (**2**)	1.7	0.5
Vinorelbine (**8**)	1.7	0.9
Vinflunine (**21**)	3.1	2.5

A more complete examination of these tubulin interactions, however, has shown that specific binding of [^{3}H]vinflunine to tubulin was undetectable by the standard centrifugal gel filtration method, unlike the other *Vinca* alkaloids, so that the comparative capacities of these molecules to interfere with tubulin binding could be classified as follows: vincristine > vinblastine > vinorelbine > vinflunine. Overall, vinflunine appears to exhibit quantitatively different tubulin-interacting properties to the classic *Vinca* alkaloids.[53]

Cytotoxicity

The new derivatives synthesized in this study have been evaluated for their cytotoxic properties first against the L1210 murine leukemia cell line, as a standard cytotoxicity screening assay. The IC_{50} values correspond to the concentration of test compound inducing 50% inhibition of cell proliferation. As shown in Table 10.2, vinflunine (**21**) exhibited a relatively low cytotoxic potency against L1210 cells ($IC_{50} = 80$ nM) compared to the other *Vinca* alkaloids. A similar profile was identified when cytotoxicities were evaluated against a panel of 11 human and murine cancer cell lines.[53]

Overall, the effects induced by vinflunine are comparable to those of compounds vinblastine (**1**), vincristine (**2**) and vinorelbine (**8**), but are achieved with 3- to 17-fold higher vinflunine concentrations. On the basis of these in vitro results, vinflunine could be rejected in a classic drug screen.

In Vivo Antitumor Properties

Antitumor activity was first evaluated against the P388 murine leukemia (IV/IP) in vivo screening model. Activity of the test compound is reflected by an increase of lifespan, defined as the (median survival of treated mice/median survival of control

TABLE 10.2 Cytotoxicity IC_{50} values (in nM) against the Panel of Cell Lines Used

Cell line	Origin	Vinblastine **1**	Vincristine **2**	Vinorelbine **8**	Vinflunine **21**
		Murine Tumor Cell Lines			
L1210	Leukemia	16.0	15.0	28.0	80
P388	Leukemia	13.0	6.7	4.4	72
		Human Tumor Cell Lines			
A549	Lung	3.6	23.0	9.2	81
DLD-1	Colon	14.0	75.0	120.0	280
DU145	Prostate	1.6	47.0	5.6	71
J82	Bladder	16000.0	10000.0	18000.0	14000
LoVo	Colon	3.3	5.9	7.9	77
MX-1	Breast	1.8	0.6	6.7	63
OVCAR-3	Ovary	2.0	2.3	5.1	72
SK-OV-3	Ovary	1.7	5.9	9.6	11000
T24	Bladder	4.1	7.6	11.0	97

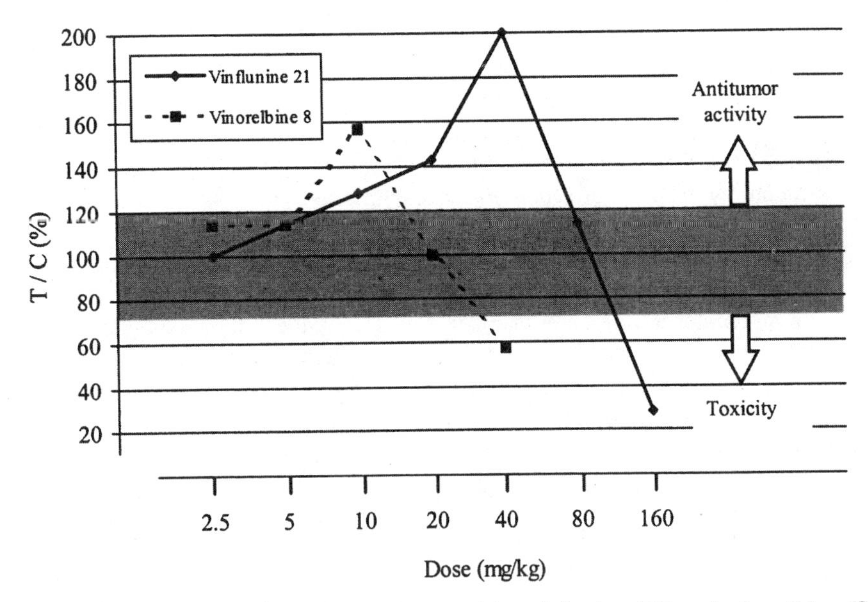

FIGURE 10.1 P388 antitumor activity exerted by vinflunine (**21**) and vinorelbine (**8**) (single dose).

mice) × 100, (%, T/C). Figure 10.1 depicts the increase of lifespan (%) versus administered dose, induced by vinflunine (**21**) and vinorelbine (**8**) administered as single doses. A markedly higher maximal effect is exerted by vinflunine, as well as being evident over a wider range of doses.[54]

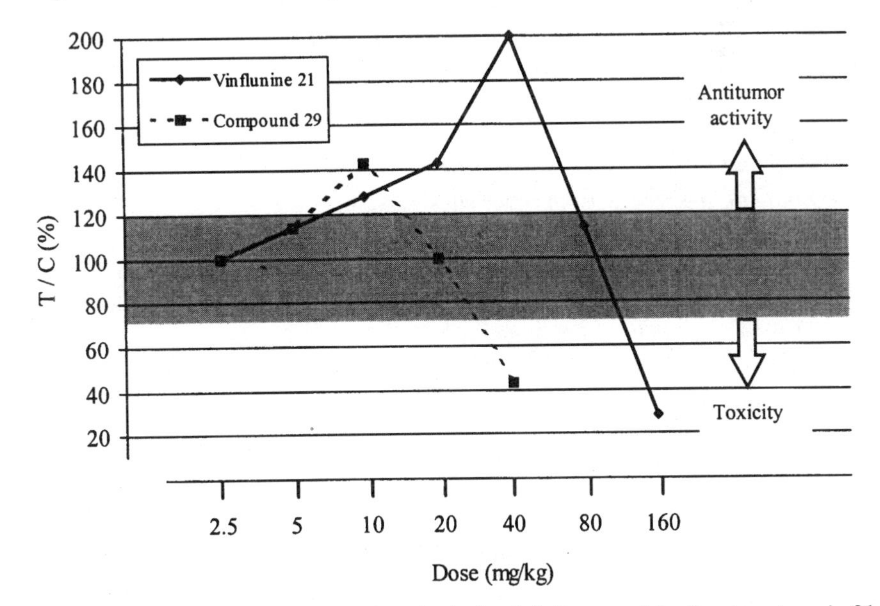

FIGURE 10.2 P388 antitumor activity (single dose): influence of the fluorine atoms in **21**.

Vinflunine exhibited marked activity in single or multiple doses according to various schedules of administration. The T/C ratios, ranging from 200% (single dose) to 457% (weekly injection over 4 weeks) for vinflunine, proved markedly superior to those of 129–186% obtained with the other *Vinca* alkaloids.[54]

To study the contribution of the fluorine atoms to the antitumor activity of **21**, experiments with the corresponding nonfluorinated derivative **29** have been conducted on this P388 (IV/IP) in vivo model. As shown in Figure 10.2, the reduced compound **29** is significantly less active than **21** under these conditions.[55]

Against a panel of eleven human tumor xenograft models, vinflunine exhibited high or moderate antitumor efficacy in 64% $\frac{7}{11}$ of the tumor xenograft models evaluated, versus moderate activity recorded for vinorelbine in only 27% $\frac{3}{11}$.[56]

In summary, the overall results recorded provide evidence of the superiority of vinflunine (**21**) over vinorelbine in each experimental tumor model system used. This enhanced efficacy could be translated into an improved spectrum of clinical activity for vinflunine in relation to the other *Vinca* alkaloids already in use.

CONCLUSIONS

Chemistry

We have shown that complex molecules can be stable in highly acidic media and have highlighted new and unexpected reactions, including halogenations and ionic hydrogenation, occurring very specifically at nonactivated sites, with high stereoselectivity. Structural modifications in this specific part of the *Vinca* alkaloid molecule were previously accessible only by total synthesis and required a large number of steps. Furthermore, optimisation of the experimental conditions has permitted us to obtain these new derivatives in high yield. The scope of these new chemical reactions is under investigation by:

- Reacting the substrates with other electrophilic reagents
- Studying the reactivity of other substrates.

Because of its promising antitumor activity, compound **21** is now synthesized on a kilogram scale by C′ ring contraction of the precursor **22** following the method used for the synthesis of vinorelbine **8**.[17]

Pharmacology

Vinflunine **21** showed an unremarkable profile in terms of both cytotoxic potency and tubulin polymerization inhibition under standard in vitro drug screening conditions. These studies were conducted by comparing the effects of vinflunine with those *Vinca* alkaloids exerting well-documented clinical efficacy. Few examples of compounds with alterations in this same area have been reported, but they also had a modified tubulin binding profile.[57] Our observations are in agreement with these results and clearly indicate that C4′ modifications dramatically alter the pharmacological properties both in vitro and in vivo.

Furthermore recent new advances in the understanding of the mechanisms of tubulin interacting agents has led to renewed interest in the therapeutic potential of such compounds.[58] It is becoming clear that the various *Vinca* alkaloids differentialy inhibit tubulin function, even if they all bind to the same site(s) on tubulin dimers.[59] Further investigations aimed at identifying a particular mechanism of action of vinflunine are in progress, namely, more precise tubulin interaction examinations. In a recent physicochemical study based on sedimentation velocity, Lobert et al. showed that vinflunine induces smaller tubulin spirals with a more rapid relaxation time. These effects are believed to be responsible in part for the toxicity of *Vinca* alkaloids. Therefore, vinflunine may exert reduced toxicity compared to the other examples of this class.[60]

Ongoing investigations aimed at defining other parameters implicated in the antitumor activity of vinflunine include monitoring its cellular uptake and accumulation, drug metabolism, and pharmacokinetic properties, as well as identifying any non-tubulin-related targets. A complete study describing the pharmacological properties of a large series of vinflunine derivatives selectively modified at C4′ and/or C20′ will be published in due course.

Finally, phase I clinical trials with vinflunine **21** started in December 1998.

ACKNOWLEDGMENTS

This chapter is dedicated to George Olah at the University of Southern California for his pioneering work in the field of carbocation chemistry.

Thanks are due to our colleagues, referred to in the references, for their contribution in this stimulating research, and to Dr. Bridget T. Hill, who provided valuable criticism during the writing of this chapter.

JCJ. thanks the CNRS and La Ligue Contre le Cancer (Comité de Charente Maritime) for their financial support.

REFERENCES

1. Blasko, G.; Cordell, G. A., in *Antitumor Bisindole Alkaloids from Catharanthus roseus* (L.). *The Alkaloids*, A. Brossi, and M. Suffness, eds., Academic Press, San Diego, 1990, Vol. 37, pp. 1–76.
2. Noble, R. L., *Biochem. Cell Biol.* 1990, **68**, 1344–1351.
3. Van Tellingen, O.; Sips, J. H. M.; Beijnen, J. H.; Bult, A.; Nooijen, W. J., *Anticancer Res.* 1992, **12**, 1699–1716.
4. (a) Potier, P.; Langlois, N.; Langlois, Y.; Gueritte, F., *J. Chem. Soc., Chem. Commun.* 1975, 670–671; Langlois, Y.; Potier, P., *J. Am. Chem. Soc.* 1976, **98**, 7017–7024 (b) Mangeney, P.; Andriamialisoa, R. Z.; Langlois, N.; Langlois, Y.; Potier, P., *ibid.* 1979, **101**, 2243–2245.

5. Kuehne, M. E.; Zebovitz, T. C.; Bornmann, W. G.; Marko, I., *J. Org. Chem.* 1987, **52**, 4340–4349; Kuehne, M. E.; Bornmann, W. G., *J. Org. Chem.* 1989, **54**, 3407–3420; Kuehne, M. E.; Matson, P. A.; Bornmann, W. G., *J. Org. Chem.* 1991, **56**, 513–528.
6. Magnus, P.; Ladlow, M.; Elliott, J., *J. Am. Chem. Soc.* 1987, **109**, 7929–7930; Magnus, P.; Stamford, A.; Ladlow, M., *ibid.* 1990, **112**, 8210–8212; Magnus, P.; Ladlow, M.; Elliott, J.; Kim, C. S., *J. Chem. Soc., Chem. Commun.* 1989, 518–520; Magnus, P.; Mendoza, J., *Tetrahedron Lett.* 1992, **33**, 899–902.
7. Atta-ur-Rahman; Iqbal, Z.; Nasir, H., *Studies in Natural Products Chemistry*; Atta-ur-Rahman, ed., Elsevier, Amsterdam, 1994; Vol. 14, pp. 805–884.
8. Magnus, P.; Mendoza, J. S.; Stamford, A.; Ladlow, M.; Willis, P., *J. Am. Chem. Soc.* 1992, **114**, 10232–10245.
9. Kuehne, M. E.; Bandarage, U. K., *J. Org. Chem.* 1996, **61**, 1175–1179.
10. Jungheim, L. N.; Shepherd, T. A.; Meyer, D. A., *J. Org. Chem.* 1992, **57**, 2334–2340.
11. Sundberg, R. J.; Bettiol, J. L.; Gadamasetti, K. G.; Marshalla, M.; Kelsh, L., *Bioorg. Med. Chem. Lett.* 1994, **4**, 1999–2004.
12. Danieli, B.; Lesma, G.; Palmisano, G.; Passarella, D., *J. Org. Chem.* 1994, **59**, 5810–5813; Danieli, B.; Lesma, G.; Martinelli, M.; Passarella, D.; Silvani, A., *ibid.* 1998, **63**, 8586–8588.
13. Brossi, A.; Suffness, M., eds., *Antitumor Bisindole Alkaloids from Catharanthus roseus* (L.). *The Alkaloids*, Academic Press, San Diego, 1990; Vol. 37.
14. Armstrong, J. G.; Dyke, R. W.; Fouts, P. J.; Hawthorne, J. J.; Jansen, C. J.; Peabody, A. M., *Cancer Res.* 1967, **27**, 221–227.
15. Barnett, C. J.; Cullinan, G. J.; Gerzon, K.; Hoying, R. C.; Jones, W. E.; Newlon, W. M.; Poore, G. A.; Robinson, R. L.; Sweeney, M. J.; Todd, G. C., *J. Med. Chem.* 1978, **21**, 88–96.
16. Sweeney, M. J.; Boder, G. B.; Cullinan, G. J.; Culp, H. W.; Daniels, W. D.; Dyke, R. W.; Gerzon, K.; McMahon, R. E.; Nelson, R. L.; Poore, G. A.; Todd, G. C., *Cancer Res.* 1978, **38**, 2886–2891.
17. Mangeney, P.; Andriamialisoa, R. Z.; Lallemand, J. Y.; Langlois, N.; Langlois, Y.; Potier, P., *Tetrahedron* 1979, **35**, 2175–2179.
18. Mathe, G.; Reisenstein, P., *Cancer Lett.* 1985, **27**, 285–293.
19. Rao, K. S. P. B.; Collard, M. P. M.; Dejonghe, J. P. C.; Atassi, G.; Hannart, J. A.; Trouet, A., *J. Med. Chem.* 1985, **28**, 1079–1088.
20. Pearce, H. L., in Ref. 13, pp. 175–181.
21. Jordan, M. A.; Himes, R. H.; Wilson, L., *Cancer Res.* 1985, **45**, 2741–2747.
22. Mullin, K.; Houghton, P. J.; Houghton, J. A.; Horowitz, M. E., *Biochem. Pharmacol.* 1985, **34**, 1975–1979.
23. Pearce, H. L., in Ref. 13, p. 190.
24. Laguzza, B. C.; Nichols, C. L.; Briggs, S. L.; Cullinan, G. L.; Johnson, D. A.; Starling, J. J.; Baker, A. L.; Bumol, T. F.; Corvalan, J. R. F., *J. Med. Chem.* 1989, **32**, 548–555.
25. Lavielle, G.; Hautefaye, P.; Schaeffer, C.; Boutin, J. A.; Cudennec, C. A.; Pierré, A., *J. Med. Chem.* 1991, **34**, 1998–2003.
26. Adenis, A.; Pion, J. M.; Fumoleau, P.; Pouillart, P.; Marty, M.; Giroux, B.; Bonneterre, J., *Cancer Chemother. Pharmacol.* 1995, **35**, 527–528.
27. Nogales, E.; Wolf, S. G.; Downing, K. H., *Nature* 1998, **391**, 199–203.

28. Erickson, H. P., *Trends Cell Biol.* 1998, **8**, 133–137.
29. Hall, N. F.; Conant, J. B., *J. Am. Chem. Soc.* 1927, **49**, 3047–3061.
30. Olah, G. A.; Prakash, G. K. S.; Sommer, J., *Superacids*, Wiley-Interscience, New York, 1985.
31. (a) Gillespie, R. J.; Peel, T. E., *Adv. Phys. Org. Chem.* 1971, **9**, 1–24; (b) Gillespie, R. J.; Peel, T. E., *J. Am. Chem. Soc.* 1973, **95**, 5173–5179.
32. Sommer J.; Canivet, P.; Schwartz, S.; Rimmelin, P., *Nouv. J. Chim.* 1981, **5**, 45–53.
33. Jacquesy, J. C., in *Stable Carbocation Chemistry*, G. K. S. Prakash and P. v. R. Schleyer, eds., Wiley-Interscience, New York, 1997, Chapter 17, pp. 549–574.
34. Berrier, C.; Jacquesy, J. C.; Jouannetaud, M. P.; Renoux, A., *Nouv. J. Chim.* 1987, **11**, 611–615.
35. Berrier, C.; Jacquesy, J.C.; Jouannetaud, M.P.; Renoux, A., *Nouv. J. Chim.* 1987, **11**, 605–609.
36. Cordell, G. A.; Saxton, J. E., in *Bis Indole Alkaloids*, R. G. A. Rodrigo, ed., Academic Press, New York, 1981, Vol. XX, pp. 1–295.
37. Berrier, C.; Jacquesy, J. C.; Jouannetaud, M. P.; Vidal, Y., *Tetrahedron* 1990, **46**, 815–826.
38. Berrier, C.; Jacquesy, J. C.; Jouannetaud, M. P.; Vidal, Y., *Tetrahedron* 1990, **46**, 827–832.
39. Berrier, C.; Jacquesy, J. C.; Jouannetaud, M. P.; Lafitte, C.; Vidal, Y.; Zunino, F.; Fahy, J.; Duflos, A., *Tetrahedron* 1998, **54**, 13761–13770.
40. Vukovic, J.; Goodbody, A. E.; Kutney, J. P.; Misawa, M., *Tetrahedron* 1988, **44**, 325–331.
41. Owellen, R. J.; Donigian, D. W.; Hartke, C. A.; Hains, F. O., *Biochem. Pharmacol.* 1977, **26**, 1213–1219.
42. Lafitte C.; Jouannetaud, M. P.; Jacquesy, J. C.; Duflos, A., *Tetrahedron* 1999, **55**, 1989–2000.
43. Neuss, N.; Fukuda, D. S.; Mallett, G. E.; Brannon, D. R.; Huckstep, L.L., *Helv. Chim. Acta* 1973, **56**, 2418–2426.
44. Fahy, J.; Duflos, A.; Ribet, J. P.; Jacquesy, J. C.; Berrier, C.; Jouannetaud, M. P.; Zunino, F., *J. Am. Chem. Soc.* 1997, **119**, 8576–8577.
45. Miller, J. C.; Gutowski, G. E.; Poore, G. A.; Boder, G. B., *J. Med. Chem.* 1977, **20**, 409–413.
46. Olah, G. A., *Angew. Chem. Internatl. Ed. Engl.* 1993, **32**, 757–788; Culmann, J.-C.; Sommer, J., *J. Chem. Soc., Chem. Commun.* 1992, 481–483; Martin, A.; Jouannetaud, M.-P.; Jacquesy, J.-C., *Tetrahedron Lett.* 1996, **37**, 2967–2970, 7731–7734.
47. Olah, G. A.; Wu, A. H.; Farooq, O., *J. Org. Chem.* 1989, **54**, 1463–1465.
48. For a recent review on the synthesis of *gem*-difluoromethylene compounds, see Tozer, M. J.; Herpin, T. F., *Tetrahedron* 1996, **52**, 8619–8683.
49. Olah, G. A.; Prakash, G. K. S.; Sommer, J., in Ref. 30, pp. 243–277.
50. Lafitte, C.; Jouannetaud, M. P.; Jacquesy, J. C.; Fahy, J.; Duflos, A., *Tetrahedron Lett.* 1998, **39**, 8281–8282.
51. Neuss, R.; Gorman, M.; Cone, N. J.; Huckstep, L. L., *Tetrahedron Lett.* 1968, **7**, 783–787.
52. Gaskin, F.; Cantor, C. R., *J. Molec. Biol.* 1974, **89**, 737–758.

53. Kruczynski, A.; Barret, J. M.; Etievant, C.; Colpaert, F.; Fahy, J.; Hill, B. T., *Biochem. Pharmacol.* 1998, **55**, 635–648.

54. Kruczynski, A.; Colpaert, F.; Tarayre, J. P.; Mouillard, P.; Fahy, J.; Hill, B. T., *Cancer Chemother. Pharmacol.* 1998, **41**, 437–447.

55. Fahy, J.; Duflos, A.; Schambel, P.; Kruczynski, A.; Etievant, C.; Barret, J. M.; Hill, B. T.; Jacquesy, J. C.; Jouannetaud, M. P.; Meheust, C., *Proc. Am. Assoc. Cancer. Res.* 1998, **39**, 1136, (New Orleans).

56. Kruczynski, A.; Fiebig, H. H.; Waud, W.; Poupon, M. F.; Clerc, X.; Ricome, C.; Colpaert, F.; Hill, B. T., *Proc. Am. Assoc. Cancer. Res.* 1998, **39**, 1139, (New Orleans).

57. Borman, L. S.; Kuehne, M. E.; Matson, P. A.; Marko, I.; Zebovitz, T. C., *J. Biol. Chem.* 1988, **263**, 6945–6948; Borman, L. S.; Kuehne, M. E., *Biochem. Pharmacol.* 1989, **38**, 715–724; Borman, L. S.; Bornmann, W. G.; Kuehne, M. E., *Cancer Chemother. Pharmacol.* 1993, **31**, 343–349.

58. Jordan, M. A.; Wilson, L., *Curr. Opin. Cell Biol.* 1998, **10**, 123–130.

59. Lobert, S.; Vulevic, B.; Correia, J. J., *Biochemistry* 1996, **35**, 6806–6814.

60. Lobert, S.; Ingram, J. W.; Hill, B. T.; Correia, J. J., *Molec. Pharm.* 1998, **53**, 908–915.

CHAPTER 11

Fluorine Substitution as a Modulator of Biological Processes

KENNETH L. KIRK*

Laboratory of Bioorganic Chemistry, National Institute of Diabetes and Digestive and Kidney Diseases, National Institutes of Health

INTRODUCTION

Medicinal chemists have used the strategy of fluorine substitution for modulation of biological processes effectively since the early 1950s. This chapter presents a brief overview of this area with selected examples. The author's own work in the area of modulation of chemical neurotransmitters through fluorine substitution is reviewed, with current research and future directions highlighted at the conclusion.

Historical

Fluoroacetate was identified in the 1940s as the primary toxic component of the South African plant *Dichapetalum cymosum* (gifblaar), a member of the genus responsible for mortality in grazing livestock. Studies by Sir Rudolf Peters in the 1950s revealed that the toxic properties of fluoroacetate resulted from its in vivo conversion to fluorocitrate. [Eq. (11.1)].[1] Fluorocitrate was shown to inhibit aconitase, the enzyme that converts citrate to aconitate in the citric acid cycle. This concept of "lethal synthesis" and the recognition of the profound biological consequences elicited by the presence of the C–F bond had great influence on subsequent research and historically has been considered a seminal event in sparking recognition of the potential of fluorinated biomolecules as antimetabolites. In fact, the competitive inhibition, the relatively high K_i and the time-dependent nature of the inactivation of aconitase were observations that were hard to reconcile with the acute and irreversible toxicity of fluorocitrate. Later research by Kun and co-workers

*This chapter was written by the author in his private capacity. No official support nor endorsement is intended nor should be inferred.

Biomedical Chemistry: Applying Chemical Principles to the Understanding and Treatment of Disease, Edited by Paul F. Torrence
ISBN 0-471-32633-x

revealed that fluorocitrate functions as a neurotoxin through inhibition of mitochondrial tricarboxylic acid transport.[2] Nonetheless, reversible inhibition of aconitase in the citric acid cycle is still cited in many texts as its primary mode of action. Indeed, the later mechanistic revelations in no way detract from the importance of the contribution of Peters.

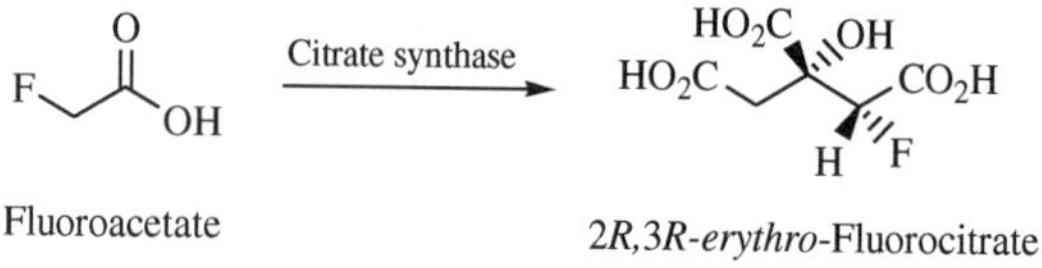

Fluoroacetate —Citrate synthase→ 2*R*,3*R*-*erythro*-Fluorocitrate (11.1)

An early demonstration of the potential medicinal benefits of fluorine substitution is found in the synthesis of fluorinated corticosteroids by Fried and Sabo in 1953, work that led to the development of potent antiinflammatory agents such as dexamethasone (**1**) (see Scheme 11.1). Heidlelberger's development in 1957 of 5-fluorourocil (**2**) as an anticancer agent was another notable early success in the use of this strategy to develop medicinal agents.[3] Since then, incorporation of fluorine into virtually every class of organic molecule has produced a plethora of fluorinated pharmaceuticals, biochemical tools, mechanistic probes, and biological tracers. The development of new fluorinating agents and new procedures that modify the reactivities of fluorinating agents are important factors behind the present rapid pace of advancement in this field. In addition, clearer understanding of the underlying biochemical mechanisms involved, coupled with knowledge of the physicochemical properties accompanying fluorine substitution, have aided the exploitation of fluorine substitution in the rational design of phamacological agents and drugs.[4]

Special Properties of Fluorine

The theoretical bases for the effectiveness of fluorine substitution in the design of analogs of biologically important molecules is well documented. A major factor is the steric similarity of the C–F bond and the C–H or C–OH bond that facilitates the acceptance of a fluorinated analog by a biological recognition site. On the other hand, the high electronegativity of fluorine often imparts altered physicochemical properties to the analog that influence the consequences of biological recognition. For example, the opposite patterns of reactivity of H^+ and F^- (fluorine as a "deceptor") are frequently exploited to design enzyme inhibitors. An example of this is seen in the mechanism of inactivation of thymidyate synthase by 5-fluorourodine

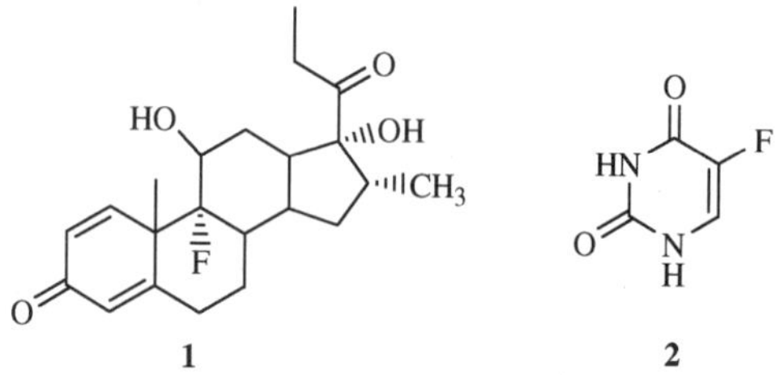

1 **2**

SCHEME 11.1

monophosphate in which inability of the ternary inhibition complex to expel F^+ blocks further reaction. The ability of fluorine to suffer elimination (as F^-) triggered by an adjacent carbanionic center is the basis for the design of a large number suicide substrates for several classes of enzymes. The pyridoxal phosphate-dependent enzymes have been targeted extensively by the strategy. The use of α-fluoroketones in the design of transition state inhibitors of hydrolytic enzymes takes advantage of the high electronegativity of fluorine that facilitates nucleophilic attack on the adjacent carbonyl group. Blockade of metabolism through introduction of fluorine, and changes in physicochemical properties such as acidity and lipophilicity are other factors that have been exploited in drug design.[4]

Fluorinated Medicinal and Pharmacological Agents

In every class of organic compound, including carboxylic acids, alcohols, aldehydes and ketones, amino acids, amines, steroids, nucleosides and nucleotides, and carbohydrates, biochemical and pharmacological studies are well represented by fluorinated analogs, often as a central theme. Typical examples of drug classes in which fluorine substitution has been critical to bioactivity include the fluoroquinolone antibiotics such as Ciprofloxacin (**3**) (see Scheme 11.2), potent inhibitors of bacterial DNA gyrase, and CNS agents such as fluoxetin (Prozac) (**4**) and the anorectic agent fenfluramine (**5**). Unabated activity toward development of new nucleoside and nucleotide-based drugs have produced promising new fluorinated antiviral agents such as (−) 2,3-dideoxy-5-fluoro-3-thiacytidine (**6**) and anticancer agents, such as Gemcitabine (**7**). These and other medicinal and pharmacological agents clearly validate the use of fluorinated anologs to effect useful modulation of biological properties. This topic has been the subject of many reviews and symposia.[4,5]

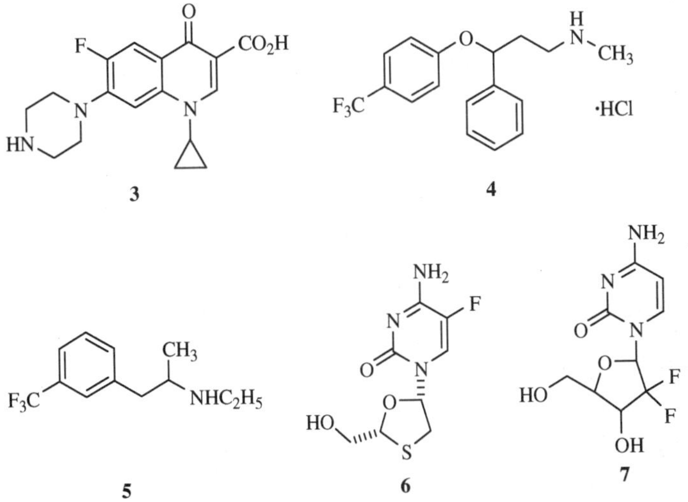

SCHEME 11.2

Fluorinated Compounds as Mechanistic Probes and Biological Tracers

Applications of fluorinated analogs as tools to study biochemical mechanisms include the use of chiral and prochiral fluorinated substrates to probe reaction stereochemistry. In early work, the prochiral fluoromethylene group of fluoropyruvate was used extensively to study the stereochemistry of many reactions that link pyruvate to carbohydrate metabolism, the Krebs cycle, and several biosynthetic schemes.[6] Examination of the effects of electronegative centers introduced by fluorine substitution on rates of enzyme-catalyzed reactions has been used to provide details of biochemical mechanisms. For example, in the elegant studies carried out by Poulter, rate retardation caused by a neighboring highly electronegative CF_3 group was used as evidence that carbonium ion formation is important in key steps of terpene biosynthesis.[7] More recently, a similar strategy was used by Walsh to delineate a stepwise E_1 mechanism in a key step of peptidoglycan biosynthesis by bacteria.[8] Many other examples could be cited.

Fluorinated analogs also are effective biochemical tracers. The NMR properties of ^{19}F make it a particularly useful probe for the study of the structure and function of fluorine-labeled biomolecules. ^{19}F has spin $\frac{1}{2}$, has a high natural sensitivity, has a much higher chemical shift range than does ^{1}H, chemical shifts are particularly sensitive to environment, and, of special importance, there is present no endogenous signal. In research on peptides and proteins, for example, there has been extensive use of fluorine-labeled amino acid residues using ^{19}F NMR to probe such parameters as local environments of individual amino acids, effects of ligands on chemical shifts in fluorine labeled receptors, intramolecular communication, and other phenomena. With increasing power of NMR instrumentation, use of ^{19}F-NMR in this and other areas of biochemistry will increase in value.

^{18}F ($t_{1/2}$ 109 min) can be made by reactor radiation and by cyclotron production. Radioactive decay of ^{18}F produces β^+ particles that are annihilated by electrons to give two 551-keV photons emitted at 180° from each other, such that the three-dimensional origin of this emission can be determined by coincidence detection. The short half-life of ^{18}F and low β^+ energy make this nuclide ideal for positron emission tomography (PET) imaging. Notable successes in this area include the use of ^{18}F-labeled fluorodeoxyglucose to study regional glucose utilization of the brain; ^{18}F-labeled steroids to image estrogen, progesterone, and androgen receptors; and ^{18}F-labeled DOPA (as a precursor to dopamine) to study CNS regional dopaminergic activity, particularly in Parkinson's disease.[10]

FLUORINE-SUBSTITUTED NEUROACTIVE COMPOUNDS

The brief discussion above gives examples from the literature of applications of fluorinated compounds in biomedical research. For several years, the author has been studying the biochemistry and pharmacology of fluorinated analogs of aromatic amino acids and the corresponding aromatic amines, many of which function as neurotransmitters. This work is reviewed in this section, with particular attention to chemistry and mechanistic principles. A status report on the author's current activity in this area is also presented.

Background

The very complex processes that combine to produce functioning neuronal networks present fascinating challenges to medicinal chemists who attempt to design molecular strategies to modulate nerve functions. Many chemical neurotransmitters are involved. The optimum levels of these are maintained by a balance of biosynthesis, feedback inhibition of biosynthesis, uptake and storage mechanisms, release to synapses to activate nerve action potentials, and deactivation through neuronal reuptake or through metabolism. Each of these several steps required for control of neurotransmitter levels can be targets for intervention.

Among the vast array of available medicinal agents and pharmacological tools related to neuroscience, fluorinated analogs have achieved noteworthy successes. Fluorinated amino acids and amines have received particular attention. Certain amino acids are themselves neurotransmitters, while others play critical roles in neurotransmission by functioning as biological precursors of amine neurotransmitters. Based on this, α-fluoromethyl- and α,α-difluoromethylene amino acids, such as analogs of DOPA and histidine (**8** and **9**, respectively; see Scheme 11.3) can affect neurotransmitter levels by inhibiting decarboxylase enzymes that convert the corresponding parent amino acid to the amine neurotransmitters.[11] Likewise, fluorinated inhibitors, for example, **10** and **11**, of monoamine oxidase enzymes can increase levels of amine neurotransmitters by blocking their metabolic clearance.[12]

Fluoroimidazoles: The Development of a New Reaction

The author's research in this area began in the late-1960s. The late Dr. Louis A. Cohen, then Chief, Section on Biochemical Mechanisms, Laboratory of Chemistry, NIAMD, recognized that ring-fluorinated imidazoles represented a very promising synthetic target. The absence of reported ring-fluorinated imidazoles seemed remarkable in view of the importance of imidazole in biological structure and function and the increasing recognition of the value of fluorine substitution. However, many attempts, using convention procedures to introduce fluorine in

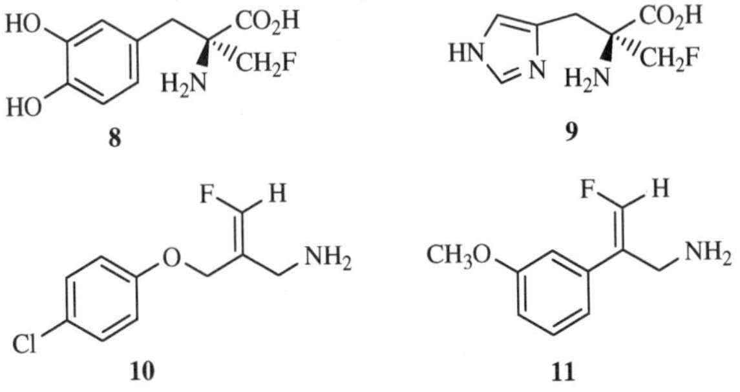

SCHEME 11.3

various imidazole precursors, failed to produce even a trace of evidence for ring fluorination [Eq. (11.2)]. Subsequently, the author found that, although imidazole diazonium fluoroborates are thermally unreactive with respect to extrusion of nitrogen under classic Schiemann conditions, photochemical decomposition of imidazole diazonium fluoroborates in fluoroboric acid cleanly leads to replacement of the diazonium group with fluorine. The procedure was made more convenient since diazotization could be carried out in the same aqueous fluoroboric acid solution, and isolation of the diazonium salt was obviated [Eq. (11.3)].[13,14]

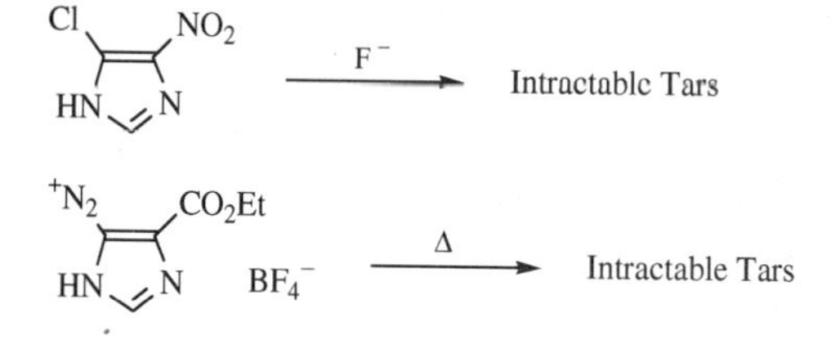

$\xrightarrow{F^-}$ Intractable Tars

BF_4^- $\xrightarrow{\Delta}$ Intractable Tars (11.2)

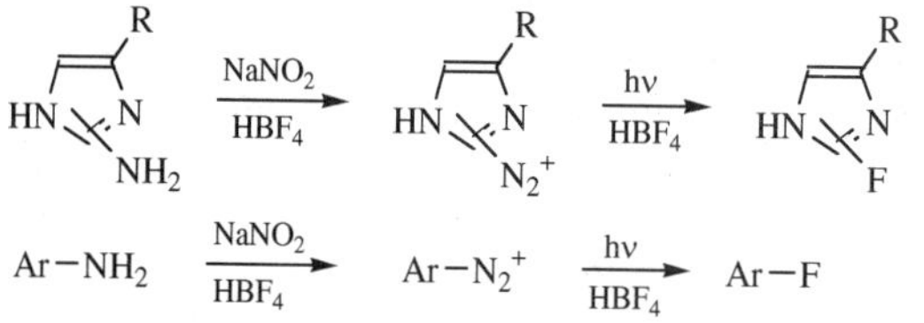

(11.3)

$$Ar{-}NH_2 \xrightarrow[HBF_4]{NaNO_2} Ar{-}N_2^+ \xrightarrow[HBF_4]{h\nu} Ar{-}F$$

This discovery led to the preparation of a large number of fluoroimidazole derivatives, including 2- and 4-fluorohistidines (**12, 13**; see Scheme 11.4), 2- and 4-fluorohistamines (**14, 15**), and fluorinated analogues of purine precursors, for

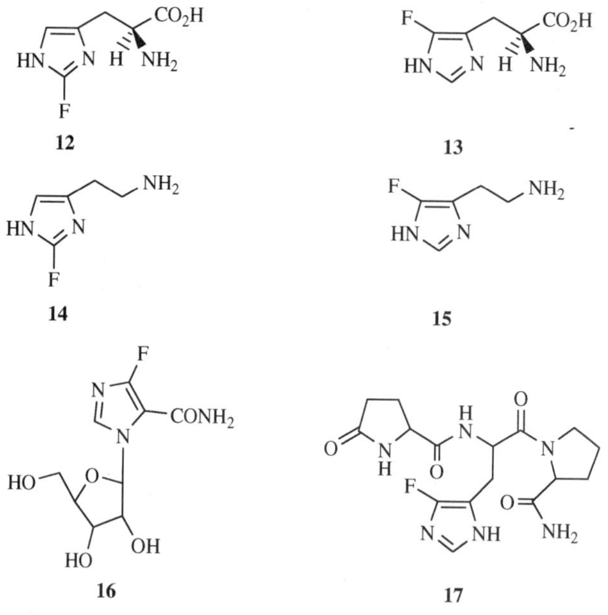

12 **13** **14** **15** **16** **17**

SCHEME 11.4

example, 4-fluoroimidazole-5-carboxamide riboside (**16**). The biochemistry and pharmacology of fluorohistidines were particularly fruitful,[15] and remains an area of research in the author's laboratory, 30 years later. The important functional role carried out by histidine in proteins and peptides was an early target for study. Ribonuclease and thyrotropin-releasing hormone (TRH) were two examples of the effective use of incorporation of fluorohistidines to probe biochemical function. With 4-fluoroimidazole TRH (**17**), for example, the first complete separation of the pituitary and cardiovascular activities of TRH was demonstrated.[16]

Fluorocatecholamines

Background A combination of factors contributed to the successes achieved in the research with fluoroimidazoles. The important functions and widespread occurrence of the imidazole ring in biological systems assured ample areas to explore with fluorinated analogs. In addition, the properties of the ionizable nitrogen atoms in the imidazole nucleus are altered significantly by the presence of the highly electronegative fluorine substituent. For example, the pK_1 of imidazole is 7.0 whereas pK_1 values of 2- and 4-fluoroimidazole determined by titration are 2.4 and 2.5, respectively. Finally, prior to the discovery of the photochemical Schiemann reaction, fluoroimidazoles were unknown, so the systems that could be investigated were plentiful.

The catechol nucleus offered the same combination of factors. The catechol ring is present in a series of natural chemical neurotransmitters, the agonist properties of which are dependent on at least one phenolic hydroxyl group. Second, the ionzable phenolic group again will be susceptible to the effects of the electronegative fluorine substituted on the aromatic nucleus. Finally, as before, the absence of reports of fluorinated catecholamines suggested ample areas of research would be available for such analogs.

The naturally occurring catecholamines—dopamine (DA) (**18a**), *R*-norepinephrine (*R*-NE) (**19a**), and *R*-epinephrine (*R*-EPI) (**20a**)—have many important biological functions. Such catecholamines originate from tyrosine, which is ring-hydroxylated to give dihydroxyphenylalanine (DOPA, **21a** (see Scheme 11.5, "series a" for structures **18–21**). Decarboxyalation of DOPA produces DA, which is both an important neurotransmitter in the central nervous system (CNS) and also the precursor of *R*-NE. DA also has actions on kidneys and heart. NE, biosynthesized from DA by β-hydroxylation, is a neurotransmitter in the CNS, and is the principal neurotransmitter of the sympathetic nervous system. *R*-EPI, which is elaborated from *R*-NE in the adrenal medulla, has potent actions on the heart, smooth muscle, and other organs.[17]

Research in the author's laboratory produced fluorinated analogs of these neuroactive amines, as well as their amino acid precursors, DOPA and tyrosine. In addition, ring-fluorinated analogs of the neurotransmitter 5-hydroxytryptamine (**22**), and the related hormone, melatonin (**23**)[18] were prepared.

Ring-Fluorinated Analogs of Dopamine In early work, the author prepared 2-, 5-, and 6-fluorodopamine (FDA; **18b–d**). 6-FDA (**18d**) was prepared from DA in

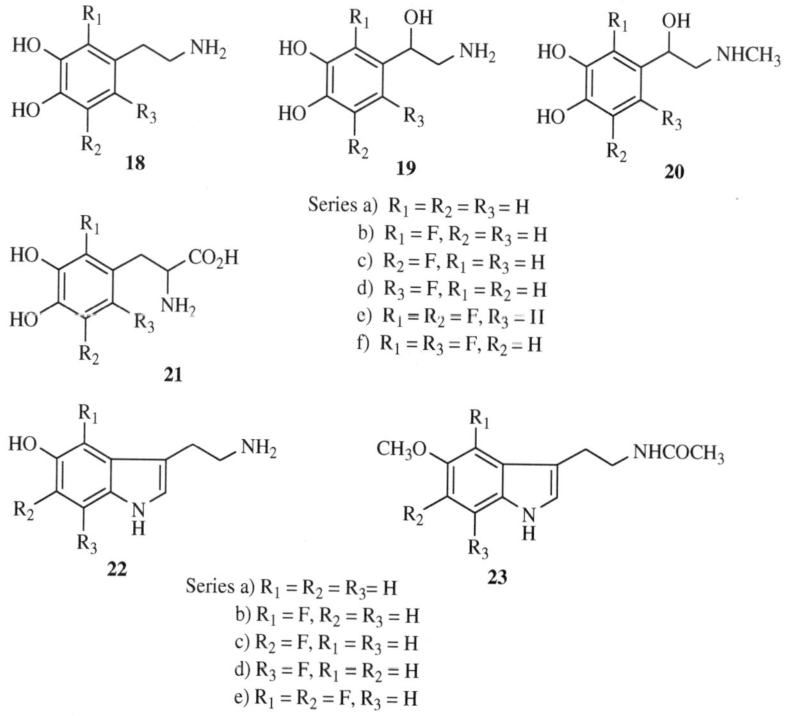

Series a) $R_1 = R_2 = R_3 = H$
b) $R_1 = F, R_2 = R_3 = H$
c) $R_2 = F, R_1 = R_3 = H$
d) $R_3 = F, R_1 = R_2 = H$
e) $R_1 = R_2 = F, R_3 = H$
f) $R_1 = R_3 = F, R_2 = H$

Series a) $R_1 = R_2 = R_3 = H$
b) $R_1 = F, R_2 = R_3 = H$
c) $R_2 = F, R_1 = R_3 = H$
d) $R_3 = F, R_1 = R_2 = H$
e) $R_1 = R_2 = F, R_3 = H$

SCHEME 11.5

a sequence that utilized the photochemical Schiemann reaction. Later, FDAs as well as other catecholamines and amino acids were synthesized by functionalization and sidechain elaboration of fluorinated veratroles. FDAs had comparable affinity as dopamine and were full agonists at dopaminergic receptors.[19] In addition, only subtle effects of fluorine substitution were observed in in vivo studies.[20] These results were not expected, in view of the rather narrow structural requirements of dopamine receptors. Nonetheless, these and other data became useful in the development of 6-fluoro-DOPA (**21d**) as a positron emission scanning agent for quantititation of dopaminergic function in the central nervous system.[10,21]

Ring-Fluorinated Analogs of Norepinephrine and Related Adrenergic Agonists NE, EPI, and other adrenergic agonists function through interactions with membrane-bound adrenergic receptors. Two classes of adrenergic receptors, designated α- and β-adrenergic receptors, have been identified in classic pharmacological studies. These receptors now have been further divided into various subclasses (e.g., α_1-,α_2-,β_1-,β_2-,β_3-) with further refinements in pharmacology and molecular biology producing additional classifications and subclassifications. Sympathetic responses regulated by adrenergic receptors include heart rate, force of cardiac contraction, vasomotor tone, blood pressure, bronchial airway tone, and fat and carbohydrate metabolism. The presence of catecholamine-containing neurons in the central nervous system increases the importance of attempts to define the molecular interactions between transmitter and receptor.[17,22]

The response of a given tissue to an adrenergic agent depends on the type and concentration of receptors present. Development of drugs that stimulate or inhibit the response of a given subset of adrenergic receptors has been the focus of a large area of medicinal chemistry. Indeed, numerous useful drugs have been marketed that depend on selectivity of adrenergic receptor response. As evidenced by the nearly absolute requirement for one or more phenolic groups in biogenic amines for agonist action, this functionality clearly plays an important part in receptor binding and activation. Fluorine substitution offered an attractive means to test effects of electronic perturbation of the phenolic system on interaction with adrenergic receptors. Accordingly, synthesis of ring-fluorinated analogs of NE, EPI, and other adrenergic agonists was initiated.

Synthesis 2-, 5-, and 6-FNEs were prepared from fluorinated dibenzyloxybenzaldehydes **24b–d** in a sequence consisting of ZnI_2-catalyzed condensation with trimethysilylcyanide (TMSCN), followed by in situ $LiAlH_4$ reduction of the trimethylsilycyanohydrin to give the fluorinated 3,4-dibenzyloxyphenethanolamines **25b–d**. Hydrogenolysis of the benzyl ethers gave FNEs **19b–d** [Eq. (11.4)].[23] 2-, and 6-FEPIs (**20b,d**) were prepared from **25b,d** by formylation with ethyl formate, $LiAlH_4$ reduction, and debenzylation [Eq. (11.5)].[24]

CH_3O, CH_3O, CHO, R_1, R_2, R_3 ⇒ BnO, BnO, CHO, R_1, R_2, R_3 **24b-d** —TMSCN/ZnI_2→ BnO, BnO, OTMS, CN, R_1, R_2, R_3

—$LiAlH_4$→ BnO, BnO, OH, NH_2, R_1, R_2, R_3 **25b,d** —H_2/Pd→ HO, HO, OH, NH_2, R_1, R_2, R_3 **19b-d** (11.4)

25b,d —HCO_2Et→ BnO, BnO, OH, NHCHO, R_1, R_2, R_3 —1) $LiAlH_4$ 2) H_2/Pd→ HO, HO, OH, $NHCH_3$, R_1, R_2, R_3 **20b,d** (11.4)

Agonist selectivities of ring-fluorinated adrenergic agonists. Initial evaluation of the agonist properties of FNEs in isolated organ preparations revealed the remarkable effects of fluorine substitution. 5-FNE was comparable to NE in evoking α_1-adrenergic (guinea pig aorta contraction) and β_1-adrenergic (guinea pig atria) responses. However, 6-FNE was much less potent as a β-adrenergic agonist but retained α-adrenergic potency. In contrast, 2-FNE had greatly reduced α-adrenergic activity, but retained β-adrenergic activity.[19,23] The agonist selectivity was related directly to receptor binding affinities[19] and later was extended to include FEPIs as well as fluorinated analogs of β-adrenergic agonist, isoproterenol, and the α-adrenergic agonist phenylephrine. To summarize, in virtually every adrenergic agonist that was examined (DA, a weak adrenergic agonist, is an important exception) fluorine in the 2 position markedly decreased α-adrenergic adrenergic activity, and fluorine in the 6 position markedly decreased β-adrenergic activity.[18]

These results had important practical consequences. The close structural similarities of FNEs to NE itself provided advantages over existing selective adrenergic agonists in pharmacological studies. Unlike other selective agonists, FNEs can be elaborated from fluorinated precursors by the normal biosynthetic pathway, and they are metabolized by the normal metabolic enzymes. For these and other reasons, FNEs and FEPIs have found many important uses as pharmacological tools, biological tracers, and mechanistic probes.

Mechanism(s) of Fluorine-Induced Adrenergic Selectivities In addition to pharmacological studies, the author also carried out extensive research designed to determine the molecular basis of these fluorine-induced selectivities. Several mechanisms were proposed to explain the "antisymmetrical" nature of the selectivity. For example, intramolecular hydrogen bonding between the benzylic OH group and an *ortho*-fluorine substituent, or dipole–dipole repulsions between the benzylic C–O and aromatic C–F bonds were considered as factors contributing to predominant formation of conformations favorable for binding to α- or β-adrenergic receptors. Alternatively, the speculation was made that fluorine-induced alterations in the electron density distributions of the catechol ring could lead to differential inhibition of binding to α- or β-adrenergic receptors. Attempts were made to examine the effects of intramolecular interactions by synthesizing several structural analogs of the fluorinated catecholamines in which such interactions, if they existed, would be expected to be quite different. For example, substitution of fluorine in the 2 position (**26b**) (see Scheme 11.6, "series b") of the potent β-adrenergic agonist **26a** had no effect on receptor affinity, fluorine in the 6 position (**26c**) abolished activity, suggesting that conformational effect may be relatively

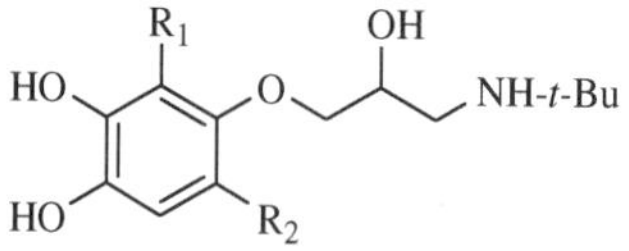

R_1 OH

HO O NH-*t*-Bu

HO R_2

26 Series a, $R_1 = R_2 = H$
b, $R_1 = F, R_2 = H$
c, $R_2 = F, R_1 = H$

SCHEME 11.6

unimportant in defining receptor selectivities. However, overall, the results of these analog studies were inconclusive in defining mechanisms of fluorine-induced selectivities.[18,25]

As an alternative approach to studying this problem, attention was turned to modifications of the structure of the receptor itself. As a first step, the selectivity profiles of these fluorinated agonists were assessed by the use of cloned adrenergic receptors, transiently or stably expressed in mammalian cells. Then, efforts were made to localize the region on the receptor which discriminates between 2-F and 6-F substitution of NE and EPI by studying various α_2-β_2 chimeric receptors. Because of the structural similarities to the natural agonists, it was felt that the fluorinated catecholamines would be sensitive tools to probe structural parameters that define α and β properties of the adrenergic receptors. Using FNEs and FEPIs, trends were noted that were suggestive of receptor regions that may represent important differences in α- versus β-adrenergic receptors. However, as the efficiency of binding to the chimeric receptors decreased, so did the selectivity of binding of 2- and 6-FNE and 2- and 6-FEPI, to the extent that interpretation of results became very difficult (unpublished results). More selective agonists clearly were desirable. The work described below addresses this issue.

Synthesis of the enantiomers of FNEs. It is important to note that although 2-, 5-, and 6-FDA (**18b–d**) did bind to adrenergic receptors, albeit with decreased affinity, the 2- and 6-fluoro analogs showed little or no discrimination between α- and β-adrenergic receptors. This implicated the benzylic OH group as an important factor in defining receptor selectivity. It also suggested that the *S* isomers might have weak affinities towards both α- and β-adrenergic receptors. In fact, classic pharmacological studies have shown that DA and *S*-NE have comparable affinities at both α- and β-adrenergic receptors. Since adrenergic selectivities of FNEs emanate from an inhibition of binding and FDAs show no such selectivities, it seemed likely that inhibition of binding requires both the presence of the ring fluorine substituent and the benzylic OH group in the *R* configuration. In other words, *R*-2-FNE could be active at the β-adrenergic receptor and virtually inactive at the α-adrenergic receptor whereas the reverse could hold true for *R*-6-FNE. Nonselective activity seen with the racemic mixture thus would be attributed to the unnatural *S* isomer. These and other considerations gave preparation of the enantiomers of FNEs and FEPIs high priority in our research program.

The enzyme-like CBS catalysts (e.g., **27**) developed by Corey[26] have been applied to the enantioselective synthesis of arylethanolamines. The remarkably high enantioselectivities obtained with these catalysts provided the basis for one approach to *R*-FNEs. Chloroketones **28b,d**, prepared from aldehydes **24b,d**, were reduced with $BH_3 \cdot THF$ in the presence of CBS catalyst **27** to give the *R*-chloroalcohols **29b** (80%) and **29d** (85%). A single recrystallization provided the chloroalcohols enriched to $\geq$95% e.e., as determined by chiral HPLC. After conversion to the azides (**30b,d**), catalytic hydrogenation in the presence of oxalic acid gave the hemioxalates of *R*-2-FNE (*R*-**19b**) (76%) and *R*-6-FNE (*R*-**19d**) (51%). In both cases, the enantiomeric excesses of the final product was identical to that of chloroalcohols **28** [Eq. (11.6)].[27]

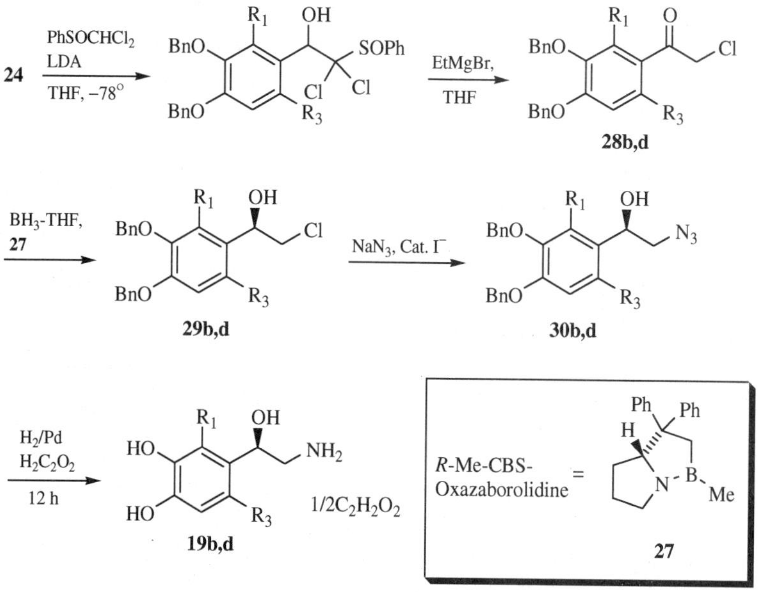

(11.6)

A second enantioselective approach to FNEs involved an enantioselective variant of the author's initial synthesis of racemic FNEs. As described above, the key step was the ZnI_2-catalyzed addition of trimethylsilylcyanide to aldehydes **24b,d**. The chiral Lewis acid (salen)Ti^{IV} complexes, such as **31**, have been used to catalyze the addition of TMSCN to aldehydes in good yield and high enantioselectivity.[28] (*R,R*)-**31**-Catalyzed addition of TMSCN to fluorinated aldehydes **24b,d** was followed by reduction in situ with $LiAlH_4$ to produce the corresponding phenethanolamines *S*-**25b,d** in 25% yield (>95% e.e.). The benzyl protecting groups were removed by hydrogenolysis to give *S*-FNEs **19b** and **19d** in 68 and 87% yields, respectively. Both of the final products were obtained in $> 95\%$ e.e.[27] [Eq. (11.7)].

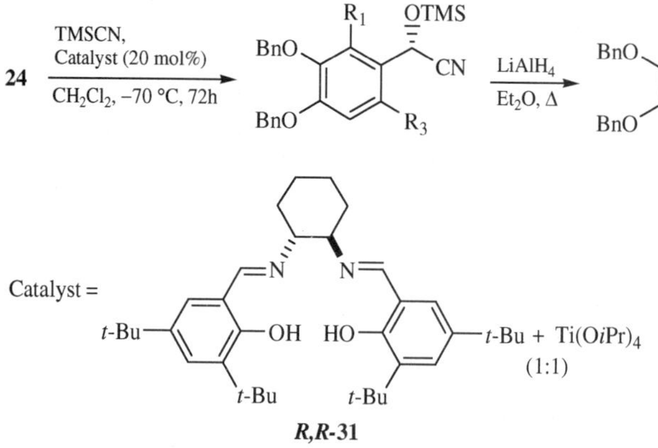

(11.7)

Through the use of the (*R,R*)-form of catalyst **31**, it was anticipated that the phenethanolamine generated would be enantiomeric to the series prepared by the enantioselective carbonyl reduction strategy. As expected, the rotations of FNEs **19b,d** produced by the two procedures were indeed opposite in sign. The result was further confirmed by direct comparison using chiral HPLC. This sequence was repeated with (*S,S*)-**31** to give *R*-FNEs. Thus, two routes to both enantiomers of FNEs **19b** and **19d** in high enantiomeric excess ($>95\%$) are now available.

Binding affinities of the enantiomers of FNEs. To assess the significance of chirality on fluorine-induced adrenergic selectivities, binding affinities of *R*-FNEs, *S*-FNEs, and racemic FNEs were carried out using procedures described.[29] *R*-NE, as expected, was 10-20-fold more active at α_1-, α_2-, β_1-, and β_2-adrenergic receptors than the unnatural *S* enantiomer. At β_1- and β_2-adrenergic receptors the β-selective *R*-2-FNE was over 100-fold more active than the S enantiomer. At the α_1- and α_2-adrenergic receptor, *R*-2-FNE and *S*-2FNE had comparable activity. At α_1- and α_2-adrenergic receptors, the α-selective *R*-6-FNE was about 20-fold more active than the *S* enantiomer. *R*-6-FNE was also several fold more active than the *S* enantiomer at the β-adrenergic receptor. Fluorine in the 2 or 6 position of *S*-NE either had little effect on activity at both α- and β-adrenergic receptors. Thus, the preliminary results confirm the expectation that the natural *R* configuration of the β-hydroxyl group of NE is essential for 2- or 6-fluorine substitution to have marked effects on the affinity of NEs at α- and β-adrenergic receptors. However, the prediction that the *S* enantiomers would be more active than the *R*-enantiomers at the "wrong" receptor appears not to be borne out.[27]

Earlier studies with racemic 2-FNE and 6-FNE had indicated that the main consequence of having a 2-fluoro substituent was a reduction in activity at α-adrenergic receptors while the main effect of having a 6-fluoro substituent was a reduction in activity at β-adrenergic receptors. The present results with the *R* enantiomers confirm that fluorine in the 2 position indeed reduces activity compared to *R*-NE at α-adrenergic receptors by about 20-fold. However, fluorine at the 2 position also enhances activity at β receptors by at least 10-fold compared to *R*-NE. Fluorine in the 6 position, as expected from earlier studies with racemates, has little effect on activity at α-adrenergic receptors but reduces activity compared to (*R*)-NE at β-adrenergic receptors by 10–20-fold.[27]

Ring-Fluorinated Catecholamino Acids

Ring-Flourinated Analogs of DOPA Ring-fluorinated analogs of DOPA have received particular attention related to the development of [^{18}F]-labeled DOPA as a PET-scanning agent for the quantitation of regional brain dopaminergic activity. For such applications, pharmakokinetic studies are critical. For example, early efforts using [^{18}F]-5-FDOPA for this purpose were thwarted because of rapid metabolism of this analog through COMT-mediated methylation. The author prepared 2-, 5-, and 6-FDOPA (**21b-d**),[18] analogs that were used to study effects of ring fluorination on COMT-catalyzed methylation, for in vivo metabolic studies, and as chromatographic standards for radiochemical syntheses of [^{18}F]-labeled analogs. Research from several groups now has demonstrated that [^{18}F]-6-fluoro-DOPA ([^{18}F]-6-FDOPA) is an effective PET-scanning agent that can be used, for example, as a diagnostic tool in Parkinson's disease. Injected [^{18}F]-6-fluoro-DOPA crosses the blood–brain barrier, is decarboxylated, and the resulting [^{18}F]-6-FDA is taken up and stored in central dopaminergic neurons. Quantitation of stored [^{18}F]-6-FDA by PET has been used to study central dopaminergic activity, in both normal and abnormal subjects.[10,21]

Fluorinated Analogs of Dihydroxyphenylserine (DOPS) The pharmacological properties and therapeutic potential of (2*S*,3*R*)-*threo*-(3,4-dihydroxyphenyl)-serine (L-*threo*-DOPS) (**32a**) have been studied. Of special interest is evidence that administered L-*threo*-DOPS crosses the blood–brain barrier and is subsequently decarboxylated to produce NE in the CNS, particularly in situations where catecholamine deficiencies are indicated. Several clinical trials suggest that L-*threo*-DOPS may be beneficial in treating disorders of both the central and sympathetic nervous systems that are characterized by NE deficiencies. For example, Tohgi and co-workers found a dose-dependent increase in cerebrospinal fluid NE concentrations in six advanced parkinsonian patients.[30] In three of six patients the "freezing phenomenon" in gait and speech, symptoms unresponsive to DOPA treatment, improved. *threo*-DOPS gave improvement of certain memory functions in patients with Korsakoff's disease (amnesia induced by chronic alcoholism). It was also effective in treatment of orthostatic hypotension associated with Shy–Drager syndrome and with familial amyloid neuropathy. There is also evidence that L-*threo*-DOPS functions directly, as the amino acid, in certain systems.[31] The biological interest and therapeutic potential of L-*threo*-DOPS suggested that ring fluorinated analogs of this amino acid, as precursors of the corresponding FNEs, would be useful tools for studying mechanistic aspects of these actions [Eq. (11.8)]. In particular, such studies could give evidence as to whether α-adrenergic or β-adrenergic innervation is involved.

32 —Dopa Decarboxylase→ ***R*-19** (11.8)

Series a: $R_1 = R_3 = H$
b: $R_1 = F, R_3 = H$
d: $R_1 = H, R_3 = F$

Synthesis of Racemic FDOPS Kellogg and co-workers have developed a *threo*-selective synthesis of 3-substituted serine derivatives using a $ZnCl_2$-catalyzed aldol condensation of aldehydes with the trimethylsilyl ketene acetal **33** derived from the benzophenone imine of glycine ethyl ester.[32] Using this approach, condensation of fluoroaldehyde **24d** in the presence of 5 mol% of $ZnCl_2$ with **33** gave a 6 to 1 mixture of *threo* and *erythro* adducts (**34d** and **35d**, respectively). A similar 7 : 1 mixture of *threo* and *erythro* products (**34b, 35b**) was obtained from **24b**. After separation of the major diastereomer in each series, the imine and silyl ether functionalities were removed with dilute acid to give **36b,d**. Saponification of the ester followed by hydrogenolysis of the benzyl groups produced 2-F- and 6-F-*threo*-DOPS (**32b,d**) [Eq. (11.9)].[31]

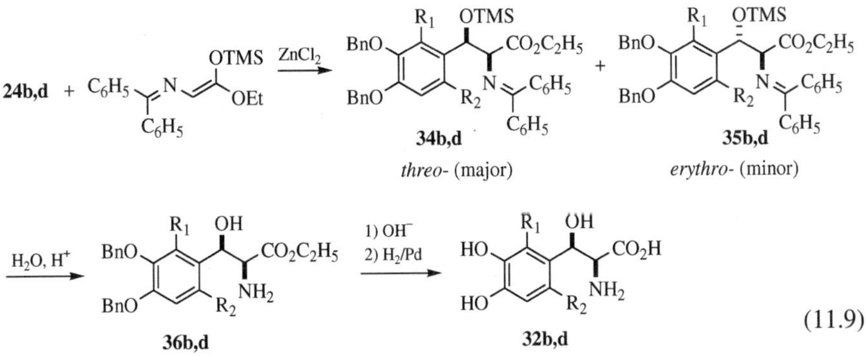

(11.9)

Synthesis of the enantiomerically pure fluorinated-2S,3*R*-*threo*-DOPS. Evans and Weber have developed the isothiocyanate **37** as a chiral glycine equivalent, and have used this for the synthesis of β-hydroxy-α-amino acids.[33] With certain modifications of the Evans conditions, oxazolidinone (**37**) was condensed with aldehyde **24d** to give thiocarbamate (**38**) [diastereomeric ratio (d.r.) 100 : 1] [Eq. (11.10)]. The chiral auxiliary was removed with methoxymagnesium bromide to

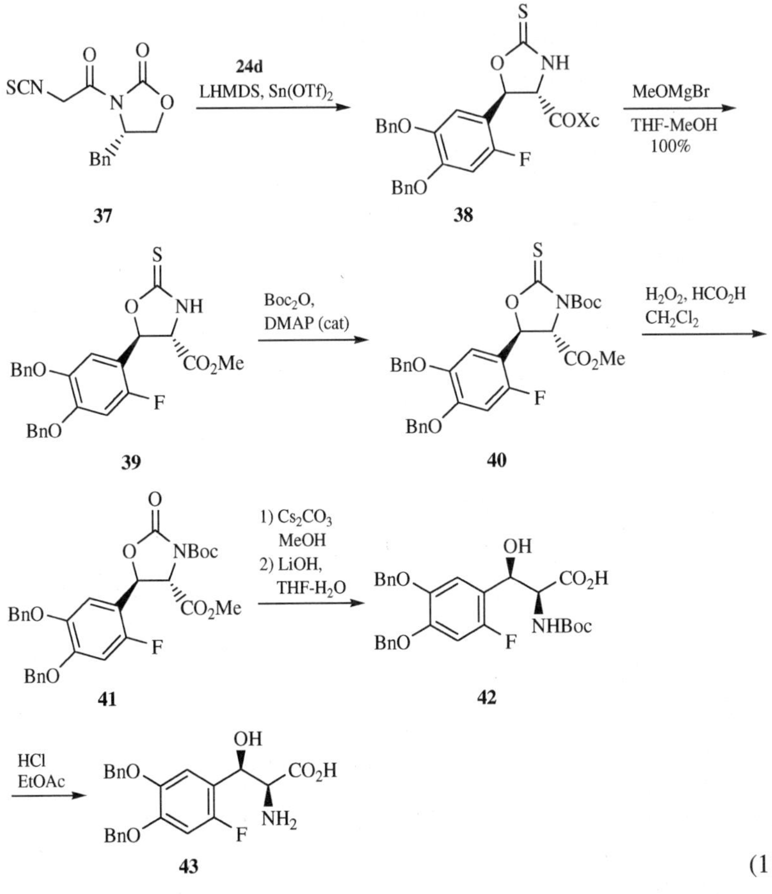

(11.10)

provide methyl ester (**39**) and the amide was converted to the *t*-butyl carbamate (**40**). Following a sulfur-to-oxygen exchange procedure (71%), the cyclic carbamate (**41**) was cleaved (Cs_2CO_3 in MeOH) and the methyl ester saponified to provide the *N*-Boc protected (3,4-dibenzyloxy-6-fluoro)phenylserine (**42**). The Boc group was cleaved gaseous HCl in ethyl acetate to furnish amino acid (**43**).[27]

These individual steps in this sequence proceed in good yield and have been found adaptable to scaleup. Further work on this route includes optimization for large-scale production of the target and the preparation of (2*S*,3*R*)-2-FDOPS (**28b**). Ready access to these analogs should provide unique tools for the study of the actions of FNEs in the CNS.

Effects of Ring Fluorination on the Metabolism of Catecholamines and Amino Acids

Fluorinated Analogs as Substrates for Monoamine Oxidases MAO is a flavin-linked mitochondrial enzyme that catalyzes the oxidation of monoamines to carbonyl compounds. An important role for MAO is the regulation of neurotransmitter amine levels. Two forms of MAO, designated MAO A and MAO B, occur in human brain. MAO A prefers more polar substrates, such as NE and 5HT, both of which are essentially pure MAO A substrates. MAO B preferentially deaminates the more lipophilic substrates such as benzylamine. DA and tyramine are effectively deaminated by both MAO A and MAO B. The influence of fluorine substitution on MAO substrate selectivity appears to reflect increased lipophilicity of fluorinated amines. For example, both 3-fluorotyramine and 3,5-difluorotyramine are better substrates for MAO B than is tyramine. Likewise, 4,6-difluoro-5HT is an excellent substrate for MAO B, while the parent 5HT is a poor substrate.[18]

Fluorinated Catecholamines as Substrates for Catechol-O-methyl Transferase Catechol-*O*-methyl transferase (COMT) catalyzes the transfer of a methyl group from *S*-adenosyl methionine (AdoMet) to one of the hydroxyl groups of a catechol. This provides an important mechanism for deactivation of biologically active catechols such as catecholamines, catecholamino acids, and steroidal catechols. FNEs were useful tools in revealing mechanistic details of catecholamine methylations. Methylation of NE, EPI, and DOPA with COMT gives predominantly *meta*-methylation, despite comparable acidities of the 3- and 4-OH groups, suggesting that side-chain orientation and other factors control regiochemistry. On the other hand, methylation of FDOPAs and FNEs indicated that the more acidic hydroxyl group is methylated preferentially. For example, the rank order of preference for *para*-methylation (5-FNE > NE > 6-FNE > 2-FNE) seen at pH 7 is even more pronounced at pH 9. This relative increase of *para*-methylation for each compound was shown to follow a titration curve corresponding to the ionization of a group with pK_a values of 8.6, 7.7, 7.9, and 8.4 for NE, 2-, 5-, and 6-FNE, respectively. These pK_a values are the same as, or similar to, the pK_a values of a phenolic group of these substrates. Similar studies with 2,5-difluoro-NE (**19e**) supported these results. To summarize, *meta*-methylation of NE is favored at physiological pH because of orientation of the side chain, but phenol ionization of the catechol favors *para*-methylation.[18,34]

Effects of Fluorination on COMT-Catalyzed Methylation of DOPA Information on the methylation of fluorinated catecholamines and FDOPAs is important with respect to the development of ^{18}F-labeled catecholamines and amino acids as PET-scanning agents (see section titled "Ring Fluorinated Analogs of DOPA"). Methylation of DOPA to produce 3-methoxytyrosine is the principal mechanism by which DOPA is cleared from the circulation. The presence of fluorine on the aromatic ring affects both the regioselectivity and rate of methylation of DOPA by COMT. Thus, the rate of methylation of 2- and 5-FDOPA by COMT is some 3–4 to four times that of DOPA, but the rate of methylation of 6-FDOPA is nearly 10 times slower. This slower rate of methylation has increased its utility as a PET-scanning agent. The differences in methylation rates reflect the relative affinites of the substrates for COMT.[18]

SUMMARY

Fluorinated analogs of amino acids and amines involved in neurotransmission have proven to be valuable tools, with important applications in pharmacology, biochemistry, and medicine. Whereas many effects of fluorine substitution on biological activity can be explained readily by steric and electronic factors, other results, such as fluorine-induced adrenergic selectivities, remain to be fully explained. Identification of new biological targets together with advances in organofluorine chemistry that make fluorinated analogs more readily available assure that organofluorine chemistry will continue to play an important role in biomedicinal chemistry.

ACKNOWLEDGMENT

The author has reviewed research from his own laboratory as part of this report. He gratefully acknowledges the vital and sustained contributions made to this research by his students, postdoctoral fellows, and collaborators.

REFERENCES

1. Peters, R., in *Ciba Foundation Symposium: Carbon-Fluorine Compounds: Chemistry, Biochemistry, and Biological Activities,* Elsevier, Amsterdam, 1972, pp. 55–76.
2. Kirsten, E.; Sharma, M. L.; Kun, E., *Molec. Pharmacol.* 1978, **14**, 172–184.
3. Goldmam, P., *Science* 1969, **164**, 1123–1133.
4. For a recent review, see Kirk, K. L.; Filler, R., in *Biomedical Frontiers of Fluorine Chemistry,* ACS Symposium Series 639, I. Ojima, J. R. McCarthy, and J. T. Welch, eds., American Chemical Society, Washington, DC, 1996, pp. 1–24.

5. Welch, J. T.; Eswarakrishnan, S., *Fluorine in Bioorganic Chemistry*, Wiley, New York, 1991.
6. For a review, see Kirk, K. L., in *Biochemistry of Halogenated Organic Compounds, Biochemistry of the Elements*, E.; Frieden, series ed., Plenum, New York, 1991, Vol. 9B, pp. 1–63.
7. For a review, see ref 6. pp. 65–103.
8. Kim, D. H.; Lees, W. J.; Haley, T. M.; Walsh, C. T., *J. Am. Chem. Soc.,* 1995, **117**, 1494–1502.
9. For reviews, see Sykes, B.; Weiner, J. H., *Magn. Res. Biol.* 1980, **1**, 171–196; Gerig, J. T., *Prog. NMR Spectrosc.* 1994, **26**, 293–370.
10. Fowler, J. S., in *Organofluorine Compounds in Medicinal Chemistry and Biomedical Applications*, R. Filler, Y. Kobayashi, and L. M. Yagupolskii, eds., Elsevier, Amsterdam, 1993, pp. 309–338.
11. Kollonitsch, J., in *Biomedicinal Aspects of Fluorine Chemistry*, R. Filler and Y. Kobayashi, eds., Kodansha, L Tokyo; Elsevier Biomedical Press, Amsterdam, pp. 93–122.
12. Palfreyman, M. G.; Bey, P.; Sjoerdsma, A., *Essays Biochem.* 1987, **23**, 28–81.
13. Kirk, K. L.; Cohen, L. A., *J. Am. Chem. Soc.* 1973, **95**, 4619–4624.
14. Kirk, K. L.; Nagai, W.; Cohen, L. A., *J. Am. Chem. Soc.* 1973, **95**, 8389–8392.
15. Kirk, K. L.; Cohen, L. A., in *Biochemistry Involving the Carbon-Fluorine Bond*, ACS Symposium Series 28, R. Filler, ed., American Chemical Society, Washington, DC, 1976, pp. 23–36.
16. Feurestein, G.; Losovsky, D.; Cohen, L. A.; Labroo, V. M.; Kirk, K. L.; Kopin, I. J.; Faden, A. I., *Neuropeptides* 1984, **4**, 303–310.
17. Cooper, J. R.; Bloom, F. E.; Roth, R. H., *The Biochemical Basis of Neuropharmacology*, Oxford Univ. Press, New York, 1996, pp. 226–351.
18. For reviews, see Kirk, K. L., *J. Fluorine Chem.* 1995, **72**, 261–266; Kirk, K. L.; Nie, J.-Y., in *Biomedical Frontiers of Fluorine Chemistry*, ACS Symposium Series 639, I. Ojima, J. R. McCarthy, and J. T. Welch, eds., American Chemical Society, Washington, DC, 1996, pp. 313–327; Kirk, K. L.; Cantacuzene, D.; Creveling, C. R., in *Biomedicinal Aspects of Fluorine Chemistry*, R. Filler and Y. Kobayashi, eds., Kodansha, L Tokyo; Elsevier Biomedical Press, Amsterdam, pp. 75–91.
19. Nimit, Y.; Cantacuzene, D.; Kirk, K. L.; Creveling, C. R.; Daly, J. W., *Life Sci.* 1980, **27**, 1577–1585.
20. Goldberg, L. I.; Kohli, J. D.; Cantacuzene, D.; Kirk, K. L.; Creveling, C. R., *J. Pharmacol. Exp. Ther.* 1980, **213**, 509–513.
21. Ding, Y.-S.; Fowler, J. S., in *Biomedical Frontiers of Fluorine Chemistry*, ACS Symposium Series 639, I. Ojima, J. R. McCarthy, and J. T. Welch, eds., American Chemical Society, Washington, DC, 1996, pp. 328–343.
22. Wiener, N., in *The Pharmacological Basis of Therapeutics* A. G. Gilman, L. S. Goodman, T. W. Rall, and F. Murad, eds., Macmillan, New York, 1985, pp. 145–180.
23. Kirk, K. L.; Cantacuzene, D.; Nimitkitpaisan, Y.; McCulloh, D.; Padgett, W. L.; Daly, J. W.; Creveling, C. R., *J. Med. Chem.* 1979, **22**, 1493–1497.
24. Adejare, A.; Gusovsky, F.; Padgett, W. L.; Creveling, C. R.; Daly, J. W.; Kirk, K. L., *J. Med. Chem.* 1988, **31**, 1972–1977.

25. Kirk, K. L., in *Selective Fluorination in Organic and Bioorganic Chemistry*, J. T. Welch, ed., ACS Symposium Series 456, American Chemical Society, Washington, DC, 1991, pp. 136–155.
26. Corey, E. J.; Link, O. J., *Tetrahedron Lett.* 1989, **30**, 6225–6287.
27. Kirk, K. L.; Herbert, B.; Lu, S.-F., Jayachandran, B.; Padgett, W. L.; Oshunleti, O. O.; Daly, J. W.; Haufe, G., in *Asymmetric Synthesis of Fluoro-Organic Compounds*, P. V. Ramachandra, ed., American Chemical Society, Washington, DC, in press.
28. Belekon', Y.; Flego, M.; Ikonnikov, N.; Moscalenko, M.; North, M.; Orizu, C.; Tararov, V.; Tasinazzo, M., *J. Chem. Soc. Perkin I* 1997, 1293–1295.
29. Nie, J.-Y.; Shi, D.; Daly, J. W.; Kirk, K. L., *Med. Chem. Res.*, 1996, 318–331; and references cited therein.
30. Tohgi, H.; Abe, T.; Takahashi, S.; Takahashi, J.; Ueno, M.; Nozaki, Y., *Neuroscience Letters*, 1990, **116**, 194–197.
31. Chen, B.-H.; Nie, J.-Y.; Singh, M.; Pike, V. W.; Kirk, K. L., *J. Fluorine Chemistry* 1995, **75**, 93–101; and references cited therein.
32. van der Werf, A. W.; Kellogg, R. M.; van Bolhuis, F., *J. Chem. Soc. Chem. Commun.* 1991, 682–683.
33. Evans, D. A.; Weber, A. E., *J. Am. Chem. Soc.* 1987, **108**, 6757–6771.
34. Thakker, D.; Boehlert, C.; Kirk, K. L.; Antkowiak, R.; Creveling, C. R., *J. Biol. Chem.* 1986, **261**, 178–184.

PART III

UNDERSTANDING THE CHEMICAL BASIS OF DRUG ACTION AND DISEASE

Malignant melanoma is the deadiest of the three recognized skin cancers that also include basal cell carcinoma and squamous cell carcinoma. In the United States alone, this disease kills more than 7000 people a year, while estimates are that more than 40,000 individuals will be diagnosed with the disease this year. Prevention and early detection remain as key elements in the battle against maligant melanoma. Once diagnosed, surgery is effective at early stages. For later-stage disease, chemotherapy (dacarbazine and nitrosoureas) and combination chemotherapy give low (20–25%) response rates. Interferon (Intron A) has been approved by the U.S. Food and Drug Administration for malignant melanoma, but again response rates are low. Thus there remains a great need for more effective chemotherapeutic options for this killer. In Chapter 12, d'Ischia and Prota begin by describing an ingenuous approach to malignant melanoma therapy that relies upon the chemical mechanistic understanding of *melanin* biosynthesis and, in particular, the enzyme *tyrosinase*.

Malaria represents an enormous drain on human prospects in much of the world. No vaccine is available, and present chemotherapeutic agents are becoming less effetive as resistant forms of *Plasmodium falciparum* develop. In Chapter 13, Posner and colleagues write on the discovery of the natural product antimalarial *artemisinin* and the elucidation of its remarkable mechanism of action. They proceed to outline this *carbon-centered free radical* mechanism at a molecular level and shown how this detailed understanding can be applied to the rational design of new antimalarials.

Parkinson's disease causes the progressive deterioration of the central nervous system and affects more than one million people in the United States alone. Its origins lie in the degeneration of pigmented neurons in the *substania nigra* (also termed *locus niger*) in the brain. This causes a reduction in *dopamine* output. The standard treatment for Parkinson's disease has been *levodopa* (in combination with another drug), which is converted to dopamine in the brain, but there are major problems, side-effects, or limitations to this therapy. Dryhurst and collaborators provide a description of our present understanding of Parkinson's disease and their hypothesis for the underlying neurotoxic mechanism. These efforts may illumine future strategies for Parkinson's disease treatment.

CHAPTER 12

Thiouracil and Related Thioureylene Compounds as Melanoma Seekers: Paradigms for a Chemical Approach to the Design of Targeting Anticancer Agents*

MARCO D'ISCHIA and GIUSEPPE PROTA

Department of Organic and Biological Chemistry, University of Naples Federico II

INTRODUCTION

Formerly regarded as an atypical and very rare neoplastic disease, malignant melanoma has become the leading cause of death from all diseases arising in the skin.[1] The dramatic rise in incidence, which continues to increase at a pace faster than that of any other malignancy among the young and middle-aged, and the often fatal outcome warrant a prominent position of melanoma in anticancer programs, especially among the red-haired, fair-skinned populations in Australia and sunny regions of the United States. In addition to clinical investigations urged by the alarming epidemiological reports, this tumor has been the focus of considerable interest also on the part of basic researchers. This derives from the biological uniqueness of the cells from which the tumor originates, the melanocytes (Fig. 12.1), which play a central role in skin homeostasis and protection against noxious environmental stimuli, and are implicated in a variety of phenomena of outstanding social, biomedical, and cosmetic relevance, ranging from racial pigmentation through skin tanning to pigmentary disorders such as albinism and vitiligo.[2]

During malignant transformation, melanocytes retain at least in part their differentiation markers, including morphology, dendriticity, and the production of

*Dedicated to the memory of Professor Bengt Larsson, an esteemed colleague and a friend, who gave a fundamental contribution to the use of thioureylene compounds as melanoma seekers.

Biomedical Chemistry: Applying Chemical Principles to the Understanding and Treatment of Disease, Edited by Paul F. Torrence
ISBN 0-471-32633-x

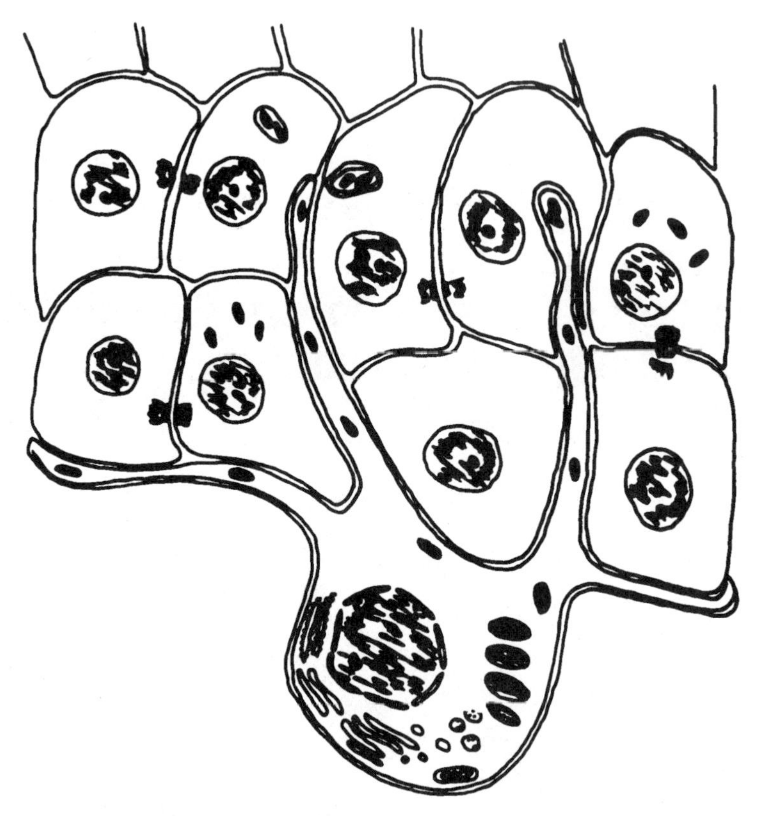

FIGURE 12.1 Schematic representation of cutaneous melanocytes highlighting their relationships with surrounding keratinocytes.

melanin pigments. Eventually, once the malignancy has taken its irreversible course, the tumor progresses rapidly with devastating destructiveness, due to its pronounced tendency to metastasize. Partial differentiation may thus result in phenotypic heterogeneity of the tumor, which may be manifest as irregular pigmentation.

Unlike the vast majority of cancers, melanoma is highly refractory to conventional therapeutic modalities, including radiotherapy, single- and double-agent chemotherapy, and immunotherapy. An attractive explanation for this unusual resistance would invoke the embryological origin of melanocytes from the neural crest, implying the existence of peculiar traits in common with most neural cells. In fact, surgical removal offers the only effective curative treatment when early diagnosis is possible, but it is clearly inadequate for disseminated melanoma, in which slowly proliferating metastatic cells are widespread throughout the body. In consequence, there is an urgent need to develop innovative targeting strategies aimed at efficiently arresting growth of primary tumor and metastatic spread.

In the 1990s, a number of approaches have been pursued in various laboratories,[3] which rely on the exploitation of the distinguishing features of malignant melanocytes. One attractive targeting modality revolves around recognition of cell surface molecules, such as receptors to α-melanocyte stimulating hormone or antigenic epitopes of other molecules characteristic of melanoma cells, by conjugates of the hormone or antibodies with cytotoxic drugs. But by far those strategies that hold most promise for the future rely on the biochemical hallmark of malignant melanocytes, specifically, the ability to retain, and even overactivate, the pathway of melanogenesis, leading to melanin pigments via the tyrosinase-catalyzed oxidation of tyrosine.[4] The very peculiarity of this pathway is the generation of a broad spectrum of reactive intermediates and transients, providing a range of options for selectively inflicting cytotoxicity to tumor cells.[5] Melanogenesis-based strategies are typically exemplified in those approaches that involve activation of phenolic prodrugs by melanogenic enzymes,[6,7] the uptake of melanin affinic drugs,[8] and the use of melanoma seekers for localizing tumor metastases and delivering radionuclides or cytotoxic moieties.[8–10] The latter approach, especially, has been the focus of much theoretical and practical interest, as it hinges on use of suitably designed molecules capable of being specifically incorporated into malignant melanocytes by reacting with appropriate intermediates in the processes of melanin formation.

In this scenario, 2-thiouracil (referred to in the following paragraphs and Scheme 12.1 as *thiouracil*) and other related thioureylene compounds have attracted a good deal of attention because of their unique affinity for newly synthesized melanin but not for preformed pigment, which allows us to discriminate between overactive pigment producing tissues, as in melanoma, and normal tissues, in which pigment production occurs at slower rates.

It is the aim of this chapter to highlight the chemical rationale underlying exploitation of thioureylene derivatives as prototype compounds for development of novel melanoma targeting agents, and to show how far a mere molecular approach can go in the design of mechanism-based drugs for melanoma scanning and for selectively delivering cytotoxicity to the tumor. We first sketch an historical profile of the key observations that led to the discovery of thiouracil incorporation into growing melanins; then, we focus on the basic studies carried out at Naples and other centers on the mechanism of interaction of thioureylene compounds with critical intermediates in the pigment pathway and the underlying rationales; finally, we suggest new possible directions of research along the routes tracked by current chemical knowledge in the field.

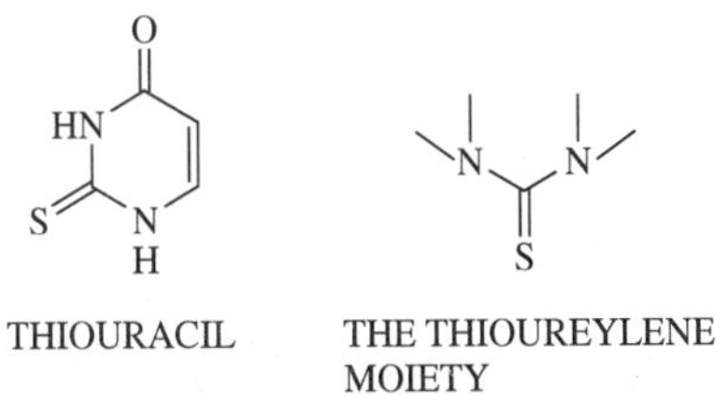

THIOURACIL

THE THIOUREYLENE MOIETY

SCHEME 12.1

HISTORICAL PERSPECTIVE

Prior to the 1960s, thiouracil and related compounds were known in the medical circles mainly for their potent antithyroid activity, as they were largely used for the treatment of hyperthyroidism and to produce goiter in laboratory animals for experimental and pharmacological purposes.

Following an earlier observation by Markert, Whittaker in 1966 provided evidence that thiouracil can cause an anomalous gigantism of the melanin granules in the otolith cells of differentiating ascidian larvae, as well as a significant lightening of their color from jet black to greenish brown.[11] By means of different experiments carried out on retinal pigment epithelium from chick embryo eyes, the author later came to the conclusion that the changes induced by thiouracil in developing pigment cells were due to the covalent incorporation of the drug into the melanin granules.[12] The degree of incorporation of thiouracil into the pigment proved to be proportional to the level of tyrosinase activity, as determined by incubation experiments with 2-(^{14}C)-radiolabeled drug in chick retinal pigment cell culture. Moreover, a significant decrease in the extent of incorporation (about 80%) was observed in the presence of phenylthiourea, an effective inhibitor of tyrosinase, the enzyme responsible for the biosynthesis of melanins.

To fully appreciate the details of the chemistry and biochemistry underlying thiouracil incorporation into melanin pigments, it seems appropriate at this stage to provide the less familiar reader with an overview of the knowledge of melanogenesis and melanin formation at the time when Whittaker"s experiments were carried out.[13] This can be readily summarized in the classical Raper scheme, initially worked out in the 1920s and later refined by Mason (Fig. 12.2).

In this scheme, the copper enzyme tyrosinase catalyzes the hydroxylation/oxidation of tyrosine to dopa and then to dopaquinone (a sequence now known to occur as a single step). This latter is a highly reactive *o*-quinone, which suffers rapid intramolecular cyclization following nucleophilic attack by the amino group to give an unstable dihydroxyindoline termed leucodopachrome. As it is an electron rich catechol, in the oxidizing reaction medium leucodopachrome is readily converted to dopachrome, the first detectable intermediate in the early stages of the process. A characteristic feature of this aminochrome is the tendency to suffer rearrangement with and/or without decarboxylation to give the thermodynamically more stable 5,6-dihydroxyindole (DHI) and/or 5,6-dihydroxyindole-2-carboxylic (DHICA). These latter are oxidatively converted to highly elusive *o*-quinones which rapidly polymerize to give dark melanin pigments insoluble in all solvents.[4] Until the early 1970s, studies of melanins were largely concentrated to the dark-to-black pigments occurring in mammalian skin, hair and eyes and in cephalopod ink.[13] Much less attention was paid to a yellow-to-reddish brown variant of melanins, the pheomelanins, originally isolated from red hair and feathers and subsequently found also in the skin of red-headed, fair complexioned individuals of Celtic origin.

Without entering into the details of pheomelanin chemistry, which has been repeatedly reviewed in papers and book chapters,[4,5,14] it will suffice here to say that these pigments contain a substantial proportion of sulfur and exhibit physical and

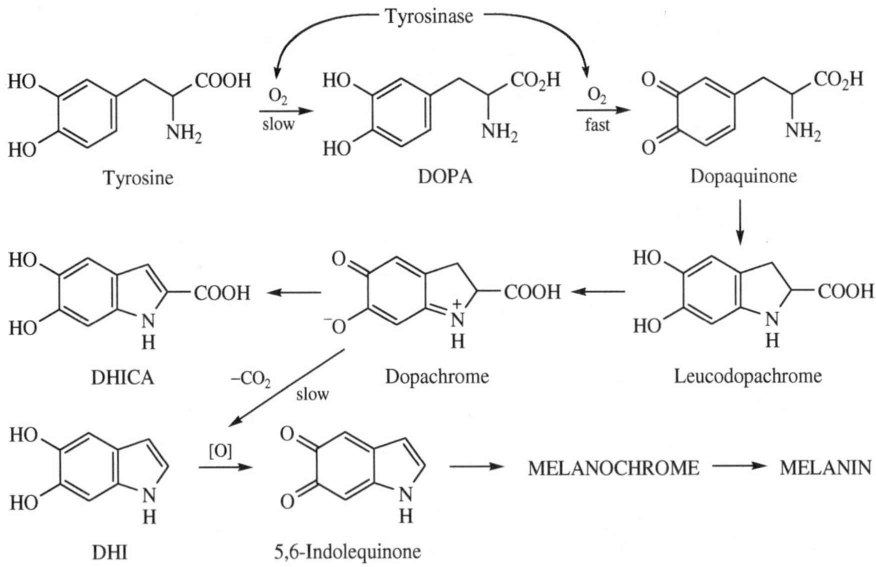

Tyrosinase

O_2 slow; O_2 fast

Tyrosine → DOPA → Dopaquinone

Leucodopachrome → Dopachrome → DHICA

$-CO_2$ slow

DHI → [O] → 5,6-Indolequinone → MELANOCHROME → MELANIN

FIGURE 12.2 The Raper–Mason scheme of melanogenesis.

chemical properties markedly different from those of their darker counterparts, including a significant solubility in aqueous alkali.

By means of biomimetic chemical experiments, Prota and co-workers succeeded in demonstrating that pheomelanins are biogenetically related to black melanins, and originate by the intervention of cysteine in the melanin pathway, via nucleophilic addition to enzymatically generated dopaquinone, to produce 5-*S*-cysteinyldopa as the main adduct (Fig. 12.3). This latter is then susceptible to oxidative cyclization to 1,4-benzothiazine intermediates that suffer polymerization to give eventually reddish-brown pheomelanins.

Taking the pheomelanin pathway as a model, Whittaker[12] suggested that the inhibitory effect of thiouracil was due to the ability of the compound to react with an intermediate product of tyrosine oxidation along the melanin pathway, to give a pigment that he referred to as thiouracil *pheomelanin.* Mainly on the basis of theoretical grounds, the author went on to propose a model for thiouracil–pheomelanin based on Hempel's model of melanoma melanin[4,13] (Fig. 12.4).

Speaking with hindsight, it is now clear that Whittaker's hypothesis, although in principle attractive, is in fact devoid of chemical foundations, as it is based on a speculative melanin model that has not withstood the test of time. Furthermore, thiouracil pheomelanin was envisaged as arising from the a posteriori incorporation of thiouracil in the preformed melanin pigment, which is in contrast with the considerably higher affinity of the drug for the growing pigment.

After Whittaker's studies, the issue of thiouracil incorporation into melanin remained dormant until 1979, when by serendipity Dencker and co-workers[15] found that the drug was strongly retained in tissues with higher rates of melanin formation compared to other pigmented tissues. Using whole-body autoradiographic tech-

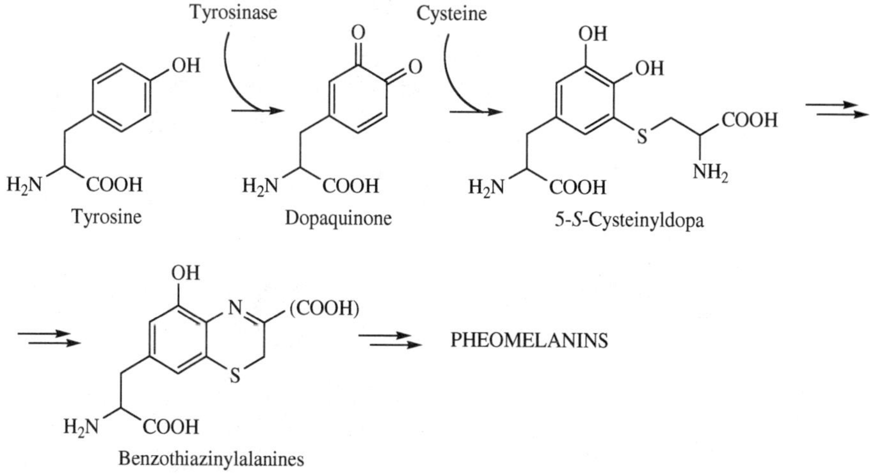

Tyrosinase
Cysteine
Tyrosine
Dopaquinone
5-*S*-Cysteinyldopa
Benzothiazinylalanines
PHEOMELANINS

FIGURE 12.3 Schematic outline of the biosynthesis of pheomelanins via the nonenzymatic conjugation of cysteine with dopaquinone to give 5-*S*-cysteinyldopa.

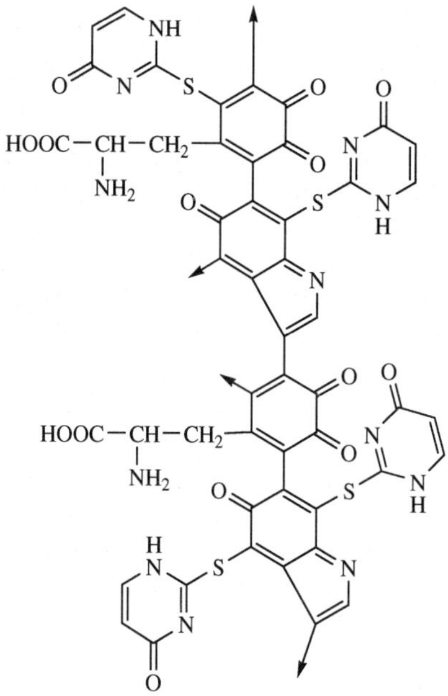

FIGURE 12.4 Part of the complex model structure of thiouracil pheomelanin proposed by Whittaker.

niques on pregnant mice, they were able to show that ^{14}C-thiouracil is selectively incorporated in the fetal eyes, where the rate of melanin synthesis is high, but not in the maternal eyes. This observation paved the way to a series of studies demonstrating that thiouracil and a number of related thioureylene compounds,

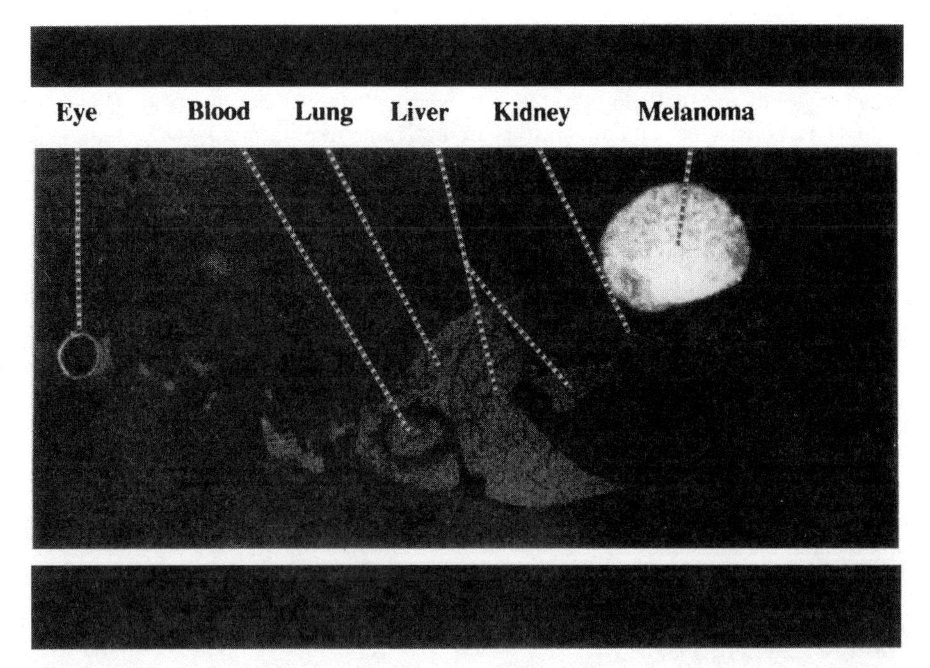

FIGURE 12.5 Whole-body autoradiogram showing selective uptake of radiolabeled thiouracil in melanoma tissue in a mouse transplanted with Harding–Passey melanoma, 4 days after intraperitoneal injection. (From Ref. 22, courtesy of Dr. U. Mars, Uppsala University and Munksgaard International Publishers Ltd., Copenhagen, with permission).

including ^{14}C- and ^{35}S-2-thiouracil, as well as radioiodine-labeled thiouracil and ^{35}S-methimazole, are specifically incorporated into growing melanin in melanotic melanoma, but not in amelanotic tumors[8–10] (Fig. 12.5).

As knowledge of the chemistry of melanogenesis improved in the 1990s, the bases were thus set to unravel the mechanism of incorporation of thiouracil into pigmented tissues and to dissect at chemical level the complex processes responsible for selective affinity for growing melanin. This issue was systematically investigated at Naples in the 1990s and has led to an improved appreciation of the potential of thiouracil as a unique, mechanism-based type of melanoma seeker.

GENERAL FEATURES OF THE INCORPORATION OF THIOURACIL INTO MELANOMA MELANIN

When administered intravenously to mice carrying subcutaneously growing B16 melanomas, [2-^{14}C]thiouracil is rapidly incorporated into the tumor, as evidenced by direct analysis of the purified pigment isolated from the excised tumors at fixed intervals of time after intraperitoneal inoculation of the label.[16] Considering the rapid circulation of the blood volume in mice, with a complete turnover approximately every 30 s, it can be concluded that significant concentrations of thiouracil are delivered to the tumor in the very first minutes following injection and, hence, that incorporation is due mainly to active mechanisms causing incorporation into newly synthesized melanin rather than to binding to preformed pigment. This time course is consistent with the results of autoradiographic studies, revealing the preferential, selective accumulation of thiouracil into tissues with high rates of melanin formation. During processing of the tumors for melanin isolation, about 33% of the radioactivity remains incorporated into the pigmented insoluble fraction, while 67% is found in the soluble fraction, indicating that the drug is initially incorporated into the melanocyte cytoplasm and is then transferred to melanogenic compartments, the melanosomes, where elevated tyrosinase activity ensures rapid polymerization of pigment precursors.

Accumulation of thiouracil into melanoma melanin is associated with a profound structural modification of the pigment, as denoted by the partial decoloration of the granules after administration of the drug, and a consistent decrease in the general absorbance. Thus, the crucial gap was at what stage and by what mechanism(s) thiouracil affects the biosynthetic pathway of melanogenesis and is retained into pigmented melanosomes.

EFFECTS OF THIOURACIL ON THE EARLY STAGES OF MELANOGENESIS

The most peculiar feature of thioureylene compounds is their pronounced nucleophilic character, which is localized for the most part on the sulfur center, competing favorably with the nitrogens in addition and/or substitution reactions. This accounts for the marked propensity of thioureas to bring about nucleophilic attack to electrophilic quinones to give *S*-substituted isothioureas. On this basis, it seemed reasonable to envisage a covalent addition with reactive quinonoid intermediates in the melanin pathways as the file d'Arianne to unravel the mechanistic intricacies underlying incorporation of thiouracil into melanoma tissues. That this prediction was correct was apparent from preliminary kinetic experiments on the effect of thiouracil on the tyrosinase catalyzed oxidation of tyrosine, revealing a pronounced ability of the drug to inhibit in a concentration-dependent manner the generation of dopachrome.[17] Since polarographic determination of oxygen consumption revealed a relatively high rate of tyrosine oxidation even under conditions of complete suppression of dopachrome formation, which argued evidently against a mere inhibition of tyrosinase, it was concluded that thiouracil could interact with an enzymatically produced melanogenic intermediate generated by oxidation of tyrosine and preceding dopachrome formation. Consistent with this view, analysis of a mixture obtained by tyrosinase-catalyzed oxidation of tyrosine in the presence of thiouracil, under conditions of complete suppression of melanogenesis, showed the formation of an aminoacidic product whose properties did not match those of any known oxidation product of tyrosine. This product was isolated and shown to be a thiouracil dopa adduct, 3,4-dihydroxy-6-(4′-hydroxypyrimidinyl-2′-thio)phenylalanine, evidently arising by nucleophilic addition of thiouracil to dopaquinone (Fig. 12.6).

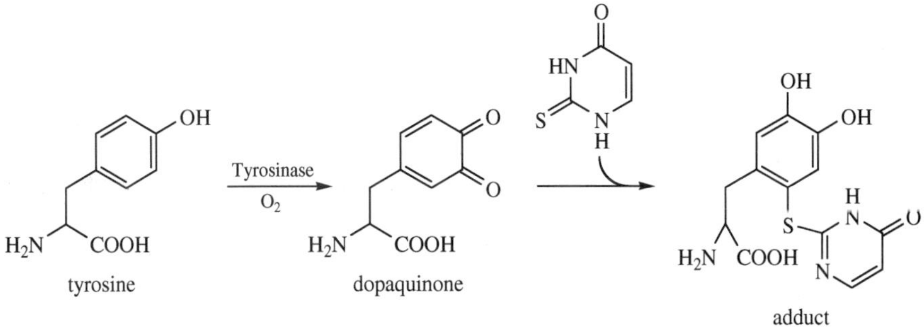

FIGURE 12.6 Structure and mechanism of formation of the thiouracil dopa adduct.

For the more chemically oriented reader, a comment seems appropriate here regarding the regiochemistry of the reaction of thiouracil with dopaquinone. Although this reaction resembles to a considerable extent the synthesis of cysteinyldopas in the pheomelanin pathway, it proceeds apparently under efficient regiochemical control to give the adduct at the 6 position, which contrasts with the preferential mode of attack of cysteine and other sulfhydryl compounds at the 5 position.[14] Such a dichotomy is intriguing and has been the subject of a longstanding issue for which no definitive explanation has been provided. One possibility is that thiouracil brings about true nucleophilic attack to the *o*-quinone moiety, whereas in the case of sulfhydryl compounds electron-transfer processes between the electron-rich sulfur centers and the quinone are actually operative. Alternatively, different degrees of nucleophilic reactivity of the attacking species may be invoked, whereby the more reactive add preferentially to the more electrophilic 5 position of the *o*-quinone ring.[17]

Whatever the actual mechanism, the significant finding that emerged is that thiouracil can effectively target dopaquinone, leading to an adduct that cannot undergo intramolecular cyclization to give eumelanin-like pigments through the usual 5,6-dihydroxyindole pathway. In line with this view, studies of the oxidation behavior of the thiouracil dopa adduct[16] revealed the rapid decay of the product, and the concomitant generation of a yellow elusive chromophore that gradually turned darker in color to afford eventually a brown insoluble pigment. On this basis, it was concluded that, following coupling with dopaquinone, incorporation of thiouracil into growing melanin results mainly from oxidation and/or inglobation of the adduct into the pigment polymer. Unfortunately, all attempts to detect the thiouracil dopa adduct in melanoma tissues from tumor-bearing mice met with failure, due probably to rapid metabolization by melanogenic enzymes.

REACTIONS OF THIOURACIL WITH QUINONOID MELANOGENIC INTERMEDIATES DISTAL TO TYROSINASE

Although dopaquinone is a primary target of thiouracil in the melanin pathway, other transient quinonoid species formed at stages distal to tyrosinase could also interact with thiouracil to form covalent adducts. This seemed realistic, given the presence of indolic precursors such as DHI and DHICA in distinct subcellular compartments, like the coated vesicles,[4,5] from which they may be released and exposed to thiouracil in an oxidative environment.

Under a variety of conditions, thiouracil failed to affect to any significant extent the rate of rearrangement of dopachrome, indicating that the latter was not a likely target of the drug. By contrast, thiouracil was able to suppress in a concentration-dependent manner the oxidative conversion of DHI and DHICA to melanin, leading to complex patterns of coupling products, which were isolated and assigned the structures shown in Figure 12.7.[18]

Coupling products of DHI with thiouracil reflect a regiochemical course of the reaction leading predominantly to 2- and 3-substituted adducts, in marked contrast with the preferential mode of coupling of sulfhydryl compounds, such as cysteine and glutathione, through the 4 position.[4,14] This is difficult to rationalize in terms of structures such as **1** (see Scheme 12.2), commonly reported for 5,6-indolequinone, and suggests either a significant contribution to the resonance hybrid of structures such as **2**, or a tautomerization step leading to the quinone methide (**3**), which is expectedly more proclive to react at the pyrrole moiety of the indole ring.

Also of particular interest is the mode of polymerization of thiouracil-5,6-dihydroxyindole(s) adducts. As apparent from the structures of oligomeric adducts, chain elongation in the case of thiouracil-DHI conjugates proceeds mainly through the 4 and 7 positions of the indole ring, without detectable involvement of the 2

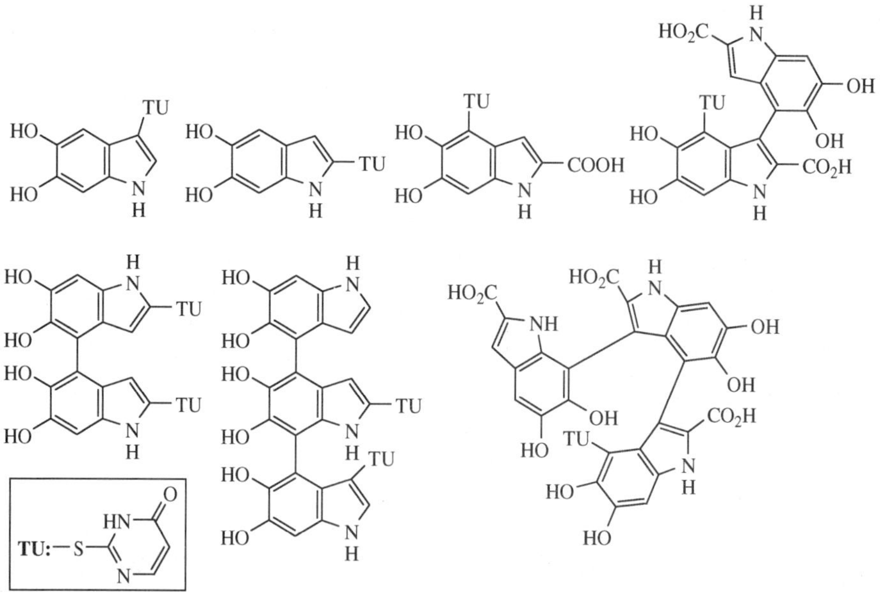

FIGURE 12.7 Structures of main products arising by oxidation of DHI and DHICA in the presence of thiouracil.

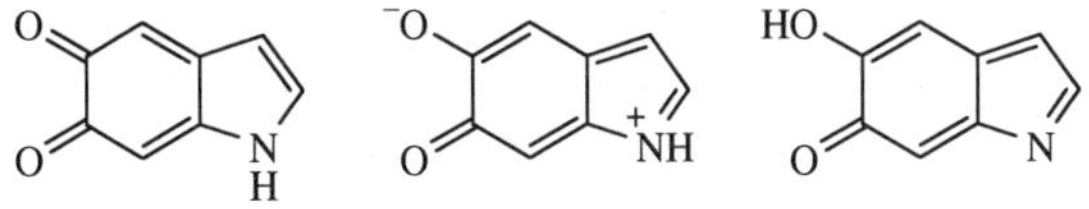

SCHEME 12.2

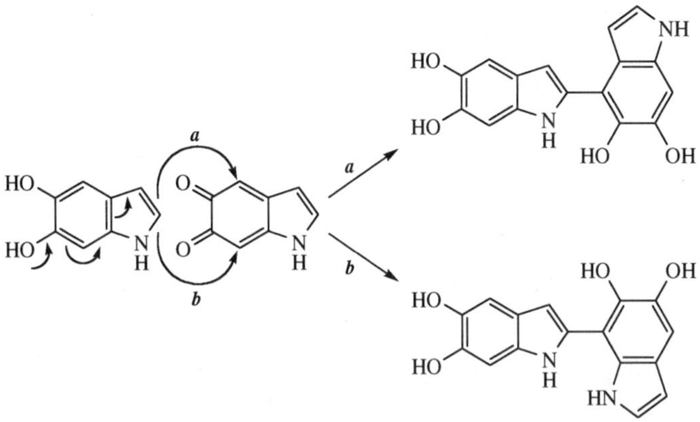

FIGURE 12.8 Suggested mechanism of polymerization of DHI via nucleophilic attack to 5,6-indolequinone.

position. This is in marked contrast with the normal regiochemical course of the polymerization of DHI, which is dictated by nucleophilic attack of DHI through the reactive 2 position to the electrophilic 4 and 7 positions of a transient 5,6-indolequinone[4,5,19] (Fig. 12.8).

A plausible explanation would envisage both alterations of electronic distribution on the indole ring caused by the thiouracil moiety and steric hindrance opposed by the substituent against the approaching indole unit in the course of dimerization.

STRUCTURAL PROPERTIES OF THIOURACIL-CONTAINING MELANINS

To gain an insight into the effects of thiouracil on the structure and general properties of melanins, various synthetic pigments were prepared by tyrosinase-catalyzed oxidation of dopa, DHI, and DHICA in the presence and in the absence of thiouracil, and were subjected to chemical analysis. A detailed account of the methods for analysis of melanins is beyond the scope of the present chapter, and the interested reader is referred to relevant reviews, book chapters, and papers.[4,5,20,21] What should be emphasized here is that melanins are untractable heterogeneous pigments of supposedly high molecular weight, virtually insoluble in all solvents and lacking well-defined chemical and physical properties. In consequence, application of modern spectral techniques for structural characterization is generally unrewarding and gives at best fragmentary information of peripheral significance. The only available approaches to assess the basic monomer composition and the degree of structural integrity of natural and synthetic melanins rest on extensive chemical degradation of the pigments, for example, by alkaline hydrogen peroxide, followed by quantitation of specific fragments of diagnostic value. These include chiefly pyrrole-2,3,5-tricarboxylic acid (PTCA) and pyrrole-2,3-dicarboxylic acid (PDCA) arising by oxidative breakdown of indolequinone units through the mechanism schematically outlined in Figure 12.9.

Of these pyrroles, PTCA has generally been taken as an index of the relative proportion of carboxylated indole units structurally related to DHICA, whereas PDCA specifically arises from DHI units unsubstituted at the 2 position.[20]

From the data in Table 12.1 it appears that incorporation of thiouracil into dopa, DHI, and DHICA melanins results in marked variations in the chemical yields and intensity of absorption of the pigment preparations, as well as in the yields of formation of PTCA and PDCA.[18] In particular, PTCA decreased with thiouracil

TABLE 12.1 Spectrophotometric and Chemical Analysis of Synthetic Melanins Prepared in Both Presence and Absence of Thiouracil

Sample[a]	A_{350}[b]	PTCA Yield, ng/mg	PDCA Yield, ng/mg
Dopa melanin	0.87	3,600	1,100
TU-Dopa melanin	0.55	4,600	7,600
DHI melanin	0.93	5,246	4,780
TU-DHI melanin	0.50	1,809	7,280
DHICA melanin	0.58	58,000	—
TU-DHICA melanin	0.42	48,000	—

[a]Prepared with a thiouracil/substrate molar ratio of 0.5.
[b]Determined on 0.1-mg/mL solutions.

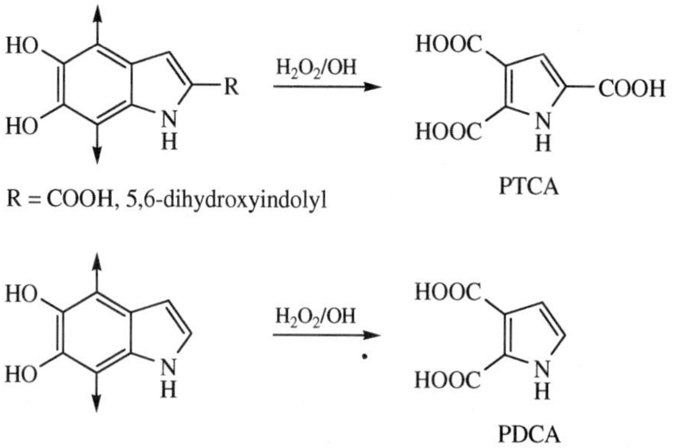

FIGURE 12.9 Structure and origin of pyrrolecarboxylic acids by oxidative degradation of indole units in melanin polymers.

incorporation in DHI and DHICA melanins, but increased in dopa melanins, whereas PDCA increased both in dopa and DHI melanins.

These effects are complex, but may be ascribed to incorporation of the rather bulky thiouracil moieties into the growing pigment, affecting the process of melanogenesis not only for what concerns the extent and mode of elongation of the pigment polymer but also with regard to the oxidizability and intrinsic absorption properties of the thiouracil-linked monomer units.[18]

STRUCTURE–ACTIVITY RELATIONSHIPS

Structure–activity relationship studies corroborate the critical role of sulfur for incorporation of thiouracil into growing melanin, since sulfur-lacking analogs, including uracil, are virtually devoid of binding properties, and argue strongly for a *free* thioureylene or thioamide moiety as the minimum essential structural requirement for the expression of melanin binding properties. An inventory of the compounds now known to be selectively incorporated into growing melanins is provided in Figure 12.10, and is due for the most part to the systematic work of Drs. Larsson, Mars, and co-workers at Uppsala.[22]

Apparently, the presence of a cycle or its dimension are not critical, as both five-membered (e.g., methimazole) and acyclic thioureylene (e.g., thiourea) compounds are effective. Inhibition of dopachrome formation as a chemical index[22] indicated that the rank order of activity was 4,5,6-triamino-2(*H*)-pyrimidinethionesulfonate > trithiocyanuric acid >2-thiouracil >4-amino-2-mercaptopyrimidine. This criterion, however, may not be entirely reliable, since in several instances inhibitory effects in vitro do not correlate with melanoma seeking properties.

A relevant case is that of thiourea.[23] Compared to its congeners, thiourea inhibits dopachrome formation to a much lesser degree than do other thiocarbonyl derivatives, including mercaptothiazoles and mercaptooxazoles, yet it is incorporated into melanoma to about the same extent as thiouracil, and only somewhat less than methimazole. However, at variance with 2-mercaptobenzothiazole and a number of thioureylene compounds, thiourea does not inhibit tyrosinase, but affects the initial stages of the tyrosinase-catalyzed oxidation of dopa to afford the expected conjugate with dopa.[24] It can thus be concluded that the lack of correlation between chemical and biological properties may result from various effects, including differences in metabolism, half-life, solubility, and inhibitory effects on tyrosinase.

THE MECHANISM OF INCORPORATION OF THIOUREYLENE COMPOUNDS INTO GROWING MELANIN

The complex of the results obtained from the abovementioned studies confirms the expected similarities in the mode of incorporation of thiouracil and related thioureylene compounds into growing melanins, and supports a rather general mechanism based on attack of the nucleophilic sulfur-containing functionality to

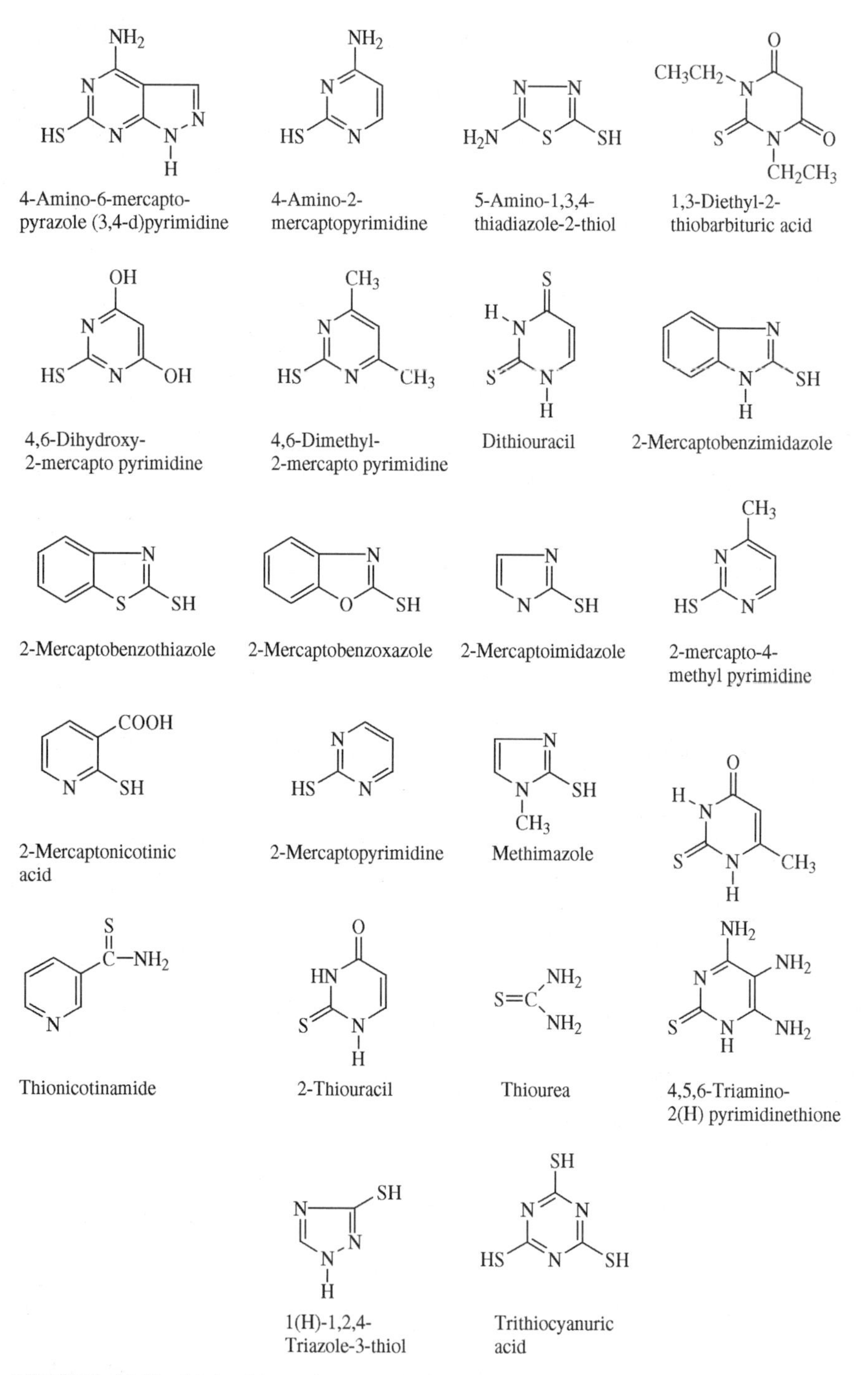

FIGURE 12.10 Main thioureylene- and thioamide-containing compounds with specific affinity for growing melanins.

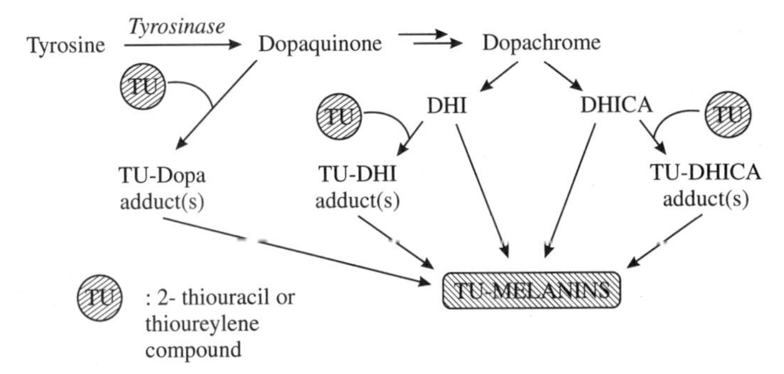

FIGURE 12.11 Schematic outline of the mechanism of incorporation of thiouracil and related compounds into growing melanin.

quinonoid intermediates of the melanin pathway to give covalent conjugates. A general overview of the proposed mechanistic scheme, highlighting the main sites of intervention of thiouracil and other thioureylene compounds in the melanin pathway, is given in Figure 12.11.

The key coupling reactions, which may occur at various stages of the pathway, lead to the formation of 1 : 1 adducts in which the melanogenic properties of the quinone acceptors are modified or inhibited, so that they can no longer give rise to melanin pigments. Nevertheless, the inherent reactivity of the catechol moieties, although partially affected by electronic and steric effects caused by the proximate thiopyrimidinyl group, is generally preserved and accounts for the generation under oxidative conditions of polymeric products probably through a sequence of catechol–quinone coupling steps not dissimilar from those occurring in melanogenesis. Thioureylene conjugates with dopa and 5,6-dihydroxyindoles can thus give rise to quinonoid derivatives that may couple with dopa, dopaquinone, and 5,6-dihydroxyindoles to give mixed-type species with altered degrees of polymerization, solubility, and UV-absorbing properties compared to the parent melanin(s). Within this general scheme, differences may be observed among various thioureylene derivatives in the relative abilities to trap the various quinonoid intermediates in the melanin pathway. This may also give rise to appreciable differences in the temporal and spatial patterns of incorporation of the label in melanoma tissues.

THIOUREYLENE-BASED COMPOUNDS AS PROTOTYPE AGENTS FOR THE TARGETING OF MELANOMA: PRINCIPLES AND STRATEGIES

The preceding survey has provided the reader with the basic chemistry underlying the selective incorporation of thiouracil and related thioureylene compounds into melanoma melanin. In this section, we focus on the main concepts directing research toward a practical exploitation of thiouracil and related compounds for the targeting of melanoma in the clinics.

Autoradiographic investigations indicate that thioureylene derivatives are incorporated with a high degree of specificity in murine melanotic melanoma, and are concentrated in those regions of the tumor where the rate of melanin synthesis is sustained. Moreover, chemical analysis has demonstrated that a substantial fraction of the admnistered drug enters the melanosomes, the subcellular organelles containing the whole biochemical machinery for the manufacture of melanins. Notably, melanotic melanoma metastases also accumulate radiolabeled thiouracil, whereas poor uptake is seen in amelanotic melanoma. This may, in principle, be a problem, given the tendency toward decreasing melanin synthesis generally observed in metastases during the course of dissemination.

One optimal utilization of thiouracil and related compounds is for melanoma scanning and as a marker for monitoring the effects of treatment.[9] To this aim, thiouracil has to be tagged with a suitable radionuclide and, in this respect, radioiodinated 5-iodo-2-thiouracil has proved to be of potential interest, since it exhibits the same degree of incorporation of the unlabeled drug in transplanted murine melanoma as well as in metastates and in cultured human melanoma cells. It may be worth mentioning here the results of two pilot clinical studies showing that injection of ^{123}I-labeled thiouracil is as useful as conventional diagnosis with ^{67}Ga-citrate for detection of ocular melanoma, and that total doses of radioactivity of 150–200 MBq are sufficient for the imaging of disseminated cutaneous melanoma with γ-scintigraphy in patients injected with ^{131}I-thiouracil.

In addition to imaging, thiouracil and related compounds may also be utilized in suitably devised strategies for the treatment of melanoma.[9,10] Three rational applications may be envisaged: (1) as carriers of therapeutic radionuclides, (2) as carriers of boron-10 for boron neutron capture therapy (BNCT), and (3) as vehicles for delivering localized cytotoxicity through appropriate cytostatic or cytotoxic agents.

Especially promising is the use of thiouracils as carriers of ^{10}B for BNCT, an intensively investigated therapeutic modality based on the emission of strongly ionizing, cell-killing α particles and lithium ions by nuclear fission of ^{10}B induced by thermal neutrons.[25,26] A number of interesting prototypes of ^{10}B-labeled thioureylenes have thus far been prepared and tested in laboratory animals. 5-Dihydroxyboryl-2-thiouracil (**4**) (see Scheme 12.3) and its 6-propyl derivative (**5**) were found to be selectively accumulated in B16 and Harding–Passey melanoma transplanted to mice, and caused sufficient concentration of ^{10}B in the tumor tissue, persisting long enough for effective application in BNCT.[26] Adducts of decaborane with 5-(diethylamino)methyl-2-thiouracil (**6**) and 1*H*-1,2,4-triazole-3-thiol have likewise been prepared and shown to be selectively incorporated in melanins in vitro and in vivo. A carboranyl thiouracil as a potential agent for BNCT has also been synthesized,[27] witnessing the variety of chemical approaches that have been pursued to covalently bind ^{10}B to a thioureylene melanoma seeker.

The use of thiouracil as a vehicle for locally delivering cytotoxicity is also under active investigation. Two variants of this strategy may be pursued, involving conjugation with (1) conventional cytostatic agents for chemotherapy or (2) specifically designed prodrugs susceptible of oxidative conversion to cytotoxic

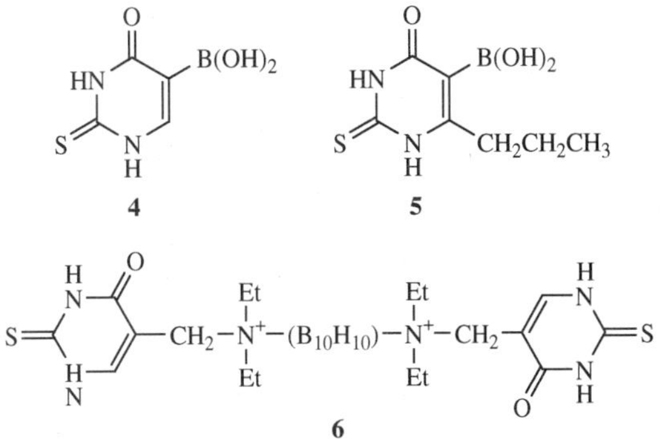

4 **5**

6

SCHEME 12.3

o-quinones in the presence of tyrosinase. In the former case, arotinoids and retinoids have been linked to the 5 position of thiouracil and have been shown to be incorporated into melanoma, without much cytotoxicity, however. Similar discouraging results have been reported with adducts of thiouracil with nitrogen mustards.[10] The latter rationale embodies a sort of "double targeting" principle. It revolves around the design of phenolic prodrugs that are delivered by the thioureylene carrier to melanoma tissues. Once there, they are specifically activated in subcellular melanogenic compartments through the action of tyrosinase, converting the phenol moiety to a potentially cytotoxic *o*-quinone.[6,7] Conjugates of tyrosinase-targeted phenolic prodrugs with thioureylene compounds have been prepared and are currently under investigation for their selectivity and antitumor activity.

CONCLUSIONS AND FUTURE PERSPECTIVES

The present overview has highlighted the chemical bases underlying selective incorporation of thiouracil and other thioureylene compounds into melanoma melanin. The actual clinical impact of the various thiouracil-based strategies is currently under assessment, and there are grounds to believe that in the near future novel exciting breakthroughs will radically modify the physician's attitude toward this tumor. The importance of thioureylene compounds for localization and scanning of melanoma is now well established, and a good deal of research is being devoted to the possible exploitation of these compounds for therapeutic modalities specifically tailored to melanoma. Although theoretical and practical difficulties have somewhat slowed down the progress of research, several approaches appear particularly valuable in perspective and deserve continued investigation at a multidisciplinary level. In any case, it should be clear from the foregoing that future developments in the field of melanoma treatment will critically depend on the understanding of the peculiar biochemical features of this tumor. In this respect, melanoma can be regarded as a unique setting in cancer research, since there are few other tumors for which a comparable knowledge of peculiar metabolic pathways is available. This fortunate circumstance offers an ever-expanding range of opportunities for translating basic chemical concepts to rational therapeutic treatments. The discovery of the mechanism of incorporation of thiouracil into growing melanins, in particular, has tracked unprecedented avenues toward tumor targeting, and has been paradigmatic in showing how critical is the understanding of biological processes at chemical level for realization of important advances in the biomedical sciences.

REFERENCES

1. Kirkham, N.; Cotton, D. W. K.; Lallemand, R. C.; White, J. E.; Rosin, R. D., eds., *Diagnosis and Management of Melanoma in Clinical Practice*, Springer-Verlag, London, 1992.
2. Nordlund, J. J.; Boissy, R. E.; Hearing, V. J.; King, R. A.; Ortonne, J. P., eds., *The Pigmentary System. Physiology and Pathophysiology*, Oxford Univ. Press, New York, 1998.
3. Riley, P. A., in *Diagnosis and Management of Melanoma in Clinical Practice*, N. Kirkham, D. W. K. Cotton, R. C. Lallemand, J. E. White, and R. D. Rosin, eds., Springer-Verlag, London, 1992, pp. 157–172.
4. Prota, G., *Melanins and Melanogenesis*, Academic Press, San Diego, 1992.
5. Prota, G.; d'Ischia, M.; Napolitano, A., in *The Pigmentary System. Physiology and Pathophysiology*, J. J. Nordlund, R. E. Boissy, V. J. Hearing, R. A. King, and J. P. Ortonne, eds., Oxford Univ. Press, New York, 1998, pp. 307–332.
6. Prota, G.; d'Ischia, M.; Mascagna, D., *Melanoma Res.* 1994, **4**, 351.
7. Jimbow, K.; Miyake, Y.; Gili, A.; Ota, M.; Chang, D.; Singh, S.; Shokravi, M.; Reszka, K. J.; Jimbow, M.; Thomas, P., in *Melanogenesis and Malignant Melanoma: Biochemistry, Cell Biology, Molecular Biology, Pathophysiology, Diagnosis and Treatment*, Y. Hori, V. J. Hearing, and J. Nakayama, eds., Elsevier, Amsterdam, 1996, pp. 257–269.
8. Larsson, B. S., in *The Pigmentary System. Physiology and Pathophysiology*, J. J. Nordlund, R. E. Boissy, V. J. Hearing, R. A. King, and J. P. Ortonne, eds., Oxford Univ. Press, New York, 1998, pp. 373–389.
9. Larsson, B. S., in *Melanogenesis and Malignant Melanoma: Biochemistry, Cell Biology, Molecular Biology, Pathophysiology, Diagnosis and Treatment*, Y. Hori, V. J. Hearing, and J. Nakayama, eds., Elsevier Science, Amsterdam, 1996, pp. 301–307.
10. Larsson, B. S., *Melanoma Res.* 1991, **1**, 85.
11. Whittaker, J. R., *Exp. Cell Res.* 1966, **44**, 1941.
12. Whittaker, J. R., *J. Biol. Chem.* 1971, **246**, 6217.
13. Nicolaus, R. A., *Melanins*. Hermann: Paris, 1968.
14. Prota, G., in *Progress in the Chemistry of Organic Natural Products*, W. Herz, G. W. Kirby, R. E. Moore, W. Steglich, and Ch. Tamm, eds., Springer-Verlag, Vienna, 1995, Vol. 64, pp. 93–148.
15. Dencker, L.; Larsson, B.; Olander, K.; Ullberg, S.; Yokota, M., *Br. J. Cancer* 1979, **39**, 449.
16. Palumbo, A.; Napolitano, A.; De Martino, L.; Vieira, W.; Hearing, V. J., *Biochim. Biophys. Acta* 1994, **1200**, 271.

17. Palumbo, A.; d'Ischia, M.; Misuraca, G.; Iannone, A.; Prota, G., *Biochim. Biophys. Acta* 1990, **1036**, 221.
18. Napolitano, A.; Palumbo, A.; d'Ischia, M.; Prota, G., *J. Med. Chem.* 1996, **39**, 5192.
19. d'Ischia, M.; Napolitano, A.; Prota, G., *Gazz. Chim. Ital.* 1996, **126**, 783.
20. Pezzella, A.; d'Ischia, M.; Napolitano, A.; Palumbo, A.; Prota, G., *Tetrahedron* 1997, **53**, 8281.
21. Ito, S., in *The Pigmentary System. Physiology and Pathophysiology*, J. J. Nordlund, R. E. Boissy, V. J. Hearing, R. A. King, and J. P. Ortonne, eds., Oxford Univ. Press, New York, 1998, pp. 439–450.
22. Mars, U.; Larsson, B. S., *Pigment Cell Res.* 1995, **8**, 194.
23. Mars, U.; Larsson, B. S., *Melanoma Res.* 1996, **6**, 113.
24. Palumbo, A.; Mars, U.; De Martino, L.; d'Ischia, M.; Napolitano, A.; Larsson, B.S.; Prota, G., *Melanoma Res.* 1997, **7**, 478.
25. Mishima, Y.; Ichihashi, M.; Honda, C.; Shiono, M.; Nakagawa, T.; Obara, H.; Shirakawa, J.; Hiratsuka, J.; Kanda, K.; Kobayashi, T.; Nozaki, T.; Aizawa, O.; Sato, T.; Karashima, H.; Yoshino, K.; Fukuda, H., in *Progress in Neutron Capture Therapy for Cancer*, B. J. Allen, D. E. Moore, and B. V. Harrington, eds., Plenum Press, New York, 1992, pp. 577–583.
26. Larsson, B. S.; Larsson, B.; Roberto, A., *Pigment Cell Res.* 1989, **2**, 356.
27. Wilson, J. G., *Pigment Cell Res.* 1989, **2**, 297.

CHAPTER 13

Carbon-Centered Radicals and Rational Design of New Antimalarial Peroxide Drugs

GARY H. POSNER, JARED N. CUMMING, and MIKHAIL KRASAVIN
Department of Chemistry, Johns Hopkins University

BACKGROUND

Today, nearly 40% of the world's population is at risk of malaria infection. Of the 300–500 million people who now have the infectious disease, 1–2 million, mostly children, die each year.[1] Not only is *prevention* of malaria not now attainable with vaccines, but also *chemotherapy* of malaria is becoming progressively more difficult using current standard drugs;[2] the *Plasmodium falciparum* malaria parasites have developed very widespread multidrug resistance to antifolates and to clinically used quinoline antimalarial drugs (e.g., chloroquine, **1**).[3] Moreover, with increasing global warming and increasing international travel, the spread of malaria to nontropical parts of the world is likely to occur more and more frequently. Even in the United States, for example, cases of malaria have been recorded recently (i.e., in the late 1990s).[4] Fundamental recent advances in understanding the details of the malaria parasite's complex life cycle promise eventually to generate new strategies to block transmission of this infectious disease and to provide new methods for effective and safe chemotherapy of infected individuals.[5,6] As one important way to find new antimalarial drugs, medicinal chemistry researchers often turn to folk remedies as a rich source of age-old wisdom about how to cure a particular human disease. For example, an ancient Chinese herbal remedy against malaria led organic chemists in China to isolate and characterize a novel class of endoperoxide antimalarial drugs.[7,8] Artemisinin (qinghaosu, **2**) is the endoperoxide found in the *Artemisia annua* plant that is responsible for this centuries-old herbal cure of malaria patients.[9] Clearly, the chemical structure of endoperoxide artemisinin (**2**) has no

Biomedical Chemistry: Applying Chemical Principles to the Understanding and Treatment of Disease, Edited by Paul F. Torrence
ISBN 0-471-32633-x

nitrogen atom, and therefore this type of trioxane (a six-membered ring containing three oxygen atoms) belongs to a completely different class of organic molecules than does chloroquine (**1**). Consequently, it was expected and has now been confirmed that malaria parasites resistant to chloroquine, fortunately, are not resistant to artemisinin.

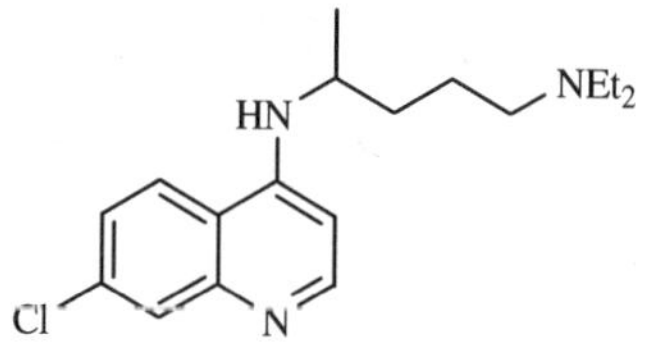

chloroquine (**1**)

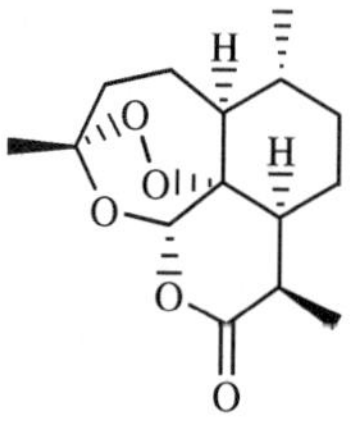

artemisinin (qinghaosu, **2**)

Indeed, since the mid-1970s, artemisinin has been used in tropical parts of the world to cure well over 1 million malaria patients.[1] Because artemisinin is very soluble neither in water nor in vegetable oils, organic chemists have reduced its lactone carbonyl group into a lactol functionality, producing dihydroartemisinin (DHA, **3**).[10,11] Ether derivatives of this lactol [e.g., artemether (**4**), arteether (**5**)] have been synthesized and used widely as oil-soluble antimalarial drugs, and also water-soluble derivatives [e.g., sodium artesunate (**6**), sodium artelinate (**7**)] are being used or considered for easy administration to malaria patients (e.g., orally or *via* steady release from suppositories).[12] Because the C10 acetal functionality in all of these DHA derivatives **4–7** is subject to chemical and enzymatic hydrolysis, all of these endoperoxides are considered to be prodrugs for the parent DHA compound **3**. Recent efforts have been directed at preparing various C10 nonacetal analogs that would be hydrolytically more stable, and this work is continuing in organic chemistry labs around the world with some promising results.[13–16]

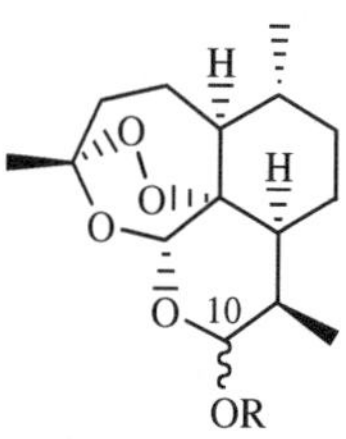

3, R = H, dihydroartemisinin

4, R = Me, artemether

5, R = Et, arteether

6, R = $COCH_2CH_2COONa$, sodium artesunate

7, R = CH_2–C_6H_4–COONa

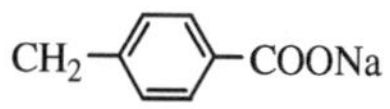

sodium artelinate

How do such endoperoxides in the artemisinin family of trioxanes kill malaria parasites selectively without harming uninfected cells? The answer is quite simple and quite astonishing. Malaria parasites develop inside human erythrocytes and digest their host's hemoglobin as a source of amino acid nutrients, releasing heme as a by-product.[17] Although the ferrous iron of heme in hemoglobin is sterically encumbered to such a large extent that it cannot come close enough to reduce peroxides such as artemisinin, the ferrous iron in free heme is sufficiently exposed so that it is capable of reducing such endoperoxides. Thus, while endoperoxide drugs such as artemisinin circulate in humans and are not disturbed by hemoglobin, when they reach erythrocytes infected with malaria parasites and therefore encounter heme, they are rapidly reduced. This selective triggering of the endoperoxide drug only within malaria parasites that reside inside infected erythrocytes sets off a cascade of chemical reactions involving various cytotoxic chemical entities that kill the malaria parasite.[18] Thus, when malaria parasites are exposed to endoperoxide drugs such as artemisinin, the parasites inadvertently but effectively kill themselves!

Much of this fascinating story of parasite self-annihilation was uncovered by the Meshnick research group during the 1990s, and a review was published in 1996.[18] This chapter summarizes research mainly at Johns Hopkins University during the 1990s that has elucidated the nature of the reactive chemical intermediates (e.g., carbon-centered free radicals) formed after heme-triggering of endoperoxides. Also discussed is how such mechanistic understanding at the molecular level has allowed rational design of promising new peroxide drug candidates for effective chemotherapy against malaria. Because this chapter, for didactic and space-limitation reasons, is not meant to be a comprehensive summary of malaria chemotherapy, the following important chemical research is not discussed: (1) the historically first discoveries of the Chinese groups about the isolation and chemistry of artemisinin (**2**) and its lactol derivatives,[19] (2) the pioneering mechanistic and synthetic work of the Geneva group,[20,21] (3) the exhaustive synthetic and molecular modeling studies by the Avery group,[22] (4) the insightful advances by the Ziffer team,[11,16] (5) the direct synthesis of tetraoxanes by the Vennerstrom team,[23] and (6) the sustained and important chemical contributions from the Jung group.[24]

CARBON-CENTERED RADICALS AND MECHANISM OF ANTIMALARIAL ACTION

Microbial metabolism of artemisinin (**2**) affords two major degradation products; ring-contracted tetrahydrofuran **8** and C4-hydroxylated dioxolane **9** [Eq. (13.1)].[25]

microbial degradation: **2** → **8** + **9** + other products (13.1)

In 1992 we modeled this natural metabolic pathway in vitro using regiospecifically ^{18}O labeled trioxane tosylate **10** (Scheme 13.1).[26] This trioxane tosylate **10** was chosen for two reasons: (1) it is comparable to artemisinin (**2**) in terms of in vitro antimalarial potency[27] and (2) it could be prepared with the ^{18}O label specifically as

SCHEME 13.1

shown in structure **10**. When exposed to ferrous ions in vitro at room temperature, trioxane tosylate **10** rapidly formed two major types of products, ring-contracted tetrahydrofurans **11′** and **11** and C4-hydroxylated dioxolane **12**, in excellent analogy to formation of products **8** and **9** via microbial degradation of artemisinin (**2**). Although ring-contracted tetrahydrofuran acetal **11′** could be detected by ^{1}H NMR spectroscopy, it quickly lost ^{18}O-labeled methyl acetate to form tetrahydrofuran aldehyde **11**. To account for formation of these isolated major products **11** and **12**, including the absence of ^{18}O in tetrahydrofuran **11** and the presence of ^{18}O in only one position of dioxolane **12**, we proposed the chemical mechanism shown in Scheme 13.1.[26] Knowing where the ^{18}O label ends up in products **11** and **12** was crucial for implicating carbon radical intermediates as shown in mechanistic Scheme 13.1. *This was the first time that carbon-centered radicals were proposed as reactive intermediates in iron-induced triggering of trioxanes like artemisinin.*

With ferrous iron acting as a single electron donor (i.e., a reducing agent) and the peroxide bond acting as an electron acceptor (i.e., an oxidizing agent), oxyanion–oxyradicals **13** and **15** could be formed; it is possible that they are in equilibrium, with iron(III) moving back and forth between O1 and O2. Carbonyl bond formation at C3 in oxyradical **13** could drive homolytic cleavage of the C3–C4 bond to form transient C4 carbon-centered radical **14** that could displace Fe(II) by cyclizing to form the tetrahydrofuran ring system in product **11′**. Indeed, this process is catalytic in iron(II).[28] Also, oxyradical **15** can undergo a 1,5-hydrogen atom abstraction, a well-known transformation, exclusively of the C4α-H atom (as shown) via a six-centered transition state to form C4 carbon-centered radical **16**; this carbon radical intermediate **16** could release iron(II) by cyclization with O2 to form epoxide **12′** that would spontaneously undergo intramolecular attack (S_Ni) by the O1 hydroxyl group to form the isolated C4-hydroxylated dioxolane **12**.

The stereospecificity of the 1,5-hydrogen atom abstraction shown in Scheme 13.1 provided an excellent opportunity to probe the relative importance of the two degradation pathways to antimalarial activity. If abstraction of H4α is essential for high antimalarial potency, then blockage of this abstraction would drastically decrease the antimalarial activity of a compound. To investigate this hypothesis, we synthesized C4-methyl trioxane alcohols **17** and **18**, as well as C4-*gem*-dimethyl trioxane alcohol **19**.[29] In vitro antimalarial testing indicated that C4β-methyl trioxane alcohol **18**, which *can* undergo a 1,5-hydrogen atom abstraction of H4α, is over 100 times *more* potent than both the C4α-methyl and C_4-*gem*-dimethyl trioxane alcohols, which *cannot* undergo such an abstraction. Not only is this system a useful stereochemical probe, but, in addition, C4β-methyl trioxane alcohol **18** is approximately twice as active as natural artemisinin (**2**) against malaria parasites in vitro.[29]

17: $R_\alpha = CH_3$, $R_\beta = H$
18: $R_\alpha = H$, $R_\beta = CH_3$
19: $R_\alpha = CH_3$, $R_\beta = CH_3$

As shown in Eq. (13.2), iron(II)-induced degradation of C4-*gem*-dimethyl trioxane benzyl ether **20**, which is also 100 times less potent in vitro than C4β-methyl trioxane alcohol **18**, gave ring-contracted ester **21** as the major product.[29] This result indicates that, despite increased steric bulk at C4, the analog is still triggered by iron, but that a 1,5-hydrogen atom abstraction ultimately leading to a C4-hydroxy dioxolane does not occur. This result indicates that formation of a C4-radical such as **16** during iron(II)-mediated triggering of trioxane antimalarials is important to their potent antimalarial activity.

Fe^{II}

20 → **21**, 38% (13.2)

To further characterize the nature of this carbon-centered radical at C4, we synthesized a series of structurally diverse C4-substituted 1,2,4-trioxanes.[28] C4β-Benzyl trioxane alcohol **22** is 200-fold more potent than its C4α epimer in vitro. C4β-[(Trimethylsilyl)methyl] trioxane alcohol **23** is at least 10 times as active as its C4α-substituted isomer.[28] These results further support the hypothesis that formation of a C4-radical via 1,5-hydrogen atom abstraction is essential for high antimalarial activity.

As with C4β-methyl trioxane alcohol **18**, C4β-benzyl trioxane alcohol **22** is a potent antimalarial agent, with in vitro efficacy comparable to that of artemisinin.[28] In addition, both of these C4β-substituted trioxanes are more than 10 times as potent as their C4-unsubstituted parent trioxane alcohol **23** against these parasites, indicating that a more stable *tertiary* C4 radical seems to surpass a *secondary* C4-radical at promoting antimalarial activity. Surprisingly, however, C4β substitution with a moiety that substantially stabilizes the resulting C4 radical above that of simple tertiary alkyl, whether with the (trimethylsilyl)methyl moiety of trioxane **23** or with the phenyl group of trioxane **25**, does not enhance the antimalarial potency of a compound. In fact, the in vitro activity of these trioxanes[28] actually drops to approximately threefold *less* than that of their C4-unsubstituted parent trioxanes, **24**[27] and **26**,[30] respectively.

All the C4β-oriented substituents in these C4-substituted trioxanes **18**, **22**, **23**, and **25** are spatially remote from the α face of the molecule where the endoperoxide moiety is situated and, according to molecular models, therefore cannot interfere with the approach of iron to the endoperoxide linkage. Instead, the loss of activity in C4β-(trimethylsilyl)methyl trioxane alcohol **23** and C4β-phenyl trioxane **25** seems to be due to shunting of the degradation pathway away from the branch that gives rise to a C4 carbon-centered radical leading to a C4-hydroxy dioxolane.[28] Indeed, as shown in Eq. (13.3), iron(II)-induced degradation of C4β-phenyl analog **25** gave predominantly ring-contracted ester **27** as a mixture of diastereomers. In contrast, as

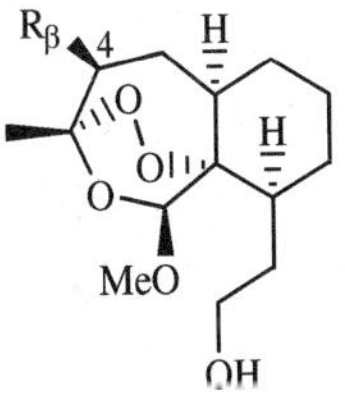

22: $R_\beta = PhCH_2$
23: $R_\beta = Me_3SiCH_2$

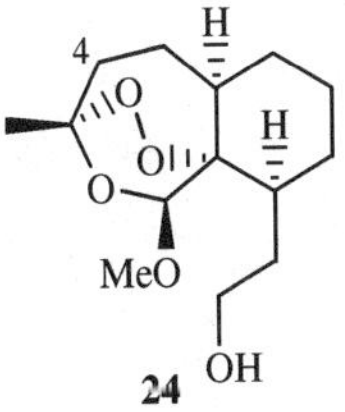

24

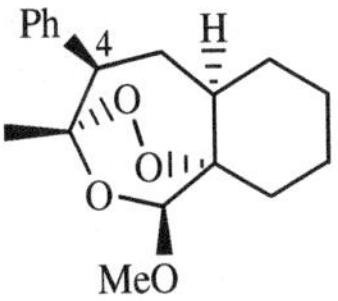

25

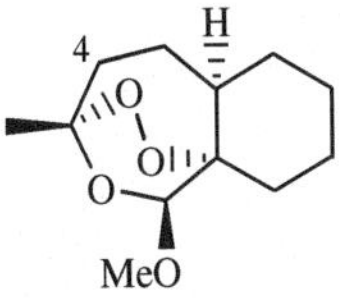

26

shown in Eq. (13.4), degradation of C4β-methyl trioxane benzyl ether **28**, which is comparably active to artemisinin in vitro, produced significant quantities of both ring-contracted esters **29** and C4-hydroxy dioxolane **30** as a single diastereomer.

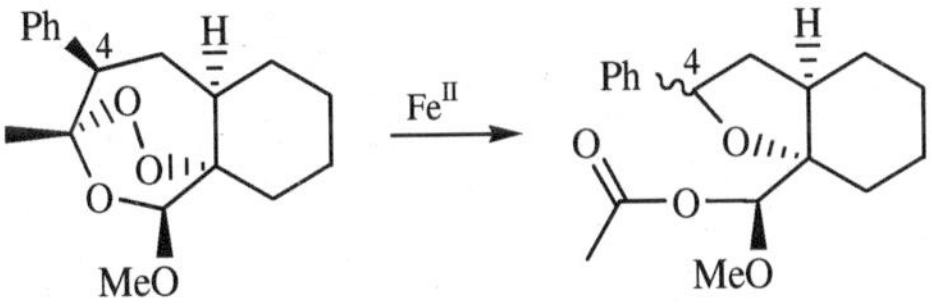

$$\mathbf{25} \xrightarrow{Fe^{II}} \mathbf{27},\ 43\% \tag{13.3}$$

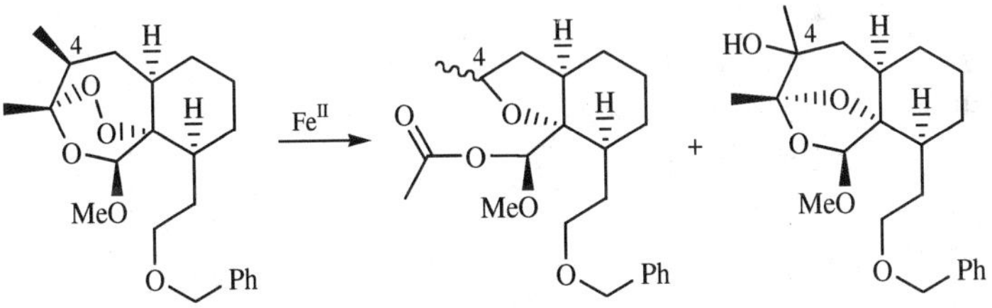

$$\mathbf{28} \xrightarrow{Fe^{II}} \mathbf{29} + \mathbf{30} \tag{13.4}$$

as a 4 : 1 mixture of products

Further evidence for a C4-carbon radical intermediate was provided by iron(II)-triggered reductive cleavage of C4β-trimethylstannylmethyl trioxane **31** (Scheme 13.2). After 1,5-H-atom shift, intermediate C4-radical **32** can undergo a β scission of the relatively stable trimethyltin radical, leading to formation of intermediate **33** bearing a characteristic exocyclic methylene group. Exocyclic methylene dioxolane **34** was actually isolated as one of several products in this reaction (Scheme 13.2), providing further support for the intermediacy of C4-carbon radical **32** and therefore also generally for C4-carbon-centered radicals via 1,5-H atom shifts in iron(II) triggering of such trioxanes.[31]

Electron spin resonance (ESR) spectroscopy has provided additional *direct* evidence for the intermediacy of carbon-centered free radicals on iron(II) triggering

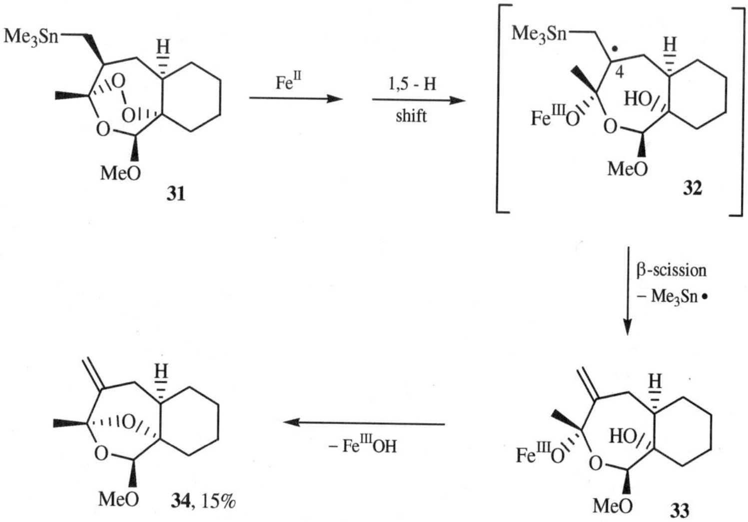

SCHEME 13.2

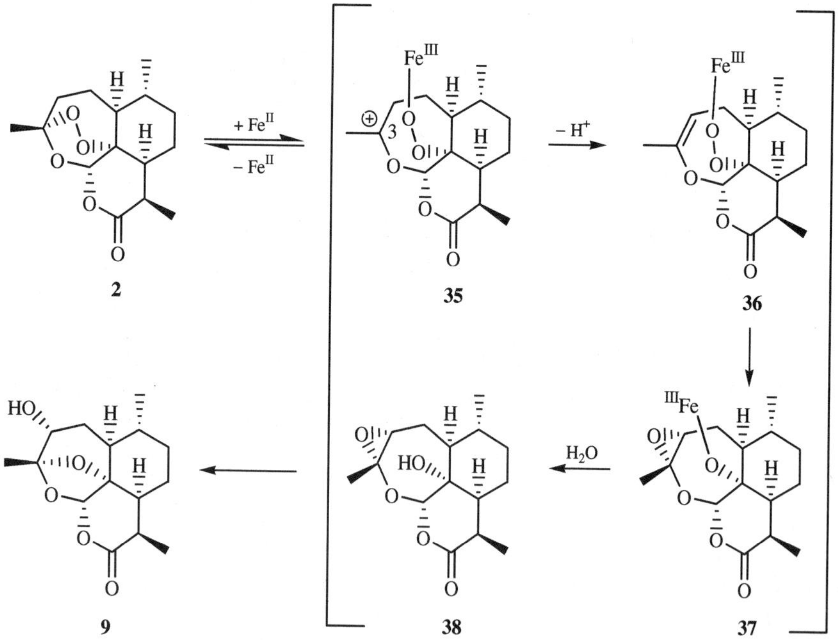

SCHEME 13.3

of trioxanes like artemisinin (**2**).[32,33] Also, isolation and characterization of a covalent adduct between the trioxane skeleton and a porphyrin unit was reported[34,35] when artemisinin (**2**) was treated with a *meso*-tetraphenylporphyrin heme model; carbon radical addition to such a porphyrin system is most likely the way such an adduct is formed. Additional evidence consistent with the intermediacy of C4-radicals such as **16** (Scheme 13.1) has come in the form of isolating a very small amount of an epoxide such as **12′** (Scheme 13.1) during iron(II) triggering of artemisinin.[33]

It has been proposed, although not generally accepted, that iron triggering of the peroxide bond in artemisinin (**2**) to form C4-hydroxy dioxolane **9** occurs via Lewis acid activation of the peroxide unit with formation of C3-carbocation **35** (Scheme 13.3).[36,37] After loss of a proton, vinyl ether **36** could then be formed, followed by epoxide **37** and then epoxy alcohol **38** and finally C4-hydroxy dioxolane **9**. To probe chemically for a C3-carbocation intermediate such as **35**, we have prepared C3-cyclopropyl trioxane **39** and have treated this active antimalarial endoperoxide with iron(II). As shown in Eq. (13.5), only cyclopropyl-containing products were isolated in the yields indicated after chromatographic purification.[38] Even in the ^{1}H NMR spectrum of the crude product mixture, there is virtually complete absence of any C3-olefinic product(s) that would be expected from spontaneous rearrangement of a C3-cyclopropyl carbinyl carbocation into a homoallylic carbocation. A similar absence of any olefinic product(s) was observed also using ferric chloride.[36] These results argue against the importance of the heterolytic mechanism outlined in Scheme 13.3 as a significant pathway.

$FeBr_2$, THF, 0-25 °C

39 → **40**, 24% + **41**, 22% + **42**, 21% (13.5)

How is the C4-radical intermediate **43** that is formed by iron(II) reduction of the peroxide bond in artemisinin (**2**) transformed into the intermediate expoxy alcohol **38** that has recently been isolated? As suggested in Scheme 13.1, direct cyclization of the C4-radical center with the adjacent oxygen atom could produce epoxy alcohol **38**. A second but indirect path also is possible, involving first a β-scission step forming vinyl ether **44** and releasing Fe(III)–O (Scheme 13.4). A subsequent step might involve a rebound epoxidation of the vinyl ether **44** by a high-valent iron–oxo species [e.g., Fe(III)–O$^{\cdot}$ ↔ Fe(IV)=O].

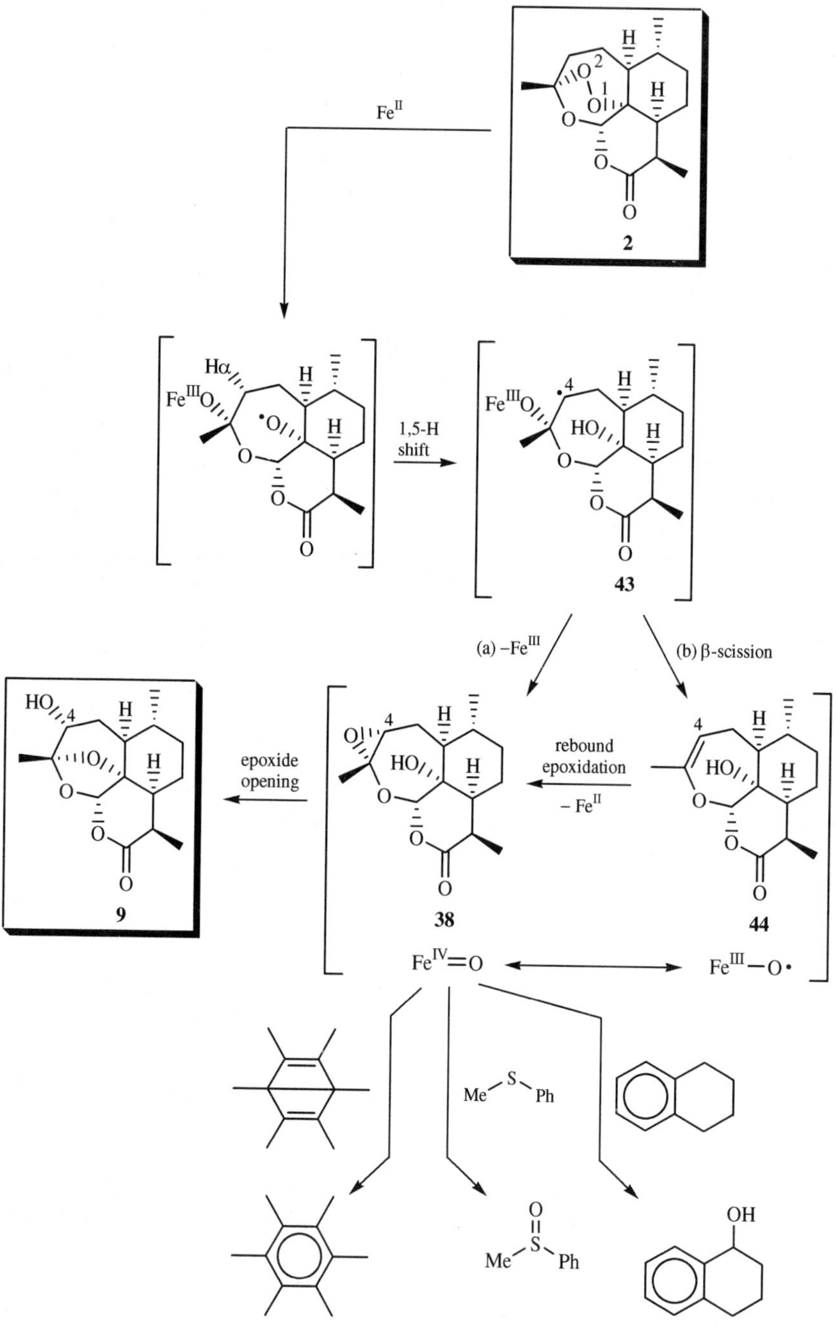

Fe^II
2
Hα
Fe^III O
1,5-H shift
43
(a) –Fe^III
(b) β-scission
epoxide opening
rebound epoxidation
– Fe^II
9
38
44
Fe^IV=O
Fe^III—O•
Me S Ph
OH

SCHEME 13.4

Evidence for the intermediacy of such a high-valent iron–oxo intermediate was provided by reporter reactions characteristic of such a species in the presence of various traps (Scheme 13.4).[39] For example, hexamethyl Dewar benzene was rearranged into hexamethylbenzene in up to 40% yield, methyl phenyl sulfide was oxidized to the corresponding sulfoxide, and tetralin (1,2,3,4-tetrahydronaphthalene) was oxidized to 1-hydroxytetralin. All three reported reactions required the presence of both artemisinin (**2**) and a source of iron(II) for significant product formation.[39] In addition, deoxygenated solvent did not affect the oxidations of methyl phenyl sulfide or of tetralin, implying that the oxygen atom in these oxygen transfer reporter reactions did not come from molecular oxygen present during the degradation. Importantly, the presence of hexamethyl Dewar benzene as Fe(IV)=O trap during chemical degradation of artemisinin (**2**) decreased the amount of C4-hydroxydeoxyartemisinin **9** by about two-thirds, implying that the high-valent iron–oxo species is required in formation of this product. A non-heme high-valent iron–oxo species has been invoked also as a reasonable intermediate in understanding some subtle SAR differences in semisynthetic antimalarial trioxane sulfides[40] and in reaction between ferrous salts and hydrogen peroxide.[41,42] Also, direct spectroscopic observation of an ^{18}O incorporation in high-valent iron–oxo species has been reported from hemin-induced reduction of ^{18}O-labelled antimalarial trioxanes.[43,44] In addition, isolation of a δ-*meso* oxidation product (i.e., a hydroxyporphyrin) from hemin-promoted reaction with artemisinin (**2**) is consistent with the action of a strong oxidizing (perhaps a high-valent iron–oxo) species.[45]

What kills the malaria parasites? Any one or a combination of the reactive and cytotoxic species shown in Scheme 13.4 [e.g., carbon-centered radical **43**, alkylating epoxide **38**, oxidizing high-valent iron–oxo Fe(IV)=O] may kill the parasites. Further research is needed to clarify the relative importance of these cytotoxic species.

Criticism of an epoxide such as **38** being a possibly lethal intermediate has been based on the observation that such a synthesized compound lacks in vitro antimalarial activity.[46] Because of its expected high chemical reactivity, however, such an alkylating epoxide would probably not survive transport through cell membranes; its cytotoxicity likely requires that it be formed inside a malaria parasite.

RATIONAL DESIGN OF NEW ANTIMALARIAL PEROXIDES

Mechanism-Based Design

On the basis of our understanding at the molecular level of the chemical cascade leading from antimalarial trioxane to the various cytotoxic intermediates already discussed in this chapter, we have rationally designed a series of structurally simple 3-*aryl*trioxanes **45**. We anticipated that the 3-aryl substituent would facilitate progression through this cascade by resonance stabilization of adjacent unsaturation after β scission (see Scheme 13.4) and, therefore, would lead to larger amounts of cytotoxic intermediates and thus to more potent antimalarial trioxanes.[47] The first

3-aryltrioxane reported was prepared in the Jefford group.[48] Of the 25 structurally simple and easily synthesized 3-aryltrioxanes we prepared, 22 have in vitro IC_{50} values of less than 100 nM, 11 have in vitro IC_{50} values of less than 50 nM, and several have in vitro IC_{50} values in the 15–30 nM range compared to the 9.2 nM IC_{50} value for structurally complex, natural artemisinin (**2**). Preliminary assay of the four 12β-methoxy-3-aryltrioxanes (**45g**, **45j**, **45k**, and **45l**) in a rodent system against chloroquine-sensitive *Plasmodium berghei* (N) resulted in the data shown in Table 13.1.[47] For convenience in comparing relative potencies of antimalarials having different molecular weights, ED_{50} and ED_{90} values are presented not only in the standard mg/kg (milligram/kilogram) terms but here also in μmol/kg (micromole/kilogram) values. Several observations emerge: (1) these four simple synthetic trioxanes are potent antimalarials; (2) high levels of in vitro antimalarial activity correlate well with high antimalarial efficacy in vivo; (3) especially noteworthy for ease of administration ultimately in clinical settings is the *oral activity* of these simple, crystalline trioxanes; and (4) structurally simple benzylic alcohol trioxane **45g**, acetate trioxane **45j**, and fluorobenzyl ether trioxane **45k** are up to twice as potent as the complex natural artemisinin (**2**). All of these orally active trioxanes are stable at 60 °C for at least 36 h. Thus, these preclinical in vivo results demonstrate the high antimalarial efficacy of simple 12β-methoxytrioxanes **45g**, **45j**, **45k**, and **45l**. These new and easily prepared trioxanes, therefore, are promising lead compounds appropriate for further preclinical evaluation.

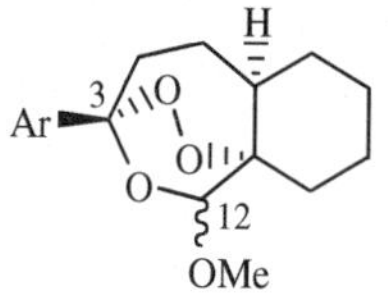

45

a, Ar = Ph
b, Ar = *p*-PhPh
c, Ar = 1-naphthyl
d, Ar = *p*-ClPh
e, Ar = *p*-MeOPh
f, Ar = 2-furyl
g, Ar = *p*-$HOCH_2Ph$
h, Ar = *p*-$MeOCH_2Ph$
i, Ar = *p*-$MeOC(O)OCH_2Ph$
j, Ar = *p*-$MeC(O)OCH_2Ph$
k, Ar = *p*-(*p*′-$FPhCH_2OCH_2$)Ph
l, Ar = *p*-FPh
m, Ar = *p*-F-*o*-MePh
n, Ar = *p*-CF_3Ph

Also on the basis of the radical cascade mechanism outlined in Schemes 13.1 and 13.4, we have designed a series of structurally simple seven-membered endoperoxides **46**. We anticipated that these seven-membered ring endoperoxides could be reduced by iron(II) to trigger sequential formation of oxyradical **47**, carbon radical

TABLE 13.1 In vivo Antimalarial Activities

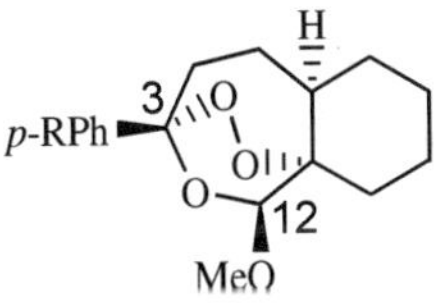

p-RPh 3 O O O 12 MeO H

$C_{12\beta}$-trioxane	R	ED$_{50}$, mg/kg (μmol/kg)[a] subcutaneous		ED$_{50}$ oral		ED$_{90}$, mg/kg (μmol/kg) subcutaneous		ED$_{90}$ oral	
45g	$HOCH_2$	3.4	(11)	5.5	(17)	6.8	(21)	12	(37)
45j	$MeC(O)OCH_2$	2.8	(7.7)	14	(39)	6.0	(17)	22	(61)
45k	p'-$FPhCH_2OCH_2$	3.5	(8.2)	3.8	(8.9)	7.1	(17)	6.8	(16)
45l	F	6.8	(22)	10	(32)	13	(42)	23	(75)
	artemisinin	3.0	(11)			8.5	(30)		
	chloroquine	1.0	(1.9)			1.2	(2.3)		

[a] Four different doses (1, 3, 10, and 30 mg/kg) were administered each day for four days to five mice per dose regimen to establish the ED values indicated above via a previously reported protocol.

48, activated olefin **49** along with Fe(IV)=O and then epoxide **50**, and finally hydroxyether **51** (Scheme 13.5).[49] This sequence of transformations matches exactly the mechanistic steps in iron(II) triggering of antimalarial trioxanes like artemisinin (**2**). One important structural characteristic of cycloheptane endoperoxide **46** is the plane of symmetry that bisects the bond connecting the two peroxide oxygen atoms; as a consequence of this symmetry element, it does not matter which oxygen atom is induced by iron to acquire radical character. Despite their extraordinary structural simplicity and facile synthesis, phenyl and *p*-anisyl endoperoxides **46** have in

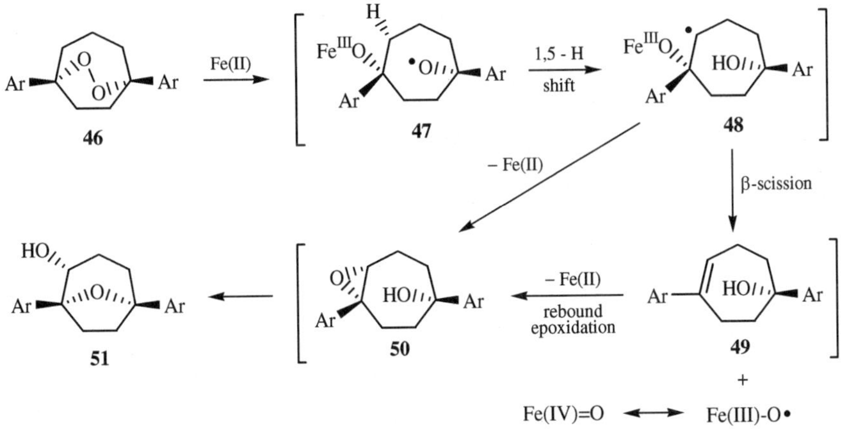

Ar O O Ar **46** —Fe(II)→ [$Fe^{III}O$ H •O Ar Ar **47** —1,5 - H shift→ $Fe^{III}O$ • HO Ar Ar **48**]

– Fe(II)

β-scission

HO Ar O O Ar **51** ← [O HO Ar Ar **50**] ← – Fe(II) rebound epoxidation — [Ar HO Ar **49**]

+

Fe(IV)=O ⟷ Fe(III)-O•

SCHEME 13.5

vitro antimalarial potencies that are about 13–18% that of natural artemisinin (**2**). This noteworthy success in predicting antimalarial activity based on chemical mechanistic understanding at the molecular level provides strong support for the fundamental correctness of mechanistic Schemes 13.1 and 13.4!

A carbon-centered radical seems to be important also as a reactive intermediate in iron(II)-triggered fragmentation of six-membered endoperoxide **52** (Scheme 13.6).[50] Iron(II)-induced reductive cleavage of the peroxide bond in endoperoxide **52**, forming oxyradical **53**, can lead to carbon-centered radical intermediate **54**, which can suffer loss of ethylene and iron(II), finally generating the observed 1,4-diketone **55**. The intermediacy of ethylene was confirmed by trapping it as 1,2-dibromoethane. Although oxyradical **53** cannot undergo a 1,5-H-atom abstraction to form a carbon-centered free radical, it apparently is prompted to form carbon radical **54** by the concomitant formation of a strong carbonyl C=O bond; a similar mechanistic step occurs in Scheme 13.1 leading to C radical **14**. Structurally simple, symmetrical, and easily synthesized cyclohexane endoperoxide **52** is about 17% as antimalarially active in vitro as natural artemisinin (**2**), and this fluorophenyl endoperoxide has measurable antimalarial activity also in vivo in a rodent antimalarial assay.[50] In the case of this type of cyclohexane endoperoxide, the mechanism shown in Scheme 13.6 involves formation of a potentially cytotoxic 1,4-diketone (e.g., **55**) as well as formation of ethylene that could be oxidized by the malaria parasite's cytochrome oxidase enzymes into ethylene oxide, an extremely reactive and cytotoxic alkylating agent.[51]

Carbon-centered radicals appear to be key intermediates also in iron(II)-triggering of cyclic peroxy ketal **56** (Scheme 13.7).[57] In Scheme 13.7, ferrous ion reduction can proceed with iron associated with either of the cyclic peroxide oxygen atoms, leading to either or both oxygen-centered radicals **57a** and/or **57b** which may be in equilibrium with each other. These oxyradicals then can cyclize to form the corresponding epoxy carbon-centered radicals **58a** and **58b**. β-Scission of a

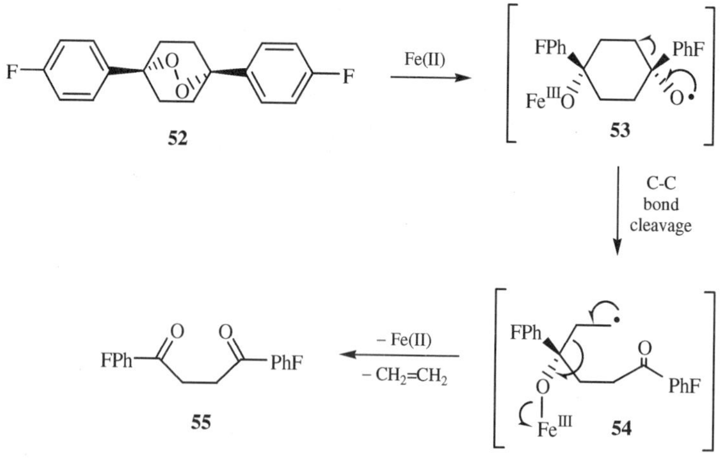

F FPh PhF Fe(II) $Fe^{III}O$ O• **52** **53** C-C bond cleavage FPh PhF – Fe(II) – CH_2=CH_2 FPh O PhF Fe^{III} **55** **54**

SCHEME 13.6

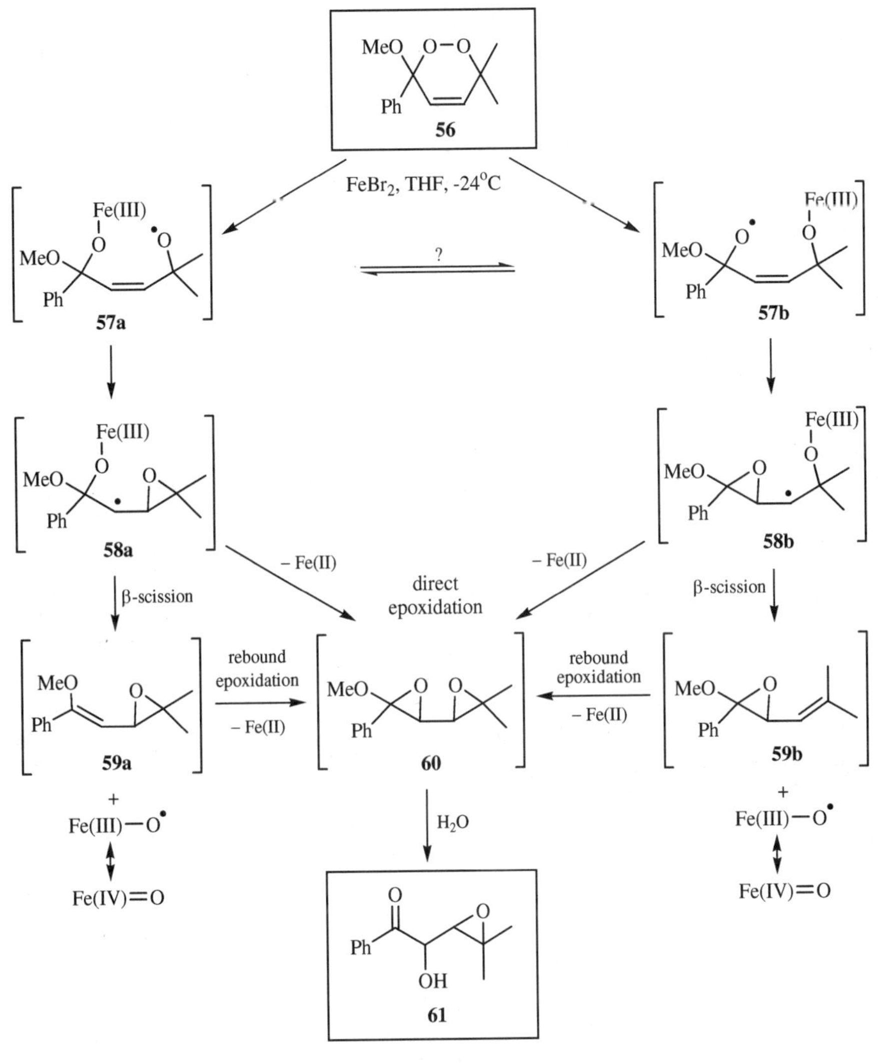

SCHEME 13.7

high-valent iron-oxo species, forming olefins **59a** and/or **59b**, and then intermolecular rebound epoxidation could produce diepoxide **60**; also diepoxide **60** might be formed via direct intramolecular epoxidation from epoxy radicals **58**. Quenching the reaction mixture with water would then rapidly hydrolyze methoxy epoxide **60** into the observed major product hydroxy ketone **61**, isolated as only one stereoisomer. If diepoxide **60** were formed inside a malaria parasite, it would likely be a strongly cytotoxic alkylating agent.

A series of more than 20 new, cyclic peroxy ketals such as **56** were prepared easily through a two-step protocol starting with readily available aryl methyl ketones.[53] Chemical structure–antimalarial activity correlations led to seven new peroxides that have in vitro antimalarial IC_{50} values ranging from about 17 to 34% that of artemisinin (**2**). The most potent among these new peroxides was sulfone **62**.

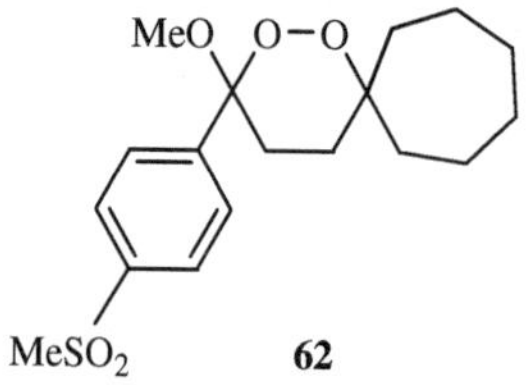

MeO O–O

$MeSO_2$ **62**

Iron(II)-mediated rearrangement of monocyclic trioxanes **63** into 1,2-diol monoesters **66** has been shown to involve radical intermediates (Scheme 13.8).[54] The initially formed oxyradical **64** undergoes a 1,5-H-atom shift to form carbon-centered radical **65**. Release of Fe(II) is accompanied by formation of the observed product diol monoester **66**. This overall sequence of reactions represents additional support for the intermediacy of carbon-centered radicals being formed on iron(II) triggering of antimalarial trioxanes such as artemisinin (**2**).

The endoperoxide arteflene (**67**) was developed by the Hoffmann-LaRoche company[55,56] as a synthetic analog of the scarce natural antimalarial yingzhaosu (**68**). Due in large part to the lengthy and costly synthesis of arteflene (**67**), however, it apparently is not being pursued currently as a practical antimalarial by the pharmaceutical industry.[12] Nevertheless, its mode of action when exposed to iron(II) seems to involve a carbon-centered radical intermediate. For example, closely structurally related endoperoxide **69** is triggered by iron(II) to form oxyradical **70**, which fragments to form carbon radical **71** and the isolated acetate ester **72** [Eq. (13.6)].[57]

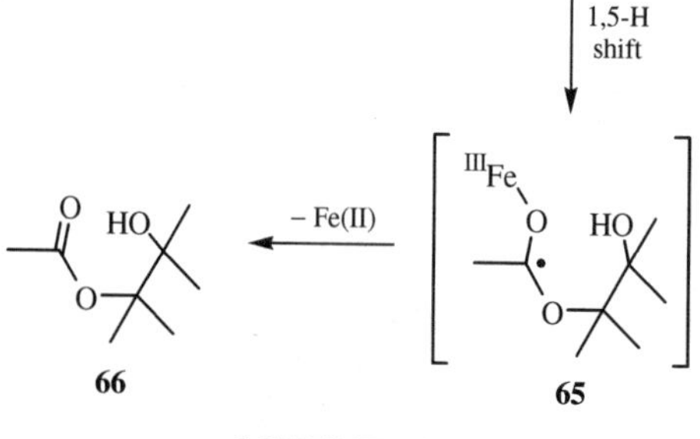

H O–O R O **63** —Fe(II)→ [^{III}Fe O •O H R O] **64**

1,5-H shift

[^{III}Fe O HO O] **65** —– Fe(II)→ O HO O **66**

SCHEME 13.8

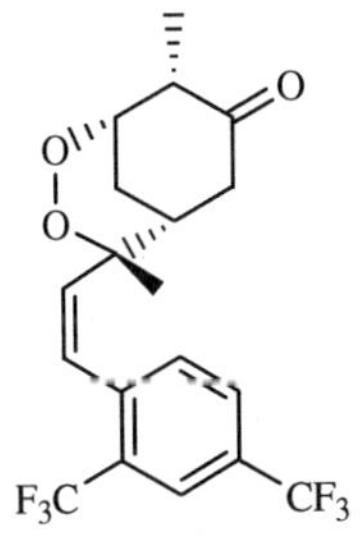

arteflene (**67**)

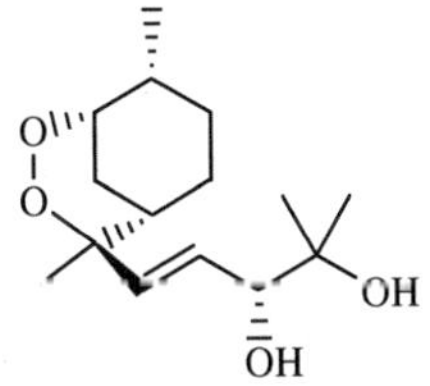

yingzhaosu (**68**)

(13.6)

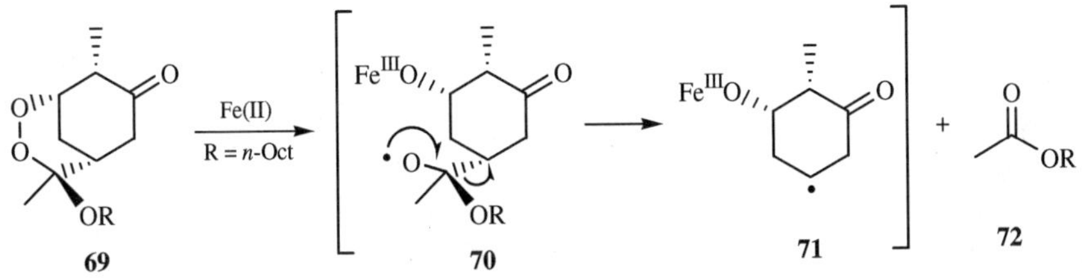

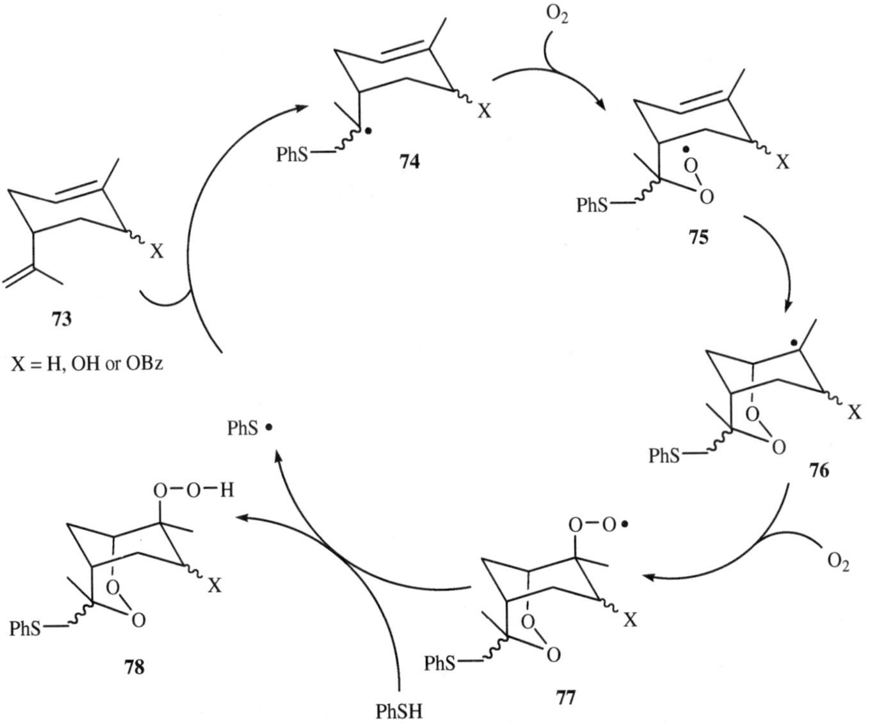

69 $\xrightarrow[R = n\text{-Oct}]{Fe(II)}$ [**70** → **71**] + **72**

$X = H$, OH or OBz

SCHEME 13.9

Radical-Based Synthesis

The possibility of using free radicals to synthesize antimalarial peroxides occurred to Professor Mario Bachi of the Weizmann Institute of Science, Israel, during his visiting professorship in the chemistry department at Johns Hopkins University in 1995. Scheme 13.9 summarizes the multistep sequence of radical addition reactions, occurring all in one reaction vessel, planned for this extremely short construction of a complex endoperoxide. Thiol radical regiospecific and chemospecific addition to only the terminal carbon atom of the exocyclic carbon–carbon double bond of chiral pool diene **73** could form carbon-centered radical **74**. Addition of molecular oxygen to carbon radical **74** could produce peroxy radical **75**, which could cyclize regiospecifically onto the remaining olefinic double bond to form tertiary carbon radical **76**. Reacting with a second molecule of oxygen, carbon radical **76** could give peroxy radical **77**, which could abstract a hydrogen atom from benzenethiol to form the desired dioxabicyclo[3.3.1]nonane target molecule **78** and also a thiol radical to keep this radical chain process going. Despite the very delicate balance of various factors required for Scheme 13.9 to work well, using diene **73a** with X = H and using triphenylphosphine to reduce the final hydroperoxide **78** into an alcohol did in fact produce endoperoxide alcohols **79a** and **80a** in an outstanding 54% overall yield![58,59] This triumph of chemical

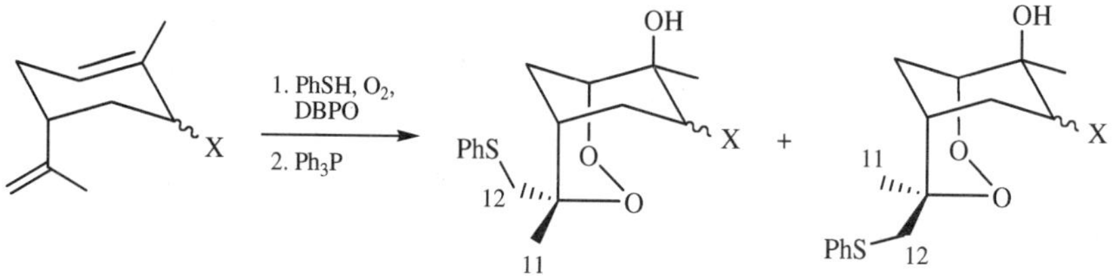

1. PhSH, O_2, DBPO
2. Ph_3P

		79	80
73a	X = H	**a**, 30%	24%
73b	X = β-OBz	**b**, 11%	11%
73c	X = α-OBz	**c**, 4%	20%
73d	X = α-OH	**d**, 12%	0

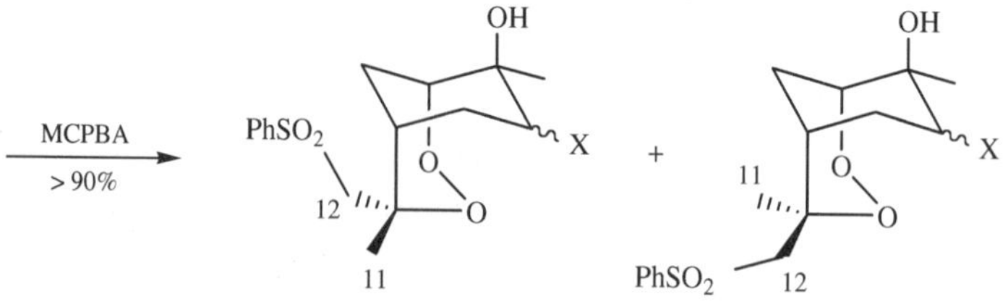

MCPBA
> 90%

SCHEME 13.10

reasoning for radical-based synthesis of dioxabicyclo[3.3.1]nonanols was used also to prepare the corresponding endoperoxides with X = OC(O)Ph and X = OH as sulfides and also as sulfones (Scheme 13.10).[59] Several of these endoperoxide sulfones are potent antimalarials even when administered to rodents orally, and they are currently under mechanistic study as well as preclinical evaluations as potential new antimalarial drug candidates.[60]

CONCLUSIONS

Applying the principles of modern organic chemistry to understanding in detail the molecular events that occur when iron(II) triggers reduction of the peroxide bond in trioxane and endoperoxide antimalarials has established firmly that carbon-centered radicals are important intermediates. This mechanistic understanding has allowed rational design of diverse new types of peroxidic antimalarials, some of which are highly efficacious even when administered to rodents orally. Whether one or more of these new peroxidic antimalarials ultimately becomes a widely used drug now depends critically on establishing a partnership between academic laboratories and pharmaceutical companies. With its vast resources, industry must now support preclinical and then clinical evaluations of these and related new lead compounds in the worldwide search for effective and safe new drugs to cure people infected with malaria.

ACKNOWLEDGMENTS

We thank the NIH (AI-34885) for generous financial support, Professor Mario Bachi for many stimulating discussions about carbon-centered radicals, Professor Theresa Shapiro for setting up and supervising our Hopkins in vitro antimalarial testing facility and for constant encouragement and good advice, and Dr. Poon Ploypradith and Ms. Suji Xie for invaluable assistance in keeping the Hopkins in vitro antimalarial testing program up and running so smoothly and reliably.

REFERENCES

1. *TDR News* (News from the WHO Division of Control of Tropical Diseases) 1994, **46**, 5.
2. Peters, W., *Chemotherapy and Drug Resistance in Malaria*; 2nd ed., Academic Press, London, 1987.
3. Ridley, R. G.; Hudson, A. T., *Exp. Opin. Ther. Patents* 1998, **8**, 121.
4. *The New York Times*, August 2, 1998, p. 23.
5. Rosenthal, P. J.; Meshnick, S. R., *Molec. Biochem. Pharm.* 1996, **83**, 131.
6. Corey, J., *Business Week*, Sept. 21, 1998, p. 70.
7. Klayman, D. L., *Science* 1985, **228**, 1049.
8. Klayman, D. L., *Nat. His.* (Oct.) 1989, p. 18.

9. Hien, T. T.; White, N. J., *Lancet* 1993, **341**, 603.
10. Zhou, W.-S.; Xu, X. X., *Acc. Chem. Res.* 1994, **27**, 211.
11. Ziffer, H.; Highet, R. J.; Klayman, D. L., *Prog. Chem. Org. Nat. Prod.* 1997, **72**, 121.
12. Posner, G. H., *Exp. Opin. Ther. Patents* 1998, **8**, 1487.
13. Posner, G. H.; Parker, M. H.; Northrop, J.; Elias, J. S.; Ploypradith, P.; Xie, S.; Shapiro, T. A., *J. Med. Chem.* 1999, **42**, 300.
14. Jung, M.; Lee, S., *Bioorg. Med. Chem. Lett.* 1998, **8**, 1003.
15. Vroman, J. A.; Khan, I. A.; Avery, M. A., *Tetrahedron Lett.* 1997, **38**, 6173.
16. Pu, Y. M.; Ziffer, H., *J. Med. Chem.* 1995, **38**, 613.
17. Meshnick, S. R.; Yang, Y.-Z.; Lima, V.; Kuypers, F.; Kamchonwongpaisan, S.; Yuthavong, Y., *Antimicrob. Agents Chemother.* 1993, **37**, 1108.
18. Meshnick, S. R.; Taylor, T. E.; Kamchonwongpaisan, S., *Microbiol. Rev.* 1996, **60**, 301.
19. Liu, J.-M.; Ni, M. Y.; Fan, J.-F.; Tu, Y.-Y.; Wu, Z. H.; Wu, Y.-L.; Chou, W.-S., *Acta Chim. Sinica* 1979, **37**, 129.
20. Jefford, C. W.; Boukouvalas, J.; Kohmoto, S.; Bernardinelli, G., *Tetrahedron* 1985, **41**, 2081.
21. Jefford, C. W., *Jpn. J. Trop. Med. Hyg.* 1996, **24** (Suppl. 1), 7.
22. Avery, M. T.; Alvim-Gaston, M.; Woolfrey, J. R., *Adv. Med. Chem.*, JAI Press, Vol. 4, 1999, 125.
23. Dong, Y.; Vennerstrom, J. L., *J. Org. Chem.* 1998, **63**, 8582.
24. Jung, M.; Lee, S., *Heterocycles* 1997, **45**, 1907.
25. Lee, I. S.; El Sohly, H. N.; Croom, E. M.; Hufford, C. D., *J. Nat. Prod.* 1989, **52**, 337.
26. Posner, G. H.; Oh, C. H., *J. Am. Chem. Soc.* 1992, **114**, 8328.
27. Posner, G. H.; Oh, C. H.; Gerena, L.; Milhous, W. K., *J. Med. Chem.* 1992, **35**, 2459.
28. Posner, G. H.; Wang, D.; Cumming, J. N.; Oh, C. H.; French, A. N.; Bodley, A. L.; Shapiro, T. A., *J. Med. Chem.* 1995, **38**, 2273.
29. Posner, G. H.; Oh, C. H.; Wang, D.; Gerena, L.; Milhous, W. K.; Meshnick, S. R.; Asawamahasakda, W., *J. Med. Chem.* 1994, **37**, 1256.
30. Posner, G. H.; Oh, C. H.; Gerena, L.; Milhous, W. K., *Heteroatom Chem.* 1995, **6**, 105.
31. Posner, G. H.; Park, S. B.; González, L.; Wang, D.; Cumming, J. N.; Klinedinst, D.; Shapiro, T. A.; Bachi, M. D., *J. Am. Chem. Soc.*, 1996, **118**, 3537.
32. Butler, A. R.; Gilbert, B. C.; Hulme, P.; Irvine, L. R.; Renton, L.; Whitwood, A. C., *Free Rad. Res.* 1998, **28**, 471.
33. Wu, W.-M.; Wu, Y.; Wu, Y.-L.; Yao, Z.-J.; Zhou, C.-M.; Li, Y.; Shan, F., *J. Am. Chem. Soc.* 1998, **120**, 3316.
34. Robert, R. A.; Meunier, B., *Chem. Eur. J.* 1998, **4**, 1287.
35. Robert, A.; Meunier, B., *Chem. Soc. Rev.* 1998, **27**, 273.
36. Haynes, R. K.; Vonwiller, S. C., *Tetrahedron Lett.* 1996, **37**, 257.
37. Haynes, R. K.; Pai, H. H.; Voerste, A., *Tetrahedron Lett.* 1999, **40**, 4715.
38. Posner, G. H.; Krasavin, M., unpublished results.
39. Posner, G. H.; Cumming, J. N.; Ploypradith, P.; Oh, C. H., *J. Am. Chem. Soc.* 1995, **117**, 5885.

40. Posner, G. H.; O'Dowd, H.; Caferro, T.; Cumming, J. N.; Ploypradith, P.; Xie, S.; Shapiro, T. A., *Tetrahedron Lett.* 1998, **39**, 2273.
41. Groves, J. T.; Van der Puy, M., *J. Am. Chem. Soc.* 1974, **96**, 5274.
42. Murata, S.; Miura, M.; Nomura, M., *J. Chem. Soc. Perkin Trans. I* **1987**, 1259.
43. Kapetanaki, S.; Varotsis, C., *J. Am. Chem. Soc.* manuscript submitted.
44. Lange, S. J.; Miyake, H.; Que, L. Jr., *J. Am. Chem. Soc.* 1999, **121**, 6330.
45. Bharel, S.; Vishwakarma, R. A.; Jain, S. K., *J. Chem. Soc. Perkin Trans I* 1998, 2163.
46. Avery, M. A.; Fan, P.; Karle, J. M.; Bonk, J. D.; Miller, R.; Goins, D. K., *J. Med. Chem.* 1996, **39**, 1885.
47. Posner, G. H.; Cumming, J. N.; Woo, S. H.; Ploypradith, P.; Xie, S.; Shapiro, T. A., *J. Med. Chem.* 1998, **41**, 940.
48. Jefford, C. W.; Velarde, J. A.; Bernardinelli, G.; Bray, D. H.; Warhurst, D. C.; Milhous, W. K., *Helv. Chim. Acta* 1993, **76**, 2775.
49. Posner, G. H.; Wang, D. W.; González, L.; Tao, X.; Cumming, J. N.; Klinedinst, D.; Shapiro, T. A., *Tetrahedron Lett.* 1996, **37**, 815.
50. Posner, G. H.; Tao, X.; Cumming, J. N.; Klinedinst, D.; Shapiro, T. A., *Tetrahedron Lett.* 1996, **37**, 7225.
51. Ortiz de Montellano, P. R., in *Bioactivation of Foreign Compounds*, M. W. Anders, ed., Academic Press, New York, 1985, Chapter 5.
52. Posner, G. H.; O'Dowd, H., *Heterocycles* 1998, **47**, 643.
53. Posner, G. H.; O'Dowd, H.; Ploypradith, P.; Cumming, J. N.; Xie, S.; Shapiro, T. A., *J. Med. Chem.* 1998, **41**, 2164.
54. Bloodworth, A. J.; Shah, A., *Tetrahedron Lett.* 1995, **36**, 7551.
55. Hofheinz, W.; Bürgin, H.; Gocke, E.; Jaquet, C.; Masciadri, R.; Schmid, G.; Stohler, H.; Urwyler, H., *Trop. Med. Parasitol.* 1994, **45**, 261.
56. Jaquet, C.; Stohler, H. R.; Chollet, J.; Peters, W., *Trop. Med. Parasitol.* 1994, **45**, 266.
57a. O'Neill, P. M.; Searle, N. L.; Raynes, K. J.; Maggs, J. L.; Ward, S. A.; Storr, R. C.; Park, B. K.; Posner, G. H., *Tetrahedron Lett.* 1998, **39**, 6065.
57b. Cazelles, J.; Robert, A.; Meunier, B., *J. Org. Chem.* 1999, **64**, 6776.
58. Bachi, M. D.; Korshin, E. E., *Synlett* 1998, 122.
59. Bachi, M. D.; Korshin, E.; Ploypradith, P.; Cumming, J. N.; Xie, S.; Shapiro, T. A.; Posner, G. H., *Bioorg. Med. Chem. Lett.* 1998, **8**, 903.
60. Bachi, M. D.; Korshin, E.; Posner, G. H., patent application filed.

CHAPTER 14

The Chemistry Of Parkinson's Disease

GLENN DRYHURST, XUE-MING SHEN, HONG LI, JILIN HAN,
ZHAOLIANG YANG, and FU-CHOU CHENG

Department of Chemistry and Biochemistry, University of Oklahoma

INTRODUCTION

Parkinson's disease (PD) was first described by the English physician James Parkinson in his short book *The Shaking Palsy* published in 1817. The cardinal symptoms of PD include resting tremor, muscular rigidity, and therefore difficulties in walking and other movements and a lack of facial expression, slowness in initiating and executing movements (bradykinesia), a stooped posture, and a propensity to falls. PD is an age-related disorder that afflicts at least one million Americans and perhaps as many as 2% of all individuals.[1] The overwhelming majority of cases of PD are termed *sporadic* or *idiopathic* to reflect the fact that they presently have no known cause.

The biochemical basis for the motor symptoms of PD is a massive loss of the neurotransmitter dopamine (DA) in two subcortical brain regions, the caudate nucleus and particularly the putamen, which together compromise the striatum.[2,3] Typically, symptoms of PD first appear when levels of striatal DA are decreased by $\geq 80\%$ compared to those in the brains of normal individuals of the same age. The loss of striatal DA is the result of the degeneration of nigrostriatal dopaminergic neurons. These neurons project from cell bodies located in a region of the midbrain known as the *substantia nigra* (black substance) pars compacta (SN_c) to axon terminals that richly innervate the striatum. A large body of evidence, based on studies of autopsied brains from patients who died with PD, strongly suggests that the pathological (neurotoxic) processes that cause the degeneration of nigrostriatal DA neurons take place in their pigmented cell bodies in the SN_c. A pathological hallmark of PD is the presence of Lewy bodies, intraneuronal occlusions in remaining SN_c cells. The fundamental mechanisms underlying the degeneration of SN_c cells and the significance of the Lewy body in PD are unknown. As a consequence,

Biomedical Chemistry: Applying Chemical Principles to the Understanding and Treatment of Disease, Edited by Paul F. Torrence
ISBN 0-471-32633-x

there is no marker for asymptomatic or early PD and no therapy available to halt or reverse the neurodegenerative processes.

BIOSYNTHESIS AND METABOLISM OF DA

Dopaminergic neurons contain the enzyme tyrosine hydroxylase that catalyzes the oxidation (hydroxylation) of the aromatic amino acid L-tyrosine to L-3,4-dihydroxyphenylalanine (L-DOPA or levodopa) (Fig. 14.1). Only very small amounts of L-DOPA are present in brain tissue owing to its rapid decarboxylation, catalyzed by L-DOPA decarboxylase (also known as aromatic amino acid decarboxylase), to DA, the final step in the biosynthesis of the neurotransmitter (Fig. 14.1). In order to function as a neurotransmitter, DA is released from neurons in response to an action potential that transiently depolarizes the neuronal membrane and interacts with one of several subtypes of dopaminergic receptors located on other neurons. In order to terminate interaction with its receptors DA is subsequently removed from the synaptic cleft by its energy-dependent reuptake into dopaminergic neurons by the DA transporter or its metabolism to inactivate metabolites. Two principal enzymes are involved in the latter biological degradation of DA: monoamine oxidase B (MAO B) and catechol-*O*-methyltransferase (COMT). MAO B, a flavoprotein enzyme, occurs both as a presynaptic enzyme and postsynaptically in astroglia and serotonergic neurons and is located in the outer membrane of mitochondria. MAO B catalyzes the oxidative deamination of DA to 3,4-dihydroxyphenylacetaldehyde acid that is rapidly oxidized to 3,4-dihydroxyphenylacetic acid (DOPAC) in a reaction catalyzed by aldehyde dehydrogenase (Fig. 14.1). COMT then catalyzes the transfer of the methyl group from *S*-adenosylmethionine to DOPAC, forming homovanillic acid (HVA). Alternatively, COMT mediates the methylation of DA to 3-methoxytyramine (3-MT), which is then oxidatively deaminated by MAO-B to 3-methoxy-4-hydroxyphenylacetaldehyde. This reactive aldehyde is then oxidized to HVA by aldehyde dehydrogenase.

CURRENT TREATMENT FOR PD

The symptoms of PD can be relieved by means of a DA replacement strategy based on the pathway shown in Figure 14.1. This is accomplished by treating patients with L-DOPA and an inhibitor of peripheral L-DOPA decarboxylase (e.g., carbidopa). The latter inhibitor is employed to block the decarboxylation of L-DOPA to DA, which cannot cross the blood–brain barrier (BBB), by L-DOPA decarboxylase in the intestinal wall, liver, kidneys, and brain capillary endothelium. Thus, L-DOPA, which is able to cross the BBB, can enter brain parenchyma, where it is decarboxylated to DA. In order to preserve brain DA, it is also common to treat PD patients with an MAO B inhibitor such as deprenyl (seleginine). Although this brain DA replacement therapy ameliorates parkinsonian symptoms, after 5–10 years most patients experience very unpleasant side effects such as dyskinesias, involuntary

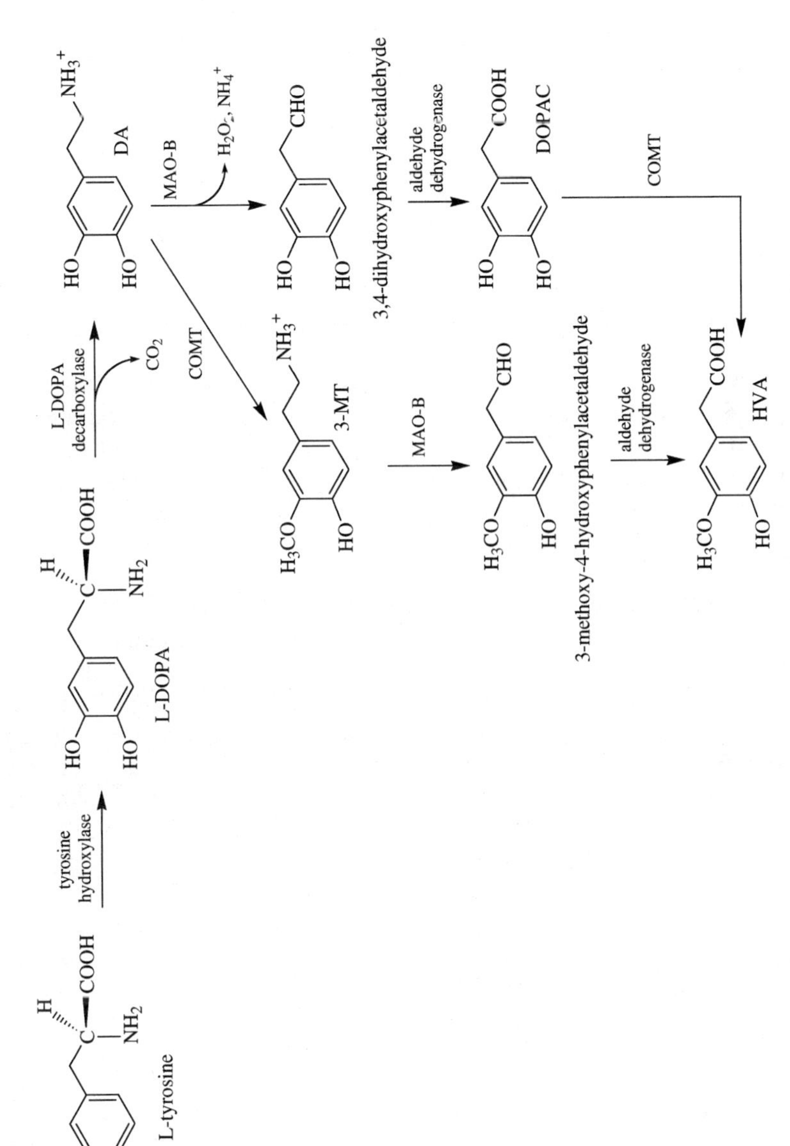

FIGURE 14.1 Biosynthesis and metabolism of dopamine in the brain.

spasms of muscle contraction causing abnormal movements and postures, that do not respond to L-DOPA. Furthermore, L-DOPA therapy fails to halt the inexorable progression of nigrostriatal neurodegeneration.

PD CHARACTERISTICS AND SYMPTOMS

Neuromelanin

The black pigment present in dopaminergic SN_c cells is known as *neuromelanin*. Although the exact structure of neuromelanin is not known, it is believed to be formed by the oxidation of cytoplasmic DA by molecular oxygen.[4] This reaction does not require an enzyme catalyst, a process known as *autoxidation*, and involves the initial oxidation of DA to DA-*o*-quinone forming superoxide ($O_2^{-\cdot}$) and thence H_2O_2 as by-products (Fig. 14.2). Deprotonation of the ethylamino sidechain of DA-*o*-quinone then permits its intramolecular cyclization to 5,6-dihydroxyindoline (**1**), which is readily further autoxidized to *p*-quinone imine (**2**), which rearranges to 5,6-dihydroxyindole (5,6-DHI, Fig. 14.1), the last known intermediate in the neuromelanin pathway.[5] Further autoxidation of 5,6-DHI gives *p*-quinone imine (**3**), which undergoes a complex sequence of reactions with precursors to give black, alkali-insoluble indolic neuromelanin polymer.

Neuromelanin in PD Brain Remaining dopaminergic SN_c cells in the PD have been reported to contain less cytoplasmic neuromelanin pigment compared to these neurons in the normal brain.[6,7] Indeed, it is common to see the statement that a characteristic change in the parkinsonian SN_c is its "depigmentation."[8] In part, such "depigmentation" probably reflects the fact that when SN_c cells die their neuromelanin granules spill out into the extracellular fluid so that the overall intensity of the pigment appears to be decreased. However, the fact that remaining SN_c cells contain less pigment than normal implies that the neuromelanin-forming reaction (Fig. 14.2) is in some way attenuated or blocked, perhaps as a consequence of the pathological processes that ultimately cause the demise of these neurons. A second interesting observation is that in certain regions of the SN_c, cells that normally have very heavy neuromelanin pigmentation may be preferentially vulnerable to degeneration in PD.[6,7] Thus, a high basal rate of DA autoxidation and hence heavy neuromelanin pigmentation appears to mark SN_c cells for preferential degeneration in PD.

Mitochondrial Dysfunction in PD

Mitochondria play a crucial role in providing chemical energy for cellular metabolism. The mitochondrial (mt) respiratory chain and oxidative phosphorylation system consists of five multisubunit protein complexes (complexes I–V) and two mobile electron carriers, coenzyme Q (CoQ) or ubiquinone and cytochrome c, located on the inner mt membrane. The respiratory chain is designed to harness the redox energy of reduced nicotinamide adenine dinucleotide (NADH) and reduced

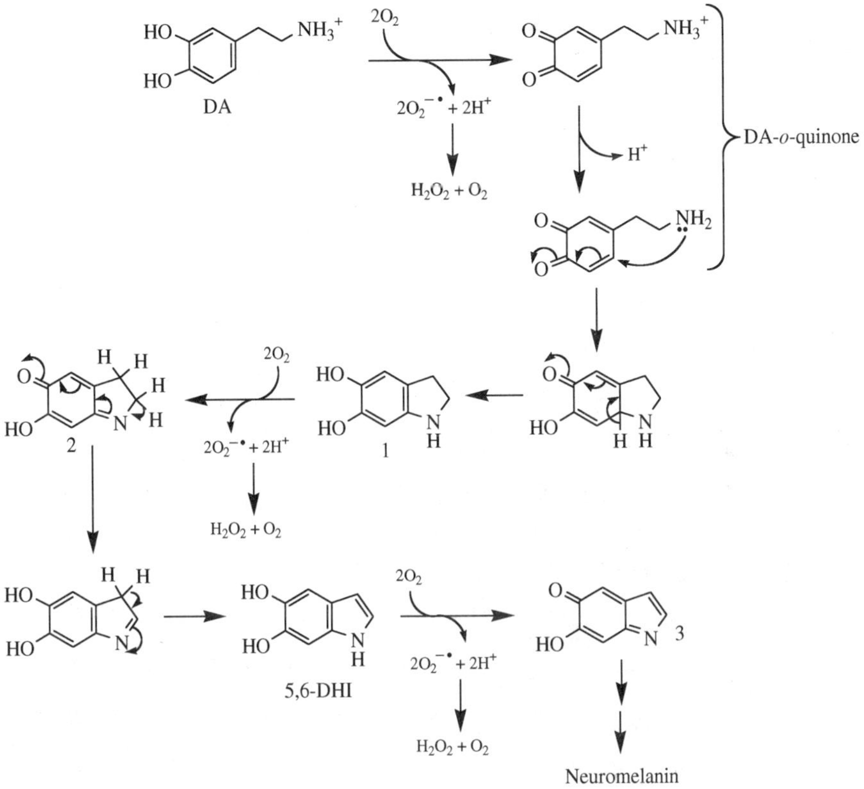

FIGURE 14.2 Autoxidation of dopamine to indolic neuromelanin polymer.

flavin adenine dinucleotide ($FADH_2$) in the form of adenosine 5′-triphosphate (ATP). Complex I, NADH–CoQ oxidoreductase, is the entry point for electrons from NADH, formed by the oxidation of substrates such as malate, pyruvate, and glutamate catalyzed by dehydrogenase enzymes (Fig. 14.3). Electrons are passed to complex III, coenzyme QH_2–cytochrome c oxidoreductase, via $CoQH_2$. $CoQH_2$ is also the entry point for electrons derived from $FADH_2$, most of which is generated via complex II, succinate–CoQ oxidoreductase. Succinate, the substrate for complex II, is produced from the oxidation of α-ketoglutarate by the α-ketoglutarate dehydrogenase complex (α-KGDH). Complex III donates electrons via the reduced form of cytochrome c (cyt c_{red}) to complex IV (cytochrome c oxidase, COX), which, in turn, donates electrons to molecular oxygen, the terminal electron acceptor, forming water. Complexes I, III, and IV also pump protons (H^+) from the inner mt matrix to the outer mt matrix, which then flow down their electrochemical gradient through ATP synthase (complex V), providing the energy to synthesize ATP from adenosine 5′-diphosphate (ADP) and inorganic phosphate (P_i) (Fig. 14.3).

Studies of homogenates of SN_c tissue from patients with PD show a specific decrease of mt complex I activity.[9,10] Furthermore, reduced immunostaining for

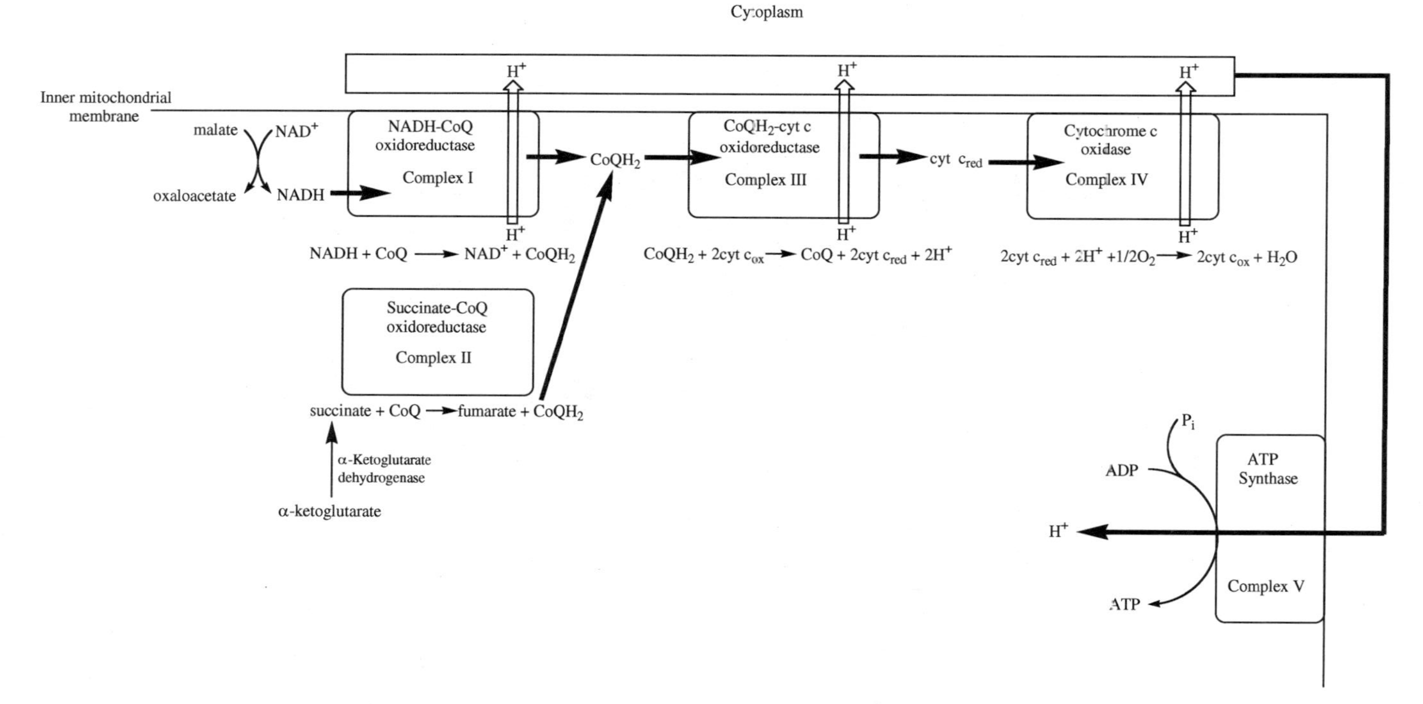

FIGURE 14.3 Schematic representation of mitochondrial energy metabolism.

complex I has been reported in remaining neuromelanin-pigmented SN_c cells in the PD brain.[11] The mt complex I deficiency in PD is not accompanied by altered complex II–IV activities, is specific to the SN_c, and is not simply a reflection of dopaminergic cell loss because it is not observed in other disorders that cause severe nigral neurodegeneration.[10] Although the activities of nonrespiratory mt enzymes have not been measured in autopsied tissue from PD patients, significantly decreased immunostaining for α-KGDH in remaining pigmented SN_c cells has been reported.[12] Inhibition of α-KGDH, which provides reducing equivalents in the form of succinate to mt complex II (Fig. 14.3), together with decreased activity of complex I in the parkinsonian SN_c, would significantly decrease ATP production in dopaminergic cells and therefore probably represent important factors in the pathogenesis of PD.

Oxidative Stress and PD

Oxidative stress refers to a condition where reactive oxygen species (ROS) such as superoxide anion radical ($O_2^{-\cdot}$), hydrogen peroxide (H_2O_2), and the highly cytotoxic hydroxyl radical ($HO^\cdot$) are formed in excess of cellular protective mechanisms resulting in tissue damage. Formation of $HO^\cdot$ depends on the presence of a transition metal ion such as Fe^{2+}, which decomposes H_2O_2 by the Fenton reaction. However, in the presence of a reducing agent, such as ascorbic acid or $O_2^{-\cdot}$, the Fe^{3+} formed in this reaction can be reduced to Fe^{2+} establishing a redox cycling (catalytic) system, known as the Haber–Weiss reaction, that dramatically elevates the rate of $HO^\cdot$ formation [Eq. (14.1)].[13] The MAO- B–mediated metabolism of DA (Fig. 14.1) and its autoxidation (Fig. 14.2) generate $O_2^{-\cdot}$ and H_2O_2 as by-products that, particularly in the presence of trace levels of free or uncomplexed iron, could lead to $HO^\cdot$ formation [Eq. (14.1)]. This raises the possibility that such processes might be responsible for SN_c cell death in PD. However, such a mechanism fails to explain why some individuals develop the disorder while most are spared. Indeed, elevated intraneuronal ROS production and oxidative stress would, in the absence of other factors, provide conditions that should increase the rate of oxidation of DA (and other catechols) leading to deposition of unusually large amounts of neuromelanin in SN_c cells prior to their death. However, as noted earlier, exactly the opposite effect occurs, namely, remaining SN_c cells contain decreased neuromelanin pigment.

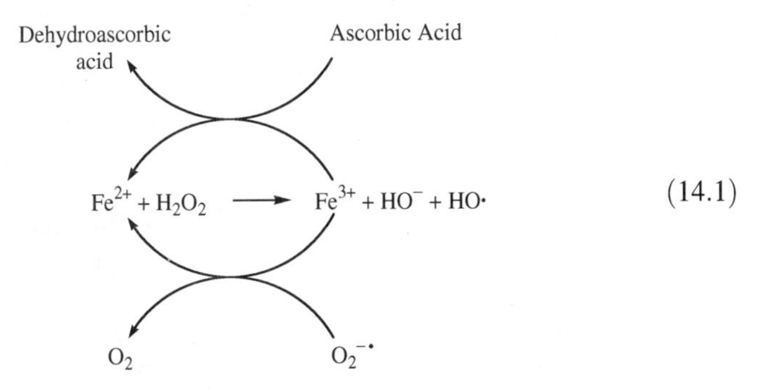

(14.1)

Nevertheless, at advanced stages of PD there is much evidence for oxidative stress in the form of elevated nigral levels of oxidation products of lipids,[14] proteins,[15] and DNA.[16] Such evidence raises the possibility that oxygen radicals may also damage the respiratory chain and inhibit the function of constituent complexes. However, in vivo studies suggest that mt complex IV, that is unaffected in PD, rather than complex I is most vulnerable to oxidative damage.[17] Furthermore, depletion of glutathione (GSH) (see later discussion) results first in a mt complex IV deficiency followed by a less pronounced decrease in complex I activity.[17] Such observations raise doubt about the role of ROS in mt complex I and α-KGDH deficiencies in the parkinsonian SN_c.

Incidental Lewy Body Disease (ILBD)

ILBD is a condition characterized by the presence of Lewy bodies and nigral degeneration, although less profound than in PD, that at post mortem appears to affect 10–15% of normal individuals over the age of 60 years.[1] This disorder is believed by many neurologists to represent an early clinically presymptomatic stage of PD,[18,19] and presumes that had these individuals lived longer, they would have developed symptoms of the disorder. However, studies of SN_c tissue from patients who died with ILBD have failed to show the presence of typical markers for oxidative stress.[15,20] Assuming that ILBD is in fact early-stage PD, these results imply that oxidative stress occurs only late in the disorder and, therefore, that it is not the cause but a (secondary) consequence of the disease. Indeed, a 1997 study indicates that at advanced stages of PD oxidative stress is not restricted to the SN_c but occurs widely in the brain,[15] perhaps as a result of the prooxidant effects of L-DOPA[21] therapy (i.e., ease of autoxidation and generation of $O_2^{-\cdot}$ and H_2O_2 as by-products).

Glutathione and PD

In addition to the mt complex I and α-KGDH deficits, another change specific to the SN_c in PD is a massive reduction (40–50%) of glutathione (GSH) compared to levels present in this brain region of age-matched control patients.[22–24] GSH is an antioxidant that serves as the major $HO^{\cdot}$ scavenger in the brain. In this reaction GSH is oxidized to glutathione disulfide (GSSG) (Fig. 14.4). Thus, the selective loss of nigral GSH in PD might again support the hypothesis that oxidative stress is a major factor in the pathogenesis of the disorder. However, arguing against this hypothesis is the fact decrements of GSH in the parkinsonian SN_c are not accompanied by corresponding increases of GSSG.[22–24] Levels of GSH in the SN_c of patients who died with ILBD are also reduced, again without corresponding increases of GSSG, but with smaller decreases of mt complex I activity than occur at advanced stages of PD. These observations, together with the fact that the activity of γ-glutamylcysteine synthetase, the rate-limiting enzyme for GSH biosynthesis, is normal in the parkinsonian SN_c,[25] imply that the loss of nigral GSH is not caused by oxidative stress, or impaired mt function and decreased production of ATP needed for its

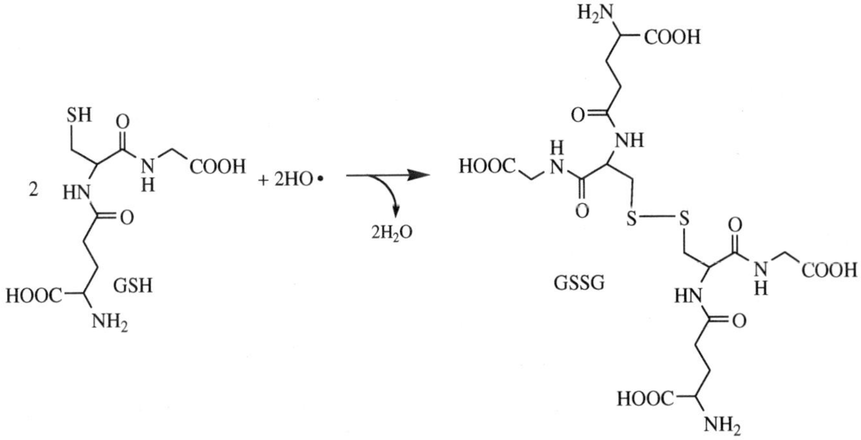

FIGURE 14.4 Oxidation of glutathione by hydroxyl radical (HO˙).

synthesis[26] (at least at early stages of the disorder), or a failure of its biosynthetic enzymes. Furthermore, they point to a connection between nigral GSH decrements and a progressive decrease of mt complex I and, perhaps, α-KGDH activity. However, experiments with experimental animals demonstrate that, by itself, depletion of brain GSH by as much as 60% or more causes neither the degeneration of nigrostriatal DA neurons[27] nor decreased activities of mt respiratory enzymes[28] although sublethal damage can occur.[29]

Using immunostaining techniques it has been demonstrated that dopaminergic SN_c cells contain cytoplasmic GSH but that much higher concentrations of the tripeptide are present in nigral glia.[30] Furthermore, remaining pigmented SN_c cells in the PD brain contain reduced concentrations of GSH.

Brain Glutathione

In vitro studies indicate that brain GSH is manufactured and stored principally in glia and that neuronal GSH is maintained by its export from glia.[31–34] However, neurons are unable to directly import GSH.[35] Rather, extracellular GSH, released from glia, is initially hydrolyzed by γ-glutamyltranspeptidase (γ-GT) to glutamate (Glu) and cysteinylglycine, which is then hydrolyzed by dipeptidases to glycine (Gly) and cysteine (CySH) (Fig. 14.5). The γ-GT-mediated step is rate-limiting in this hydrolytic sequence, which occurs exclusively extracellularly.[36] CySH is then translocated into neurons by neutral amino acid transporters[37], where it is employed for intraneuronal GSH biosynthesis. It is of potential importance, on the basis of the preceding discussion, that the activity of γ-GT is significantly increased in the parkinsonian SN_c.[25] This raises the possibility that the loss of nigral GSH in PD might in some way be related to it release from pigmented dopaminergic SN_c cells (perhaps accounting for decreased intraneuronal concentrations of the tripeptide)

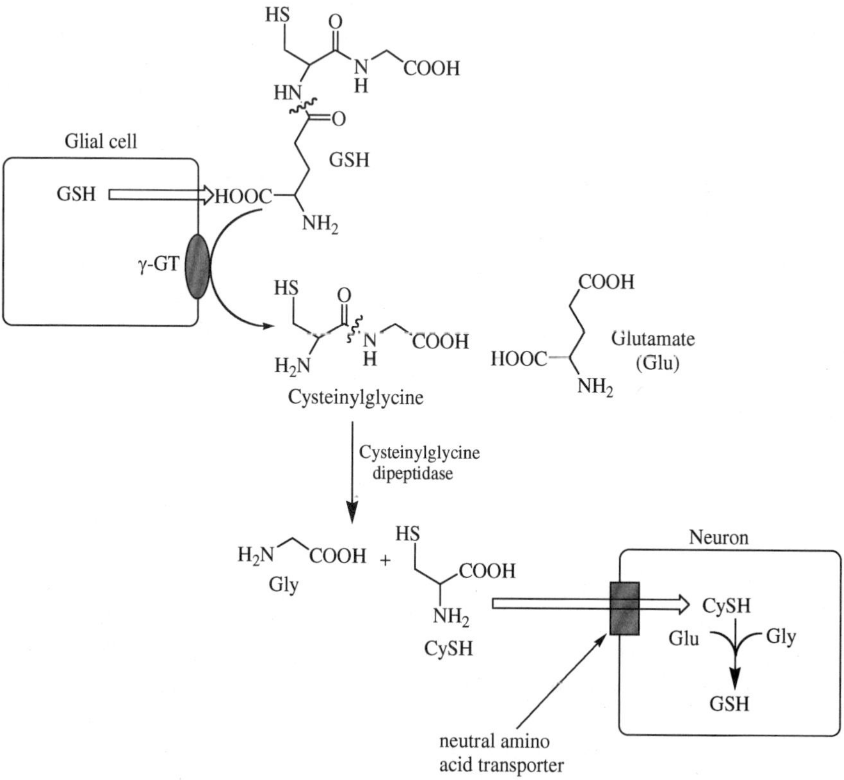

FIGURE 14.5 Maintenance of neuronal glutathione (GSH) by glia.

and surrounding glia followed by its hydrolysis by γ-GT/dipeptidases to Glu, Gly, and CySH.

Other Antioxidants and Antioxidant Enzymes in PD

Dopaminergic SN_c cells possess a weak antioxidant system.[38,39] The most obvious evidence for this is the fact that cytoplasmic DA is autoxidized to DA-*o*-quinone leading to neuromelanin pigment deposition (Fig. 14.2). Furthermore, in vitro, excess GSH blocks formation of indolic melanin polymer from DA by scavenging DA-*o*-quinone to give initially 5-*S*-glutathionyldopamine (5-*S*-Glu-DA; Fig. 14.6) and then more highly substituted glutathionyl conjugates of DA.[40] Thus, it is likely that concentrations of GSH in pigmented SN_c neurons are lower than those of cytoplasmic DA.

Levels of ascorbic acid[22] and α-tocopherol[41] are normal in the parkinsonian SN_c. The activity of nigral catalase, an enzyme that catalyzes the reduction of H_2O_2 to water, is normal in PD.[42] Similarly, the activity of GSH peroxidase is normal[25] or slightly reduced[43] whereas GSSG reductase activity is normal[25] in the parkinsonian

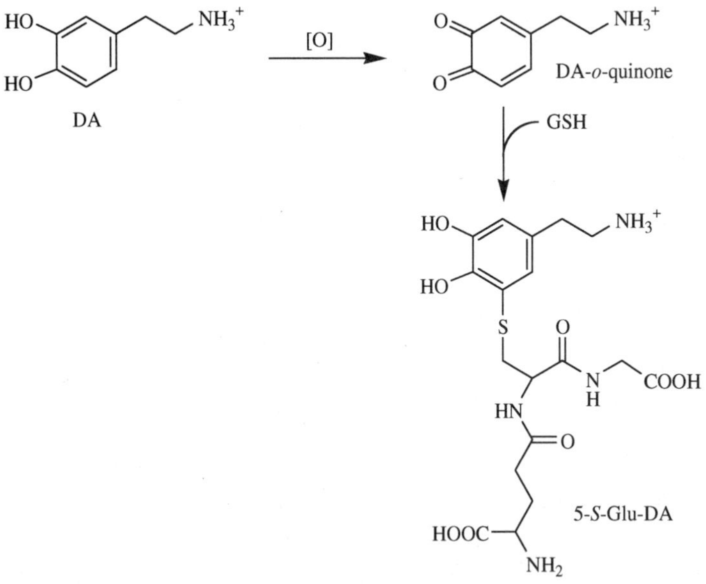

FIGURE 14.6 Initial reaction of dopamine-*o*-quinone with glutathione.

SN_c. The GSH peroxidase/GSSG reductase system is also employed to reduce H_2O_2 to water (Fig. 14.7). Thus, with the exception of GSH decrements, compromised antioxidant capacity does not appear to be related to SN_c cell pathology in PD. However, superoxide dismutase (SOD) activities are significantly increased in the parkinsonian SN_c, including the inducible manganese-dependent mt form of this enzyme.[44,45] SOD is the enzyme responsible for catalyzing the conversion of $O_2^{-\cdot}$ to H_2O_2 (Fig. 14.8). Thus, increased SOD activity in the parkinsonian SN_c implies elevated intraneuronal generation of $O_2^{-\cdot}$.

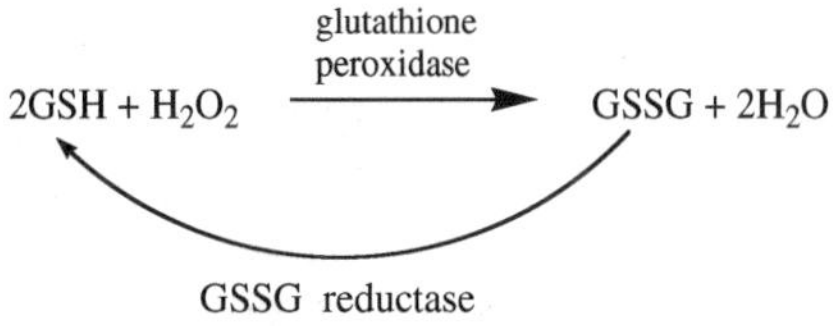

FIGURE 14.7 The glutathione peroxidase/GSSG reductase system.

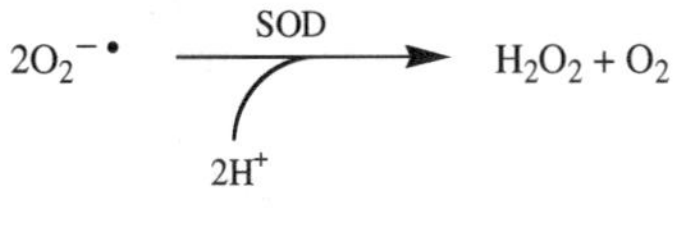

FIGURE 14.8 Superoxide dismutase (SOD)-catalyzed decomposition of superoxide anion radical ($O_2^{-\cdot}$).

Alterations in Iron in PD

Another change specific to the SN_c in PD is an increase of iron levels.[22,46] Increased iron is bound largely to neuromelanin granules[47,48] that have a high affinity for this metal.[49] Because both free or unbound iron[13] and the iron–neuromelanin complex[49] can catalyze formation of $HO^{\cdot}$ from H_2O_2 by Fenton/Haber–Weiss chemistry [Eq. (14.1)], elevated iron in the parkinsonian SN_c appears to provide support for the oxidative stress hypothesis for the pathogenesis of PD.[1] However, iron levels in the SN_c of ILBD patients are normal.[20] Thus, again assuming that ILBD represents a presymptomatic form of PD, this suggests that increased levels of nigral iron observed in the parkinsonian SN_c occur only late in the disorder and are secondary.

Although some controversy remains, the preponderance of presently available information suggests that levels of ferritin, the major iron storage protein in the brain, are not significantly altered in the parkinsonian SN_c (see Ref. 1 for a more detailed review).

The source of increased iron in the parkinsonian SN_c is unknown. However, in view of increased SOD activity in the parkinsonian SN_c (implying elevated intraneuronal $O_2^{-\cdot}$ generation), it is of interest that $O_2^{-\cdot}$ is able to release Fe^{2+} from a number of iron-containing proteins[50–53]. These observations are not inconsistent with the $O_2^{-\cdot}$-mediated mobilization of Fe^{2+} from such proteins that is subsequently scavenged and complexed by neuromelanin and that this process is in some way involved in the pathogenesis of SN_c cell death in PD.

Genetic Factors and PD

Genetic factors appear to be of major importance in early-onset cases of PD in which clinical symptoms appear at or before the age of 50 years.[54] However, no clear genetic component is evident for the more common sporadic form of PD when symptoms typically appear after age 50 years. Evidence has been provided for genetically determined mt complex I[55–57] and α-KGDH[58] defects in sporadic PD. However, even if PD patients have systemic complex I and α-KGDH deficiencies, such defects fail to explain why dopaminergic SN_c cells degenerate whereas other neurons are spared. This implies that other factors must underlie the particular vulnerability of SN_c cells in PD. Perhaps of relevance in this respect are reports that PD patients, presumably for genetic reasons, have low activity of cysteine dioxygenase (CDO).[59,60] In the liver, CDO is a major rate-limiting enzyme associated with a pathway by which CySH is converted into inorganic sulfate (SO_4^{2-}), which is employed to conjugate, detoxify, solubilize, and excrete certain xenobiotics and environmental toxicants. Such a CDO deficiency would be expected to permit such toxicants to enter the circulatory system in unusually high concentrations.[58] Furthermore, because the BBB is ineffective or even absent in the SN,[61] such bloodborne toxicants should be able to selectively enter brain parenchyma in this structure. Indeed, epidemiological studies suggest a link between exposure to high levels of agricultural, industrial, and other environmental chemicals and an increased incidence of PD.[62–66] In particular, an organochlorine pesticide, dieldrin, which inhibits mt respiration,[67,68] induces ROS and increases SOD activity,[69] depletes brain monoamines in a number of species, and is a DA neurotoxin[70] has been case-implicated with PD.[71–74] Similarly, diethyldithiocarbamate fungicides inhibit mt respiration[75] and have been case-implicated with PD.[76–77] Indeed, numerous toxins that interfere with mt respiratory chain function give rise to the degeneration of nigrostriatal DA neurons.[78]

Accelerated Oxidation of DA in the Parkinsonian SN_c

A possible connection between the loss of nigral GSH without corresponding increases of GSSG, increased γ-GT activity (implying increased expression of this enzyme and an increased rate of hydrolysis of extracellular GSH to Glu, Gly, and CySH), and selectively increased intraneuronal $O_2^{-\cdot}$ generation (implied by the rise of SOD activity) is provided by dramatically elevated concentrations of 5-*S*-cysteinyldopamine (5-*S*-CyS-DA) *relative to remaining DA* in the parkinsonian SN_c.[79,80] In the normal neuromelanin-pigmented SN_c, 5-*S*-CyS-DA is only a trace metabolite of DA. However, in the "depigmented" parkinsonian SN_c, 5-*S*-CyS-DA becomes a major metabolite, reaching concentrations 20% or more of remaining DA. This marked elevation of 5-*S*-CyS-DA, which can be formed by the $O_2^{-\cdot}$-mediated oxidation of DA in the presence of CySH[80] (Fig. 14.9), implies not only an accelerated rate of DA oxidation but also increased availability of CySH, normally present at very low concentrations in neurons. This, in turn, raises the possibility that the increased availability of CySH results from the hydrolysis of GSH by γ-GT/dipeptidases. However, in view of this accelerated rate of DA oxidation, it is of interest to recall that remaining SN_c cells in postmortem PD brains, that would presumably have degenerated had the patient lived longer, contain significantly less neuromelanin than normal. Thus, the apparently accelerated rate of DA oxidation in the parkinsonian SN_c clearly does not lead to neuromelanin formation (Fig. 14.2), but rather a new metabolic pathway appears to be established in which DA-o-quinone is scavenged by CySH to give 5-*S*-CyS-DA (Fig. 14.9).

Time Course of PD

The time from initiation of the neurodegenerative processes in PD to the first appearance of symptoms ($\geq$ 80% depletion of striatal DA) has been estimated to be approximately 5 years.[19] It is during the early stages of this clinically presymptomatic phase, when nigral concentrations of DA and GSH are high, that dopaminergic SN_c cells are lost most rapidly.[19] The long presymptomatic phase of PD implies that the degeneration of SN_c cells is probably not the result of a single massive neurotoxic insult but rather that an endogenous toxin is selectively formed in pigmented neurons, highly specific in its action, specifically, inhibition of mt complex I and α-KGDH, which is produced over a number of years and for which tolerance does not develop.

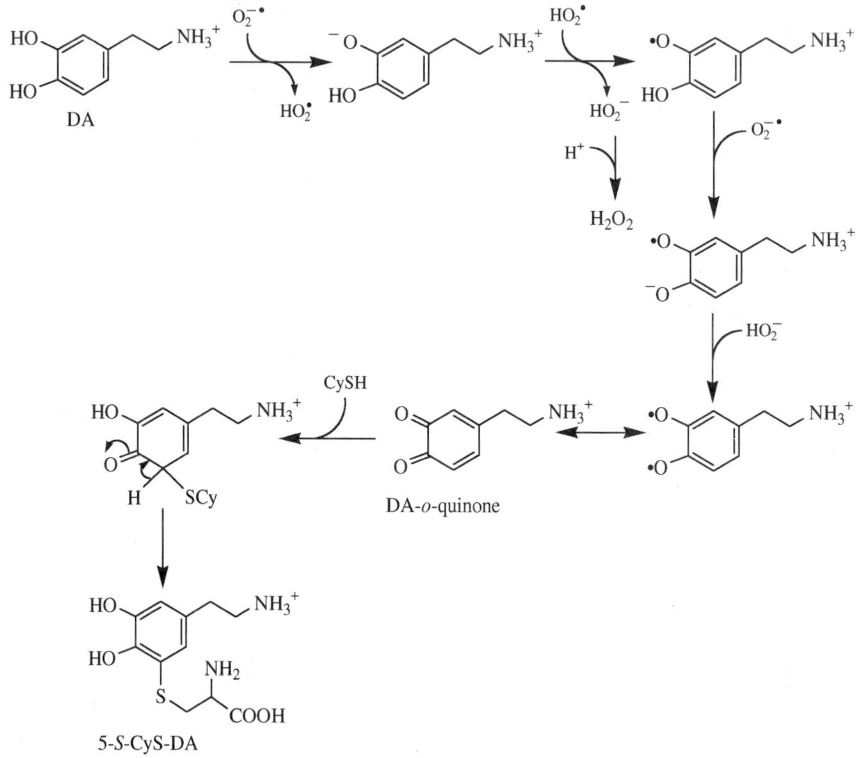

FIGURE 14.9 Superoxide ($O_2^{-\cdot}$)-mediated oxidation of dopamine in the presence of L-cysteine.

Summary

PD MODELS

Important factors that appear to be associated with the pathological mechanism underlying the degeneration of pigmented dopaminergic SN_c cells include (1) decreased concentrations of nigral GSH without corresponding increases of GSSG, possibly a very early event in SN_c cell pathogenesis; (2) compromised xenobiotic metabolism and exposure to environmental chemicals that interfere with mt respiration; (3) elevated intraneuronal $O_2^{-\cdot}$ generation; (4) apparent mobilization of unbound iron from protein stores that is scavenged by neuromelanin; and (5) an accelerated rate of DA oxidation that does not lead to neuromelanin deposition but rather to elevated concentrations of 5-*S*-CyS-DA. Each of these factors appears to represent a component of a cascade of chemical and/or biochemical processes that lead to the formation of endotoxins responsible for mt complex I and α-KGDH deficiencies, reduced ATP production, and ultimately the death of SN_c cells in PD. The early loss of nigral GSH in PD suggests that this change represents a fundamental step in the initiation and progression of the neuropathological mechanism leading to mt dysfunction and SN_c cell death. What remains unknown about PD is how this cascade is triggered and organized such that it leads to mt dysfunction, and the nature of the toxin(s) involved.

Animal Models of PD

Much of the available information bearing on PD, summarized above, is based on studies of brain tissue obtained postmortem from patients who died with the disorder. Because $\geq 80\%$ of striatal DA is lost before symptoms of PD first appear and patients may live for a decade or more longer, albeit with progressively increasing disability, the information derived from such studies reflects the terminal stage of a disorder that was initiated 20 or more years earlier. In order to gain possible insights into the earliest stages of the neurodegenerative processes in PD, it has been necessary to turn to experiments with laboratory animals using toxins that evoke the degeneration of nigrostriatal DA neurons. The two most promising neurotoxins employed in such studies are 1-methyl-4-phenyl-1,2,3,6-tetrahydropyridine (MPTP) and methamphetamine (MA). These compounds are believed to best simulate the most significant histopathological and pathobiochemical changes that occur in PD.[81]

The MPTP Model of PD

In the early 1980s, a group of young heroin abusers in northern California developed a classic parkinsonian syndrome following intravenous injection of new so-called designer drugs. These analogs of meperidine (synthetic heroin) were subsequently found to be contaminated with MPTP, the agent responsible for the rather selective degeneration of nigrostriatal DA neurons.[82] The selectivity of MPTP, which readily crosses the BBB, can be traced to its oxidation by MAO B in glial cells to 1-methyl-4-phenylpyridinium (MPP^+),[83] which is then translocated into dopaminergic neurons by the DA transporter[84] (Fig. 14.10). MPP^+ reaches concentrations of only approximately 30 μM in the cytoplasm of DA neurons but is then transported by an energy-dependent process into mitochondria, where it is concentrated to ≥ 20 mM.[85] Such high intramitochondrial concentrations of MPP^+ inhibit complex I[85] and α-KGDH,[86] causing rapid depletion of ATP.[87] Interestingly, in addition to evoking the degeneration of DA neurons, in mice MPTP/MPP^+ also causes the loss of nigrostriatal GSH without corresponding increases of GSSG.[88–90]

Important insights into the neurotoxic mechanism evoked by MPTP/MPP^+ have been obtained using a technique known as *microdialysis*, which permits measurements of extracellular concentrations of low-molecular-weight neurochemicals in the brain.[91] Using microdialysis, it has been demonstrated that perfusion of neurotoxic concentrations of MPP^+ into the rat striatum or SN_c evoke a massive instantaneous release of DA from dopaminergic neurons (Fig. 14.11*a*).[92,93] This release of DA reflects the rapid depletion of ATP as a result of the inhibition of mt complex I and α-KGDH by MPP^+, with resultant depolarization of dopaminergic neurons. However, MPP^+ is a reversible inhibitor of complex I.[78] Thus, when MPP^+

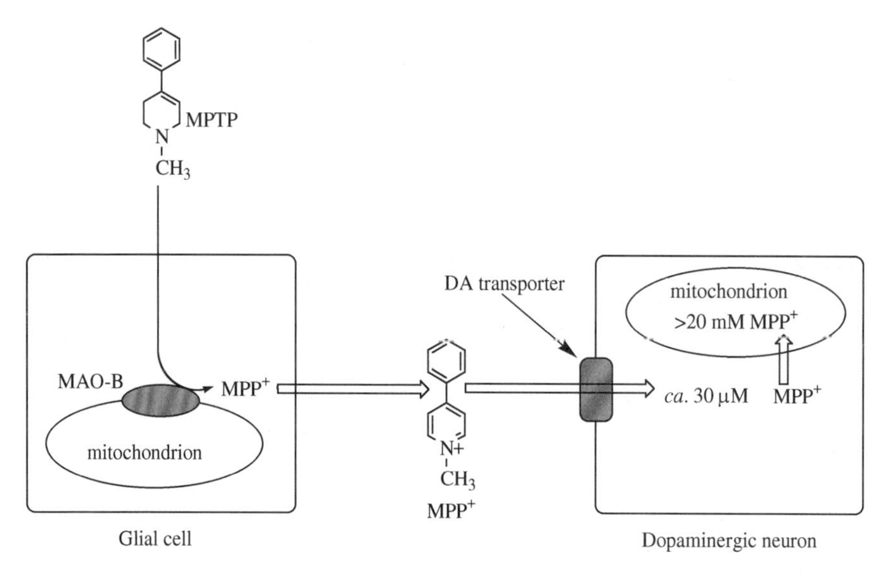

FIGURE 14.10 Metabolism of 1-methyl-4-phenyl-1,2,3,6-tetrahydropyridine (MPTP) to 1-methyl-4-phenylpyridinium (MPP^+) in glia and uptake of MPP^+ into dopaminergic neurons.

perfusion is discontinued, neuronal ATP production begins to increase, initiating reuptake of released DA so that extracellular concentrations of the neurotransmitter decrease (Fig. 14.11*a*). Interestingly, *during* perfusion of high neurotoxic concentrations of MPP^+ into the rat striatum, for example, extracellular concentrations of Glu increase only very modestly.[93] However, when such MPP^+ perfusion is discontinued and the dopaminergic neuron energy impairment begins to subside and ATP production starts to rise, extracellular concentrations of Glu exhibit a massive and prolonged increase (Fig. 14.11*b*). Because this massive elevation of extracellular Glu occurs only when perfusions are discontinued and MPP^+ is not a glutamatergic neurotoxin[94] raises important questions concerning the origin of this Glu and why it is elevated only after the dopaminergic neuron energy impairment begins to subside. The fact that MPTP/MPP^+ causes loss of GSH without corresponding increases of GSSG raises the possibility that the elevation of extracellular Glu as the energy impairment subsides might result from the γ-GT/dipeptidase-mediated hydrolysis of GSH (Fig. 14.5). This would necessarily require release of GSH from neurons and/or glia. Indeed, in vitro studies indicate that neuronal depolarization evokes the release of GSH.[95]

The Methamphetamine Model of PD

MA also evokes neurotoxic effects on nigrostriatal DA neurons that are not caused directly by the drug or its known metabolites.[96,97] Recent evidence implicates perturbations of DA neuron energy metabolism and rapid ATP depletion

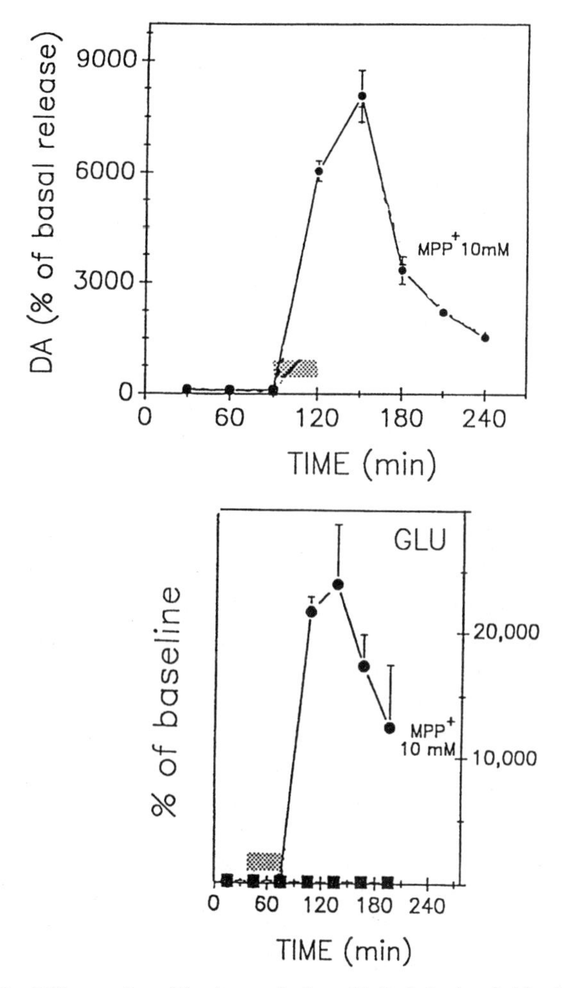

FIGURE 14.11 Effects of a 30-min perfusion (dotted bar) of 10 mM MPP^+ via a microdialysis probe into the rat striatum on the release of dopamine (DA) and glutamate (Glu). Points represent mean ± standard deviation from three to four replicate experiments. (Reprinted from Ref. 93 with permission of Elsevier Publishers, Ltd.)

with MA-induced neurotoxicity.[98,99] This energy impairment contributes to a massive transient release of DA.[100] Neurotoxic doses of MA also evoke significant decreases of striatal GSH.[101] Furthermore, MA evokes a delayed but prolonged increase of extracellular concentrations of Glu[102] that is believed to be an important component of the underlying neurotoxic mechanism.[103] However, drugs that block Glu release provide no protection against MA-induced damage to DA neurons,[104] again suggesting that the delayed elevation of extracellular Glu may result from the γ-GT/dipeptidase-mediated hydrolysis of released GSH.

PD TOXICITY AND PATHOGENESIS

Glutamate and Excitotoxicity

Before discussing other factors associated with MPTP/MPP^+ and MA neurotoxicity, it is necessary to consider the neurotransmitter Glu and the concept of excitotoxicity. Glu is an excitatory neurotransmitter that can, under certain conditions, cause damage to neurons by a process known as *excitotoxicity*. Both the excitatory and excitotoxic properties of Glu are mediated by neuronal receptors. Of particular interest in connection with the present discussion is the *N*-methyl-D-aspartate (NMDA) receptor (named for its nonphysiological agonist, *N*-methyl-D-aspartate, that most specifically activates this receptor). This is a so-called ligand-gated ion channel ionotropic Glu receptor. Under appropriate conditions, when Glu binds to the NMDA receptor, the ion channel opens and permits an influx of Na^+ or Ca^{2+} ions.

Under normal circumstances, the membrane potential of neurons is maintained at about $-70\,mV$, and the outer surface is positively charged with respect to the neuronal cytoplasm. Factors that make the cytoplasm more positive, and hence decrease the transmembrane potential, are said to depolarize the membrane. The ability to maintain the normal membrane potential depends on membrane-bound ion pumps that, in turn, are dependent on neuronal production of ATP to maintain a consistent high level of activity. These ion pumps include Na^+/K^+ ATPase and ATP-dependent Ca^{2+} pumps. The latter, for example, maintains intraneuronal concentrations of Ca^{2+} several orders of magnitude lower than extracellular concentrations. At the normal resting potential of a neuron the NMDA receptor calcium channel is blocked by an extracellular magnesium ion (Mg^{2+}).[105] With this Mg^{2+} blockade in place, the NMDA receptor is unable to transduce a signal even if stimulated by an agonist such as the excitatory amino acid (EAA) neurotransmitters Glu or aspartate (Asp). However, the Mg^{2+} blockade or gating of the NMDA receptor is voltage-dependent. Thus, when the neuron becomes depolarized (i.e., the transmembrane potential decreases) the Mg^{2+} blockade is relieved, the NMDA receptor can be activated, and Ca^{2+} can flow through the channel. Clearly, therefore, inadequate production of neuronal ATP can impair the ability of neurons to maintain normal membrane potential. Such an inability to maintain the membrane potential not only relieves the Mg^{2+} blockade of the NMDA receptor but also facilitates its activation by Glu leading to an influx of Ca^{2+}. Furthermore, the larger the neuron energy impairment, the lower the extracellular concentrations of EAAs necessary for NMDA receptor activation. A sufficiently large neuron energy impairment can essentially completely relieve the Mg^{2+} blockade of the NMDA receptor such that normal (so-called basal) extracellular concentrations of Glu and Asp can activate these receptors. A massive and prolonged influx of Ca^{2+} into neurons under such conditions can ultimately kill these cells by initiating a cascade of autodestructive processes that together are termed *excitotoxicity*[106–108] (see later discussion).

Other Factors Associated with MPTP/MPP$^+$ and MA Neurotoxicity

As noted previously, both MPP$^+$ and MA evoke a delayed, prolonged, and massive elevation of extracellular concentrations of Glu. Such an effect, particularly under conditions of reduced dopaminergic neuron ATP production evoked by MPP$^+$[87] and MA,[98] should result in excessive activation of NMDA receptors and excitotoxicity,[109] a process known to involve intraneuronal generation of $O_2^{-\cdot}$.[110] This is of interest because $O_2^{-\cdot}$ is a key species associated with damage to DA neurons evoked by MPTP/MPP$^+$ and MA, based on the observation that transgenic mice that overexpress cytoplasmic Cu/Zn SOD are resistant to the neurotoxic effects of these drugs,[111,112] whereas SOD inhibitors exacerbate such damage.[113] MPTP/MPP$^+$[114] and MA[115] also induce increased activity of striatal SOD, implying an adaptive response to elevated intraneuronal $O_2^{-\cdot}$ generation. Furthermore, NMDA receptor antagonists attenuate MPTP/MPP$^+$-induced neuronal damage,[116,117] although controversy exists,[94] as well as that evoked by MA.[118,119] An interesting property of $O_2^{-\cdot}$ is its ability to release Fe^{2+} from a number of iron-containing proteins.[50–53] Furthermore, the neurotoxicity evoked by MPTP/MPP$^+$[120] and MA[121] involves free (uncomplexed) iron because iron chelators are neuroprotective. Elevation of extracellular Glu by MPTP/MPP$^+$ or MA under conditions of neuron energy impairment with resultant NMDA receptor activation and then intraneuronal $O_2^{-\cdot}$ generation and release of Fe^{2+} from protein stores in the presence of ascorbate, implicated in both MPTP/MPP$^+$[94,122] and MA[123] neurotoxicity, and extracellular H_2O_2,[124] together provide ideal conditions for $HO^\cdot$ formation [Eq. (14.1)]. Indeed, MPTP/MPP$^+$[125] and MA[126] mediate extracellular $HO^\cdot$ formation and $HO^\cdot$ scavengers are neuroprotective.[127,128]

DA is also a key participant in the mechanism underlying the neurotoxicity evoked by MPTP/MPP$^+$ and MA because its synthesis and uptake inhibitors are neuroprotective.[129,130] Furthermore, while Glu evokes excitotoxicity by mechanisms primarily mediated by NMDA receptor activation, emerging evidence suggests that elevation of extracellular concentrations of this EAA alone, at least to levels that could occur in vivo, may not be sufficient to cause neuronal damage.[94,131] It may be of relevance, therefore, that both the uptake and intraneuronal oxidation of DA present in serum, normally a constituent of the media employed to culture neurons in vitro, are essential steps associated with Glu oxidative toxicity and, perhaps, NMDA receptor-mediated excitotoxicity.[132] Furthermore, neurons cultured in serum exhibit excitotoxic damage when exposed to Glu but fail to respond to high concentrations of this EAA when cultured under serum-free conditions.[133] This again raises the possibility that serum-derived DA may play a role in NMDA receptor-mediated excitotoxicity.

Although the ultimate causes of nigrostriatal DA neurodegeneration evoked by MPTP/MPP$^+$ and MA are not known, the underlying neurotoxic mechanisms clearly involve many factors implicated in the pathogenesis of dopaminergic SN_c cell death in PD. These include (1) loss of GSH without corresponding increases of GSSG; (2) elevated intraneuronal $O_2^{-\cdot}$ generation that, in turn, suggests NMDA receptor activation by increased extracellular concentrations of Glu under conditions of neuron energy impairment; (3) mobilization of iron and extracellular $HO^{\cdot}$ formation; and (4) a key role for DA.

A Chemical Hypothesis for the Pathogenesis of PD

The preceding lines of evidence drawn from postmortem studies of brains from both PD and ILBD patients and factors implicated with the dopaminergic neurotoxicity of MPTP/MPP^+ and MA in animals, have led us to propose a new hypothesis for the underlying neurotoxic mechanism, presented in schematic form in Fig. 14.12. The essential first step in this mechanism is a large but transient impairment of DA neuron energy metabolism. This could be caused, for example, by exposure to an inhibitor of mt respiration such as MPP^+ or indeed many other compounds.[78] In the case of MA, doses should be sufficient to induce a profound neuron energy impairment by a combination of metabolic perturbations, exchange with DA such that its transporters operate in a highly activated ATP-consuming state, disruption of ion gradients and severe hyperthermia, an important factor associated with the neurotoxic mechanism.[134] The normal decline of neuron energy metabolism with aging,[135] a major risk factor for sporadic PD, might be a predisposing factor but, by itself, fails to explain the selective vulnerability of dopaminergic SN_c cells to degeneration in this disorder. However, exposure to environmental toxicants/xenobiotics[62–77] that interfere with mt respiratory enzymes, such as certain pesticides and fungicides discussed previously, particularly in view of defective xenobiotic metabolism[59,60] and an ineffective BBB[61] in the SN_c, superimposed on the effects of aging and systemic mt complex I[55–57] and α-KGDH[58] defects could represent a combination of factors that mediate a profound selective dopaminergic neuron energy impairment. Regardless of cause, the importance of the transient DA neuron energy impairment is twofold. First, the energy impairment must be sufficiently large to profoundly depolarize dopaminergic neurons evoking not only a massive release of DA but also, equally important, GSH.[95] It will subsequently be demonstrated that intraneuronal GSH would be expected to scavenge and detoxify putative intraneuronal toxicants, thus necessitating its depolarization-mediated release. The second important role of the transient energy impairment is that when neurons are depolarized, the Mg^{2+} blockade of NMDA receptors is relieved so that they can be activated by extracellular EAAs such as Glu and Asp. The larger the energy impairment, the lower the concentration of these EAAs necessary for NMDA receptor activation until, ultimately, even basal extracellular levels of Glu and Asp can activate these receptors[136] and mediate intraneuronal $O_2^{-\cdot}$ generation.[110]

During the profound DA neuron energy impairment, intraneuronal $O_2^{-\cdot}$ generation, mediated by NMDA receptor activation by basal levels of Glu and Asp, in excess of the scavenging capacity of SOD would be expected to release Fe^{2+} from iron-containing proteins, some of which is extracellular. Released Fe^{2+} in the presence of extracellular ascorbate catalytically decomposes H_2O_2 to $HO^{\cdot}$ [Eq. (14.1)]. Being the major $HO^{\cdot}$ scavenger in the brain, extracellular GSH (basal plus that released from DA neurons during the energy impairment) should therefore be

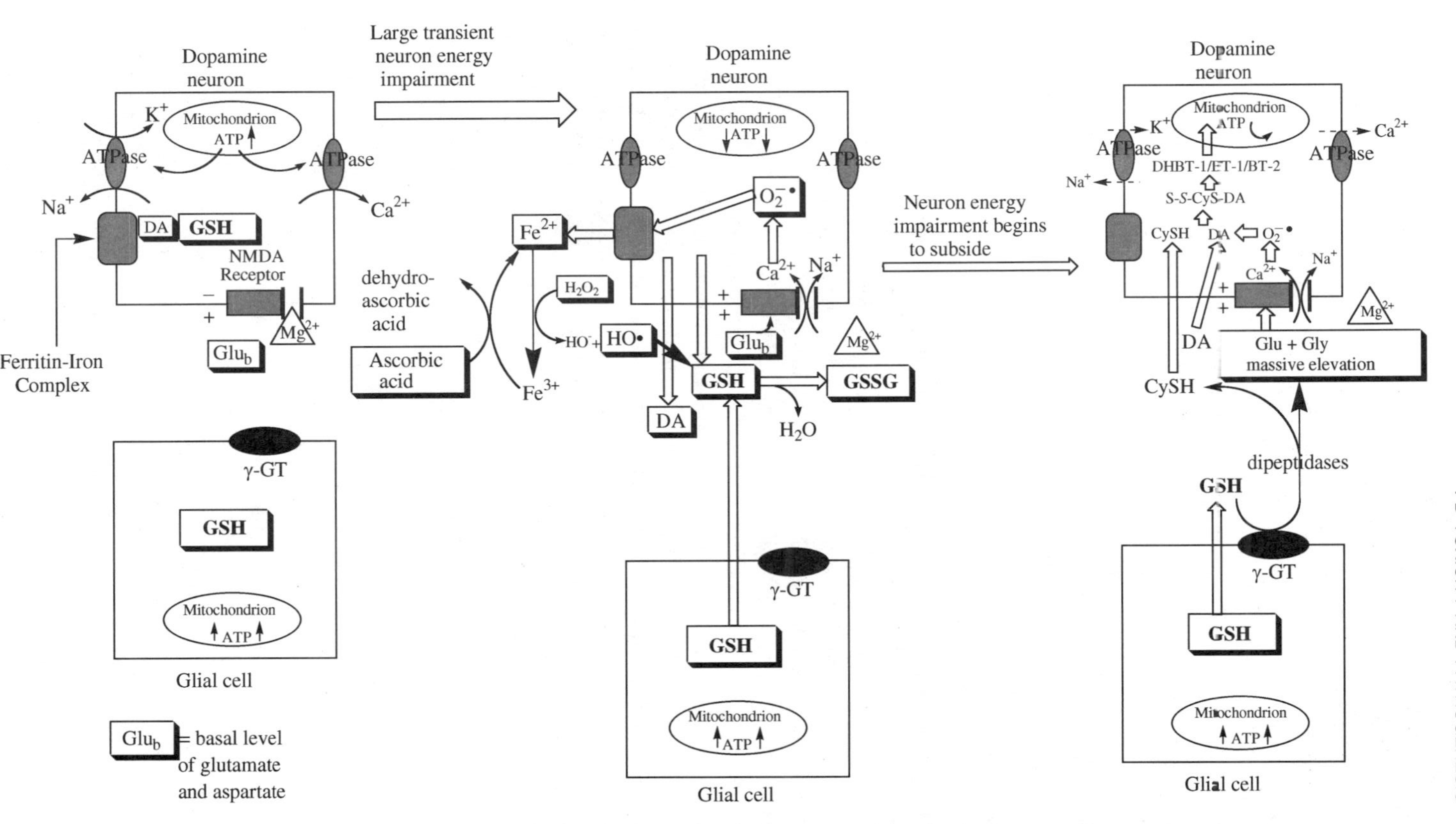

FIGURE 14.12 Cascade of biochemical and chemical processes hypothesized to contribute to the degeneration of dopaminergic neurons. (Reprinted from Ref. 140 with permission from the *Journal of Neurochemistry.*)

immediately oxidized to GSSG.[137] This, in turn, should trigger the release of GSH from glia to replenish extracellular GSH as a mechanism to protect neighboring neurons against direct damage by highly cytotoxic $HO^{\cdot}$.[138,139]

When the neuron energy impairment begins to subside ATP stores would start to increase, initiating DA reuptake by its transporter. Increasing neuronal ATP production should also begin to reinstate the Mg^{2+} blockade of NMDA receptors, thus decreasing intraneuronal $O_2^{-\cdot}$ generation, Fe^{2+} mobilization, and extracellular $HO^{\cdot}$ formation. More importantly, however, is the fact that DA is oxidized by $O_2^{-\cdot}$ (Fig. 14.9).[80] Thus, as DA reuptake commences it should be oxidized by and hence scavenge intraneuronal $O_2^{-\cdot}$ and completely block Fe^{2+} mobilization from storage proteins and extracellular $HO^{\cdot}$ formation. As a consequence, the $HO^{\cdot}$-mediated oxidation of extracellular GSH to GSSG should cease. However, in order to replenish intraneuronal GSH would require, as discussed in connection with Figure 14.5, continued release of the tripeptide from glia and its hydrolysis by γ-GT/dipeptides to provide the CySH that can be translocated into neurons and utilized for GSH biosynthesis. This γ-GT/dipeptidase hydrolysis of released glial GSH, however, forms not only CySH but also Glu and Gly (Fig. 14.5). It might therefore be this hydrolysis of released glial GSH that accounts for the massive elevation of extracellular Glu concentrations as a transient DA neuron energy impairment evoked by neurotoxic concentrations of MPP^+ (Fig. 14.11*b*)[93] and MA[102] subsides.

We have begun to explore the ideas just presented using microdialysis to measure extracellular concentrations of GSH and CySH in the rat striatum and SN_c before, during, and after perfusions of neurotoxic concentrations of MPP^+. Typical results are presented in Fig. 14.13.[140] Normal (basal) extracellular concentrations of GSH and CySH in rat brain are approximately equimolar at 2 μM.[140,141] During the 30-min period when MPP^+ enters the rat striatum or SN_c, extracellular GSH and CySH remain at basal levels (Fig. 14.13). Because rapid depletion of neuronal ATP evoked by MPP^+ causes an instantaneous and massive release of DA (Fig. 14.11*a*), it would be expected that a similar depolarization-mediated release of neuronal GSH should also occur.[95] The fact that extracellular GSH concentrations do not change during the large MPP^+-induced DA neuron energy impairment is consistent with the conclusion that released neuronal GSH is oxidized to GSSG by extracellular $HO^{\cdot}$, generated by the pathway conceptualized in Fig. 14.12. This, we propose, triggers additional release of GSH from surrounding glia in amounts sufficient to scavenge excess $HO^{\cdot}$ and maintain extracellular GSH at basal levels. This interpretation is supported by experiments in which Fe^{2+} was perfused into the rat striatum as a method to directly generate extracellular $HO^{\cdot}$ [Eq. (14.1)].[142] As with MPP^+, during perfusion of Fe^{2+} ($HO^{\cdot}$) into rat brain extracellular GSH remains at basal levels. Because GSH is very easily oxidized by $HO^{\cdot}$ (Fig. 14.4),[137] this observation implies that during Fe^{2+} perfusion the tripeptide must be released from a brain compartment in quantities sufficient to scavenge $HO^{\cdot}$ and maintain basal extracellular concentrations. In view of the fact that DA or another neurotransmitter, 5-hydroxytryptamine (5-HT), are not released during Fe^{2+} ($HO^{\cdot}$) perfusions into rat brain implies that GSH is probably released only from glia. The observation that extracellular concentrations of CySH (Fig. 14.13) and Glu (Fig. 14.11*b*) remain at or

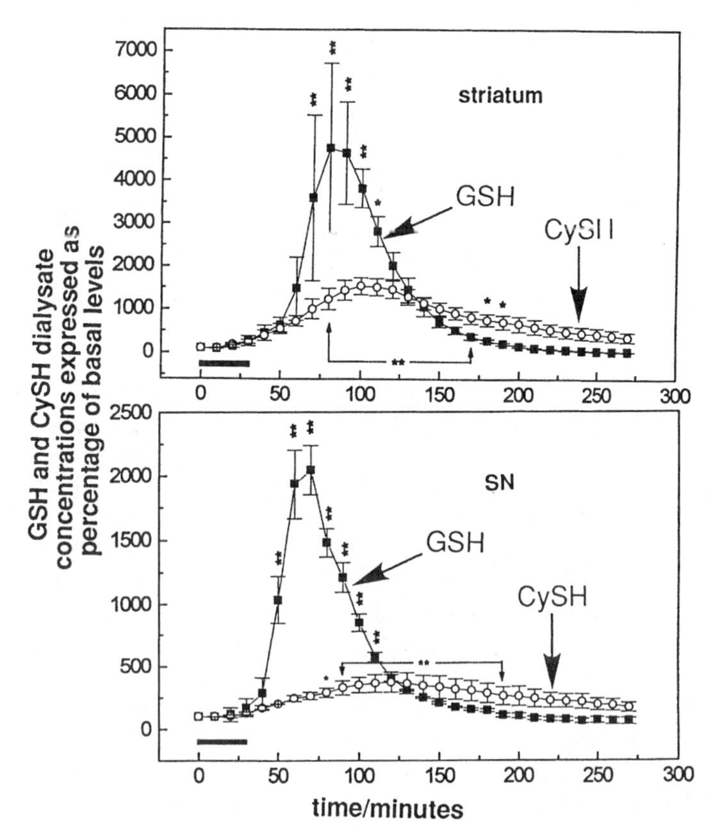

FIGURE 14.13 Time-dependent effects of a 30-min perfusion (horizontal black bar) of MPP^+ (2.5 mM) into the rat striatum and substantia nigra (SN) on microdialysate concentrations of GSH and CySH. Data are mean ± SEM (vertical bars) percentages of basal GSH and CySH levels. (Reprinted from Ref. 140 with permission from the *Journal of Neurochemistry.*)

close to basal levels during a large MPP^+-induced neuron energy impairment further supports the conclusion that GSH, released first from neurons and then from glia, is principally oxidized by $HO^\cdot$ to GSSG and is not therefore available for hydrolysis by γ-GT/dipeptidases.

When MPP^+ perfusions are discontinued and the neuron energy impairment begins to subside, DA reuptake is initiated (Fig. 14.11*a*). The hypothesis we propose suggests that as DA returns via its transporter to its parent neurons, it is oxidized and scavenges intraneuronal $O_2^{-\cdot}$ and therefore blocks the release of Fe^{2+} from iron-containing proteins and extracellular $HO^\cdot$ formation and hence oxidation of extracellular GSH. Indeed, microdialysis experiments demonstrate that as extracellular levels of DA begin to decline (because of its reuptake) (Fig. 14.11*a*), extracellular levels of GSH begin to increase massively (Fig. 14.13). However, after reaching peak extracellular concentrations GSH rather rapidly declines back to basal levels. After discontinuing MPP^+ perfusions, extracellular concentrations of CySH (Fig. 14.13) and Glu (Fig. 14.11*b*) also increase. In the case of CySH, it is clear that extracellular levels increase more slowly and peak later than those of GSH before declining toward basal levels. Such observations suggest that as the MPP^+-induced neuron energy impairment subsides, elevation of extracellular CySH and Glu concentrations result from the γ-GT/dipeptidase-mediated hydrolysis of GSH released from glia. This has been confirmed by microdialysis experiments in which γ-GT was inhibited by acivicin[143] prior to perfusion of MPP^+ into the rat striatum and SN_c. Under such conditions, when perfusions of MPP^+ are discontinued, extracellular concentrations of GSH increase massively but then decline only very slowly whereas extracellular CySH remains at very low concentrations.[140] Thus, the increase of extracellular concentrations of CySH and therefore Glu (and presumably Gly) that occur as an MPP^+-induced DA neuron energy impairment subsides appears to reflect the γ-GT/dipeptidase-mediated hydrolysis of GSH released from glia. It is significant, however, that the total release of GSH during this period is significantly greater than accumulated extracellular CySH (Fig. 14.13). This, according to the hypothetical scheme presented in Fig. 14.12, may reflect the rapid import of extracellular CySH into DA neurons as the first step in a process designed normally to replenish intraneuronal GSH. During the period of recovering but still reduced neuron ATP production, the Mg^{2+} blockade of NMDA receptors should remain partially relieved. Under these circumstances, massively elevated extracellular levels of Glu (formed by hydrolysis of released glial GSH) should continue NMDA receptor activation and intraneuronal $O_2^{-\cdot}$ generation far in excess of the scavenging capacity of SOD and therefore oxidation of DA in the presence of CySH as they enter dopaminergic neurons. The $O_2^{-\cdot}$-mediated oxidation of DA forms DA-*o*-quinone that in the presence of CySH, a thiol resistant to oxidation by $O_2^{-\cdot}$, generates initially 5-*S*-CySDA (Fig. 14.9),[80] the conjugate found in elevated concentrations in the parkinsonian SN_c[79,80] and in the rat striatum following neurotoxic doses of MA.[144] Thus, as the large transient DA neuron energy impairment subsides, translocated CySH (formed by hydrolysis of released glial GSH) would not be available for intraneuronal GSH synthesis but rather would be scavenged by DA-*o*-quinone forming 5-*S*-CyS-DA.

Possible $O_2^{-\cdot}$-Mediated Intraneuronal Oxidation Chemistry of 5-*S*-CyS-DA

Although 5-*S*-CyS-DA is the initial in vitro product of the $O_2^{-\cdot}$-mediated oxidation of DA in the presence of CySH, this conjugate is further oxidized.[80] Indeed, in vitro studies reveal that 5-*S*-CyS-DA is appreciably more easily oxidized than DA.[145–147] The initial step in this reaction is the oxidation of 5-*S*-CyS-DA to *o*-quinone **4** (Fig. 14.14). Depending on conditions, the subsequent reactions of **4** can be extremely complex. However, a major pathway involves the intramolecular cyclization (condensation) of *o*-quinone **4** to *o*-quinone imine **5**. This unstable intermediate can directly oxidize 5-*S*-CyS-DA to give radical **6** that disproportionates to 5-*S*-CyS-DA and *o*-quinone **4**, and radical **7** that disproportionates to the dihydrobenzothiazine

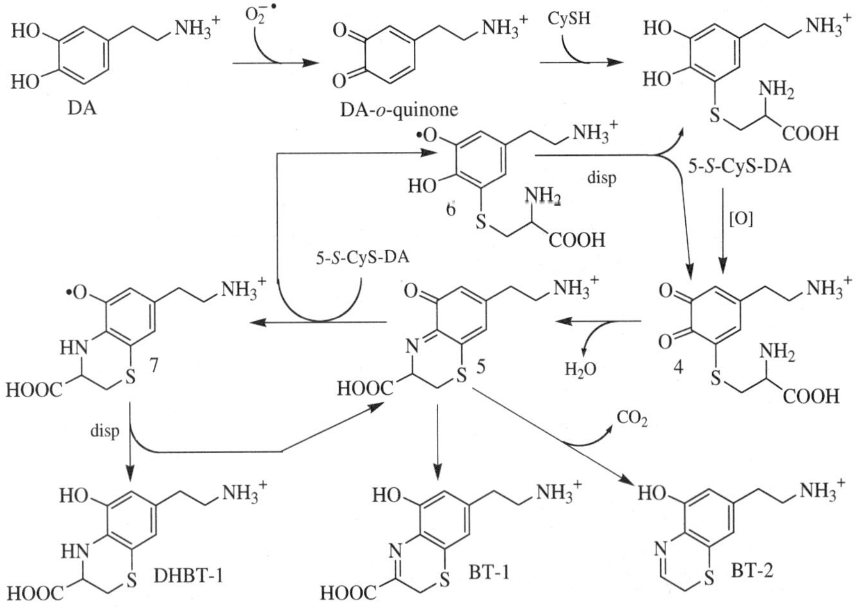

FIGURE 14.14 Oxidation of DA in the presence of CySH in buffered aqueous solution (pH 7.4) leading to 5-*S*-CySH-DA and then dihydrobenzothiazine and benzothiazine metabolites.

(DHBT) 7-(2-aminoethyl)-3,4-dihydro-5-hydroxy-2*H*-1,4-benzothiazine-3-carboxylic acid (DHBT-1) and *o*-quinone imine **5** (Fig. 14.14).[148] *ortho*-Quinone imine **5** can also rapidly rearrange (tautomerize) to the benzothiazine BT-1 and decarboxylate to BT-2. Whether, in fact, this oxidation chemistry leading from DA to 5-*S*-CyS-DA and then DHBT-1, BT-1, and BT-2 actually occurs in the cytoplasm of pigmented dopaminergic SN_c cells in PD or of nigrostriatal DA neurons as the energy impairment evoked by MPP^+ or MA subsides remains to be established. Nevertheless, the sequence of events summarized in Figure 14.12, namely, massive release and γ-GT/dipeptidase-mediated hydrolysis of GSH and subsequent reaction of translocated CySH with DA-*o*-quinone, provides one plausible explanation for the loss of nigral GSH in PD and of nigrostriatal GSH evoked by MPTP/MPP^+ and MA.

Possible Roles of 5-*S*-CyS-DA, DHBT-1, BT-1, and BT-2 in the Pathogenesis of PD

Elevation of 5-*S*-CyS-DA in the parkinsonian SN_c and the ease of oxidation of this conjugate to DHBT-1, BT-1, and BT-2 raises the possibility that these compounds might participate in the pathogenesis of PD. While DHBT-1 evokes a profound neurobehavioral response when injected into the rat or mouse brain,[145–147] neither this putative metabolite nor 5-*S*-CyS-DA causes the degeneration of nigrostriatal DA neurons (unpublished results). However, recalling that 5-*S*-CyS-DA and its DHBT/BT oxidative metabolites should be formed intraneuronally, it is of considerable significance that when incubated in vitro with intact rat brain mitochondria for 5 min, DHBT-1 evokes a dose-dependent inhibition of malate/pyruvate-supported complex I respiration (oxygen consumption) with an IC_{50} of approximately 0.8 mM (Fig. 14.15).[149] Addition of succinate restores oxygen consumption, indicating that DHBT-1 does not interfere with complex II respiration. When incubated with freeze-thawed rat brain mitochondria (mt membranes), DHBT-1 evokes a slower, time-dependent inhibition of NADH–CoQ_1 reductase (Fig. 14.16). Furthermore, when DHBT-1 is incubated with mt membranes for 60 min and then the membranes are washed (to remove DHBT-1), the activity of NADH–CoQ_1 reductase remains reduced.[149] These results indicate that DHBT-1 evokes a time-dependent irreversible inhibition of complex I.

The time dependence of the irreversible inhibition of NADH–CoQ_1 reductase by DHBT-1 is related to the oxidation of this putative intraneuronal metabolite in a reaction catalyzed by a presently unknown constituent of the inner mt membrane.[150] In this reaction, DHBT-1 is initially oxidized to *o*-quinone imine **5** that, as noted previously, rapidly rearranges to BT-1 and BT-2 (Fig. 14.17). However, BT-1 and BT-2 also evoke a time-dependent irreversible inhibition of NADH–CoQ_1 reductase when incubated with rat brain mt membranes.[150] This time dependence is also related to the oxidation of BT-1 and BT-2 in a reaction catalyzed by mt membranes. However, while HPLC analysis throughout the course of these reactions reveal the time-dependent decrease of BT-1 and BT-2 concentrations, no low-molecular-weight (LMW) metabolites have yet been detected. These observations suggest that BT-1 and BT-2 are oxidized to intermediates that covalently bind to mt membrane proteins. Support for this suggestion derives from the fact

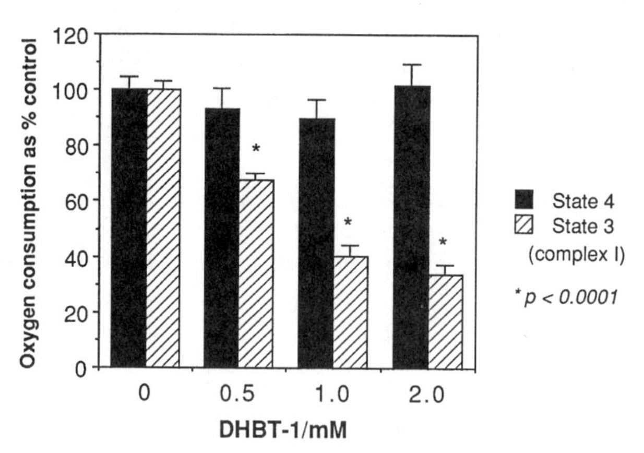

FIGURE 14.15 Effects of DHBT-1 on state 4 and state 3 (complex I) oxygen consumption (respiration). DHBT-1 was incubated for 5 min at 27 °C with intact rat brain mitochondria prior to addition of malate (2.5 mM) and pyruvate (2.5 mM). State 4 oxygen consumption was then measured (2 min), and then adenosine 5′-diphosphate was added to stimulate state 3 (complex I) respiration.

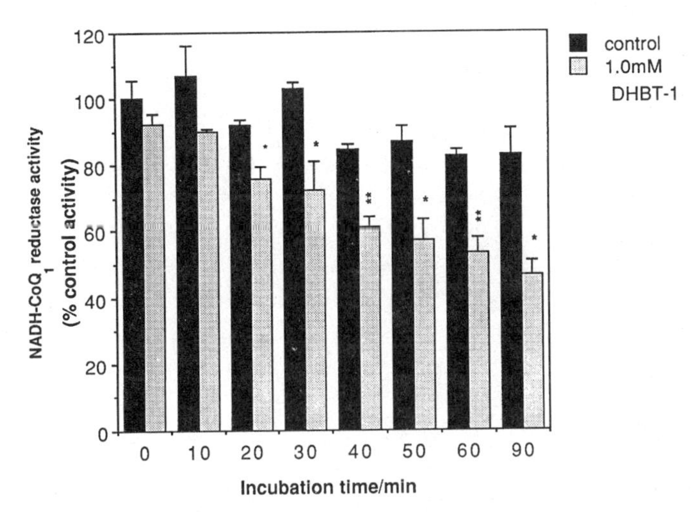

FIGURE 14.16 Time-dependent inhibition of NADH-CoQ1 reductase by DHBT-1 (1 mM) when incubated with freeze–thawed rat brain mitochondria (mt membranes) in 20 mM potassium phosphate buffer (pH 8.0).

that > equimolar concentrations of GSH completely block the irreversible inhibition of NADH–CoQ_1 reductase by DHBT-1 and BT-1/BT-2 without affecting the rate of their mt membrane-catalyzed oxidations. Furthermore, following incubations of DHBT-1 or BT-1/BT-2 with mt membranes in the presence of GSH, new metabolites are formed that include epimers **10** and **11**, 2-*S*-glutathionyl conjugates of BT-1 (Fig. 14.17). According to such lines of evidence, it appears that a component of the inner mt membrane catalyzes the oxidation of BT-1 and BT-2 to the highly electrophilic intermediates **8** and **9**, respectively, that covalently bind to key sulfhydryl (–SH or cysteinyl) residues at or close to the active site of complex I, thus evoking irreversible inhibition of NADH–CoQ_1 reductase.[150]

DHBT-1 and BT-1/BT-2 also evoke a time-dependent inhibition of the α-KGDH complex when incubated with rat brain mt membranes but have no effect on the activity of cytochrome c oxidase (complex IV) activity.[151] Again, the time dependence of the inhibition of α-KGDH is related to the oxidation of DHBT-1 to BT-1 and BT-2 that are further oxidized to **8** and **9**, respectively, which probably covalently bind to –SH residues at the active site of this enzyme complex (Fig. 14.17). Support for this suggestion is provided by the fact that inhibition of α-KGDH by DHBT-1 and BT-1/BT-2 is blocked by GSH with resultant formation of metabolites that include epimers **10** and **11** (Fig. 14.17). Indeed, it is the ability of GSH to scavenge putative electrophilic intermediates **8** and **9**, formed by the mt membrane-catalyzed oxidation of DHBT-1, BT-1/BT-2, that necessitates the massive release of the tripeptide during the initial neuron energy impairment in order to ensure that these metabolites can subsequently inhibit complex I and α-KGDH.

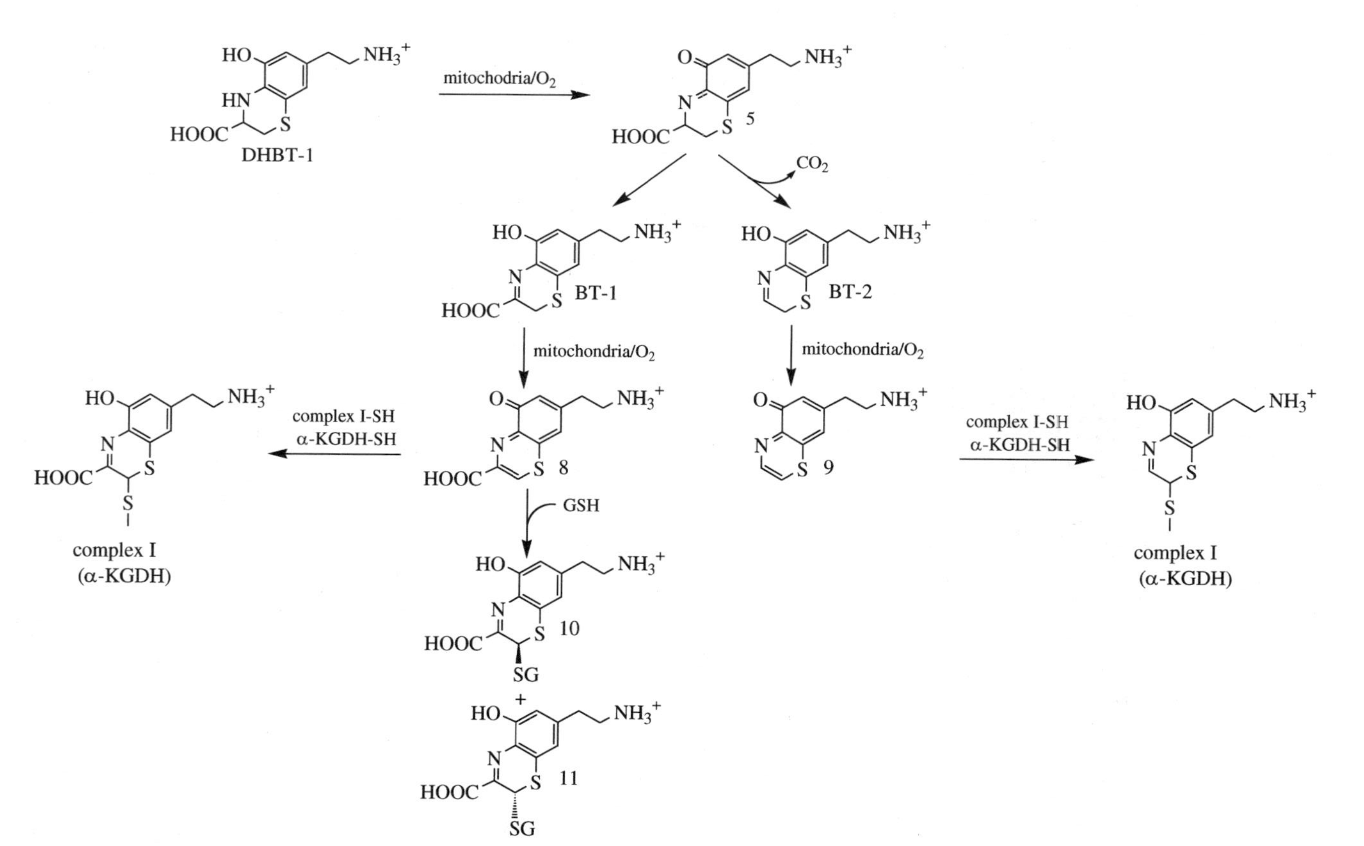

FIGURE 14.17 Proposed mechanism for the oxidations of DHBT-1, BT-1, and BT-2 catalyzed by rat brain mitochondrial membranes with resultant inhibition of complex I and α-KGDH.

CONCLUSIONS

It is important to emphasize that it remains to be experimentally established whether the degeneration of pigmented dopaminergic SN_c cells in PD results from the cascade of chemical and biochemical processes outlined in Figure 14.12 and the intraneuronal chemistry described in Figures 14.14 and 14.17. Nevertheless, the loss of nigral GSH without increased GSSG and increased concentrations of 5-*S*-CyS-DA in the parkinsonian SN_c are both consistent with the hypothetical mechanism proposed. Furthermore, the ease of oxidation of 5-*S*-CyS-DA suggests that under intraneuronal conditions where the oxidation of DA is accelerated, this conjugate would be expected to undergo oxidation to DHBT-1, BT-1 and BT-2, putative metabolites that in vitro are selective irreversible inhibitors of mt complex I and α-KGDH. There is presently no evidence that the latter compounds are formed in the parkinsonian SN_c or in the brains of humans, other primates, or rodents exposed to the nigrostriatal dopaminergic neurotoxins MPTP or MA. However, the fact that DHBT-1 and BT-1/BT-2 can be accumulated by brain mitochondria and oxidized to electrophilic intermediates that appear to covalently bind to mt proteins suggests that efforts to detect these putative metabolites in the free (unbound) state are likely to be difficult or unsuccessful.

Dopaminergic neuron energy impairment evoked by typical neurotoxic doses of MPTP/MPP^+ or MA employed in experiments with laboratory animals are probably so large that they evoke a rapid and severe depletion of intraneuronal GSH together with a massive release of DA. The subsequent reuptake of DA and CySH and massive intraneuronal $O_2^{-\cdot}$ generation, mediated by NMDA receptor activation by elevated extracellular Glu, would, according to the hypothesized mechanism, lead to formation of DHBT/BT mt toxicants in such high concentrations that neuron death would be rapid. However, the slow progression of dopaminergic SN_c cell death in PD implies that many smaller SN_c cell energy impairments over a number of years occur rather than a single massive neurotoxic insult. However, the initial neuron energy impairment necessary to ultimately lead to the demise of SN_c cells must, according to the hypothesis, be sufficiently large to evoke release of DA and GSH and permit NMDA receptor activation by endogenous extracellular levels of EAAs with resultant intraneuronal $O_2^{-\cdot}$ generation in excess of the scavenging capacity of SOD. Because putative DHBT/BT endotoxins irreversibly inhibit mt complex I and α-KGDH, the initial neuron energy impairment would cause a small but permanent decrease of SN_c cell energy metabolism. Subsequent neuron energy impairments triggered, for example, by periodic exposure to certain environmental toxicants, would progressively further permanently decrease SN_c energy metabolism. Such a permanent and growing bioenergetic defect would, in turn, permit the hypothesized neurotoxic mechanism to be triggered by quite minor environmental exposures or even, ultimately, by the progressive age-dependent decline of neuron energy metabolism such that SN_c cell death would occur.

Numerous other lines of evidence appear to provide some additional, albeit indirect, support for the pathological cascade outlined in Figure 14.12. For example, both intraneuronal $O_2^{-\cdot}$ and extracellular $HO^{\cdot}$ are implicated as key factors in the mechanism, and oxygen radical species have been shown to trigger an increased γ-GT gene transcription rate and elevation of intracellular mRNA,[152] which could account for increased γ-GT activity in the parkinsonian SN_c. Astroglia are suggested to play a key role in the neuropathological mechanism presented in Figure 14.12. Thus, it can be predicted that ablation of astroglia should protect DA neurons. Indeed, astroglial ablation prevents MPTP-induced degeneration of nigrostriatal DA neurons.[153] This is generally believed to reflect the fact that MPTP must be metabolized by MAO B to MPP^+ in glia (Fig. 14.10).[83] However, it is possible that glia play a second, more important role by serving as the source of GSH that is hydrolyzed by (upregulated) γ-GT/dipeptidases to Glu, Gly, and CySH as the transient DA neuron energy impairment subsides.

Periodic, transient SN_c neuron energy impairment occurring over months or years would, according to the hypothetical pathologic cascade, evoke periods of accelerated intraneuronal oxidation of DA. However, DA-*o*-quinone, formed by both the normal autoxidation and abnormal $O_2^{-\cdot}$-mediated oxidation of DA, would be scavenged by translocated CySH rather than forming neuromelanin, an effect that might account for the "depigmentation" of SN_c cells in the parkinsonian brain. Furthermore, SN_c cells that contain heavy neuromelanin pigmentation must normally sustain particularly high basal levels of DA autoxidation. This could reflect unusually high cytoplasmic concentrations of DA, a particularly weak endogenous antioxidant system, including relatively low concentrations of GSH. Thus, as a transient SN_c energy impairment subsides, translocated CySH would be expected to react with DA-*o*-quinone formed both from the autoxidation and abnormal $O_2^{-\cdot}$-mediated oxidation of DA. This, in turn, would lead to high intraneuronal concentrations of 5-*S*-CyS-DA and then DHBT/BT toxicants. As a consequence, heavy pigmentation would mark SN_c cells for preferential degeneration in PD.

Finally, the pathological cascade presented in Figure 14.12 implicates a key role for NMDA receptor activation by Glu both during the DA neuron energy-impairment and as the energy impairment subsides. Indeed, recent reports suggest that NMDA receptor antagonists may be useful therapeutic agents for PD.[154–156]

Much remains to be learned about the mechanisms that underlie dopaminergic SN_c cell death in PD and of nigrostriatal DA neurons evoked by neurotoxins such as MPTP and MA. Nevertheless, in the event that some key features of the neuropathological mechanism are accurately reflected in the scheme proposed, there are clearly a number of potential strategies that might be employed to interrupt one or more key steps. These include antioxidants (ROS scavengers), NMDA receptor antagonists, drugs that interfere with intraneuronal enzymes responsible for $O_2^{-\cdot}$ generation, and inhibitors of γ-GT. It is anticipated that a variety of drugs designed for such purposes will be evaluated in the coming years. However, as noted in earlier sections, the clinical symptoms of PD become apparent only at a stage when the disorder has nearly run its course. Thus, regardless of therapeutic strategies that are ultimately developed, they will be regarded as truly successful only if they can be employed before symptoms appear. Thus, a major challenge for neuroscientists will be to discover some type of behavioral, clinical, or probably most realistically, a chemical, marker for SN_c degeneration in the early presymptomatic stages of PD. The use of agents that can slow or halt the neurodegenerative processes at this stage provides the best hope for rescuing remaining SN_c cells.

ACKNOWLEDGMENTS

The work in the author's laboratory described was supported by National Institutes of Health grants NS29886 and GM32367, with additional support provided by the Vice President for Research and Research Council at the University of Oklahoma.

REFERENCES

1. Jenner, P.; Olanow, C. W., "Pathological Evidence for Oxidative Stress in Parkinson's Disease and Related Degenerative Disorders," in *Neurodegeneration and Neuroprotection in Parkinson's Disease*, C. W. Olanow, P. Jenner, and M. Youdim, eds., Academic Press, New York, 1996, pp. 23–45.
2. Ehringer, H.; Hornykiewicz, O., *Klin. Wochenschr.* 1960, **38**, 1236–1239.
3. Kish, S. J.; Shannak, K.; Hornykiewicz, O., *N. Engl. J. Med.* 1988, **318**, 876–880.
4. Graham, D. G., *Molec. Pharmacol.* 1978, **14**, 633–643.
5. Zhang, F.; Dryhurst, G., *Bioorg. Chem.* 1993, **21**, 392–410.
6. Mann, D. M. A.; Yates, P. O., *Mech. Ageing Dev.* 1983, **21**, 193–203.
7. Kastner, A.; Hirsch, E. C.; Lejeune, O.; Javoy-Agid, F.; Rascol, O.; Agid, Y., *J. Neurochem.* 1992, **59**, 1080–1089.
8. Fornstedt, B.; Brun, A.; Rosengren, E.; Carlsson, A., *J. Neural Transm.* (*Park. Dis. Dementia Sect.*) 1989, **1**, 279–295.
9. Schapira, A. H. V.; Cooper, J.; Dexter, D.; Clark, J. B.; Jenner P.; Marsden, C. D., *J. Neurochem.* 1990, **54**, 823–827.
10. Schapira, A. H. V.; Mann, V. M.; Cooper, J. N.; Dexter, D.; Daniel, S. E.; Jenner, P.; Clark, J. B.; Marsden, C. D., *J. Neurochem.* 1990, **55**, 2142–2145.
11. Hattori, N.; Tanaka, M.; Ozawa, T.; Mizuno, Y., *Ann. Neurol.* 1991, **30**, 563–571.
12. Mizuno, Y.; Matuda, S.; Yoshino, H.; Mori, H.; Hattori, N.; Ikebe, S.-J., *Ann. Neurol.* 1994, **35**, 204–210.
13. Gutteridge, J. M. C., *Ann. New York Acad. Sci.* 1996, **738**, 201–213.
14. Dexter, D. T.; Carter, C. J.; Wells, F. R.; Javoy-Agid, F.; Agid, Y.; Lees, A.; Jenner, P.; Marsden, C. D., *J. Neurochem.* 1989, **52**, 381–389.
15. Alam, Z. I.; Daniel, S. E.; Lees, A. J.; Marsden, D. C.; Jenner, P.; Halliwell, B., *J. Neurochem.* 1997, **69**, 1326–1329.
16. Alam, Z. I.; Jenner, A.; Lees, A. J.; Cairns, N.; Marsden, C. D.; Jenner, P.; Halliwell, B., *J. Neurochem.* 1997, **69**, 1196–1203.
17. Benzi, G.; Curti, D.; Pastoris, O.; Marzatico, F.; Villa, R. F.; Dagani, F., *Neurochem. Res.* 1991, **16**, 1295–1302.
18. Gibb. W. R.; Lees. A. J., *J. Neurol. Neurosurg. Psychiatry* 1988, **51**, 745–752.

19. Fearnley, J. M.; Lees, A. J., *Brain* 1991, **114**, 2283–2301.
20. Dexter, D. T.; Sian, J.; Rose, H.; Hindmarsh, J.-G.; Mann, V. G.; Cooper, J. M.; Wells, F. R.; Daniel, S. E.; Lees, A. J.; Schapira, A. H. V.; Jenner, P.; Marsden, C. D., *Ann. Neurol.* 1994, **35**, 38–44.
21. Spencer, J. P. E.; Jenner, A.; Aruoma, O. I.; Evans, P. J.; Kaur, H.; Dexter, D. T.; Jenner, P.; Lees, A. J.; Marsden, C. D., *FEBS Lett.* 1994, **353**, 246–250.
22. Riederer, P.; Sofic, E.; Rausch, W.-D.; Schmidt, B.; Reynolds, G. P.; Jellinger, K.; Youdim, M. B. H., *J. Neurochem.* 1989, **52**, 515–520.
23. Sofic, E.; Lange, K. W.; Jellinger, K.; Riederer, P., *Neurosci. Lett.* 1992, **142**, 128–130.
24. Sian, J.; Dexter, D.; Lees, A.; Daniel, S.; Agid, Y.; Javoy-Agid, F.; Jenner, P.; Marsden, C. D., *Ann. Neurol.* 1994, **36**, 348–355.
25. Sian, J.; Dexter, D. T.; Lees, A. J.; Daniel, S.; Jenner, P.; Marsden, C. D., *Ann. Neurol.* 1994, **36**, 356–361.
26. Mithhöfer, K.; Sandy, M. S.; Smith, M. T.; Di Monte, D., *Arch. Biochem. Biophys.* 1992, **295**, 132–136.
27. Toffa, S.; Kunikowska, G. M.; Zeng, B.-Y.; Jenner, P.; Marsden C. D., *J. Neural Transm.* 1997, **104**, 67–75.
28. Seaton, T. A.; Jenner, P.; Marsden, C. D., *Biochem. Pharmacol.* 1996, **52**, 1657–1663.
29. Andersen, J. K.; Mo, J. Q.; Hom, D. G.; Lee, F. Y.; Harnish, P.; Hamill, R. W.; MacNeil, T. H., *J. Neurochem.* 1996, **67**, 2164–2171.
30. Pearce, R. K. B.; Owen, A.; Daniel, S.; Jenner, P.; Marsden, C. D., *J. Neural Transm.* 1997, **104**, 661–677.
31. Raps, S. P.; Lai, J. C. K.; Hertz, L.; Cooper, A. J. L., *Brain Res.* 1989, **493**, 398–401.
32. Sagara, J.-L.; Miura, K.; Bannai, S., *J. Neurochem.* 1993, **61**, 1667–1671.
33. Sagara, J.; Miura, K.; Bannai, S., *J. Neurochem.* 1993, **61**, 1672–1676.
34. Yudkoff, M.; Pleasure, C.; Cregar, L.; Lin, Z.-P.; Nissim, I.; Stern, J.; Nissim, I., *J. Neurochem.* 1990, **55**, 137–145.
35. Anderson, M. E.; Meister, A., *FASEB J.* 1989, **3**, 1632–1636.
36. Inoue, M.; Saito, Y.; Hirata, E.; Morino, Y.; Nagase, S., *J. Protein Chem.* 1987, **6**, 207–225.
37. Bannai, S., *Biochim. Biophys. Acta* 1984, **779**, 289–306.
38. Calabrese, V.; Fariello, R. G., *Biochem. Pharmacol.* 1988, **37**, 2287–2288.
39. Fariello, R. G.; Ghirardi, O.; Peschechera, A.; Ramucci, M. T.; Angelucci, T., *Neuropharmacology* 1988, **27**, 1077–1088.
40. Zhang, F.; Dryhurst, G., *J. Electroanal. Chem.* 1995, **398**, 117–128.
41. Dexter, D. T.; Ward, R. J.; Wells, F. R.; Daniel, S. E.; Lees, A. J.; Peters, T. J.; Jenner, P.; Marsden, C. D., *Ann. Neurol.* 1993, **32**, 591–593.
42. Marttila, R. J.; Lorentz, H.; Rinne, U. K., *J. Neurol. Sci.* 1988, **86**, 321–331.
43. Kish, S. J.; Morito, C.; Hornykiewicz, O., *Neurosci. Lett.* 1985, **58**, 343–346.
44. Saggu, H.; Cooksey, J.; Dexter, D.; Wells, F. R.; Lees, A. J.; Jenner, P.; Marsden, C. D., *J. Neurochem.* 1989, **53**, 692–697.
45. Marttila, R. J.; Lorentz, H.; Rinne, U. K., *J. Neurol. Sci.* 1988, **86**, 321–331.
46. Sofic, E.; Riederer, P.; Heisen, H.; Bechmann, H.; Reynolds, G. P.; Habenstreit, G.; Youdim, M. B. H., *J. Neural Transm.* 1988, **74**, 199–205.

47. Jellinger, K.; Kienzl, E.; Rumpelmair, G.; Riederer, P.; Stachelberger, H.; Ben-Shachar, D.; Youdim, M. B. H., *J. Neurochem.* 1992, **59**, 1168–1171.
48. Good, P. F.; Olanow, C. W.; Perl, D. P., *Brain Res.* 1992, **593**, 343–346.
49. Ben-Shachar, D.; Riederer, P.; Youdim, M. B. H., *J. Neurochem.* 1991, **57**, 1609–1614.
50. Yoshida, T.; Tanaka, M.; Somomatsu, A.; Hirai, S., *Neurosci. Lett.* 1995, **190**, 21–24.
51. Thomas, C. E.; Morehouse, L. A.; Aust, S. D., *J. Biol. Chem.* 1985, **260**, 3275–3280.
52. Brieland, J. K.; Fantone, J.C., *Arch. Biochem. Biophys.* 1991, **284**, 78–83.
53. Flint, D. H.; Tuminello, J. F.; Emptage, M. H., *J. Biol. Chem.* 1993, **268**, 22369–22376.
54. Tanner, C. M.; Ottman, R.; Goldman, S. M.; Ellenberg, J.; Chan, P.; Mayeux, R.; Langston, J. W., *JAMA* 1999, **281**, 341–346.
55. Ikebe, S.; Tanaka, M.; Ozawa, T., *Nature* 1995, **334**, 345–348.
56. Swerdlow, R. H.; Parks, J. K.; Miller, S. W.; Tuttle, J. B.; Trimmer, P. A.; Sheehan, J. P.; Bennett, J. P.; Davis, R. E.; Parker, W. D., *Ann. Neurol.* 1996, **40**, 663–671.
57. Sheehan, J. P.; Swerdlow, R. H.; Parker, W. D.; Miller, S. W.; Davis, R. E.; Tuttle, J. B., *J. Neurochem.* 1997, **68**, 1221–1233.
58. Kobayashi, T.; Matsumme, H.; Matuda, S.; Mizuno, Y., *Ann. Neurol.* 1998, **43**, 120–123.
59. Ho, S. L.; McCann, K. P.; Bennett, P.; Kapadi, A. L.; Waring, R. H.; Ramsden, D.B.; Williams, A. C., *Adv. Neurol.* 1996, **69**, 53–60.
60. Steventon, G. B.; Heafield, M. T. E.; Waring, R. H.; Williams, A. C., *Neurology* 1989, **39**, 883–887.
61. Ross, H. R.; Romrell, L. J.; Kaye, G. I., *Histology. A Text and Atlas*, Williams Wilkins, Baltimore, 1995, Chapter 11, p. 282.
62. Semchuk, K. M.; Love, E. J.; Lee, R. G., *Neurology* 1992, **42**, 1328–1335.
63. Rybicki, B. A.; Johnson, C. C.; Uman, J.; Gorell, J. M., *Mov. Disord.* 1992, **1**, 87–92.
64. Meco, G.; Bonifati, M. G.; Vanacore, N.; Fabrizio, E., *Scand. J. Environ. Health* 1994, **20**, 301–305.
65. Gorell, J. M.; Johnson, C. C.; Rybicki, B. A.; Peterson, E. L.; Kortsha, G. X.; Brown, G. G., *Neurology* 1997, **48**, 650–658.
66. Gorell, J. M.; Johnson, C. C.; Rybicki, B. A.; Peterson, E. L.; Richardson, R. J., *Neurology* 1998, **50**, 1346–1350.
67. Mehrotra, B. D.; Bansal, S. K.; Desaiah, D., *J. Appl. Toxicol.* 1982, **2**, 278–283.
68. Singh, G. J.; Singh, B., *Neurobehav. Toxicol. Teratol.* 1984, **6**, 201–213.
69. Pedrajas, J. R.; Lopez-Barea, J.; Peinado, J., *Comp. Biochem. Physiol. C. Pharmacol. Toxicol. Endocrinol.* 1996, **115**, 125–131.
70. Sanchez-Ramos, J.; Facca, A.; Basit, A.; Song, S., *Exp. Neurol.* 1998, **150**, 263–271.
71. Corrigan, F. M.; Murray, L.; Wyatt, C. L.; Shore, R. F., *Exp. Neurol.* 1998, **150**, 339–342.
72. Fleming, L.; Mann, J. B.; Bean, J.; Briggle, T.; Sanchez-Ramos, J. R., *Ann. Neurol.* 1994, **36**, 100–103.
73. Menegon, A.; Board, P. G.; Blackburn, A. C.; Mellick, G. D.; Le Couteur, D. G., *Lancet* 1998, **352**, 1344–1346.
74. Seidler, A.; Hellenbrand, W.; Robra, B. P.; Vieregge, P.; Nischan, P.; Joerg, J.; Oertel, W. H.; Ulm, G.; Scheider, E., *Neurology* 1996, **46**, 1275–1284.

75. Bachurin, S. O.; Shevtzova, E. P.; Lermontova, N. N.; Serkova, T. P.; Ramsay, R. R., *Neurotoxicology* 1996, **17**, 897–903.

76. Hoogenraad, T. U., *Lancet* 1988, **1**, 767.

77. Meco, G.; Bonifati, M. G.; Vanacore, N.; Fabrizio, E., *Scand. J. Environ. Health* 1994, **20**, 301–305.

78. Cooper, J. M.; Schapira, A. H. V., *J. Bioenerg. Biomembr.* 1997, **29**, 175–183.

79. Fornstedt, B.; Brun, A.; Rosengren, E.; Carlsson, A., *J. Neural Transm. (Park. Dis. Dement. Sect.)* 1989, **1**, 279–295.

80. Spencer, J. P. E.; Jenner, P.; Daniel, S. E.; Lees, A. J.; Marsden, C. D.; Halliwell, B., *J. Neurochem.* 1998, **71**, 2112–2122.

81. Gerlach, M.; Riederer, P., *J. Neural Transm.* 1996, **103**, 987–1041.

82. Langston, J. W.; Ballard, P. A.; Tetrud, J. W.; Irwin, I., *Science* 1983, **219**, 979–980.

83. Ransom, B. R.; Kunis, D. M.; Irwin, I.; Langston, J. W., *Neurosci. Lett.* 1987, **75**, 323–328.

84. Javitch, J. A.; D'Amato, R. J.; Strittmatter, S. M.; Snyder, S. H., *Proc. Nat. Acad. Sci.* (USA) 1985, **82**, 2173–2177.

85. Ramsay, R. R.; Dadgar, J.; Trevor, A.; Singer, T. P., *Life Sci.* 1986, **39**, 581–588.

86. Mizuno, Y.; Saitoh, T.; Sone, N., *Biochem. Biophys. Res. Commun.* 1987, **143**, 971–976.

87. Chan, P.; Delanney, L. E.; Irwin, I.; Langston, J. W.; Di Monte, D., *J. Neurochem.* 1991, **57**, 348-351.

88. Ferraro, T. N.; Golden, G. T.; DeMattei, M.; Hare, T. A.; Fariello, R. G., *Neuropharmacology* 1986, **25**, 1071–1074.

89. Yong, V. W.; Perry, T. L.; Krisman, A. A., *Neurosci. Lett.* 1986, **63**, 56–60.

90. Oishi, T.; Hasegawa, E.; Murai, Y., *J. Neural Transm.* (*Park. Dis. Dement. Sect.*) 1993, **6**, 45–52.

91. Robinson, T. E.; Justice, J. B., *Microdialysis in the Neurosciences*, Elsevier Science, Amsterdam, 1992.

92. Rollema, H.; Damsma, G.; Horn, A. S.; De Vries, J. B.; Westerink, B. H. C., *Eur. J. Pharmacol.* 1986, **126**, 345–346.

93. Carboni, S.; Melis, F.; Pani, L.; Hadjiconstantinou, M.; Rossetti, Z. L., *Neurosci. Lett.* 1990, **117**, 129–133.

94. Matarredona, E. R.; Santiago, M.; Machado, A.; Cano, J., *Br. J. Pharmacol. Chemotherap.* 1997, **121**, 1038–1044.

95. Zängerle, L.; Cuenod, M.; Winterhalter, K. H.; Do, K. Q., *J. Neurochem.* 1992, **59**, 181–189.

96. Sonsalla, P. K.; Jochnowitz, N. D.; Zeevalk, G. D.; Oostveen, J. A.; Hall, E. D., *Brain Res.* 1996, **738**, 172–175.

97. Gibb, J. W.; Hanson, G. R.; Johnson, M., "Neurochemical Mechanisms of Toxicity," in Amphetamine and Its Analogs, Psychopharmacology, Toxicology and Abuse, A. K. Cho and D. S. Segal, eds., Academic Press, New York, 1994, pp. 269–289.

98. Chan, P.; Di Monte, D. A.; Luo, J.-J.; Delanney, L. E.; Irwin, I.; Langston, J. W., *J. Neurochem.* 1994, **62**, 2484–2487.

99. Albers, D. S.; Zeevalk, G. D.; Sonsalla, P. K., *Brain Res.* 1996, **718**, 217–220.

100. O'Dell, S. J.; Weihmuller, F. B.; Marshall, J. F., *Brain Res.* 1991, **564**, 256–260.

101. Moszczynska, A.; Turenne, S.; Kish, S. J. *Neurosci. Lett.* 1998, **255**, 49–52.
102. Nash, J. F.; Yamamoto, B. K., *Brain Res.* 1992, **581**, 237–243.
103. Stephens, S. E.; Yamamoto, B. K., *Synapse* 1994, **17**, 203–209.
104. Boireau, A.; Bordier, F.; Dubédat, P.; Doble, A., *Neurosci. Lett.* 1995, **195**, 9–12.
105. Nowak, L.; Bregestovski, P.; Ascher, P.; Herbet, A.; Prochiantz, A., *Nature* 1984, **307**, 462–465.
106. Choi, D. W., *Neurosci. Lett.* 1985, **58**, 293–297.
107. Rothman, S. M.; Thurston, J. H.; Hanhart, R. E., *Neuroscience* 1987, **22**, 471–480.
108. Olney, J. W., *Annu. Rev. Pharmacol. Toxicol.* 1990, **30**, 47–71.
109. Novelli, A.; Reilly, J. A.; Lysko, P. G.; Henneberry, R. C., *Brain Res.* 1998, **451**, 205–212.
110. Lafon-Cazal, M.; Pletri, S.; Culcasi, M.; Bockaert, J., *Nature* 1993, **364**, 535–537.
111. Przedborski, S.; Kostic, V.; Jackson-Lewis, V.; Naini, A. B.; Simonettá, S.; Fah, S.; Carlson, E.; Epstein, C. J.; Cadet, J. L., *J. Neurosci.* 1992, **12**, 1658–1667.
112. Cadet, J. L.; Sheng, P.; Ali, S.; Rothman, R.; Carlson, E.; Epstein, C., *J. Neurochem.* 1994, **62**, 380–383.
113. Corsini, G. U.; Pintus, S.; Chiueh, C. C.; Weiss, J. F.; Kopin, I. J., *Eur. J. Pharmacol.* 1985, **119**, 127–128.
114. Thiffault, C.; Aumont, N.; Quirion, R.; Poirier, J., *J. Neurochem.* 1995, **65**, 2725–2733.
115. Acikgoz, O.; Gonenc, S.; Kayatekin, B. M.; Uysal, N.; Pekcetin, C.; Semin, I.; Gre, A., *Brain Res.* 1998, **813**, 200–202.
116. Chan, P.; Langston, J. W.; Di Monte, D. A., *J. Pharmacol. Exp. Therap.* 1993, **267**, 1515–1520.
117. Santiago, M.; Venero, J. L.; Machado, A.; Cano, J., *Brain Res.* 1992, **586**, 203–207.
118. Sonsalla, P. K.; Nicklas, W. J.; Heikkila, R. E., *Science* 1989, **243**, 398–400.
119. Johnson, M.; Hanson, G. R.; Gibb, J. W., *Eur. J. Pharmacol.* 1989, **165**, 315–318.
120. Santiago, M.; Matarredona, E. R.; Granero, L.; Cano, J.; Machado, A., *J. Neurochem.* 1997, **68**, 732–738.
121. Yamamoto, B.; Zhu, W., *J. Pharmacol. Exp. Ther.* 1998, **287**, 107–114.
122. Revuelta, M.; Romero-Ramos, M.; Venero, J. L.; Millan, F.; Machado, A.; Cano, J., *Neuroscience* 1997, **77**, 167–174.
123. Matsuda, L. A.; Schmidt, C. J.; Gibb, J. W.; Hanson, G. R., *Brain Res.* 1987, **400**, 176–180.
124. Hyslop, P. A.; Zhang, Z.; Pearson, D. V.; Phebus, L. A., *Brain Res.* 1995, **671**, 181–186.
125. Chiueh, C. C.; Wu, R.-M.; Mohanakumar, K. P.; Sternberger, L. M.; Krishma, G.; Obata, T.; Murphy, D. L., *Ann. New York Acad. Sci.* 1994, **738**, 25–36.
126. Giovanni, A.; Liang, L. P.; Hastings, T. G.; Zigmond, M. J., *J. Neurochem.* 1995, **64**, 1819–1825.
127. De Vito, M. J.; Wagner, G. C., *Neuropharmacology* 1989, **28**, 1145–1150.
128. Aubin, N.; Curet, O.; Deffois, A.; Carter, C., *J. Neurochem.* 1998, **71**, 1635–1642.
129. Martin, F. R.; Sanchez-Ramos, J.; Rosenthal, M., *J. Neurochem.* 1991, **57**, 1340–1346.
130. Marek, G. J.; Vosmer, G.; Seiden, L. S., *Brain Res.* 1990, **513**, 274–279.
131. Massieu, L.; Morales-Villagran, A.; Tapia, R., *J. Neurochem.* 1995, **64**, 2262–2272.
132. Maher, P.; Davis, J. B., *J. Neurosci.* 1996, **16**, 6394–6401.

133. Erdö, S. L.; Michler, A.; Wolff, J. R.; Tytko, H., *Brain Res.* 1990, **526**, 328–332.
134. Huether, G.; Zhou, D.; Rüther, E., *J. Neural Transm.* 1997, **104**, 771–794.
135. Beal, M. F., *Mitochondrial Dysfunction and Oxidative Damage in Neurodegenerative Diseases,* Landes, Austin, TX, 1995.
136. Zeevalk, G. D.; Nicklas, W. J., *J. Pharmacol. Exp. Ther.* 1991, **257**, 870–878.
137. Shaw, C. A., "Multiple Roles of Glutathione in the Nervous System" in *Glutathione in the Nervous System,* C. A. Shaw, ed., Taylor & Francis, Washington, DC, 1998, pp. 3–23.
138. Cooper, A. J. L., "Role of Astrocytes in Maintaining Cerebral Glutathione Homeostasis and Protecting the Brain Against Xenobiotics and Oxidative Stress" in *Glutathione in the Nervous System,* C. A. Shaw, ed., Taylor & Francis, Washington, DC, 1996, pp. 91–115.
139. Dringen, R.; Kranich, O.; Hamprecht, B., *Neurochem. Res.* 1997, **22**, 727–733.
140. Han, J.; Cheng, F.-C.; Yang, Z.; Dryhurst, G., *J. Neurochem.*, **73**, 1683–1695.
141. Lada, M. W.; Kennedy, R. T., *J. Neurosci. Meth.* 1997, **72**, 153–159.
142. Xie, C. X.; St. Pyrek, J.; Porter, W. H.; Yokel, R. A., *Neurotoxicology* 1995,**16**, 489–496.
143. Stole, E.; Smith, T. K.; Manning, J. M.; Meister, A., *J. Biol. Chem.* 1994, **269**, 21435–21439.
144. LaVoie, M. J.; Zigmond, M. J.; Hastings, T. G., *Soc. Neurosci. Abstr.* 1996, **22**, 221.
145. Zhang, F.; Dryhurst, G., *J. Med. Chem.* 1994, **37**, 1084-1098.
146. Shen, X.-M.; Dryhurst, G., *Chem. Res. Toxicol.* 1996, **9**, 751–763.
147. Shen, X.-M.; Zhang, F.; Dryhurst, G., *Chem. Res. Toxicol.* 1997, **10**, 147–155.
148. Zhang, F.; Dryhurst, G., *Bioorg. Chem.* 1995, **23**, 193–216.
149. Li, H.; Dryhurst, G., *J. Neurochem.* 1997, **69**, 1530–1541.
150. Li, H.; Shen, X.-M.; Dryhurst, G., *J. Neurochem.* 1998, **71**, 2049–2062.
151. Shen, X.-M.; Li, H.; Dryhurst, G., *J. Neural Transm.* Submitted.
152. Liu, R. M.; Shi, M. M.; Giulivi, C.; Forman, H. J., *Am. J. Physiol.* 1998, **274**, L330–L336.
153. Takada, M.; Li, Z.; Hattori, T., *Brain Res.* 1990, **509**, 55–61.
154. Greenamyre, J. T.; O'Brien, C. F., *Arch. Neurol.* 1991, **48**, 977–981.
155. Fischer, G.; Bourson, A.; Kemp, J.; Lorenz, H., *Soc. Neurosci. Abstr.* 1996, **22**, 1760.
156. Verhagen, M. L.; Del Dotto, P.; Blanchet, P. J.; van der Munckhof, P.; Chase, T. N., *Amino Acids* 1998, **14**, 75–82.

PART IV

NOVEL CHEMISTRIES IN THE VANGUARD OF BIOTECHNOLOGY AND DRUG DISCOVERY

This section may be introduced best by a quote (italics added) from an article in *Nature Genetics*[1] by Eric. S. Lander:

> *Genomics* aims to provide biologists with the equivalent of chemistry's Periodic Table—an inventory of all genes used to assemble a living creature, together with an insightful system for classifying these building blocks.[1]

To discover how these genes relate to one another—which are expressed and which are not expressed in a specific disease, or in different stages of development, or in response to a given drug or environmental stimulus—requires the ingenuity of chemists. *DNA microarrays* (also known as *DNA chips*) promise an approach to a solution of the foregoing challenge. In Chapter 15, Garnier expands on this topic of *DNA sensors* and elaborates his own novel approach based on a biological model.

As an outgrowth of the foregoing efforts, *functional genomics*, evolving from the *human genome project*,[2] identifies gene function and associates specific genes with specific disease, thereby providing a host of new targets for drug discovery and development.[3–5] All the subdisciplines and tools of bioorganic–medicinal chemistry contribute to this new opportunity, including natural products, combinatorial approaches, rational drug design, small-molecule chemistry, and *antisense* and *antigene* strategies.

According to one traditional approach to drug design and discovery, one should first identify a particular target protein that is central to the pathogenic process of interest. Then, for instance, by rational drug design, combinatorial strategies, or random screening and subsequent modifications to a lead structure, specific drug candidates may be identified. In a wholly different tack, *antisense therapeutics* relies on the central dogma of molecular biology, namely, that DNA makes RNA, which in turn, makes protein. With the messenger RNA (mRNA) defined as the entity that makes *sense* by embodying the message that translates DNA's genetic code into the final product protein, *antisense* defines an entity (an oligonucleotide or an analog thereof) that can bind specifically to the "sense" mRNA and somehow inhibit its function. Substantial advantages accrue to this strategy, including the fact that nothing need be known about the chosen protein target. All that must be known is the sequence (or portion thereof) of the mRNA or the corresponding gene's DNA nucleotide sequence.

Antigene (also known as *triplex*) strategies rely on a different code carried in double-stranded nucleic acids. The potential of the antigene approach is illustrated in a 1998 article[6] in *Nature Genetics* in which a specific gene could be excised from a cell's genome to give a progeny cell devoid of that genetic component.

This developing field of antisense–oligonucleotide therapeutics holds great promise. For instance, one antisense oligonucleotide has received FDA approval (ISIS 2922 for AIDS CMV retinitis) and is on the market.[7]

In a related application, as the sequencing of the human genome progresses, antisense molecules can be of great use in functional genomics, specifically, the relationship of nucleotide sequence to gene purpose. The effective use of such *knockout* reagents will depend on improvements the ability of antisense oligonucleotides and their analogs to find and destroy their RNA targets.

The final two contributions to this section covey the movement and potential of the *antisense* and *ribozyme* concepts. In Chapter 16, Nielsen outlines his work on a fascinating class of macromolecules called *peptide nucleic acids*, which have extremely high affinites for their targets and are resistant to destruction by the enzymes that normally degrade nucleic acids in the cell. These nucleic acid analogs have particular relevance to the triplex or antigene strategy to prevent gene expression. In Chapter 17, Osiek, Putnam, and Bashkin introduce the promise and problems of antisense and ribozymes, and then go on to provide a description of *ribozyme mimics*, are grounded in a chemical catalytic approach to RNA destruction.

REFERENCES

1. Lander, E. S., *Nat. Genet.* 1999, **21**, 3–4.
2. Collins, F.; Galas, D., *Science* 1993, **262**, 43–46.
3. Metcalf, B. W., *Pure Appl. Chem.* 1998, **70**, 359–363.
4. Sadee, W., *Pharm. Res.* 1998, **15**, 959–963.
5. Haseltine, W. A., *Sci. Am.* 1997, **276**, 92–97.
6. Majumdar, A.; Khorlin, A.; Dyatkina, N.; Lin, F.-L. M.; Powell, J.; Liu, J.; Fei, Z.; Khripine, Y.; Watanabe, K.; George, J.; Glazer, P. M.; Seidman, M. M., *Nat. Genet.* 1998, **20**, 212–214.
7. Crooke, S. T., *Antisense Nucleic Acid Drug. Des.* 1998, **8**, vii–viii.

CHAPTER 15

New Generation of DNA Chips Based on Real-Time Biomimetic Recognition

FRANCIS GARNIER

Laboratoire des Matériaux Moléculaires, CNRS

The challenge to sequence the whole of the human genome attracted great attention and hope throughout the 1990s. Although this objective once appeared out of reach,[1] the recent (late 1990s) sequencing of five bacterial genomes and of many human genes, together with the considerable progress achieved by many biotechnology companies in the field of DNA sensors, has led some scientists to consider that the postgenomic era has now started. In fact, DNA sensors have recently enabled the analysis of gene sequences, gene mutations, and gene expression.[2–4] These achievements underline the potential use of DNA sensors for fundamental research and practical applications in medical sciences.

One of the first application domains of these biodevices concerned the sequencing of nucleic acids specific to infectious disease testing. After the introduction by Gen-Probe of the first test for Legionella disease in the mid-1980s, nucleic acid probe testing has become one of the most promising tools in the clinical laboratory; moreover, its sensitivity has been greatly increased by the development of new amplification techniques, such as the polymerase chain reaction (PCR), which offers a real molecular zoom with an amplification power of the order of 10^5.[5] Amplified DNA probe assays have been developed for numerous infectious diseases, including chlamydia, cytomegalovirus, hepatitis B and C viruses, and HIV-1 virus. Another major new application is the screening of donated blood for infectious agents. Increased sensitivity of the screening process should allow the identification of infected blood products from recently infected donors who have not yet generated antibodies to the infectious agent at the time of blood donation.

A second considerable area for application of DNA sensors is mutation analysis. Many diseases are linked to the mutation of genes, which may occur via various mechanisms such as replacement, duplication, insertion, or deletion of one or many

Biomedical Chemistry: Applying Chemical Principles to the Understanding and Treatment of Disease, Edited by Paul F. Torrence
ISBN 0-471-32633-x

base units that constitute the gene. Genetic tests have become one of the most powerful tools for screening the presence of pathogenic mutations that can induce severe diseases. Prenatal diagnosis now enables a search for the presence of genetic alterations in fetal cells, allowing an early diagnosis of hereditary diseases such as hemophilia, mucoviscidosis, and various myopathies. Owing to the high sensitivity of DNA sensors, a quantity as small as 10 mL of maternal plasma is sufficient for the reliable detection of fetal abnormalities. Preventive medical care dedicated to monogenic diseases, namely, the alteration of one gene,is also being developed. No doubt it will be further expanded to diseases involving many risk factors, such as cardiovascular diseases, some cancers, or diabetes.

Another significant area of application concerns the analysis of gene expression. For those human genes that have already been characterized, the technique is so sensitive that the presence of mRNA at the level of 1 in 300,000 can be quantified, and mRNA at the level of two to three copies per cell have been clearly identified. When applied to an organism in which the whole of the genome has been sequenced, genotype sequencing provides the necessary basic data for relating genetic information and biologic function, such as which genes are turned on under a specific set of conditions. The recent sequencing of bacterial and some human genes has thus produced the first information needed for a deeper understanding of the intricate regulations correlating genetic expression and functions.

The recent large-scale miniaturization achieved for these sensors has led to the terminology of DNA chips, which have opened the fields of gene sequencing, gene mutations, and complex gene expression. The relevance of this postgenomic era is confirmed by the enormous activity of many biotech companies (Affymetrix, Nanogene, Hyseq, Hoffmann La Roche, Incyte Pharmaceuticals, Beckman Intsruments, Vysis, OncorMed, etc). In the following paragraphs, we present a short overview of the state of art of DNA chips, and describe a new research strategy, that the CNRS laboratory recently (late 1990s) developed in this field.

HOW A DNA SENSOR FUNCTIONS

The mode of operation of DNA sensors is based on the hybridization phenomenon occurring between the complementary sequences of two single DNA strands. Hybridization builds on the specific interaction established through hydrogen bonds between the elemental bases that constitute nucleic acids, specifically, between adenine (A), and thymine (T), (or uracile), and between cytosine (C), and guanine (G), respectively. For instance, if the sequences 5′ ATT GGC TAT 3′ and 5′ TAA CCG ATA 3′ are respectively present in two single DNA strands, these two complementary sequences will hybridize and form a duplex. The stability of the duplex is controlled by the cohesive energy originating from these base associations, and is expressed by the melting temperature T_m, defined as the temperature at which 50% of the double strand is dissociated. The stability of the duplex is also largely affected by an eventual mismatch involving one or more bases in the base sequences of both single strands. This hybridization occurs under strict experimental conditions concerning the medium, the temperature, the ionic force, and other parameters, The mapping of an unknown DNA sequence, considered as a DNA target, is performed by screening it in presence of a large set of short oligonucleotide (ODN) probes, of some 10–25 base-pair of known base sequence, as shown in Figure 15.1.

The characterization of the complementary ODN probes that have led to hybridization permits the mapping of the whole sequence of the DNA target. It is clear that this DNA sequencing, schematized in Figure 15.2, will be all the more accurate as the number of ODN probes on the sensor increases. This raises the experimental problem of the ODN anchoring site density, which can be realized on a sensor substrate. The concept of DNA chip is intended to answer this question. Millions of ODN probes have been implanted on small silicon chips; one of the highest-performing devices has been advanced by the Affimetrix Company.[6] Matricial arrays of $n \times n = n^2$ sensing sites, each associated with a specific ODN probe, have been realized that display ODN densities up to approximately 10^5 sites/cm^2.

The efficiency of such a DNA sensor is directly related to the density of ODN probes, which raises the issue of what limits ODN probe density.

DETERMINING THE ACTUAL STATE OF THE ART OF DNA CHIPS

Many parameters have to be taken into account for considering this issue:

1. A *writing* problem: how to anchor different oligonucleotides on small spots on a substrate
2. A *hybridization reaction* problem: how to ensure experimental conditions allowing the precise analysis of perfect and partially perfect hybrids of different stability (problem of matching vs. mismatching)
3. A *reading* problem: once the hybridizations have occurred, how to "read," in a simple, efficient, fast, and sensitive way, which ODN probes have been engaged in complementary hybrids with the DNA target

Many solutions to these issues have already been given in the literature. Different types of substrates, probes, writing, and reading techniques have been proposed that

```
.....ATT GGC TAT.....     (base sequence in a DNA target)

  TAA CC
   AA CCG
    A CCG A          (overlapping base sequences of 5 ODN probes)
      CCG AT
       CG ATA
```

FIGURE 15.1 The unknown base sequence (ATT GGC TAT) in the DNA target is mapped from hybridization data showing that it forms complementary duplexes with 5-base-long ODN probes, with overlapping sequences.

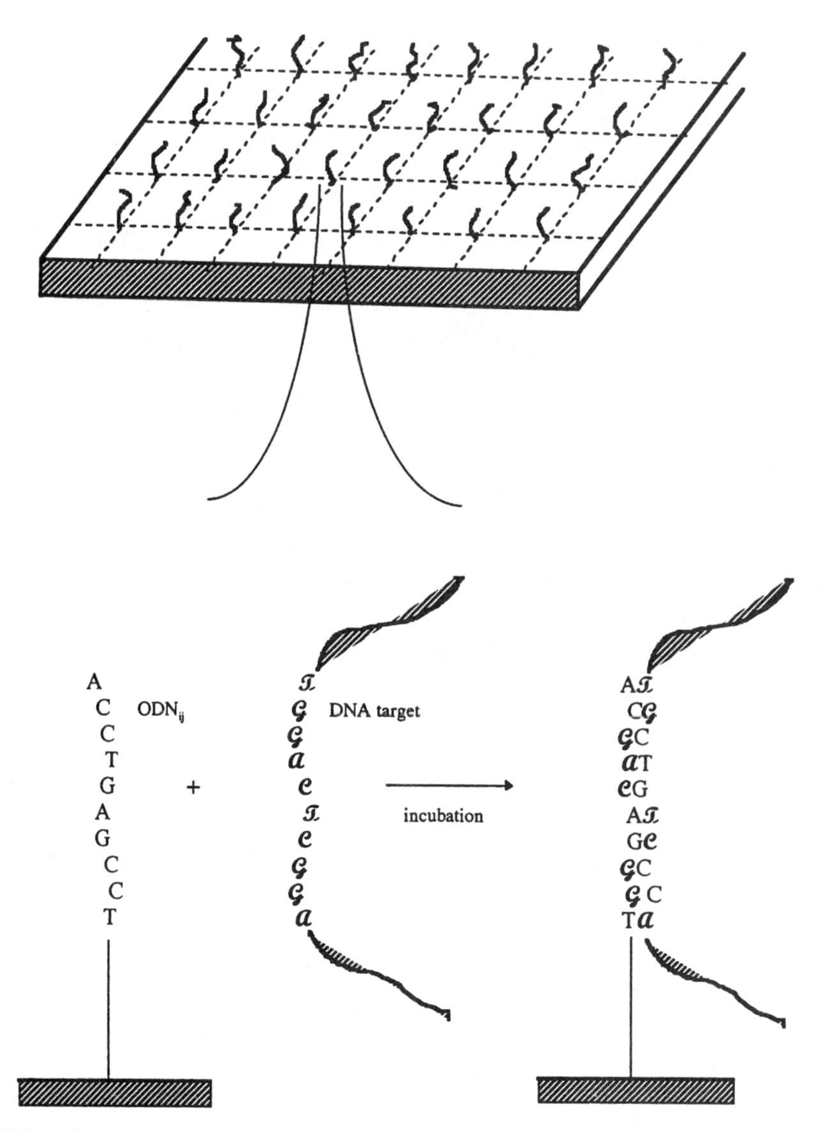

FIGURE 15.2 Schematic representation of the mode of operation of a DNA chip, involving a matricial array of $n \times n$ sensing dots, each associated with a defined ODN probe. Incubation of DNA target with 10-mer ODN$_{ij}$ probe leads to hybridization ODN$_{ij}$/DNA, owing to the complementarity of base sequences in ODN$_{ij}$ and DNA target.

are worthy of description. Various substrates have been employed, depending on the writing procedure and the experimental conditions under which these devices are used.[7] Considerations of pH, chemical stability, and inertness toward nucleic acids have been taken into account, which led to the predominant use of glass, silicon, and some polymers, where the substrate surface was either flat or porous. To prevent chemical interference between the substrate surface and the ODN probe, a molecular *spacer* is used between the substrate and the ODN.

The anchoring of ODN probes on this solid substrate is then accomplished using either already available ODNs, or through an in situ synthesis. The first route, described in the late 1980s, involved the insertion of small ODNs in polyacrylamide gels, deposited as small dots on a glass substrate.[8] The addressing of the dots, measuring approximately 100 μm, was carried out mechanically by the use of micropipettes containing the ODN solution, leading to an achievable density of sensing sites of about 10^4 cm^{-2}. A second route for the deposition of ODNs, proposed by the Affymetrix Company,[6] involved an original photochemical addressing of sensing dots, associated with an in situ synthesis of the desired ODN probes by the use of an automated ODN synthesizer, which derives from the well-known automated peptide synthesizer. A series of photochemical protection and deprotection steps are performed on an activated glass surface, by the use of shadow masks, which allows the precise chemical coupling of functionalized ODN probes on the activated surfaces. This deposition technique, termed *very large-scale immobilized polymer synthesis* (VLSIPS), combines photochemically addressed surface activation and combinatory nucleotide synthesis. Since a nucleotide is composed of 4 bases, A, C, T, and G, the number of possible combinations increases rapidly. For instance, 8 bases are mapped to 4^8, creating 65,536 possible sequences. The achieved density of sensing dots is very high, about 10^5 cm^{-2}. In the procedures described above, the hybridization reactions are optimized by controlling the experimental reaction parameters, such as temperature, electrolyte balance, and pH.[9]

A new interesting procedure, referred to as an *automated programmable electronic matrix* (APEX) proposed by the Nanogen Company, is based on an electronic addressing of the hybridization reaction between the DNA target and its complementary ODN probe.[10] Each ODN probe deposited on a microelectrode is independently electronically addressed, and the imposition of an electric field at the vicinity of this probe induces an increased electrophoretic concentration of the DNA targets, and hence an increase of the hybridization reaction. Afterward, the inversion of polarity leads to repulsion of the nonhybridized DNA targets from the ODN probe microelectrode, which brings an elegant solution to the problem of mismatching. Another significant route recently described by the Cis BioInternational Company for the deposition of ODN probes involves electrochemical addressing.[11] An array of micrometer-sized electrodes, of about 50 μm size, is patterned on a substrate by the use of well-controlled photolithographic techniques. The ODN probe is chemically derivatized and covalently bonded to a pyrrole heterocycle. A mixture of ODN-substituted pyrrole and unsubstituted pyrrole is then electrochemically polymerized onto one of the microelectrodes, individually addressed by the application of the electrochemical potential required for polymerization. This affords a thin copolymer film, poly(pyrrole, ODN-pyrrole) on the electrode. This procedure is repeated for each [electrode/ODN$_{ij}$ probe], specifically, n^2 times, leading to a DNA chip with an achievable density of sensing sites of about 5×10^4. This electrical addressing for the deposition of ODNs is reliable and efficient, although its expected cost is high, because the electropolymerization conditions require a fairly high concentration (10^{-4} M/L) of ODN-substituted monomer solution. In addition, this polymerization solution is effective practically at about half of its monomer content, due to the formation of by-products that hinder electropolymerization.

The last issue concerns the "reading" of the ODN_{ij} probes that, among the n^2 probes of the DNA sensing chip, have been subjected to hybridization with the DNA target. These interacting ODN_{ij} probes constitute the real *fingerprint*, from which signal processing and computational data analysis allow the mapping of the DNA analyte. In the systems so far described in the literature of DNA sensing, this reading process is always performed by an indirect method involving the use of a tag, which has first to be chemically bonded to the DNA target before carrying out the hybridization reaction. Various types of tags have been proposed in the literature; the first ones were based on radioactive elements (^{32}P).[11,12] Once the hybridization is performed, the reading of the chip is accomplished by the screening of the ODN sites presenting a radioactive signal. Although a high degree of sensitivity has been obtained, this technique is hampered by its low spatial resolution, which corresponds to a density of some 1000 detecting sites/cm^2. In addition to this low resolution, the environmental problems associated with the use of radioactive labeled material has led research groups to envision other types of tags. Since fluorescence is a highly sensitive detection method, various fluorophores, such as fluorescein, have been covalently bonded as tags onto the DNA target and then submitted to hybridization with the ODN probes.[6] Once hybridization has occurred, the DNA chip is then read by scanning its surface with an argon laser under a confocal microscope. The detection of excited fluorophores allows one to characterize the sites where tagged DNA have hybridized with ODN probes.[13] The emitted light is amplified with a photomultiplier, or recorded by a CCD camera,[12] affording a series of pixels associated with the reacted ODN_{ij} probe sites. The spatial resolution of such reading is about 30 μm, which corresponds to a density of some 10^5 sites/cm^2. The reading sensitivity reaches about 10^{-9} M of DNA target, and the reading time, if accomplished with high resolution, is of the order of 20 min.

Another detection technique is based on the fact that hybridization of the ODN probe by its binding with the DNA target induces a modification of the electrical charge density on the ODN, principally associated with the phosphate groups of the nucleic acids. When using a layer of these ODN probes as the grid of a field-effect transistor, $Si/SiO_2/ODN$, the hybridization-induced variation of charge at the insulating SiO_2 surface leads to a corresponding variation of charge at the Si/SiO_2 interface.[14] The response of the transistor is directly related to the variation of this charge density, and hence to the concentration of reacted ODNs. The interest in this approach relates to the absence of tag on the DNA target, the direct electrical reading of the result, and the amplification of the signal brought by the transistor. Although sensitivities up to 10^{-9} g of DNA have been quoted with these devices, their sensitivity to nonspecific adsorption of DNAs limits their potential of application.

In summary, the potential of these DNA chips has already been largely confirmed by many impressive results, obtained, for instance, in the case of the analysis of the mutations in the genes of reverse transcriptase and the protease of the HIV virus, and also in the case of sequencing of the gene responsible of the suppression of the P53 tumor, which is involved in a large proportion of cancers. When analyzing the technology used in the mode of operation of the most advanced DNA sensing devices, it appears that the problems associated with the writing of ODN probes on DNA chips has been largely solved. Indeed, the systems proposed by Affimetrix and by Nanogen involve a precise optical addressing for the writing, or an electronic addressing for the control of the hybridization reaction, respectively. The electrochemically controlled ODN deposition proposed by Cis BioInternational is also appealing for constructing DNA chips, as it offers a very large-scale immobilization of ODN probes. On the other hand, the reading of the hybridized sensing sites still raises many questions. In fact, the currently described DNA chips require a first fixation of a tag, either a radioactive or a fluorescent one, involving a supplementary chemical step that costs time and effort. Second, the subsequent reading of the chip requires an expensive equipment, such as a laser associated with a confocal microscope, which limits the spatial resolution of sensing sites achievable with these chips. If a high resolution is desired, a rather long reading time, of some tenths of minutes, is needed. Finally, the transducing of an optical signal into an electric one results in a loss of the initial sensitivity. There is still a great need for improvement in the sensitivity, the ease of reading, and the time required for screening these DNA chips. Can new approaches be found for constructing biosensors? When considering the intrinsic specificity, sensitivity, and speed of recognition processes occurring in biological systems, it appears obvious that we have very much to learn from biology for building efficient sensors. In a quest for models, let us consider how a living body is organized for accomplishing the sensing of its environment.

HOW TO BUILD A SENSING ARCHITECTURE IN THE IMAGE OF BIOLOGICAL SYSTEMS

Living beings possess an ultimate degree of organization, showing on one hand the most elaborate distribution of recognition functions able to sense the chemical and the physical environment and, on the other hand, a large variety of acting centers that allow them to constantly adapt to this changing environment. The sensing activity is carried out by a complex system involving highly specialized entities operating in a multistep process: (1) receptacles formed by designed cell assemblies (e.g., retina, synapses) that act as transducers of external information (light, heat, sound, etc.), in form of pulses and (2) a nervous system that forwards these pulses to the brain, where (a) the information is stored and processed, leading to a (b) feedback message through the same information transfer route, the nervous system, to (c) various chemical actuators that define the response of the living being to external stimuli. In fact, one of the keys of this system is formed by the existence of a nervous system, which is able to establish an exchange of information between the brain and multibillions of receptacles distributed as molecular assemblies ranging from nanometers to micrometers throughout the living being.

A comparable sensing of the environment has raised a continuing interest among scientists, and various routes have been proposed for duplicating these functions.

How might one build a sophisticated architecture able to transport information, in the image of the nervous system, and connected at one end to various synapse analogs able to sense the environment, and at the other end to a substrate able to store and process the information, and eventually to deliver a feedback message to the interface with the environment? To answer this puzzling question, physicists have for a long time proposed a macroscopic approach through the construction of various large-sized devices, from very simple primitive sensors such as the mercury-based thermometer, to highly sophisticated systems such as the Coulter cell counter, which combine sensors, logic, and memories capable of accomplishing discrimination and eventually feedback operations. Macroscopic sensors and actuators mimicking the basic operations accomplished by living beings have been thus developed by physicists, and the recent progress in electronics technology has allowed the further miniaturization of these devices, together with an integration of logic and memory functions in the likeness of human body. This approach, however, encounters increasing technological problems, and it is now believed that the progress in scale down and miniaturization of these devices will not range below the thousand angstrom range. In addition, it must also be pointed out that, because of the lack of relevant detection system, no satisfactory answer has been brought by this physical approach for the duplication of some important basic biological functions such as olfaction and taste.

On the other hand, *biology* is the privileged discipline that has developed the concepts and experimental tools for understanding the fundamental processes associated with the behavior of living beings. From this knowledge, intriguing progress has been accomplished at the interface of biology and physical chemistry, based on the remarkable selectivity and activity shown by biological entities for designing new sensors and actuators. Biologically active entities, such as enzymes, have been successfully used for building gas sensors, and also as potent catalysts, such as as chemical actuators. This latter approach to sensors, however, is hampered by many factors, including (1) the difficulty of recovering a signal from the enzyme probe, which introduces the problems of long-term stable grafting; (2) the limited lifetime of most biological entities; and finally (3) the decrease of activity often observed for enzymes when not in synergy with their biological partners.

These considerations raise the issues as to whether it is possible to overcome the problems of miniaturization into which physics is running, to bypass the inherent complexity of biology that addresses highly organized and fragile entities, and to build a new class of nanoscale sensing architectures operating at the molecular level, in the image of biologically active systems. *Chemistry* appears a promising discipline for bringing a solution to this challenging aim, as it involves the engineering of molecular and macromolecular assemblies with specifically designed properties, from the microscopic to the mesoscopic and up to the macroscopic level. In fact, chemists know how to synthesize elemental "bricks," and how to assemble them in an ordered way into materials and how to shape specific chemical and physical properties. Chemistry may thus offer new privileged routes for building sensors in the image of biology.

In the following paragraphs, we report on the advances that have been accomplished in the design of new materials able to perform, at the molecular level, the basic functions of recognition, signal transduction, and even chemical actuation, which underlie intelligent structures.[15] The basic concept of these new class of sensors is associated with a knowledge recently acquired by chemists, who know how to engineer macromolecules that possess intrinsic electric conductivity. These can be considered as macromolecular wires in the image of nerves, and they can be functionalized with various prosthetic groups able to perform sensing and actuating functions.

CONJUGATED POLYMERS AS SYNTHETIC MACROMOLECULAR WIRES MIMICKING A NERVOUS SYSTEM

The concept of macromolecular wires stems from the early 1980s, when it was shown that the individual macromolecular chains of organic conjugated polymers (see Fig. 15.3) possess a high intrinsic electrical conductivity.[16] These organic conducting polymers have raised much interest, and a deep knowledge base on their physical and chemical properties has since been gathered. These polymers can be chemically synthesized, affording a powder that presents less than ideal processing properties. Much more preferable is a versatile electrochemical route, which permits generation of a thin film on an electrode. The deposition thickness can be controlled easily from several nanometers to several millimeters. The macromolecular chains of these polymers show a regular alternation of single and double bonds, as exemplified by polyacetylene, the prototype of these conjugated polymers. A large variety of conjugated polymers have been since proposed in the literature, either chemically synthesized (e.g., polyacetylene, polyphenylenes) or electropolymerized as thin films onto an electrode (e.g., polypyrrole, polythiophenes).

All the carbon atoms of the macromolecular chains possess an unpaired atomic *p* orbital. Hybridization of these orbitals creates a supramolecular π-type orbital that extends throughout the skeleton of the polymer chain, leading to the formation of two electronic bands: a fully occupied one, the valence band; and a completely empty one, the conduction band, reminiscent of the highest occupied molecular orbital (HOMO) and the lowest unoccupied molecular orbital (LUMO), respectively. These polymers can be easily oxidized (or reduced), either chemically or

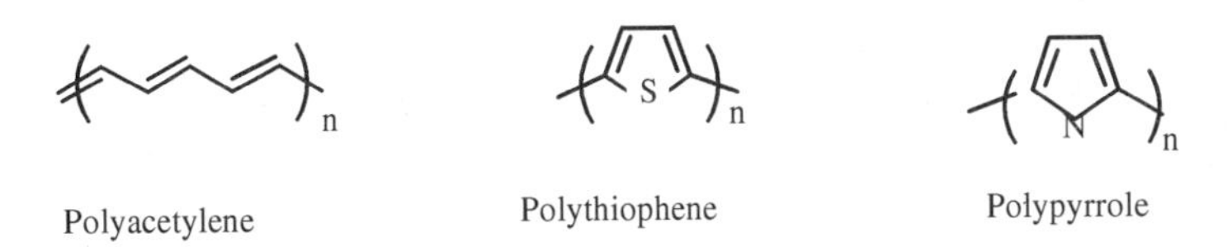

FIGURE 15.3 Some examples of conjugated conducting polymers.

electrochemically, which means that electrons can be injected in the conduction band, or extracted from the valence band. These extraneous charges are free to move, thereby leading to an intrinsic electrical conductivity along the macromolecular chain. A precise description of the oxidation–reduction process would acknowledge the contribution of localized states within the gap, known as *polaronic states*; however, the given simplification is sufficient for understanding the origin of conduction. The experimentally obtained electrical conductivities of conjugated polymers range from 1 to 10^4 S/cm.[16] This is 2–3 orders of magnitude lower than that of metals but high enough for considering these macromolecular chains of several-micrometer length as molecular wires, able to transport information. When these polymers are deposited as thin films on an electrode, the reversible oxidoreduction process can be driven electrochemically, by applying a potential sweep to the electrode, which results in a cyclic voltammogram representing the electrochemical signature of the polymer. The obtained oxidation peak value depends on the steric constraints associated with the variation of the internal polymer chain structure occurring on (electro)chemical oxidation. In its neutral state, a polyheterocycle consists of heterocycle units, such as pyrrole nuclei, linked together through a sigma bond. Free rotation around this sigma bond allows the polypyrrole chain structure to accommodate steric hindrances that might originate from the presence of bulky groups substituted on the pyrrole nuclei. On oxidation, the injection of a positive charge onto the chain results in the transformation of the polyaromatic structure into a polyquinoid one, which requires all pyrrole nuclei of the polypyrrole chain to become coplanar. Depending on the steric hindrance existing in the pendent substituents along the polymer chains, the oxidation for passing from the polyaromatic to the polyquinoid structure will require more or less energy. The electrochemical oxidation potential of a functionalized conjugated polymer thus forms a very sensitive criterion that reflects the steric modifications occurring at the vicinity of the polymer chains.

In conclusion, conjugated polymers can be described as a tridimensional network of conducting macromolecular chains, of about 10 Å diameter, characterized by an electrochemical signature that is very sensitive to their nearby environment. Any modification appearing on the pendent groups along the chains will be electrically transduced to the supporting electrode, in the form of a modification of the electrochemical signature of the polymer.

If these polymers can now be considered analogs of nerves, how can corresponding synapses be constructed? The answer is obtained by the functionalization of these polymers with various prosthetic groups that can be covalently bonded as pendent groups (R) along these conducting chains.

Different functionalities can thus be inserted either as homo- or as plurisubstituted polymers, and many examples have been given in the literature on the introduction of various chemical, catalytic, optical, and prochiral properties.[15] Among these functionalities, a particularly interesting class concerns molecules or entities that are known to specifically interact with species in solution. In fact, chemists have long designed molecules and materials that show selectivity toward cations, enzymes, DNAs, or antibodies by the use of chelates,[17] inhibiting peptides,[18] complementary oligonucleotides,[19] and antigens, respectively. Such reversible interactions are at the basis of affinity chromatography. Specific chemical recognition toward various chemical or biological species can be built through the proper functionalization of conjugated polymers. These sensing groups, which operate at the molecular level, represent the desired analogs of synapses.

Functionalized conjugated polymers can thus be considered as a three-dimensional network of macromolecular conducting wires bearing various sensing groups that mimic the sensorial organization of living bodies.[15] Some examples of such new molecular-based sensing devices have been described in the literature. These have been designed for specific chemical (ion) or biological (enzyme, DNA) analytes. Besides the beauty of synthesizing sophisticated architectures able to perform recognition processes at the image of biological entities, there is also a tremendous interest related to their application in some sensing areas, such as that linked to DNA recognition. The mapping of DNA sequence, the characterization of genetic diseases, the diagnosis of viruses, the determination of human lymphocyte activity (HLA) groups[20] obviously confirm the very large application potential of DNA sensing devices.

In the following paragraphs, we first describe the route used for building polymer-based sensors, then present the first results obtained in the sensing of DNA, and finally envision the potential of these macromolecular systems in the frame of biosensing devices.

HOW TO CONSTRUCT MOLECULE-BASED DNA SENSING ARCHITECTURES

The scenario for constructing such a sensing device appears relatively simple as these polymers can be easily electropolymerized onto a metallic (micro)electrode, previously patterned on a substrate. By counting the number of coulombs involved in the electropolymerization reaction of the monomer, a precise control can be exerted on the total quantity of monomer units (and consequently the thickness) of the polymer film that will be obtained on the electrode metal surface.

When considering the recognition of biological species, some further requirements have to be taken into account. Biological recognition is most often realized in aqueous solutions, which means that the conjugated polymer must be electroactive (i.e., hydrophilic) in water. In addition, some future attempts can be envisioned for the implantation of such an electrode in vivo, which means that the polymer should be biocompatible. In this regard, polypyrrole appears one of the best adapted polymer candidates, and the majority of work has focused on this polymer.

A third consideration involves the availability and the fragility of the R-sensing group that must be grafted onto the polymer chain. The introduction of R can first be envisioned on the monomer, in form of a R-substituted pyrrole molecule. However, we have learned from the chemistry of conjugated polymers that the electropolymerization requires substantial concentrations of highly pure monomer (around 10^{-1} M/L). Furthermore, the appearance of some degradation products during the electropolymerization, has led to the general observation that a polymerization medium does not proceed beyond a 50% consumption of the contained monomer. This means that a large quantity of R-substituted pyrrole monomer is needed, and that a significant amount of it is decomposed and lost. All these considerations appear to be incompatible with the use of costly and fragile R substrates, such as oligonucleotides. So another route has been developed, involving the first electropolymerization of a precursor polymer possessing an activated ester group, ester*, on the 3 position of the pyrrole nuclei to produce a poly(3-ester* pyrrole). The precursor polymer film is then substituted in a second step with a nucleophilic R group, which displaces the labile ester, leading to the desired functionalized polymer poly(3-R pyrrole). The substitution reaction for grafting the R group can be realized by a simple contact of the polymer film surface with a (micro)pipette containing the desired R reacting solution. The tiny necessary quantity of R will be consumed in a chemical substitution step without loss. The validity of this precursor polymer approach has been confirmed by the use of an electrochemical probe where R represents aminoferrocene. This substitution has been carried out on a poly(3-hydroxysuccinimide* pyrrole). Such substitution occurs with a very high yield, approaching 100%, even for large concentrations of about 10^{-6} mol/cm^2-area of electrode. It has also been observed that the grafted ferrocene retains a high degree of electroactivity, which confirms the validity of this precursor polymer route approach.[21]

The last point concerns the respective sizes of the probe and target entities. In fact, the sensing function R (e.g., an ODN probe) and most importantly the target entity (e.g., a DNA single strand) can easily reach a geometric size that largely exceeds that of the pyrrole monomer unit. A partial functionalization of the pyrrole monomer units within the polymer film then becomes sufficient. This can be achieved by the use of a precursor copolymer, containing a statistical distribution of (3-ester* pyrrole) units together with unreactive pyrrole units, but bearing other interesting functionalities such as hydroxyl groups, which are useful for preventing the organic electrode from nonspecific binding.

A multisensing matricial device for DNA recognition can thus be easily constructed as follows. The first step involves the patterning of the substrate (which can be a silicon chip or a plastic sheet) for establishing the electric microcircuitry. The deposition, by the use of conventional microlithography techniques, of [n columns $\times$ n lines] of gold stripes affords $n \times n$ gold micrometer-sized electrodes at the column-line intersections. Each ij electrode is individually addressed by imposing the required half-potential at line i and column j.

The subsequent deposition of the poly(ODN pyrrole) can be achieved following two procedures, corresponding to either an electrical addressing or a mechanical addressing. The first procedure involves the only deposition of one precursor polymer spot at a time, which can be achieved through the electrochemical control of the polymerization. When addressing electrically line i and column j, a precursor polymer film will be deposited only on the ij electrode, which is subsequently immersed in the medium containing the desired oligonucleotide probe ODN_{ij}. This procedure allows an electrochemical "writing" of the sensing device, electrode by electrode, and hence an electrical addressing of ODNs. A second procedure that can be employed involves the electrochemical deposition of a film of the precursor polymer, poly[3-ester* pyrrole] on all microelectrodes at the same time, by addressing all columns and lines in an electropolymerization medium. This step is then followed by the chemical substitution of defined oligonucleotide probes on each of these $n \times n$ precursor polymer coated electrodes, by the use of adapted micropipettes with tips maintained in contact with the precursor polymer. This step corresponds to the mechanical addressing already described in the case of polyacrylamide gels.[6] Standard experimental conditions allow this easy substitution to be achieved, and each electrode is then covalently bonded with a defined ODN_{ij} probe. The electrically controlled procedure for ODN deposition should afford a much higher resolution than the second one, based on a mechanical approach.

REALIZATION OF THE OLIGONUCLEOTIDE-MODIFIED ELECTRODE

The main synthetic problem concerned the generation of a conducting polypyrrole film functionalized with such large and bulky groups as oligonucleotides (ODNs); however, the polypyrrole film must also show aqueous medium electroactivity in aqueous medium, which is a prerequisite for allowing an electrical reading of a possible recognition of the ODN probe with external complementary DNA target. With this aim, the precursor copolymer poly[(3-acetic acid pyrrole), (3-*N*-hydroxyphthalimide pyrrole)][19] (**1**), was realized bearing the activated ester *N*-hydroxyphthalimide (Fig. 15.4), which can be further easily substituted by a prosthetic group bearing a terminal amino function.

This precursor copolymer (**1**) was first electropolymerized as a film of about 200 nm thickness, onto a platinum electrode, of 0.7 cm^2 area, in acetonitrile solution containing both monomers: 3-acetic acid pyrrole and 3-*N*-hydroxyphthalimide pyrrole. The precursor polymer film obtained (**1**) shows a high intrinsic conductivity, compatible with the concept of macromolecular wires for these materials.[15]

In subsequent advance, an amino-substituted oligonucleotide, ODN, has been grafted on (**1**) by a direct chemical substitution of the facile leaving group, *N*-hydroxyphthalimide, following a conventional route in dimethylformamide (DMF) containing 10% acetate buffer, at pH = 6.8 (Fig. 15.5). The oligonucleotide, 5′ CCT AAG AGG GAG TG 3′ (**2**), bearing an amino group on its 5′ phosphorylated position, was provided by the Biomerieux Company. This substitution led to the final functionalized polypyrrole film, poly[(3-acetic acid pyrrole), (3-ODN acetamido pyrrole)] (**3**). This polymer film was carefully washed with DMF and water, in order to remove any trace of ungrafted ODN, and also to hydrolyze the unreacted *N*-hydroxyphthalimide leaving group to acetic acid groups.

The presence of the oligonucleotide (**2**) as grafted pendent groups onto the polypyrrole chains (**3**) was confirmed by its signature in infrared spectroscopy and by the use of X-ray fluorescence which confirmed the presence of phophorus atoms in polymer (**3**).

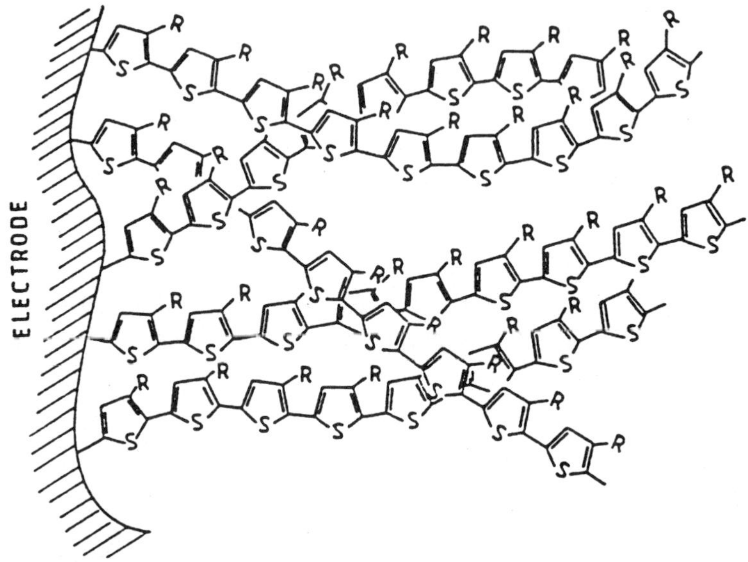

FIGURE 15.4 Macromolecular chains of a conjugated polythiophene deposited onto an electrode, and bearing prosthetic groups R on each monomer unit.

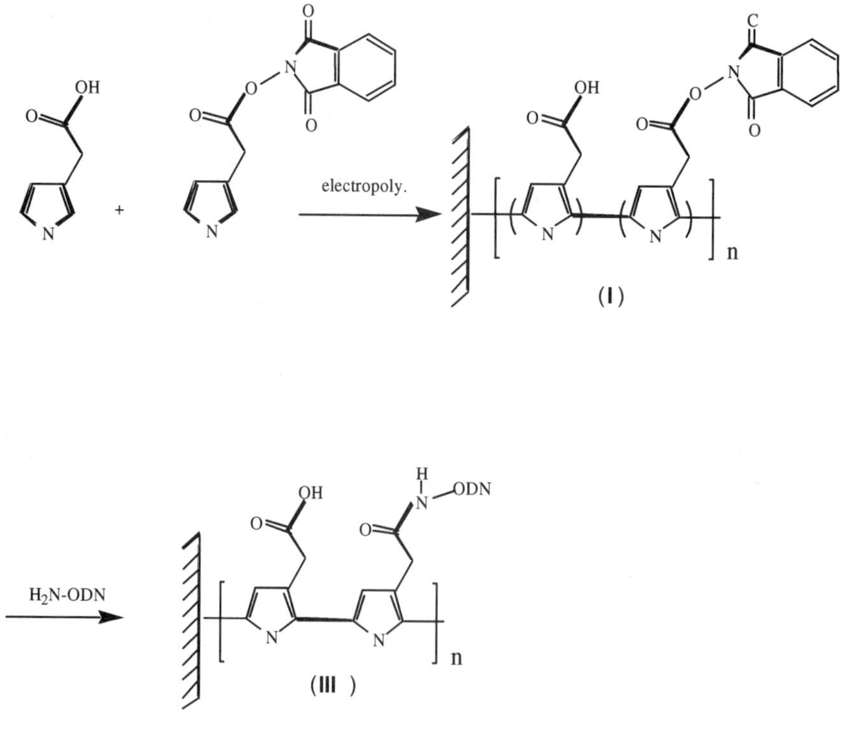

FIGURE 15.5 Electropolymerization of the precursor copolymer (**I**), bearing an activated *N*-hydroxyphthalimide ester*, followed by the substitution with an amino-functionalized oligonucleotide ODN (**II**), leading to the ODN-bearing polypyrrole (**III**).

ELECTRICAL SENSING OF DNA TARGET

The relevance of these electrodes for electrical DNA recognition has been tested using various grafted ODN probes and ODN targets in solution. The modified electrode poly[(COOH-pyrrole), (ODN-pyrrole)] (**3**), has been electrochemically characterized both in H_2O–NaCl 0.5 M, and in a biological buffered aqueous medium, involving a mixture of salmon DNA, sodium phosphate, and Tween buffer, at pH 7. The voltammograms of (**3**) in these media were identical, Figure 15.6*a*, showing an oxidation peak at −0.2 V/SCE which corresponds to the oxidation (or doping) of the poypyrrole chains. This low oxidation potential value, together with the symmetry of the redox waves, confirmed the high electroactivity of the ODN-substituted polypyrrole film. Also, the absence of any effect due to salmon DNA suggested that no interaction occurred between the grafted ODN probe and this noncomplementary DNA, and furthermore, that the electrochemical response of polymer (**3**) was not affected by the presence of such nonspecific DNA in the electrolytic medium.

In a second step, the modified electrode (**3**) was incubated during 2 h at 37 °C, in buffered aqueous solution of various oligonucleotides, used as targets for hybridiza-

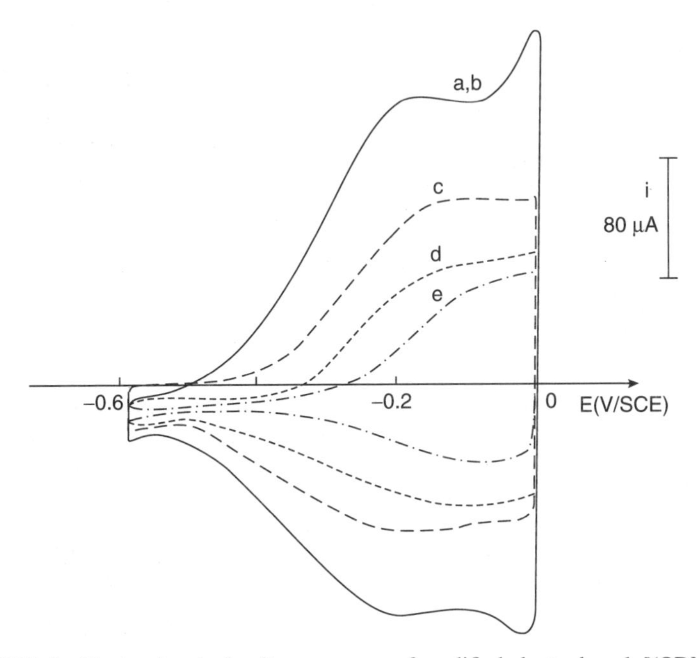

FIGURE 15.6 Electrochemical voltammograms of modified electrode poly[(ODN pyrrole) (COOH pyrrole)] (**III**) after 2 h of incubation (37 °C) in buffered aqueous solution containing (a) H_2O–NaCl 0.5 M or PEG, salmon DNA; (b) noncomplementary ODN target (**V**); 6c–e: complementary ODN target (**IV**) in concentrations of 66 nmol (c), 165 nmol (d), and 500 nmol (e).

tion with the grafted oligonucleotide probe (**2**). Among these, an oligonucleotide target constituted of 14 bases, bearing the sequence 5′ CAC TCC CTC TTA GG 3′ (**4**) has been used, the structure of which was complementary to the probe (**2**). Another noncomplementary oligonucleotide target consisting of 15 bases, with the sequence 5′ GGT GAT AGA AGT ATC 3′ (**5**), has also been analyzed. The incubation solution was 5 mL, in which different amounts of **4** and **5** were used, varying between 0 and 10^3 nmol. After incubation, these electrodes were washed using the same buffer, and electrochemically characterized in a buffered 0.5 M NaCl aqueous medium.

The results showed that when incubated in presence of the noncomplementary oligonucleotide (**5**), the electrochemical response of the modified electrode, (Fig. 15.6*b*) remained unchanged as compared to Figure 15.6*a*, whatever the concentration of (**5**). This indicated that no interaction occurred between the ODN probe (**2**), grafted on polypyrrole, and the noncomplementary oligonucleotide target (**5**) in solution. On the other hand, when the complementary ODN target (**4**) (which is known to undergo a specific hybridization reaction under the incubation conditions used in this experiment) was used, a significant modification in the voltammogram was observed (Fig. 15.6*c*–*e*). The results obtained showed on one hand a decrease of the intensity of the oxidation wave, and on the other hand a shift of the oxidation wave to higher potential.

To confirm the generality of this recognition process, and to analyze the effect of base sequence length, another ODN probe comprising 25 bases, 5′ TCA ATC TCG GGA ATC TCA ATG TTA G 3′ (**6**), has been substituted onto the precursor polymer poly (**1**), and the resulting ODN-functionalized polymer, poly-(**7**), has been analyzed in presence of a complementary ODN target bearing 25 bases, 5′ CTA ACA TTG AGA TTC CCG AGA TTG A 3′ (**8**). The results (Fig. 15.7) confirmed the occurrence of a specific recognition process, and, furthermore, the detection sensitivity appeared to largely increase with the hybridization base sequence length. The oxidation potential of the 25-mer electrode underwent a remarkable potential shift, induced by a much lower concentration, of ~25 nmol, for the 25-mer ODN target in solution. As described above, these electrochemical features are reminiscent of the behavior of functionalized conjugated polyheterocycles when engaged in a recognition process. Thus polypyrroles and polythiophenes, substituted with various crown ether derivatives, have shown an increase in oxidation potential specific insertion of a cation from the electrolytic medium into the pendent crown ether cages. Peptide-substituted polypyrroles have also shown a significant increase in oxidation potential when the peptide inhibitor ligand underwent complexation with an enzyme.[18] These shifts in oxidation potential of functionalized conjugated polyheterocycles, observed complexation, are attributed to an increase in the bulkiness and the stiffness of the pendent ODN group, which follows the complexation reaction occuring at these recognition sites.

This interpretation is supported by the fact that the shift in the voltammogram increases with the recognition length of base sequences, as observed when passing from a 15-mers to a 25-mers (Figs. 15.6 and 15.7). These results clearly confirm that oligonucleotide-functionalized polypyrroles are able to specifically recognize their

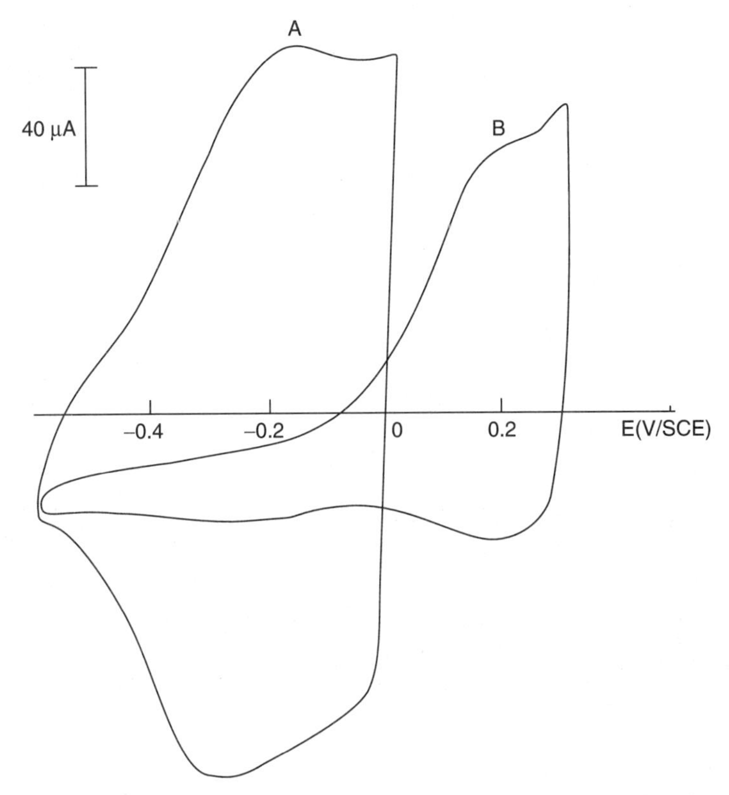

FIGURE 15.7 Electrochemical voltammograms of modified electrode poly[(COOH pyrrole), (ODN pyrrole)] (**VII**), with ODN probe (**VI**), after 2 h incubation (37 °C) in buffered aqueous solution containing the complementary ODN target (**VIII**) in concentration of 25 nmol.

complementary ODN targets, and to transduce this information into an electrochemical signal that can be easily read by an electrode.

This electrochemical modification can be further used for analyzing, in a quantitative way, the sensitivity of such bioelectrochemical sensors. The shape of the voltammograms shows that a potential "window" exists around −0.2 V/SCE, where the shift in oxidation potential from Figure 15.6*b* to 15.6*d*–*e* is accompanied by a large variation of electrode oxidation current intensity, allowing an amperometric analysis of the hybridization.

When applying a potential of −0.2 V/SCE to the electrode, no variation of the electrode oxidation current is observed on addition of the noncomplementary ODN (**5**) to the electrolytic medium (Fig. 15.8*a*). On the other hand, when adding the complementary ODN (**4**), the electrode current decreases continuously as a function of the ODN target concentration in solution (Fig. 15.8*b*). The ODN concentration variation and the asymptotic value agree with a bimolecular hybridization

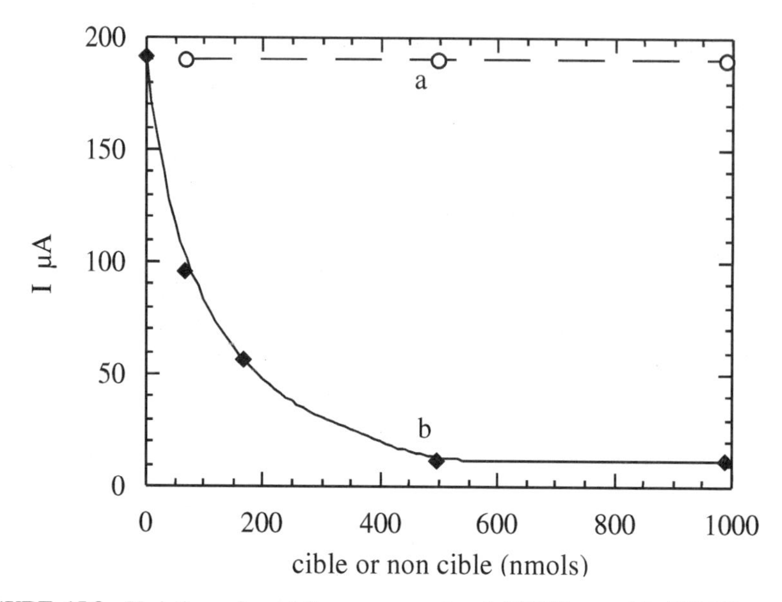

FIGURE 15.8 Variation of oxidation current of poly[(ODN pyrrole) (COOH pyrrole)] electrode (**III**) at constant potential $E = -0.2\,\text{V/SCE}$, as function of oligonucleotide target concentration used during incubation: 8(*a*) (*a*) noncomplementary ODN (**V**); (*b*) complementary ODN (**IV**).

equilibrium, with a constant in the range of $10^7\,\text{L/mol}$, in agreement with literature data for this ODN. This relationship allows thus a quantitative determination of the concentration of the complementary ODN existing in solution. The sensitivity of this electrochemical sensing process can be evaluated by the slope at the origin of this curve, of about 1 μA/nmol of complementary ODN. Taking account of the reproducibility observed for a set of four synthesized electrodes, a value of 10 nA for the electrochemical oxidation current intensity variation would correspond to an ODN target concentration of about 10^{-2} nmol as detection limit, without any signal treatment, for such an amperometric biosensing modified electrode. When compared to the sensitivity achieved when using a fluorescent probe, on the order of 10^{-10} mol,[22] the value of 10^{-11} mol obtained in this work, together with the easiness associated with a direct electrical reading technique, appears very promising.

This sensitivity is, however, still too low for practical applications in the field of gene diagnoses, but two approaches can be employed to increase sensitivity. A first improvement has been brought by the use of an internal electrochemical probe, ferrocene, which is known to present a very sharp electrochemical signal. Thus, the substitution at the 3 position of a ferrocene probe has allowed improvement of the detection limit of these ODN functionalized polypyrroles down to 10^{-13} mol of ODN target. A further increase of sensitivity currently under study involves the use of an electrochemical transistor configuration,[23] which should bring an on-site amplification of the signal, and hence an increased sensitivity.

These results confirm that ODN-functionalized polypyrroles are able to specifically recognize complementary DNA strands in aqueous media, and to transduce this recognition into a molecular signal that is sent to the supporting electrode through the conducting polypyrrole chain.

WHAT'S NEW

In conclusion, when compared to the already existing DNA chips described in the literature, what are the main potential interests offered by the present molecule-based sensing architectures operating through electrochemical recognition?

1. These ODN-bearing polymers are obtained from an electrochemically deposited polymer precursor, which means that they offer an *electrical addressing* during the deposition, or "writing" step of the $n \times n$ sensing dots. Electronics has a long history and experience in constructing matricial arrays of conducting stripes of a given micrometer width. When applying the half-potential bias at column i and line j, electropolymerization will occur only at the crossing of these two conducting stripes, affording a micrometer-sized electrode on which the precursor polymer, bearing an activated ester* group, can be electrochemically grown. The contact of the chip with a solution containing the amino functionalized ODN_{ij} will result in the controlled grafting of the ODN_{ij} on that precise electrode. An alternative way could consist in the simultaneous electropolymerization of the precursor polymer on all the micrometer-sized n^2 electrodes, followed by a mechanical adressing, using micropipetting, of each electrode ij with a solution containing the desired amino-functionalized ODN_{ij}, as already performed in the construction of some DNA chips.

2. The "reading" of the hybridized DNA chip is provided by the *transduction of the recognition process into an electrical signal*, which is transmitted directly to the supporting electrode. *No tag* is necessary, which allows bypass of the chemical step needed for the fixation of a fluorescent species onto the DNA target. Electrical signals can be easily processed and stored, allowing a real-time acquisition and computational analysis of data, and hence a very fast and clean delivery of information. These DNA chips thus provide a fast, reliable, and sensitive response.

3. Since the electrical reading of the electrodes is performed by the simple application of a potential bias, a significant *scale down* can be expected for these DNA chips. Resolution of the order of 10 μm are achievable, which means that sensing site density exceeding 10^6 can be envisioned.

4. As described previously, this approach of functionalized conjugated polymers has already been successfully applied in the field of enzyme characterization by the use of peptide inhibitors as probes, which suggests that this concept can be further generalized to the case of *antibody sensing*, by the grafting of the associated antigen onto the conjugated polymer. Initial results in this direction are very encouraging and confirm the generality of this approach.

As presented in the introduction of this chapter, this work focuses on the building of molecule-based sensing architectures that mimic recognition processes occurring in living bodies, specifically, receptacles delivering through the nervous system messages to the brain, where they are processed. These features have been synthetically designed through the engineering of ODN-functionalized conducting polymers. The molecular signals obtained from the ODN probes are transduced through the conducting polymer chains to the supporting electrode, where they can be processed by the use of external logic and memory functions. The analogy of the features shown by these synthetic architectures with the processes associated with living bodies can be even further developed when considering the whole scope of the chemical functionalization of conducting polymers. In fact, in addition to sensing functions, *actuating functions* can also be introduced in these polymers as already largely demonstrated in the literature. Electrocatalytic properties have been obtained for these polymers by the grafting of organometallic derivatives, or by the inclusion of nanometer-scale aggregates of metals such as platinum.[24,25] Electrochemically controlled delivery of negatively charged species (including drugs) or of protons have also been reported. Multifunctionalized conjugated polymers could thus perform the entire successive steps of (1) sensing the biological environment, (2) transmitting the informations to the supporting electrode, where (3) the data can be processed by the use of logics, at the image of the brain, and, most importantly, (4) deliver a feedback message to some actuating groups associated with the polymer chain, such as to trigger a reaction, or release drugs or protons, defining the response to the sensing operation. All these partial activities have been individually demonstrated, and their association into highly sophisticated interactive (implanted) biochips will be the subject of a future story.

ACKNOWLEDGMENT

The Biomérieux Company (Drs. B. Mandrand, and T. Delair) Lyon, France, is acknowledged for the support of this work.

REFERENCES

1. Hodgson, J., *Biotechnology*, 1995, **13**, 231.
2. Cotton, R. G. H.; Edkins, E.; Forrest S., *Mutation Detection, a Practical Approach*, Oxford Univ. Press, Oxford, 1998.
3. Singer, M.; Berg, P., *Exploring Genetic Mechanisms*, CRC Press, New York, 1997.
4. Rehm, H.J.; Reed, G., *Biotechnology: Recombinant Proteins, Monoclonal Antibodies and Therapeutic Genes*, Wiley-VCH, Weinheim, 1999.
5. Ehrlich H. A.; Saikiki, R. K.; Walsh, P. S.; Levenson, C. H., *Proc. Natl. Acad. Sci.* (USA) 1989, **86**, 6230.
6. Sheldon, E. L., *Clin. Chem.* 1993, **39**, 718.
7. Matson, R. S.; Rampal, J. B.; Coassin, P. J., *Analyt. Biochem.* 1994, **217**, 306.

8. Yershov, G.; Barsky, V.; Belgovskiy, E.; Kirillov, E.; Kreindlin, E.; Ivanov, I.; Parinov, S.; Gushin, D.; Drobishev, A., *Proc. Natl. Acad. Sc.* (USA), 1996, **93**, 4913.
9. Maskos, U.; Southern, E. M., *Nucleic Acids Res.* 1992, **20**, 1675.
10. Merel, P., *Biofutur*, 1995, **142**, 46.
11. Livache, T.; Roget, A.; Dejean, E.; Barthet, C.; Bidan, G.; Theoule, R., *Nucleic Acids Res.* 1994, **22**, 2915.
12. Mirzabekov, A. D., *Trends Biotechnol.* 1994, **12**, 27
13. Lipshutz, R. J.; Morris, D.; Chee, M.; Hubbell, E.; Kozal, M. J.; Shah, N.; Shen, N.; Yang, R.; Fodor, S. P. A., *Biotechniques* 1995, **19**, 442.
14. Souteyrand, E.; Martin, J. R.; Martelet, C., *Sens. Actuators B,* 1994, **20**, 63.
15. Garnier, F., *Angew. Chem.* 1989, **101**, 529.
16. Skotheim, T. A., *Handbook of Conducting Polymers*, Marcel Dekker, New York, 1986.
17. Korri-Youssoufi, H.; Hmyene, M.; Garnier, F.; Delabouglise, D., *J. Chem. Soc. Chem. Commun.* 1993, 1550.
18. Garnier, F.; Korri-Youssoufi, H.; Srivastava, P.; Yassar, A., *J. Am. Chem. Soc.* 1994, **116**, 8813.
19. Garnier, F.; Korri Youssoufi, H.; Srivastava, P.; Godillot, P.; Yassar, A., *J. Am. Chem. Soc.* 1997, **119**, 7388.
20. Van Ness, J.; Kalbfleisch, S.; Petrie, C. R.; Reed, M. W.; Tabone, J. C.; Vermeulen, N. M. J., *Nucleic Acids Res.* 1991, **19**, 3345.
21. Godillot, P.; Korri Youssoufi, H.; Garnier, F.; El Kassmi, A.; Srivastava, P., *Synth. Met.* 1996, **83**, 117.
22. Morrison, L. E.; Stols, L. M., *Biochemistry* 1993, **32**, 3095.
23. Wrighton, M. S., *Science* 1986, **231**, 32.
24. Deronzier, A.; Moutet, J. C., *Acc. Chem. Res.* 1989, **22**, 249.
25. Tourillon, G.; Garnier, F., *Phys. Chem.* 1984, **88**, 5281.

CHAPTER 16

Peptide Nucleic Acids (PNA): Toward Gene Therapeutic Drugs

PETER E. NIELSEN

Center for Biomolecular Recognition, Department of Medical Biochemistry & Genetics, The Panum Institute

INTRODUCTION

The rapid accumulation of genomic sequence information concerning human decease related genes, and the soon expected total sequencing of the human genome, as well as our steadily increasing understanding at the molecular level of the causes of human deceases—not the least cancer—have made gene therapeutic drug approaches very attractive. These approaches rely on the antisense principle (Fig. 16.1), in which the mRNA is targeted, and the antigene principle (Fig. 16.2), in which the gene itself, the genomic DNA, is targeted.

ANTISENSE DRUGS

The first antisense experiments were reported in 1978,[1] but the extremely attractive principle that in essence allows one to target (down-regulate) any (un)desired gene just from knowing part of its sequence and designing a nucleobase–complementary oligonucleotide to this (Fig. 16.1), did not really catch on for another 10 years. This was chiefly due to lack of efficient techniques for chemically synthesizing oligonucleotides. However, following the development of solid phase DNA synthesis and also taking advantage of the rapid progress in molecular biology, antisense technology experienced a major boost in the late 1980s, especially within the pharmaceutical industry. Unfortunately, the technology was initially dealt with as a chemical problem of synthesizing biologically stable analogs of DNA, thereby essentially neglecting important issues, such as cell biology, bioavailability, and

Biomedical Chemistry: Applying Chemical Principles to the Understanding and Treatment of Disease, Edited by Paul F. Torrence
ISBN 0-471-32633-x

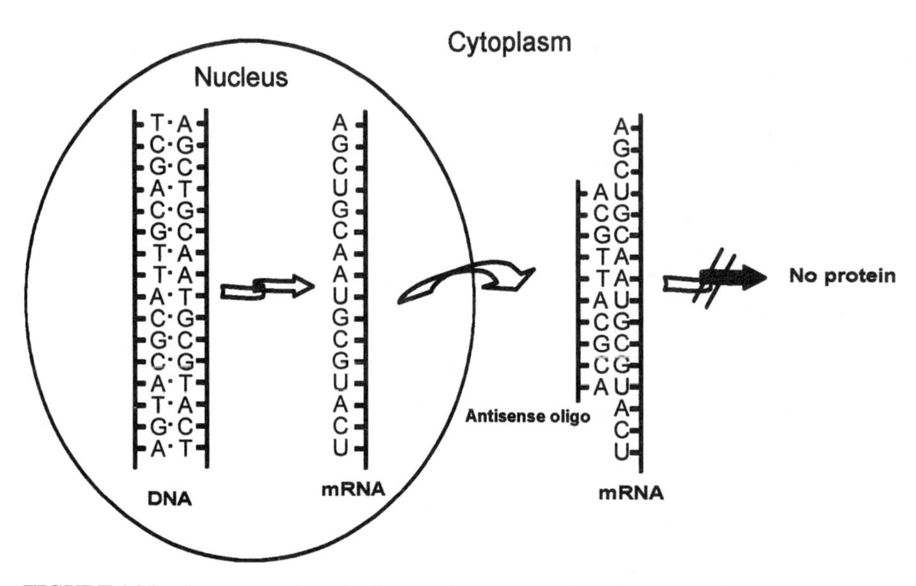

FIGURE 16.1 Antisense principle. Schematic drawing of a eukaryotic cell showing how the gene (DNA) is transcribed into mRNA in the nucleus, and where the mRNA is transported into the cytoplasm where it would finally be translated into the protein gene product. However, binding of a sequence complementary antisense oligonucleotide interferes with the translation process and inhibits the synthesis of the protein. The total length of the DNA in human cells is 2 m, and it contains 3 billion base pairs. It is estimated that less than 10% is actually transcribed into RNA, corresponding to ~100,000 genes. Thus in an antisense approach one is aiming at one gene out of 100,000 and at 15–20 bases out of 100 million.

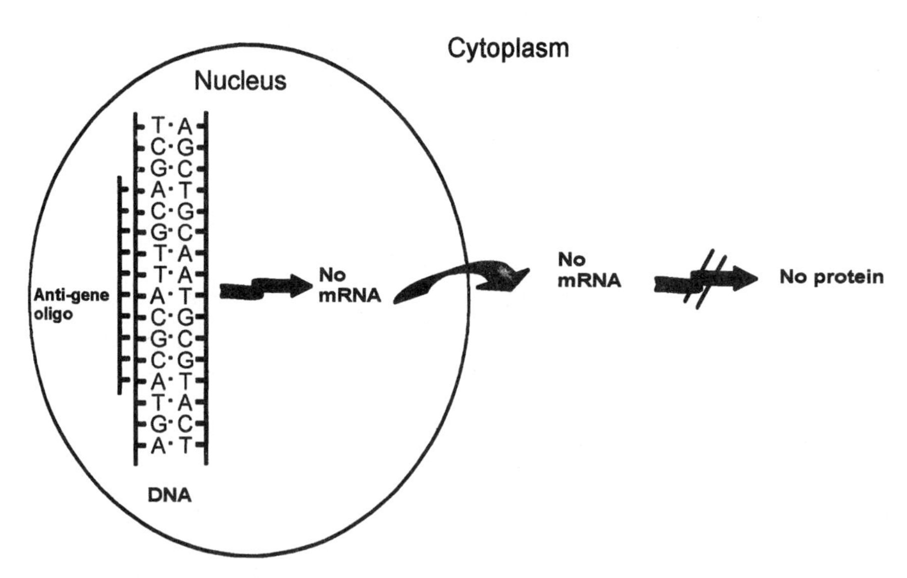

FIGURE 16.2 Antigene principle. Schematic drawing of a eukaryotic cell showing how the expression of a gene (DNA) is blocked by a reagent that sequence specifically binds to the DNA.

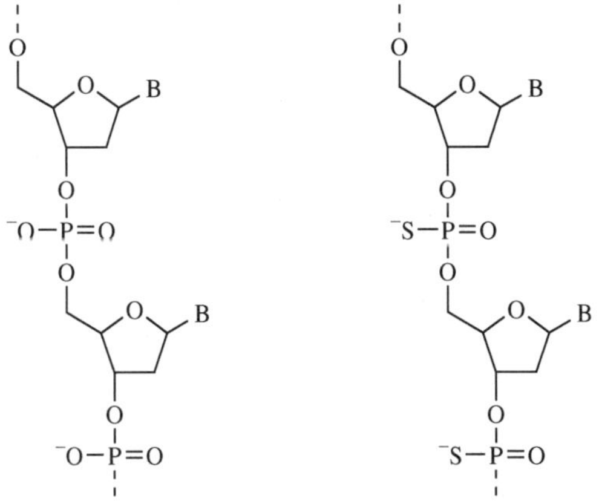

DNA Phosphorothioate

FIGURE 16.3 Chemical structures of DNA and the analogs phophorothioate, which is used in most antisense drugs undergoing clinical trials today (B is one of the nucleobases, adenine, cytosine, guanine, and thymine).

pharmacokinetics. However, quite quickly a promising DNA analog, the phosphorothioate (Fig. 16.3), in which one of the nonbridging oxygens in the phosphodiester linkage was exchanged with sulfur was developed. Virtually all the antisense drugs that are presently undergoing clinical trials are phosphorotioates. This also includes the drug vitravene against CMV induced retinitis in AIDS patients, which was recently approved by the U.S. FDA as the first oligonucleotide-based drug. However, an impressive number of oligonucleotide analogs have since been developed,[2–4] including profound alterations of either (or both) the phosphodiester linkage or (and) the (deoxy)ribose sugar, and some of these are also beginning to enter clinical trials.

ANTIGENE DRUGS

While sequence-specific recognition of mRNA is straightforward via simple Watson–Crick base-pairing, sequence-specific recognition of double-stranded DNA is much more difficult. Basically three options are presently available to the (medicinal) chemist.[5] Triplex forming oligonucleotides that recognize DNA via the major groove and for which the targets are essentially restricted to homopurine sequences. Small hairpin polyamides that are analogs of distamycin recognize DNA via the minor groove.[6] Finally, PNA may bind to double-stranded DNA via triplex or double-duplex invasion (see the following paragraphs). In terms of drug development, the antigene technology is less advanced than the antisense technology, and clinical trials have not even been initiated.

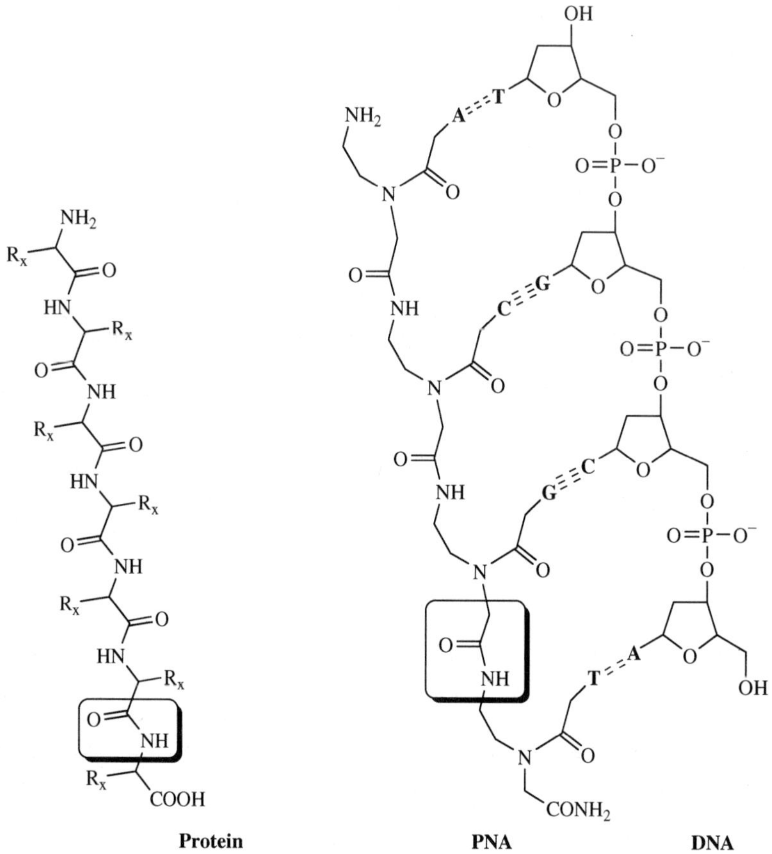

Protein **PNA** **DNA**

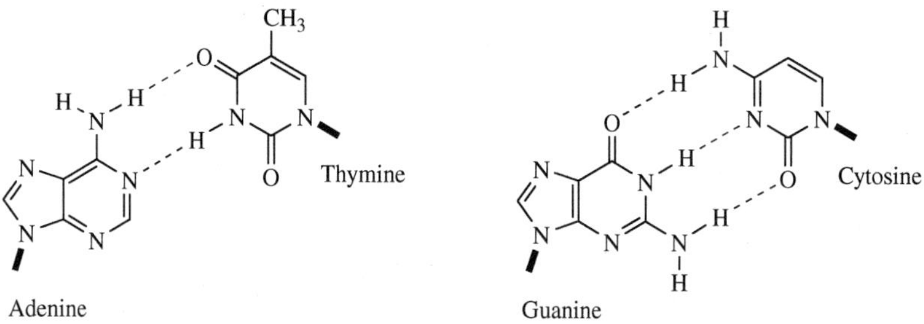

Adenine Thymine Guanine Cytosine

FIGURE 16.4 (*a*) Chemical structures of a protein (peptide) (where Rx is an amino acid side chain), a PNA and a DNA molecule. The amide (peptide) bond characteristic for both PNA and protein is boxed in. (*b*) Chemical structures of the four nucleobases, adenine (A), cytosine (C), guanine (G), and thymine (T), showing how they recognize each other (A-T and G-C) by Watson–Crick hydrogen bonding.

PEPTIDE NUCLEIC ACID (PNA)

PNA was originally conceived and designed as an oligonucleotide mimic for sequence specific recognition of double-stranded DNA via triplex formation.[7–11] Chemically, PNA is a polyamide or a pseudopeptide and it is therefore more closely related to proteins than to nucleic acids (Fig. 16.4). Nonetheless, PNA shares many structural properties with DNA. In fact, PNA may be regarded as a (pseudo)peptide with DNA-like properties in relation to nucleobase recognition.

Chemistry

PNA oligomers are synthesized by conventional solid-phase peptide chemistry using *nucleobase amino acid* monomers, such as those shown in Figure 16.5. Thus, making milligram–gram quantities of 10–20-mer PNAs is now routine, and both monomers and even oligomers are commercially available. The straightforward chemistry of PNA has inspired chemists to synthesize a large number of derivatives and analogs of PNA in order to elucidate structure–activity relations, and eventually obtain reagents with improved antisense properties.[9,11,12] Chimeric molecules between PNA and DNA have also been developed. In contrast to oligonucleotides, PNAs can easily be modified by conjugation to either peptides or molecular tags

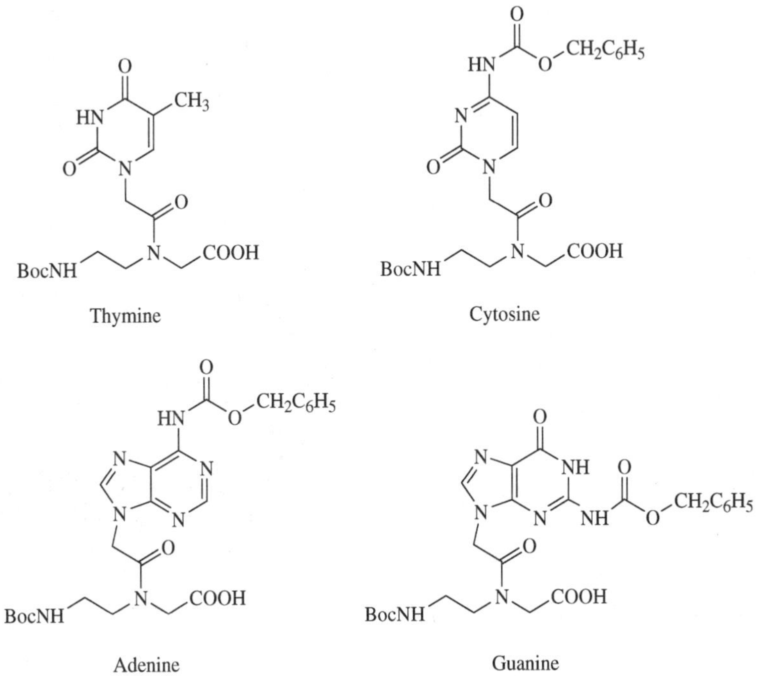

Thymine

Cytosine

Adenine

Guanine

FIGURE 16.5 Chemical structures of Boc-PNA monomers of the A, C, G and T nucleobases.

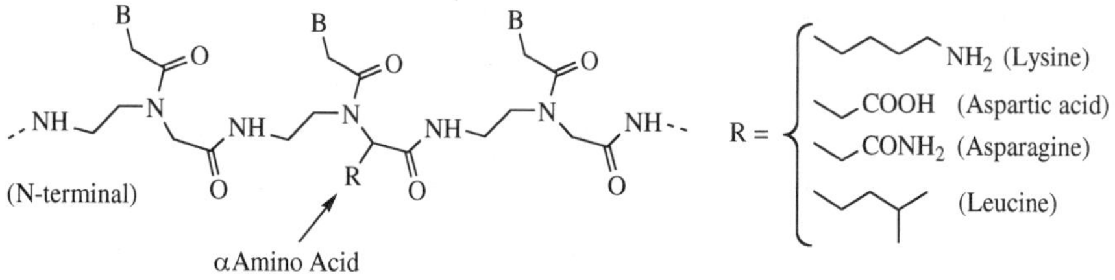

B B B
O O O
NH N NH N NH N NH
(N-terminal) O R O O
αAmino Acid
R = NH_2 (Lysine)
COOH (Aspartic acid)
$CONH_2$ (Asparagine)
(Leucine)

FIGURE 16.6 Example of a PNA oligomer containg a backbone-functionalized unit.

(fluorescein, biotin, etc.), and by employing amino acids other than glycine in the backbone, a large variety of chemically functionalized PNA oligomers are accessible (Fig. 16.6), essentially without compromising the PNA/DNA hybridization efficiency of the oligomer.[13,14]

Structure

As mentioned above, PNA is a structural mimic of DNA, and the three dimensional structure of several PNA complexes that have recently been determined[15–18] emphasize this contention (Fig. 16.7). Nonetheless, the PNA oligomers prefer to adopt a helical conformation, the P form, which, despite similarities to DNA and RNA helices, is distinctly different, characterized by a very large diameter (28 vs.

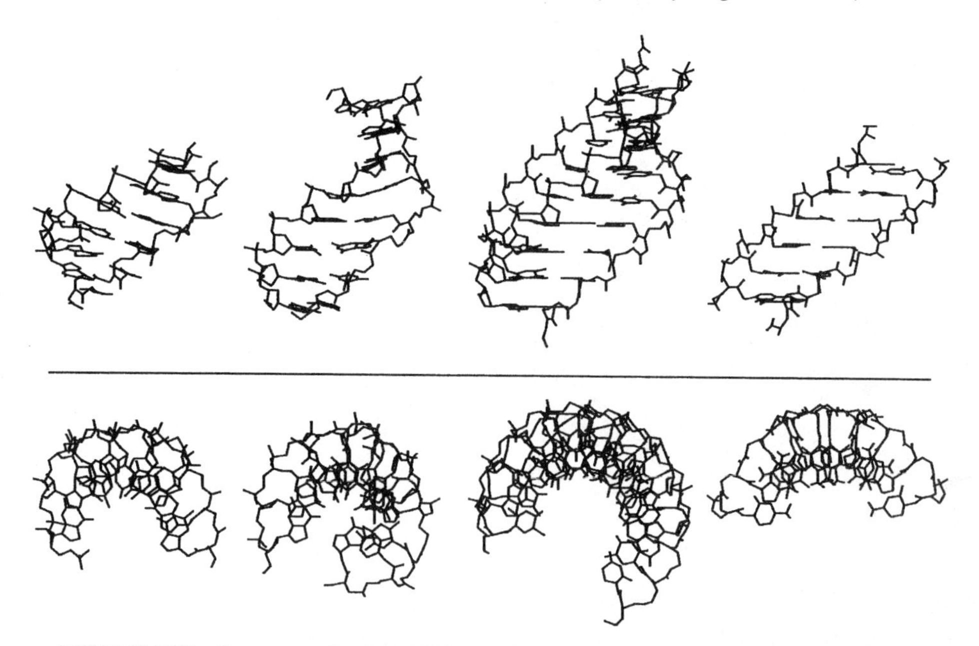

FIGURE 16.7 Structures of various PNA complexes. From left to right: PNA–RNA duplex, PNA–DNA duplex, PNA–DNA–PNA triplex, and PNA–PNA duplex. (FIGURE adapted from Eriksson, M.; Nielsen, P. E., *Quart. Rev. Biophys.* 1996, **29**, 369.)

20 Å for a B-form DNA helix) and a large pitch [18 bp (base pairs) vs. 10 bp for B-DNA]. However, due to the relatively higher flexibility of the PNA backbone, PNA oligomers are able to a large degree to adapt to the helical A form preferred by RNA[15] as well as to the helical B form preferred by DNA.[16] It is also worth recalling that PNA oligomers are achiral, and thus, as expected, PNA–PNA duplexes can equally well adapt a right- as well as a left-handed conformation. Indeed, crystals of a self-complementary PNA hexamer is built from coaxially stacked alternating right- and left-handed duplexes.[18] Most interestingly, however, a single chiral amino acid attached to the carboxy end of a PNA can be sufficient to induce, by "chiral communication," a preferred handedness to the PNA double helix.[19]

Antisense

Mammalian Several early studies using in vitro, cell free translation and cellular microinjection systems demonstrated the ability of PNA bound to mRNA to inhibit the function of the translational machinery, the ribosomes,[20–22] but studies on live cells or whole animals were lacking because of poor cellular uptake of PNAs.[23,24] However, it has been reported that various membrane penetrating peptides, such as that of the third helix of the homeo domain of the Antennapedia protein from *Drosophila*,[25] are quite efficient vehicles for cellular delivery of PNAs,[26–28] and with some cells (nerve cells in particular), a carrier may not even be required.[28,29] Furthermore, down-regulation of gene expression with such antisense PNA–peptide conjugates was reported with mammalian cells in culture;[26,28] and on injection of these conjugates[26] or even naked PNAs[29] targeted to the mRNA of neuroreceptor directly into the brain of rats, a reduction of receptor activity compatible with antisense down-regulation was observed.

Other Targets Studies on cell free systems have also identified other RNA targets for potential PNA drugs. There is increasing evidence that telomerase activity to maintain the size of the chromosomal telomers is required for malignant cancer cells. Since telomerase contains an RNA that is crucial for enzymatic activity, this RNA is an interesting target for an antisense strategy, and it has indeed been demonstrated that PNAs complementary to part of the telomerase RNA are potent inhibitors of this enzyme and therefore potential anti-cancer drug leads.[30]

Reverse transcriptase produces a complementary DNA strand using an RNA strand as template, and this reaction is indispensable in the life cycle and proliferation of retrovirus, such as HIV, whose genome is made up of RNA. PNAs targeted to HIV RNA are very potent inhibitors of the reverse transcription reaction by HIV reverse transcriptase,[31,32] and such PNA could therefore form the basis for development of efficient anti-HIV drugs for treatment of AIDS.

Bacteria Contrary to expectations, the bacterium *Escherichia coli* is somewhat permeable to PNAs,[33] and using a "leaky" mutant strain (AS19), it has been found that antisense down-regulation of bacterial genes in live bacteria is possible with PNA.[34] Most excitingly, it was found that by PNA antisense down-regulation of the

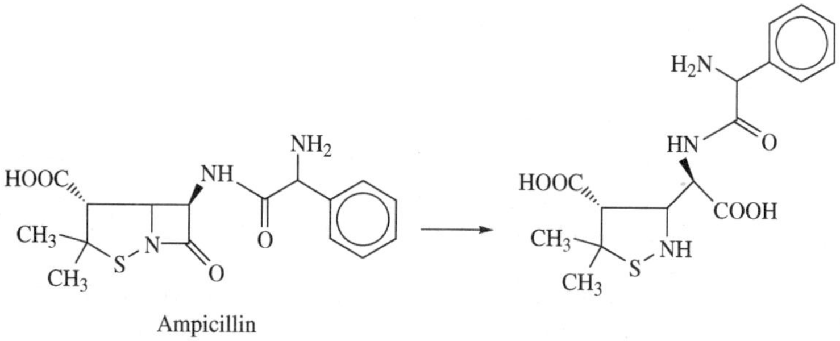

FIGURE 16.8 Hydrolytic inactivation of ampicillin by the enzyme β-lactamase.

enzyme β-lactamase that confers bacterial resistance to penicillins (Fig. 16.8), the bacteria could be resensitized to penicillin and the resistance thereby overcome.[34]

Furthermore, PNAs targeted to certain regions of the ribosomal RNA, a major and crucial component of ribosomes (Fig. 16.9) were toxic to *E. coli*,[33] and in this respect mimic bacterial antibiotics, such as tetracyclin and chloramphenicol, which also bind to rRNA. These results pave the road for the development of novel antibacterial agents that are very much needed in light of the increasing threat by multiresistant strains of pathogenic bacteria. Indeed, it is feared—and unfortunately with ample cause—that many infections in the near future will no longer be treatable by existing antibiotics because of acquired resistance by the bacteria.

Antigene

PNAs also bind very strongly to complementary homopurine sequence targets in double-stranded DNA by a process termed *triplex invasion*, in which an internal PNA_2–DNA triplex is formed (Fig. 16.10).[35] Such PNA triplex invasion complexes are very efficient inhibitors of transcription initiation as they prohibit protein (transcription factor/RNA polymerase) binding to the promotor DNA.[36,37] However, the complexes also have sufficient stability to arrest elongating RNA polymerases,[20,38] and therefore targets in the coding region of the genes can also be exploited. This should dramatically increase the probability of identifying suitable homopurine targets.

The PNA antigene approach is less well developed than the antisense approach because several issues apart from cellular uptake must be addressed. In vitro studies have shown that the strand invasion binding is very sensitive to increasing ionic strength, and using simple PNAs binding to a double-stranded target is extremely slow at physiologically relevant ionic conditions (e.g., 140 mM KCl). Nonetheless, using modified PNAs, it is possible to obtain efficient binding under these conditions,[39] and other studies have shown that cellular processes related to transcription greatly catalyze the binding.[40,41] Furthermore, a few studies[42,43] have indicated that binding of PNA to DNA targets in the cell nucleus may indeed be much more efficient than predicted from the in vitro results. Since PNA

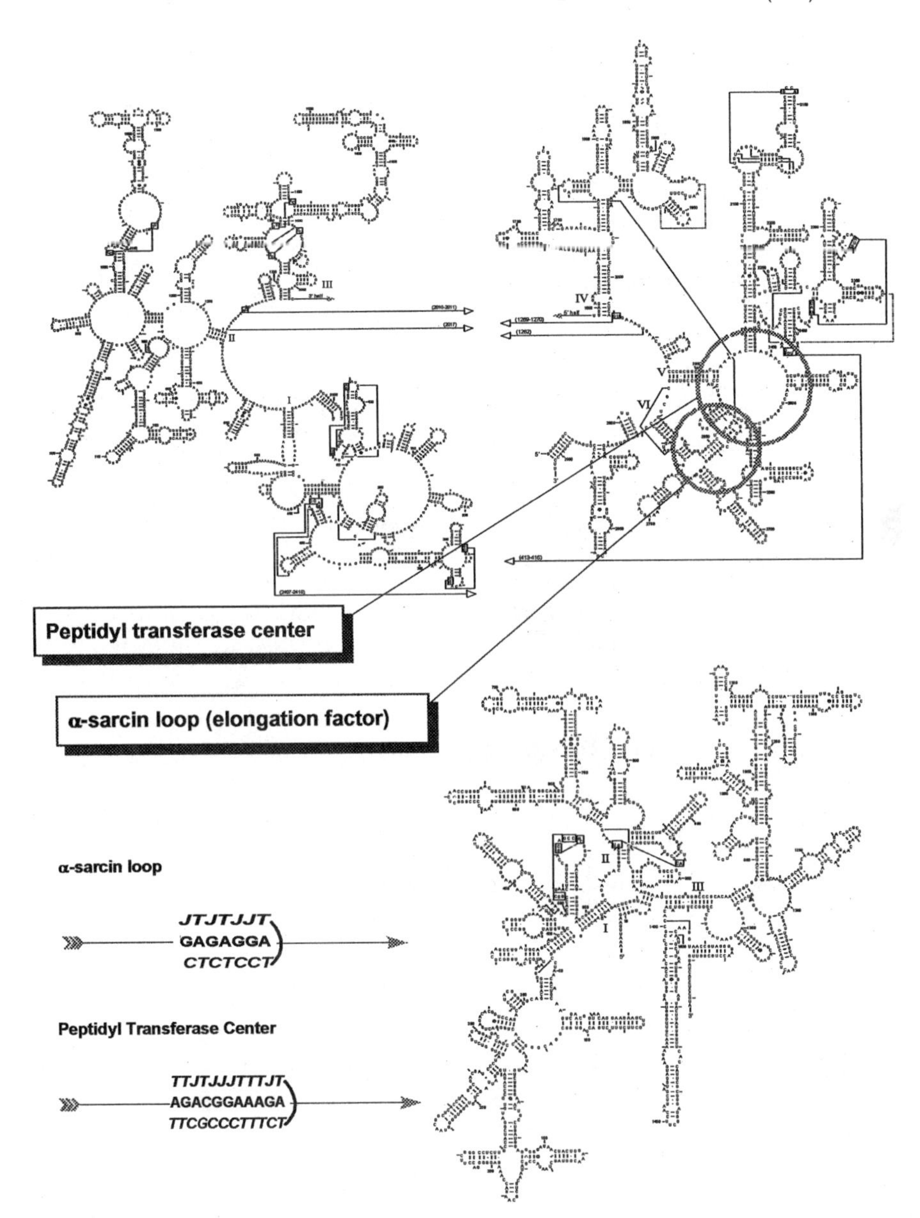

FIGURE 16.9 Sequence and secondary structure of 16*S* and 23*S* *E. coli* ribosomal RNA. PNA targets in the peptidyl transferase center and in the α-sarcin loop are indicated in the RNA and shown in the lower left corner. The PNAs in which the two strands are chemically linked (bis-PNA) are shown in italic.

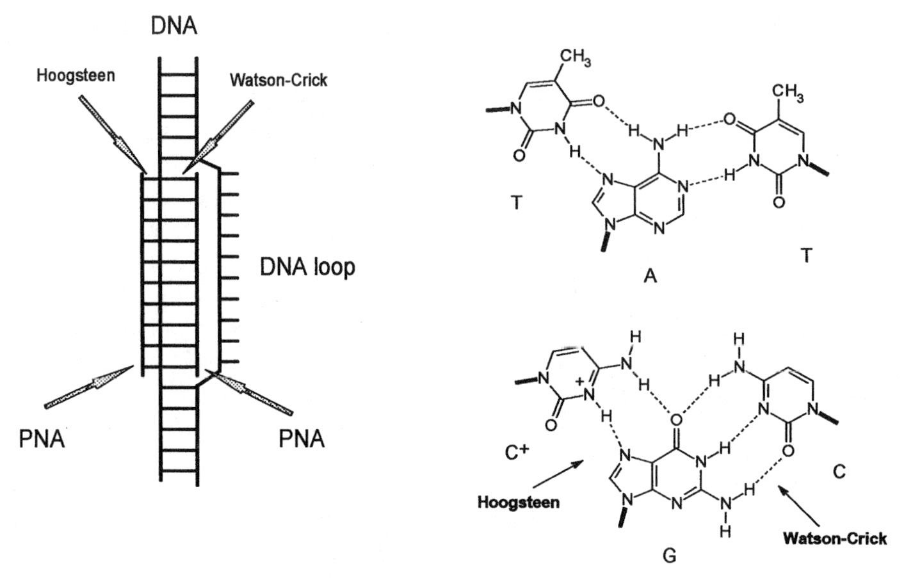

FIGURE 16.10 Schematic drawing of a triplex invasion PNA complex, and the Watson–Crick/Hoogsteen base-pairing involved in the recognition of DNA–adenine by two PNA–thymines and of DNA–guanine by two PNA–cytosines.

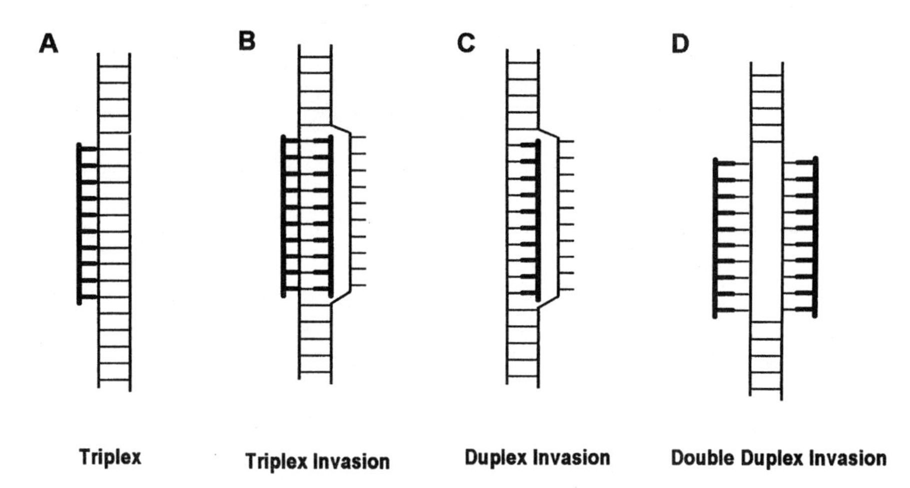

FIGURE 16.11 Schematic drawing of complexes formed on targeting double-stranded DNA with PNA. The triplex invasion complex is the more stable and is formed by homopyrimidine PNAs binding to a homopurine DNA target. The duplex invasion complex can be formed with homopurine PNAs, whereas the double duplex invasion complex requires nonstandard nucleobases in the PNAs (see Fig. 16.12). A conventional triplex seems to be formed only with cytosine-rich homopyrimidine PNAs.

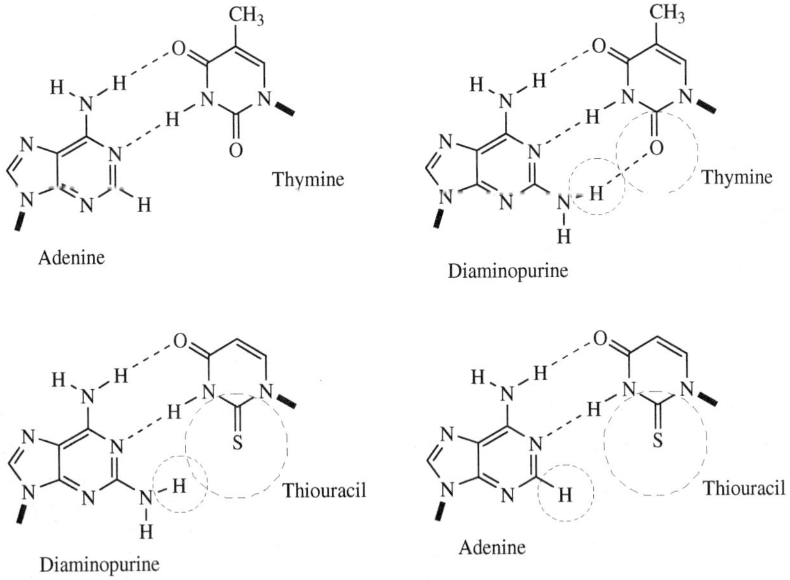

FIGURE 16.12 Chemical structures showing how adenine–thymine, diaminopurine–thymine, and adenine–thiouracil can be formed whereas steric hindrance interferes with the formation of the diaminopurine–thiouracil base pair. Thus sequence complementary PNA oligomers in which adenines have been replaced by diaminopurine and thymines have been replaced by thiouracil will not bind to each other but will still be able to bind complementary DNA.

triplex invasion binding is kinetically controlled and the complexes essentially do not dissociate once formed, a slow accumulation is expected. PNA triplex invasion complexes also inhibit replication and also appear to be somewhat mutagenic (42).

Several binding modes apart from the triplex invasion have been characterized for PNA (Fig. 16.11). In particular, the double-duplex invasion mode—in which two pseudocomplementary PNA oligomers containing diaminopurine/thiouracil substitutions (Fig. 16.12) both bind their complementary DNA strand—is interesting and can be used to target mixed purine/pyrimidine sequences in double-stranded DNA.[43]

FUTURE PROSPECTS

The first phosphorothioate oligonucleotides drug has already made its way to the clinic, and others will soon follow. It is, however, clear that phosphorothioates are not optimal as gene therapeutic drugs and that second- and third-generation chemistries will replace them. Although many issues still have not been thoroughly addressed concerning PNA—in particular relating to bioavailability, pharmacokinetics, and dynamics—the properties described so far show good promise that safe and efficient drugs against a wide variety of diseases may be developed using PNA chemistry.

REFERENCES

1. Stephenson, M. L.; Zamecnik, P. C., *Proc. Natl. Acad. Sci.* (USA) 1978, **75**, 285–8.
2. Milligan, J. F.; Matteucci, M. D.; Martin, J. C., *J. Med. Chem.* 1993, **36**, 1923–1937.
3. De Mesmaeker A.; Altmann, K. H.; Waldner, A.; Wendeborn, S., *Curr. Opin. Struct. Biol.* 1995, **5**, 343–55.
4. Freier, S. M.; Altmann, K. H., *Nucl. Acids Res.* 1997, **25**, 4429–4443.
5. Nielsen, P. E., *Chem. Eur. J.* 1997, **3**, 505–508.
6. White, S.; Szewczyk, J. W.; Turner, J. M.; Dervan, P. B., *Nature* 1998, **391**, 468–471.
7. Nielsen, P. E.; Egholm, M.; Berg, R. H.; Buchardt, O., *Science* 1991, **254**, 1497–1500.
8. Egholm, M.; Buchardt, O.; Christensen, L.; Behrens, C.; Freier, S. M.; Driver, D. A.; Berg, R. H.; Kim, S. K.; Nordén, B.; Nielsen, P. E., *Nature* 1993, **365**, 556–568.
9. Hyrup, B.; Nielsen, P. E., *Bioorg. Biomed. Chem.* 1996, **4**, 5–23.
10. Good, L.; Nielsen, P. E., *Antisense Nucleic Acid Drug Devel.* 1997, **7**, 431–437.
11. Uhlmann, E.; Peyman, A.; Breipohl, G.; Will, D. W., *Angew. Chem. Internatl. Ed.* 1998, **37**, 2796–2823.
12. Nielsen, P. E.; Haaima, G., *Chem. Soc. Rev.* 1997, **26**, 73–78.
13. Haaima, G.; Lohse, A.; Buchardt, O.; Nielsen, P. E., *Angewandte Chemie* 1996, **35**, 1939–1941.
14. Püschl, A.; Sforza, S.; Haaima, G.; Dahl, O.; Nielsen, P. E., *Tetrahedron Lett.* 1998, **39**, 4707–4710.
15. Brown, S. C.; Thomson, S. A.; Veal, J. M.; Davis, D. G., *Science* 1994, **265**, 777–780.
16. Eriksson, M.; Nielsen, P. E., *Nat. Struct. Biol.* 1996, **3**, 410–413.
17. Betts, L.; Josey, J. A.; Veal, J. M.; Jordan, S. R., *Science* 1995, **270**, 1838–1841.
18. Rasmussen, H.; Kastrup, J. S.; Nielsen, J. N.; Nielsen, J. M.; Nielsen, P. E., *Nat. Struct. Biol.* 1997, **4**, 98–101.
19. Wittung, P.; Nielsen, P. E.; Buchardt, O.; Egholm, M.; Nordén, B., *Nature* 1994, **368**, 561–563.
20. Hanvey, J. C.; Peffer, N. C.; Bisi, J. E.; Thomson, S. A.; Cadilla, R.; Josey, J. A.; Ricca, D. J.; Hassman, C. F.; Bonham, M. A.; Au, K. G.; Carter, S. G.; Bruckenstein, D. A.; Boyd, A. L.; Noble S. A.; Babiss, L. E., *Science* 1992, **258**, 1481–1485.
21. Knudsen, H.; Nielsen, P. E., *Nucleic Acids Res.* 1996, **24**, 494–500.
22. Gambacorti-Passerini, C.; Mologni, L.; Bertazzoli, C.; Marchesi, E.; Grignani, F.; Nielsen, P. E., *Blood* 1996, **88**, 1411–1417.
23. Bonham, M. A.; Brown, S.; Boyd, A. L.; Brown, P. H.; Bruckenstein, D. A.; Hanvey, J. C.; Thomson, S. A.; Pipe, A.; Hassman, F.; Bisi, J. E.; Froehler, B. C.; Matteucci, M. D.; Wagner, R. W.; Noble, S. A.; Babiss, L. E., *Nucleic Acids Res.* 1995, **23**, 1197–1203.
24. Wittung, P.; Kajanus, J.; Edwards, K.; Nielsen, P. E.; Nordén, B.; Malmström, B. G., *FEBS Lett.* 1995, **365**, 27–29.

25. Derossi, D.; Chassaing, G.; Prochiantz, A., *Trends Cell Biol.* 1998, **8**, 84–87.
26. Simmons, C. G.; Pitts, A. E.; Mayfield, L. D.; Shay, J. W.; Corey, D. R., *Bioorg. Med. Chem. Lett.* 1997, **7**, 3001–3006.
27. Pooga, H.; Soomets, U.; Hällbrink, M.; Valkna, A.; Saar, K.; Rezaei, K.; Kahl, U.; Hao, J.-X.; Xu, X.-J.; Wiesenfeld-Hallin, Z.; Hökfelt, T.; Bartfai, T.; Langel, Ü., *Nat. Biotechnol.* 1998, **16**, 857.
28. Aldrian-Herrada, G.; Desarménien, M. G.; Orcel, H.; Boissin-Agasse, L.; Méry, J.; Brugidou, J.; Rabie, A., *Nucleic Acids Res.* 1998, **26**(21), 4910–4916.
29. Tyler, B. M.; McCormick, D. J.; Hoshall, C. V.; Douglas, C. L.; Jansen, K.; Lacy, B. W.; Cusack, B.; Richelson, E., *FEBS Lett.* 1998, **421**, 280–284.
30. Norton, J. C.; Piatyszek, M. A.; Wright, W. E.; Shay, J. W.; Corey, D. R., *Nat. Biotechnol.* 1996, **14**, 615–618.
31. Koppelhus, U.; Zachar, V.; Nielsen, P. E.; Liu, X.; Eugen-Olsen, J.; Ebbesen, P., *Nucleic Acids Res.* 1997, **25**, 2167–2173
32. Lee, R.; Kaushik, N.; Modak, M. J.; Vinayak, R.; Pandey, V. N., *Biochemistry* 1998, **37**(3), 900–910.
33. Good, L.; Nielsen, P. E., *Proc. Natl. Acad. Sci.* (USA) 1998, **95**, 2073–2076.
34. Good, L.; Nielsen, P. E., *Nat. Biotechnol.* 1998, **16**, 355–358.
35. Nielsen, P. E.; Egholm, M.; Buchardt, O., *J. Molec. Recogn.* 1994, **7**, 165–170.
36. Vickers, T. A.; Griffith, M. C.; Ramasamy, K.; Risen, L. M.; Freier, S. M., *Nucleic Acids Res.* 1995, **23**, 3003–3008.
37. Praseuth, D.; Grigoriev, M.; Guieysse, A. L.; Pritchard L. L.; Harel-Bellan, A.; Nielsen, P. E.; Helene, C., *Biochim. Biophys. Acta* 1996, **1309**, 226–238.
38. Nielsen, P. E.; Egholm, M.; Buchardt, O., *Gene* 1994, **149**, 139–145.
39. Kurakin, A.; Larsen, H. J.; Nielsen, P. E., *Chem. Biol.* 1998, **5**, 81–89.
40. Bentin, T.; Nielsen, P. E., *Biochemistry* 1996, **35**, 8863–8869.
41. Larsen, H. J.; Nielsen, P. E., *Nucleic Acids Res.* 1996, **24**, 458–463.
42. Faruqi, A. F.; Egholm, M.; Glazer, P. M., *Proc. Natl. Acad. Sci.* (USA) 1998, **95**, 1398–1403.
43. Lohse, J.; Dahl, O.; Nielsen, P. E. *Proc., Natl. Acad. Sci (USA)* 1999, (in press).

CHAPTER 17

Ribozyme Mimics: The Evolution of Gene-Specific Chemotherapy

TODD A. OSIEK, WILLIAM C. PUTNAM and JAMES K. BASHKIN

Department of Chemistry, Washington University

INTRODUCTION

Typical pharmaceutical agents suffer from a lack of specificity at the molecular level, resulting in side effects that can be severe in many cases. This occurs because the molecular recognition employed in conventional chemotherapy is merely selective, rather than specific. New pharmaceutical strategies that capitalize on the highly specific information in the genetic code should provide a remarkable improvement in antiviral, antifungal, and anticancer treatment, with a consequent dramatic decrease in side effects. This genomic approach to medicine has special promise because many diseases can be broadly described as the expression of harmful proteins. The development of a gene-specific, catalytic approach to chemotherapy is the overall goal in the design of the ribozyme mimics described herein.[1–5] Ribozyme mimics were invented to be catalytic antisense reagents, meaning that one drug molecule would destroy many copies of harmful genetic information. The first antisense reagents have recently been approved for use by the FDA,[6] and this ground-breaking development in antisense technology serve to validate the entire field. We refer to these as conventional antisense reagents, to distinguish them from the catalytic drugs whose design we review here. The specific form of the genetic information that is targeted by the antisense method is messenger RNA (mRNA) for harmful genes, including viral, bacterial, fungal mRNA, and certain types of mRNA associated with cancer. Ribozyme mimics are designed to be catalytic antisense reagents that recognize harmful RNA sequences and cleave the RNA phosphodiester backbone through the natural transesterification/hydrolysis pathways.[7] We refer to the combined transesterification/hydrolysis mechanisms for RNA cleavage as *nucleophilic cleavage* to distinguish them from oxidative

Biomedical Chemistry: Applying Chemical Principles to the Understanding and Treatment of Disease, Edited by Paul F. Torrence
ISBN 0-471-32633-x

cleavage processes such as the Fenton reaction.[8] The cleavage of mRNA disrupts the natural procession of genetic material through the cell (from DNA to mRNA to protein), thus preventing the synthesis of a harmful protein, such as a viral coat protein. This chapter discusses the evolution of the antisense technique and the emergence of ribozyme mimics as catalytic antisense reagents. Another example of a catalytic drug strategy includes the superoxide dismutase (SOD) mimic invented by Monsanto scientists.[9–18]

ANTISENSE REGULATION OF GENE EXPRESSION

An emerging methodology, the "antisense" technique,[19–30] has the potential to suppress or completely stop the production of undesired proteins with great specificity. It uses the innate specificity of Watson–Crick base-pairing to make gene-specific suppressors of protein synthesis. The technique intercepts harmful messenger RNA (mRNA) with a complementary nucleic acid strand (often referred to as the *antisense probe*). Natural examples of the antisense method have been identified in prokaryotic[31] and eukaryotic[32] organisms. In its naturally occurring form,[28,31,32] the target mRNA is bound by a complementary RNA transcript. The resulting RNA–RNA duplex preventing protein synthesis by making the mRNA unavailable for binding to the protein synthesis machinery in the ribosomes (Fig. 17.1). In the early medicinal chemistry version of the antisense technique, the intercepting strand was often a deoxyribonucleic acid strand (DNA) that was introduced to the cell from the extracellular environment (Fig. 17.1).

Antisense deoxyoligonucleotides were first reported to have an inhibitory effect on protein synthesis in 1978, by Zamecnik and Stephenson, when a 13-mer deoxyoligonucleotide (DNA) was used to inhibit the proliferation of the Rous sarcoma virus.[29,33] Tests in mammalian cells by Izant and Weintraub revealed that specific genes could be suppressed by incorporating antisense genes into the cells[34].

The chemical analog to the antisense technique does not necessarily directly mimic the naturally occurring antisense mechanism in the obvious manner. Instead, in at least some cases, the cellular enzyme ribonuclease H (RNase H) mediates the suppression of protein synthesis by antisense probes (Fig. 17.2).[27,35,36]

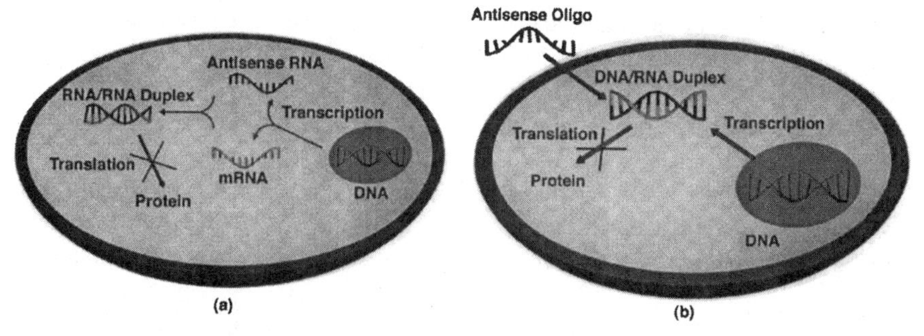

FIGURE 17.1 (*a*) The native antisense method; (*b*) a short, synthetic analog of the antisense molecule may also be effective at blocking protein synthesis.

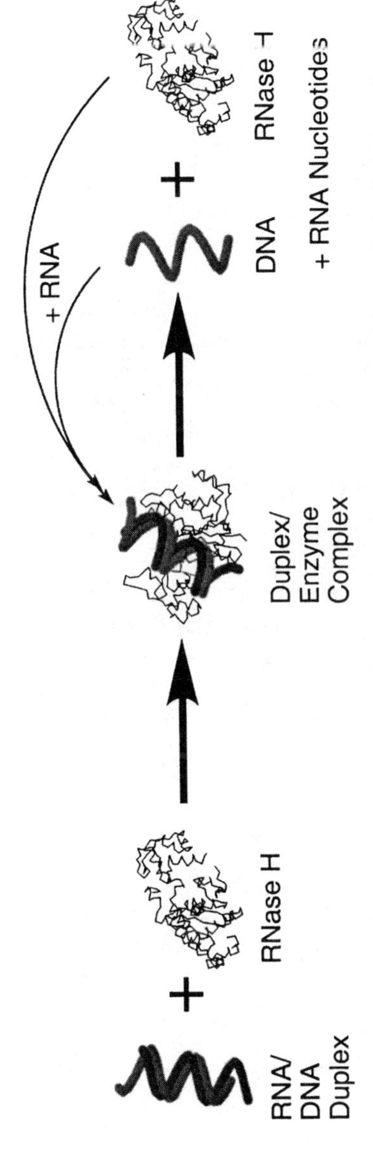

FIGURE 17.2 RNase H enzymatic activity allows the sequence-specific, catalytic digestion of specific, harmful mRNA molecules.

RNase H is known to digest the RNA strand of an RNA–DNA duplex, and it digests mRNA that is hybridized to the DNA drug. Digestion of the mRNA destroys the message and prevents synthesis of the specifically targeted protein. RNase H digestion also allows the DNA to disengage from the cleaved RNA, since cleavage products are small nucleotide products and the binding constant between DNA and RNA is dependent on the length of the double-stranded region. The disengaged DNA is free to bind another mRNA target, so many RNA molecules may be destroyed by a single DNA drug molecule acting in concert with RNase H. In principle, the result is a drug that can act catalytically.

Although the antisense technique using native DNA probes was initially very promising, it has not generally proved successful in animal studies. Furthermore, some initial studies that were successful were later attributed to other biological effects. Problems with this method include (1) high concentrations of probe are required, resulting in high cost; (2) difficulty exists in transporting the DNA probes across the cellular membrane; (3) duplex formation involves a reversible equilibrium; (4) it is difficult to localize the drug in the proper cellular region; and (5) nuclease enzymes degrade the DNA probe.

To overcome the problems of nuclease degradation and transport across the cellular membrane, many chemical modifications have been made to antisense probes.[37–39] Some of these modifications include lipophilic methylphosphonate derivatives (Fig. 17.3*a*), DNA phosphothioates (Fig. 17.3*b*), 2′-OMe RNA derivatives (Fig. 17.3*c*), and more dramatic chemical changes such as peptide nucleic acids (PNA) (Fig. 17.3*d*).

These and other chemical modifications to the phosphodiester linkage and the deoxyribose ring have often shown marked improvement in nuclease resistance and cellular uptake. Unfortunately, *most chemical changes to the natural DNA structure prevent the duplex formed between the antisense probe and the mRNA target from being a substrate for RNase H,*[40–44] *and the important path for catalytic RNA destruction is lost.* When this pathway provided by RNase H is shut down, antisense reagents can inactivate only a single RNA transcript, and they act merely as stoichiometric drugs. This can reintroduce the problem of high dosages and the consequent issues of high cost and side effects.

The idea of catalytic antisense probes was developed in our laboratory to allow medicinal chemistry to overcome the problems of uptake and stability while retaining catalytic activity. This idea is schematically shown in Figure 17.4.

Functional mimics of ribozymes constitute a nucleophilic RNA cleavage agent covalently linked to an internal nucleotide or nucleotide analog of an antisense probe. Because of the presence of the 2′-OH in RNA and its absense in DNA, RNA is cleaved by nucleophilic reactions at least one million times faster than DNA.[7] Taking advantage of this inherent chemoselectivity, ribozyme mimics can destroy target RNA strands without destroying the DNA strand to which they are attached and use for molecular recognition.

There are many important aspects to the design of a competent, active ribozyme mimic. The proper molecular placement of the cleavage agent is essential for catalysis (Fig. 17.5): a cleavage agent located at an end of the antisense probe can

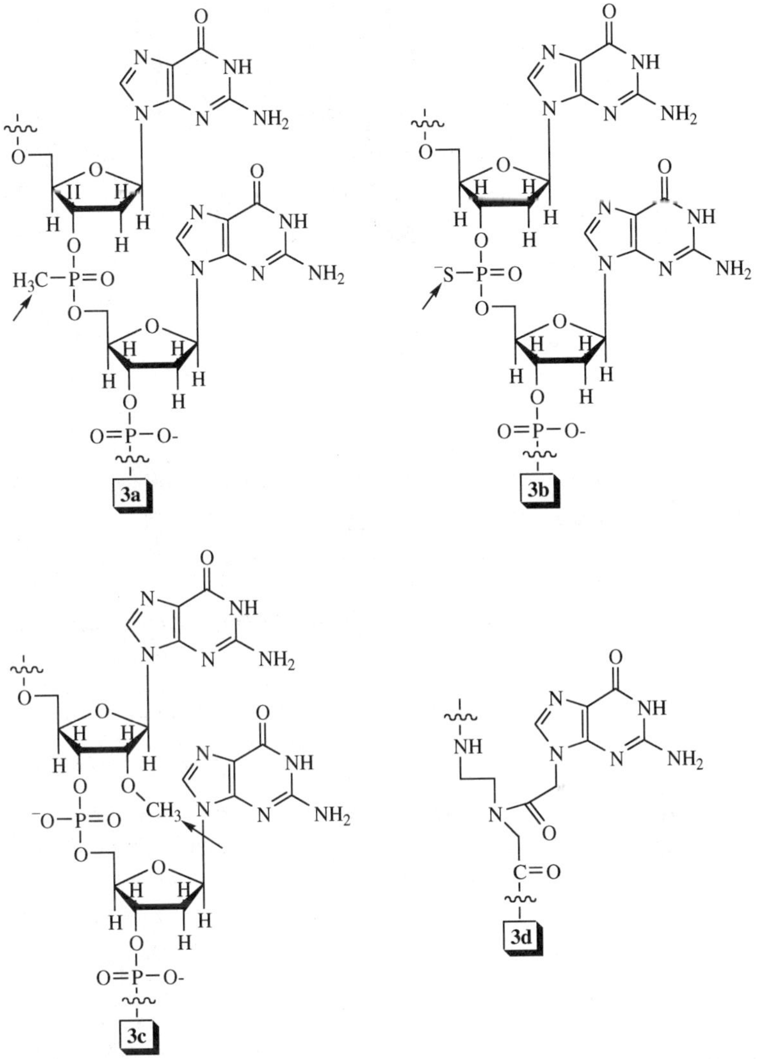

FIGURE 17.3 Chemical modifications to antisense probes; key changes are indicated by arrows.

cleave the RNA target, but there is no mechanism for disengagement of the probe from the RNA because the RNA–DNA binding constant is undiminished after cleavage (Fig. 17.5*a*). Catalytic turnover depends on a decrease in drug–RNA binding constant that is concomitant with RNA cleavage. Placing the cleavage agent in the middle of the duplex allows catalytic behavior due to the length dependence of the binding constants for nucleic acids (Fig. 17.5*b*).

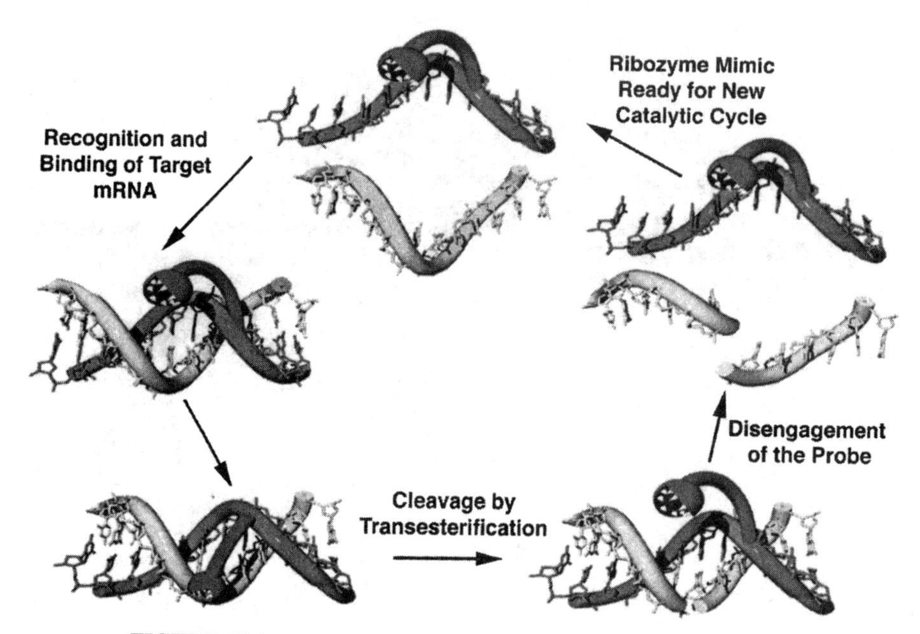

FIGURE 17.4 Schematic representation of ribozyme mimic action.

OXIDATIVE VERSUS NUCLEOPHILIC CLEAVAGE OF DNA AND RNA

Oxidative and nucleophilic cleavage are the two major pathways for cleavage of DNA and RNA. Oxidative cleavage reagents include compounds such as copper(II) *ortho*-phenanthroline (Cu-*o*-phen),[45] bleomycin,[46] and other reagents[47] that abstract hydrogen atoms from the sugar and base portions of nucleosides, which may lead to strand scission.[48] Figure 17.6 shows an iron-oxo mechanism for oxidative cleavage in which a radical is generated at C4′ by hydrogen abstraction. Breakdown of this radical, in the presence of oxygen occurs, via a hydroperoxide and leads to strand scission.

Because oxidative cleavage pathways work by attacking the sugar portion of nucleosides, they display little inherent preference for cleavage of RNA over DNA (unless this is provided by a specific binding component).

Many metal ions,[49,50] metal complexes,[51–56] enzymes, and ribozymes[20,57] catalyze nucleophilic cleavage of nucleic acids. The term *nucleophilic cleavage* arises from the fact that this mechanism involves nucleophilic attack on a phosphodiester moiety. Nucleophilic cleavage is known as *transesterification* if the nucleophile is an alcohol or alkoxide, and termed as *hydrolysis* if the nucleophile is water or hydroxide. As mentioned previously, the presence of the 2′-OH confers on RNA a dramatic susceptibility to nucleophilic cleavage versus DNA. The intramolecular transesterification mechanism that employes the 2′-OH as the nucleophile is shown in Figure 17.7.

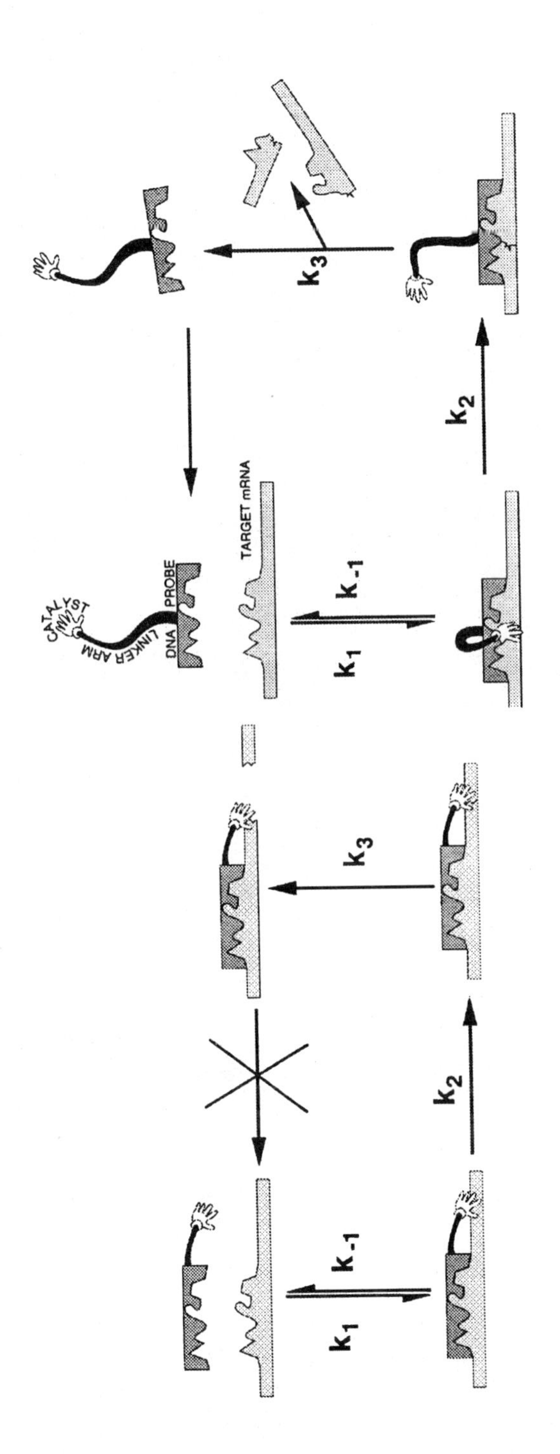

FIGURE 17.5 Proper placement of the cleavage agent is essential for catalysis: (*a*) cleavage outside the duplex gives product inhibition; (*b*) cleavage within the duplex allows catalytic turnover.

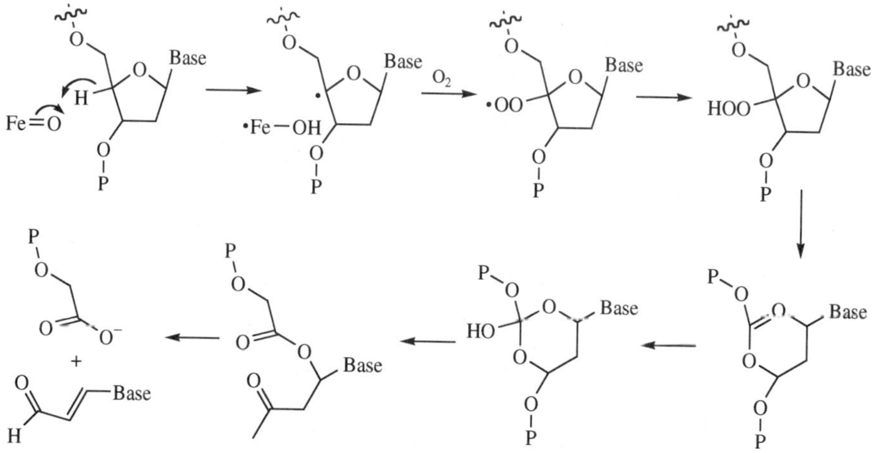

FIGURE 17.6 Example of oxidative strand scission.

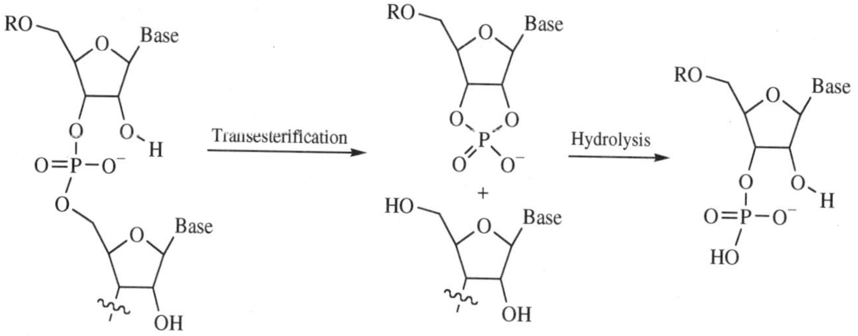

FIGURE 17.7 Transesterification and hydrolysis pathways.

When RNA is cleaved by the transesterification route shown in Figure 17.7 it leaves a 2′, 3′-cyclic phosphate and a 5′-OH group. The cyclic phosphate is rapidly hydrolyzed to a terminal phosphate monoester under typical physiological conditions.

THE ROLES OF METALS IN RNA TRANSESTERIFICATION AND HYDROLYSIS

Two major roles of metals in nucleophilic cleavage of nucleic acids have been postulated in the literature.[58] When metal ions coordinate water, they can dramatically lower the pK_a of the bound water.[59,60] This effect varies depending on the charge, ionic radius, and hardness or softness of the metal in question.[61] By varying the metal and its ligands, one can control the pK_a of coordinated water. The coordinated water (or hydroxide) can then act as a nucleophile, attacking the

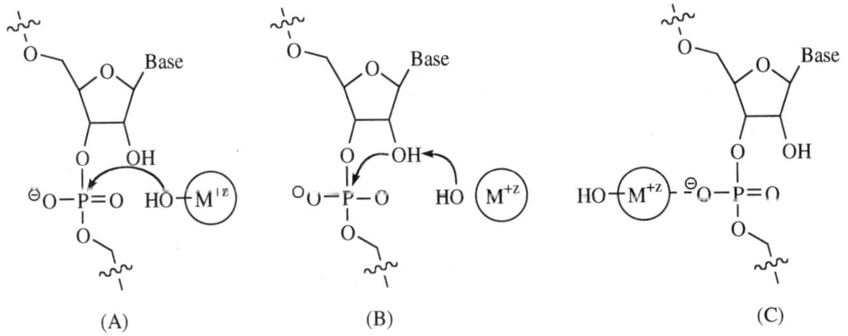

FIGURE 17.8 Proposed metal roles in nucleophilic cleavage of RNA.

phosphodiester backbone directly (Fig. 17.8*a*),[49] or as a base, deprotonating the 2′-OH to generate an intramolecular nucleophile that attacks an adjacent phosphate (Fig. 17.8*b*).

Additionally, cationic metal ions and complexes bind electrostatically to the anionic phosphodiester backbone of nucleic acids (Fig. 17.8*c*). Lewis acid or electrostatic catalysis by metal cations can reduce the electrostatic repulsion between the alkoxide nucleophile and phosphate electrophile, which are both anionic.[7] Thus, metals can act to stabilize the rate-determining transition state for nucleophilic attack by neutralizing the otherwise high negative charge that builds up in the activated complex. Several reviews of RNA cleavage mechanisms have appeared in the literature.[58]

CLEAVAGE OF RNA BY FREE METAL IONS

Interest in the ability of metals to cleave RNA dates back as far as 1939. Bamann used Ce(III) and La(III) salts to cleave RNA at 37°C.[62] Currently, metal ions spanning the periodic table have been shown to be nucleophilic cleavage catalysts. These include Mg(II), Ca(II), Fe(III), Ni(II), Cu(II), Zn(II), Pb(II), lanthanides, UO_2^{2+},[63] and thorium salts.[64] Cleavage studies have been done on a variety of substrates, including cyclic phosphates,[65–68] dinucleotides,[69] oligomers,[52] and polymers. Lanthanides and actinides have shown the highest rates of cleavage of RNA. However, a lack of standard substrates and conditions makes direct comparison difficult, although some general trends can be drawn. Breslow and Huang showed that Eu(III) is 15 times faster than Zn(II) cleavage of UpU.[70] Eu(III) also has been reported by Kuusela and Lönnberg to hydrolyze 2′, 3′-cyclic AMP 1000 times faster than Zn(II).[68]

CLEAVAGE OF RNA BY METAL COMPLEXES

Free metals have been widely studied for the cleavage of RNA in vitro, but their lack of specificity precludes their use as pharmacological agents. In the late 1980s, we and others recognized that metal complexes, which could be tethered to molecular recognition agents, would provide a vital step in building potentially therapeutic RNA cleavage agents. Tethered complexes allow us to deliver metal ions to a specific site, to tailor pK_a values by adjusting the nature of the ligands, and to control the coordination number around the metals. Much work has been done to develop and discover metal complexes that cleave RNA.[5,48,71,72] In the first report of such activity, we found that copper(II) terpyridine, copper(II) bipyridine, and a zinc(II) macrocycle cleaved RNA by nucleophilic pathways.[52] Häner[4] and Morrow[5] have synthesized many macrocycles to encapsulate lanthanides for the cleavage of RNA substrates. Like metal ions, the free metal complexes that have been developed have little or no inherent sequence specificity for RNA cleavage. However linking metal complexes to strands of DNA (or DNA analogs) provides the desired specificity, and gives us functional mimics of ribozymes, the naturally occurring RNA enzymes. These ribozyme mimics use the specificity of the genetic code to deliver metal complexes to particular substrates, according to the principles of the chemical antisense method.

DESIGN OF RIBOZYME MIMICS

Various reports have appeared describing different approaches to the sequence-specific cleavage of RNA. The tethering of nuclease enzymes enzymes to DNA strands provided the first examples of such cleavage. Zuckermann and Schultz covalently attached a mutant staphylococcal nuclease to the 3′ end of a DNA 14-mer, and in a later publication attached ribonuclease S to the same 14-mer.[73] Both of the hybrids achieved cleavage of the target RNA regions in vitro. The covalent attachment of RNase H to the 5′ termini of a 9-mer by Kanaya and co-workers[74,75] also attained sequence-specific cleavage of an RNA transcript in vitro.

The use of small molecule catalysts (instead of enzymes) linked to nucleic acid recognition elements has become a major focus of research into the de novo design of ribozyme mimics. The first example of a wholly synthetic ribozyme mimic was published in 1994, using terpyridine covalently linked to an internal nucleoside of a 17-mer DNA strand[76] (Fig. 17.9). When incubated with copper(II), the well-known nucleophilic RNA cleavage catalyst Cu-terpyridine was formed in situ, attached to the DNA probe. The resulting molecule, which we called a *ribozyme mimic*, cleaved a 159-mer RNA strand from a portion of the *gag* gene on RNA of HIV in a sequence-specific manner. In a 72 h in vitro experiment, the ribozyme mimic (5 μM) cleaved 11% of the target RNA (1 nM) at 37 °C and 18–25% of the RNA at 45 °C. This initial molecule proved the concept that a metal catalyst could be elaborated into a ribozyme mimic by covalent attachment to DNA.

Several other groups have embraced this chemistry and reported the nucleophilic cleavage of RNA using small molecule catalysts. Trawick[77] and Lönnberg[58] have reported extensive, critical reviews of ribozyme mimics. A few key examples of additional mimics are mentioned here. Much of the work described here was based on the important work of Morrow, who had previously demonstrated the catalytic

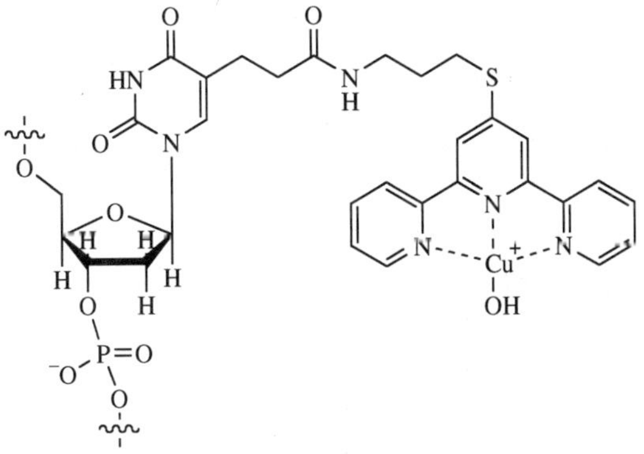

FIGURE 17.9 The active site of the first ribozyme mimic at the 6 position of a 17-mer oligonucleotide probe.

RNA cleavage activity of a lanthanide-based hexaaza macrocycle[5] (Fig. 17.10*a*). Continuing in the lanthanide arena, Häner and co-workers[78–80] synthesized terpyridine-based lanthanide macrocycles, related to Morrow's compounds, that were attached to DNA recognition elements in several ways (Fig. 17.10*b*). When Häner et al. attached their lanthanide macrocycle to the 5′ end of the DNA, they achieved 88% cleavage of their target RNA after 16 h. Komiyama and co-workers[81] prepared a lanthanide chelate using iminodiacetate that was covalently attached to the 5′ end of a 15-mer DNA probe (Fig. 17.10*c*). After the addition of the lanthanide ions (Lu, Eu, and Th), the conjugate cleaved a 39-mer of RNA (7.3% for 4 h and 17% for 8 h). Magda and co-workers[72] prepared a ribozyme mimic by using a lanthanide–texaphyrin conjugate attached to a DNA recognition agent (Fig. 17.10d). In 1997, locating a dysprosium(III) texaphyrin complex at an internal position of a DNA probe, Magda and co-workers *achieved catalytic turnover with a ribozyme mimic.*[82] Lönnberg and co-workers[83] used a 5′-imidazole DNA conjugate to direct the cleavage of zinc to an RNA strand (Fig. 17.10*e*). This ribozyme mimic cleaved 2–5% of the target RNA after 19 h at room temperature. Baker and co-workers[84] achieved decapitation of the 5′ cap of mRNA through the sequestering of Cu by a peptide derivative *N*-(2-mercaptopropionyl) gylcine (Fig. 17.10*f*): 20% of the target RNA was decapitated after 120 h at 37 °C.

Wholly organic (metal free) ribozyme mimics have also been synthesized; Lönnberg recently described this chemistry in an elegant review.[58] These reagents are based on concepts and compounds reported around 1990 by Stein,[30] Cohen,[3] and Bashkin,[85,86] and on much earlier physical organic chemistry studies of RNA cleavage by Breslow.[70] Many of these compounds utilize imidazole residues or free amine residues to act as general bases and/or general acids in the sequence-specific scission of RNA.[87,88] Komiyama and co-workers reported the cleavage of RNA by an ethylenediamine linked DNA[89] (Fig. 17.10*g*). This mimic cleaved 10% of the target RNA after 4 h at 50 °C. Reynolds and co-workers used a diimidazole construct to cleave slightly less than 10% of their target RNA after 5 days at 25 °C

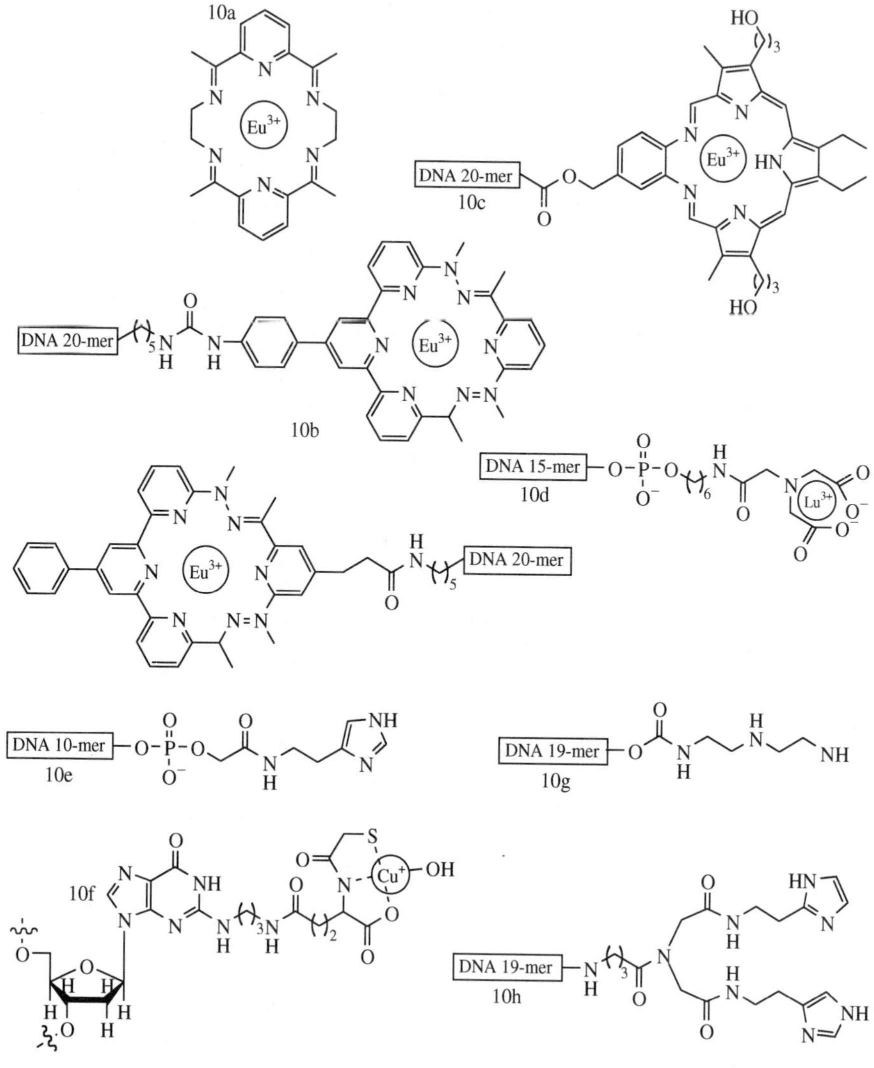

FIGURE 17.10 Other ribozyme mimic constructs.

(Fig. 17.10*h*).[90] A dihistamine construct has been shown by Vlassov and co-workers[91] to be efficient at RNA scission (60% cleavage after 8 h at 37 °C).

The next important step for the future development of ribozyme mimics was the improvement of RNA cleavage efficiencies. One possible approach to more efficient nucleophilic cleavage was to increase the residence time of the metal complexes near the molecular target (2′-OH). Since the 2′-OH resides in the minor groove of RNA–DNA duplexes, the ability to direct complexes to specific grooves of an A-form RNA/DNA duplex seems important in the design of more efficient ribozyme mimics. Groove placement has been controlled experimentally by covalently linking complexes to different positions on various nucleosides. As Figure 17.11 shows, catalysts attached via the C5 position of thymidine are directed to the major groove, while catalysts incorporated at C1′ or C2′ attack across the minor groove. The first ribozyme mimic utilized a groove-directing design to place Cu(II) terpyridine (Cutrpy) in the major groove by attachment to the C5 position of deoxyuridine.[71] It may access the 2′-OH by intercalation into the duplex, allowing access to the minor groove. Metal terpy complexes are known as intercalators of duplexes.

Functionalzation of thymidine to include a trpy catalyst pendent from the C1′ position gave an example of a ribozyme mimic directed toward the minor groove (Fig. 17.12).[71] Häner and co-workers investigated specific groove attack by lanthanide based complexes that were covalently attached to the 2′-oxygen of a ribonucleotide, C1′, and C5.[78]

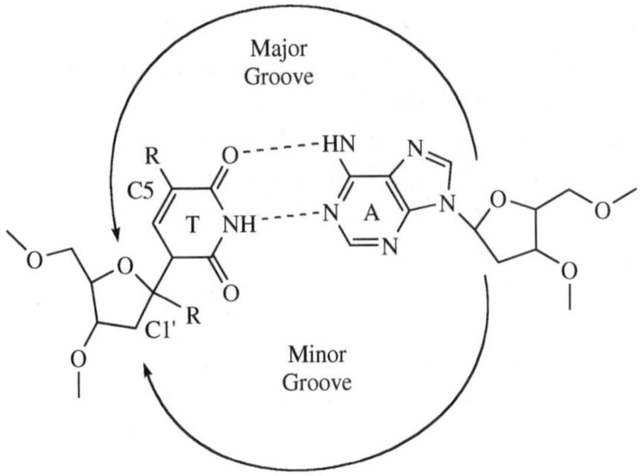

FIGURE 17.11 Major versus minor groove attack for a A : T base pair.

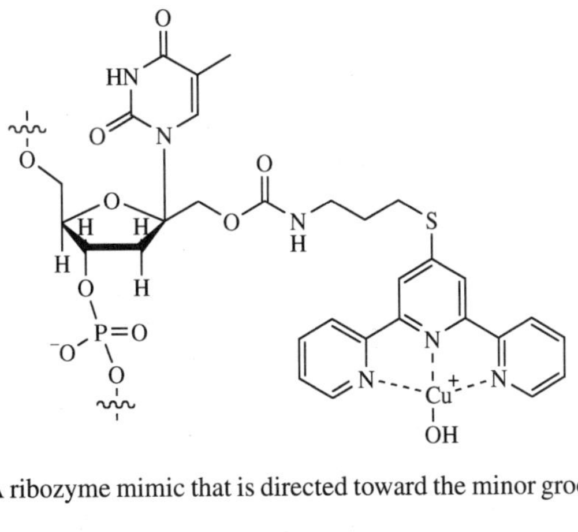

FIGURE 17.12 A ribozyme mimic that is directed toward the minor groove of a DNA/RNA duplex.

CLEAVAGE ENHANCEMENT VIA DISTORTION OF THE DUPLEX

As discussed previously, part of the development of *catalytic* antisense reagents was the recognition that cleavage must occur within a duplex to allow subsequent release of the antisense probe. Morrow's group investigated the cleavage of RNA/DNA duplexes by metal complexes.[92] They found that, when a DNA probe was hybridized to a section of $tRNA^{Phe}$, the double-stranded portion of the RNA sequence was protected from cleavage. This protection occurred with complexes of Eu^{3+} and La^{3+} that were known to be efficient nucleophilic cleavage catalysts for the cleavage of single-stranded RNA. Morrow hypothesized that cleavage was inhibited within the duplex because of its rigidity, which hampered adoption of a 5-coordinate phosphorane transition state at a tetrahedral phosphodiester site. Häner and co-workers devised an approach to circumvent this problem by creating a bulge in the DNA/RNA duplex.[78] They found that RNA bulges are more susceptible to nucleophilic cleavage than are fully base-paired RNA. Huesken concluded that the observed increase in cleavage was a result of the greater flexibility of the bulged RNA compared to duplexed RNA.[93] He argued that bulged RNA can achieve a conformation appropriate for the desirable in-line attack of the 2′-OH on the adjacent phosphorus. This conformation is consistent with an efficient, concerted nucleophilic cleavage reaction. From this recognition of the importance of flexibility at the cleavage site came the idea of attaching metal complexes to abasic sites. Each abasic site eliminates a Watson-Crick base-pair, introducing flexibility into a DNA/RNA duplex. Serinol, a reduced form of serine, was chosen by the Bashkin[94] group to create this abasic site. Others have used abasic sites to increase cleavage efficiency using both serinol and threoninol (the reduced form of threonine).[90,95,96] Because serinol mimics the three-carbon spacing of the normal backbone in DNA, minimal disruption of the duplex was expected. One serinol residue eliminated one Watson–Crick base-pair, thus allowing the RNA strand to possess enhanced flexibility opposite of the modified site. Results consistent with this concept were obtained with a Cutrpy-based ribozyme mimic (Fig. 17.13) cleaved 85% of its RNA target at 45°C after 3 days (a threefold rate increase over its fully base-paired Cutrpy counterpart).[94]

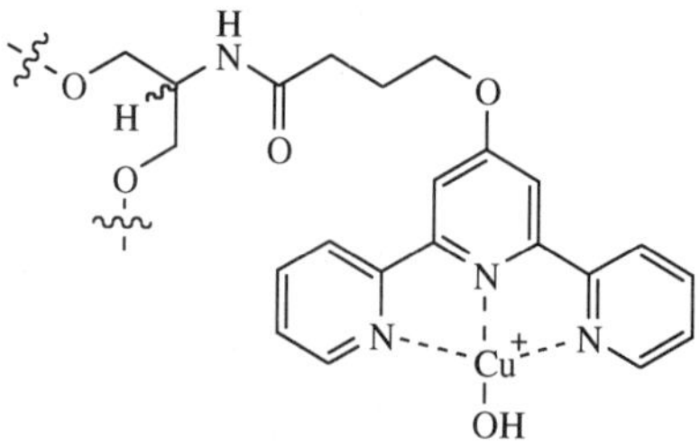

H N O O H O O N N N Cu + OH

FIGURE 17.13 A ribozyme mimic that confers an abasic site on the duplex.

OTHER RIBOZYME AND GENE THERAPY APPROACHES

Naturally occurring ribozymes[2,97,98] can also destroy RNA with the full specificity of the genetic code in a catalytic manner. Ribozyme mimics are synthetic counterparts of ribozymes in which the large catalytic domain of the natural ribozyme has been replaced by a small-molecule catalyst. The catalytic domain of a ribozyme mimic is low in molecular weight (MW), which provides mimics with an advantage over their naturally occurring, high-MW counterparts (drug delivery is generally difficult for high-MW species). Another advantage of the mimics is that natural ribozymes are difficult to modify for in vivo stability because most chemical changes to their backbone diminish or destroy their cleavage ability.[99–105] However, natural ribozymes can be expressed in vivo, which is an advantage that synthetic counterparts do not possess. There are many advantages to therapy via both natural ribozymes and ribozyme mimics, since they both promise unprecedented, gene-specific chemotherapy. Therefore, we believe that both approaches should be vigorously pursued to ensure the best chance that catalytic antisense reagents will reach the clinic.

The development of genome-based therapies such as the antisense approach represents some of the largest and most rapidly expanding areas of biomedical science.[106–111] Another type of genomic therapy works by correcting genetic problems by the replacement of damaged DNA. This approach must meet different challenges from those facing antisense therapy, which is a hybrid between medicinal chemistry and genomic strategies. Recent advances in the biochemical and medicinal areas allow human somatic cells to be transfected via a modified viral vector that carries a therapeutic gene of interest. By homologous recombination, this gene may insert into the human genome, causing the development of a new, transgenic cell. These cells can be selected for and reinserted back into the original patient, thereby introducing the gene of interest. Many aspects of this approach still must be addressed experimentally. Problems include the length constraint of transfected genes imposed by viral vectors and possible tumorogenicity of many viral vectors. With the emergence of these and other gene therapy techniques, new treatments for currently incurable or untreatable diseases will be found.

ACKNOWLEDGMENTS

The authors thank the National Science Foundation (CHE-9802660) and the donors of the Petroleum Research Fund, administered by the American Chemical Society, for partial support.

REFERENCES

1. Bashkin, J. K.; Xie, J.; Daniher, A. T.; Jenkins, L. A.; Yeh, G. C., *Nato Asi Ser., Ser. C* 1996, **479**, 355–366.

2. Cech, T. R., *JAMA* 1988, **260**, 3030–3034.
3. Cohen, J. S., ed., *Oligodeoxynucleotides: Antisense Inhibitors of Gene Expression*; CRC Press: Boca Raton, FL, 1989.
4. Häner, R.; Hall, J., *Antisense Nucleic Acid Drug Devel.* 1997, **7**, 423–430.
5. Morrow, J. R.; Buttrey, L. A.; Shelton, V. M.; Berback, K. A., *J. Am. Chem. Soc.* 1992, **114**, 1903–5.
6. "Isis Pharma," in *Dow Jones Newswire*, Sept 16, 1998.
7. Westheimer, F. H., *Science* 1987, **235**, 1173–1178.
8. Bertini, I.; Gray, H. B.; Lippard, S. J.; Valentine, J. S., *Bioinorganic Chemistry*, University Science Books, 1994.
9. Weiss, R. H.; Riley, D. P., *Drugs Fut.* 1996, **21**, 383–389.
10. Weiss, R. H.; Fretland, D. J.; Baron, D. A.; Ryan, U. S.; Riley, D. P., *J. Biol. Chem.* 1996, **271**, 26149.
11. Kasten, T. P.; Settle, S. L.; Misko, T. P.; Riley, D. P.; Weiss, R. H.; Currie, M. G.; Nickols, G. A., *Proc. Soc. Exp. Biol. Med.* 1995, **208**, 170–171.
12. Hardy, M. M.; Flickinger, A. G.; Riley, D. P.; Weiss, R. H.; Ryan, U. S., *J. Biol. Chem.* 1994, **269**, 18535–18540.
13. Kilgore, K. S.; Friedrichs, G. S.; Johnson, C. R.; Schasteen, C. S.; Weiss, R. H.; Riley, D. P.; Ryan, U. S.; Lucchesi, B. R., *J. Molec. Cell. Cardiol.* 1994, **26**, 995–1006.
14. Black, S. C.; Schasteen, C. S.; Weiss, R. H.; Riley, D. P.; Driscoll, E. M.; Lucchesi, B. R., *J. Pharmacol. Exp. Ther.* 1994, **270**, 1208–1215.
15. Weiss, R. H.; Flickinger, A. G.; Rivers, W. J.; Hardy, M. M.; Aston, K. W.; Ryan, U. S.; Riley, D. P., *J. Biol. Chem.* 1993, **268**, 23049–23054.
16. Riley, D. P.; Weiss, R. H., *J. Am. Chem. Soc.* 1994, **116**, 387.
17. Riley, D. P.; Henke, S. L.; Lennon, P. J.; Weiss, R. H.; Neumann, W. L.; Jr., R., *Inorg. Chem.* 1996, **35**, 5213.
18. Riley, D. P.; Henke, S. L.; Lennon, P. J.; Weiss, R. H.; Neumann, W. L., *J. Am. Chem. Soc.* 1997, **119**, 6522.
19. Uhlmann, E., *Prog. Biotechnol* 1994, **9**, 667–73.
20. Altmann, K.-H.; Dean, N. M.; Fabbro, D.; Freier, S. M.; Geiger, T.; Haener, R.; Huesken, D.; Martin, P.; Monia, B. P.; et al., *Chimia* 1996, **50**, 168–176.
21. De Mesmaeker, A.; Haener, R.; Martin, P.; Moser, H. E., *Acc. Chem. Res.* 1995, **28**, 366–74.
22. Seliger, H.; Krist, B.; Roesch, H.; Roesch, R.; Rueck, A.; Steiner, R.; Gupta, K. C.; Ortigao, J. F. R., *Prog. Biotechnol.* 1994, **9**, 681–4.
23. Crooke, S. T., *Cancer Ther.* 1997, 299–336.
24. Iversen, P. L.; Bayever, E., *Clin. Gene Anal. Manipulation* 1996, 372–390.
25. Nicot, A.; Pfaff, D. W., *J. Neurosci. Meth.* 1997, **71**, 45–53.
26. Oberbauer, R., *Wien. Klin. Wochenschr* 1997, **109**, 40–46.
27. Uhlmann, E.; Peyman, A., *Chem. Rev.* 1990, **90**, 544.
28. Inouye, M., *Gene* 1998, **72**, 25–34.
29. Stephenson, M. L. Z., P. C., *Proc. Natl. Acad. Sci. USA* 1978, **75**, 285–288.
30. Stein, C. A.; Cohen, J. S., *Cancer Research* 1988, **48**, 2659–2668.

31. Simons, R. W.; Kleckner, N., *Cell* 1983, **34**, 683.
32. Heywood, S. M., *Nucleic Acids Res.* 1986, **14**, 6771.
33. Zamecnik, P. C.; Stephenson, M. L., *Proc. Natl. Acad. Sci.* (USA) 1978, **75**, 280–284.
34. Izant, J. G.; Weintraub, H., *Science* 1985, **229**, 345.
35. Toulme, J. J.; Boiziau, C.; Larrouy, B.; Frank, P.; Albert, S.; Ahmadi, R., in *DNA and RNA Cleavers and Chemotherapy of Cancer and Viral Diseases*, B. Meunier, ed., Kluwer, The Netherlands, 1996; Vol. 479, 1996, pp. 271–288.
36. Ogawa, T.; Okazaki, T., *Molec. Gen. Genet.* 1984, **193**, 231–237.
37. S. Agrawal, ed., *Protocols for Oligonucleotides and Analogs*, Humana: Totowa, NJ, 1993, Vol. 20.
38. Baum, R., *Chemical and Eng. News*, April 14, 1994, pp. 21–22.
39. Eckstein, F., ed., *Oligonucleotides and Analogues: A Practical Approach*, Oxford Univ. Press, Oxford, 1991.
40. Imbach, J. L.; Rayner, B.; Morvan, F., *Nucleosides Nucleotides* 1989, **8**, 627–48.
41. Monia, B. P.; Lesnik, E. A.; Gonzalez, C.; Lima, W. F.; McGee, D.; Guinosso, C. J.; Kawasaki, A. M.; Cook, P. D.; Freier, S. M., *J. Biol. Chem.* 1993, **268**, 14514–14522.
42. Davies, J.; Hostomka, Z.; Hostomsky, Z.; Jordan, S.; Matthews, D., *Science* 1991, **252**, 88–95.
43. Hostomsky, Z.; Hostomska, Z.; Matthews, D. A., in *Nucleases*, S. M. Linn, R. S. Lloyd, R. J. Roberts, eds., Cold Spring Harbor Laboratory Press, Cold Spring Harbor, NY, 1993, pp. 341–376.
44. Crooke, S. T.; Lemonidis, K. M.; Neilson, L.; Griffey, R.; Lesnik, E. A.; Monia, B. P., *Biochem. J.* 1995, **312**, 599–608.
45. Sigman, D. S., *Acc. Chem. Res.* 1986, **19**, 180–186.
46. Absalon, M. J.; Wu, W.; Kozarich, J. W.; Stubbe, J., *Biochemistry* 1995, **34**, 2076–2086.
47. Dervan, P. B., *Nature* 1992, **359**, 87–88.
48. Meijler, M. M.; Zelenko, O.; Sigman, D. S., *J. Am. Chem. Soc.* 1997, **119**, 1135–1136.
49. Bashkin, J. K.; Jenkins, L. A., *Comments Inorg. Chem.* 1994, **16**, 77–93.
50. Morrow, J. R., *Met. Ions Biol. Syst.* 1996, **33**, 561–592.
51. Linkletter, B.; Chin, J., *Angew. Chem. Int. Ed. Engl.* 1995, **34**, 472–474.
52. Stern, M. K.; Bashkin, J. K.; Sall, E. D., *J. Am. Chem. Soc.* 1990, **112**, 5357.
53. Shelton, V. M.; Morrow, J. R., *Inorg. Chem.*1991, **30**, 4295–4299.
54. Young, M. J.; Chin, J., *J. Am. Chem. Soc.* 1995, **117**, 10577–10578.
55. Yashiro, M.; Ishikubo, A.; Komiyama, M., *J. Chem. Soc., Chem. Commun.* 1997, **1**, 83–84.
56. Chu, F.; Smith, J.; Lynch, V. M.; Anslyn, E. V., *Inorg. Chem.* 1995, **34**, 5689–5690.
57. Cech, T. R., *Science* 1987, **236**, 1532.
58. Oivanen, M.; Kuusela, S.; Lönnberg, H., *Chem. Rev.* 1998, **98**, 961–990.
59. Shiiba, T.; Komiyama, M., *Nucleic Acids Symp. Ser.* 1992, **27**, 39–40.
60. Yeh, G. C.; Beatty, A. M.; Bashkin, J. K., *Inorg. Chem.* 1996, **35**, 3828–3835.
61. Shriver, D. F.; Atkins, P.; Langford, C. H., *Inorganic Chemistry*; 2nd ed., Freeman, New York, 1994, pp. 191–192.
62. Bamann, E., *Angew. Chem.* 1939, **52**, 186–188.

63. Moss, R. A.; Bracker, K.; Zhang, J., *Chem. Commun.* 1997, 563–564.
64. Ihara, T.; Shimura, H.; Ohmori, K.; Takeuchi, J.; Takagi, M., *Chem. Lett.* 1996, 687–688.
65. Kuusela, S.; Lonnberg, H., *Nucleosides Nucleotides* 1996, **15**, 1669–1678.
66. Kuusela, S.; Lonnberg, H. J., *J. Phys. Org. Chem.* 1993, **6**, 347–356.
67. Chin, J.; Zou, X., *Can. J. Chem.* 1987, **65**, 1882–1884.
68. Kuusela, S.; Lonnberg, H., *J. Phys. Org. Chem.* 1992, **5**, 803–811.
69. Komiyama, M.; Matsumura, K.; Matsumoto, Y., *J. Chem. Soc., Chem. Commun.* 1992, 640–641.
70. Breslow, R.; Huang, D.-L., *Proc. Natl. Acad. Sci.* (USA) 1991, **88**, 4080–4083.
71. Bashkin, J. K.; Xie, J.; Daniher, A. T.; Sampath, U.; Kao, J. L.-F., *J. Org. Chem.* 1996, **61**, 2314–2321.
72. Magda, D.; Miller, R. A.; Sessler, J. L.; Iverson, B. L., *J. Am. Chem. Soc.* 1994, **116**, 7439–7440.
73. Zuckermann, R. N.; Schultz, P. G., *Proc. Natl. Acad. Sci.* (USA) 1989, **86**, 1766–1770.
74. Kanaya, S.; Nakai, C.; Konishi, A.; Inoue, H.; Ohtsuka, E.; Ikehara, M., *J. Biol. Chem.* 1992, **267**, 8492–8498.
75. Uchiyama, Y. I., H.; Ohtsuka, E.; Naki, C.; Kanaya, S.; Ueno, Y.; Ikeheara, M., *Bioconjugate Chem.* 1994, **5**, 327–332.
76. Bashkin, J. K.; Frolova, E. I.; Sampath, U., *J. Am. Chem. Soc.* 1994, **116**, 5981–5982.
77. Trawick, B. N. D., A. T.; Bashkin, J. K., *Chem. Rev.* 1998, **98**, 939–960.
78. Hall, J.; Huesken, D.; Haener, R., *Nucleic Acids Res.* 1996, **24**, 3522–3526.
79. Hall, J.; Huesken, D.; Pieles, U.; Moser, H. E.; Haener, R., *Chem. Biol.* 1994, 185–190.
80. Hall, J. H., D.; Haner, R., *Nucleosides Nucleotides* 1997, **16**, 1357–1368.
81. Matsumura, K.; Endo, M.; Komiyama, M., *J. Chem. Soc., Chem. Commun.* 1994, 2019–2020.
82. Magda, D.; Crofts, S.; Lin, A.; Miles, D.; Wright, M.; Sessler, J. L., *J. Am. Chem. Soc.* 1997, **119**, 6947–6948.
83. Hovinen, J.; Guzaev, A.; Azhayeva, E.; Azhayev, A.; Lönnberg, H., *J. Org. Chem.* 1995, **60**, 2205–2209.
84. Baker, B. F., *J. Am. Chem. Soc.* 1993, **115**, 3378–3379.
85. Bashkin, J. K.; Gard, J. K.; Modak, A. S., *J. Org. Chem.* 1990, **55**, 5125.
86. Bashkin, J. K.; McBeath, R. J.; Modak, A. S.; Sample, K. R.; Wise, W. B., *J. Org. Chem.* 1991, **56**, 3168.
87. Barbier, B.; Brack, A., *J. Am. Chem. Soc.* 1988, **110**, 6880–6882.
88. Barbier, B.; Brack, A., *J. Am. Chem. Soc.* 1992, **114**, 3511–3515.
89. Komiyama, M.; Inokawa, T., *J. Biochem.* 1994, **116**, 719–720.
90. Reynolds, M. A.; Beck, T. A.; Say, P. B.; Schwartz, D. A.; Dwyer, B. P.; Daily, W. J.; Vaghefi, M. M.; Metzler, M. D.; Klem, R. E.; Arnold, L. J., Jr., *Nucleic Acids Res.* 1996, **24**, 760–765.
91. Vlassov, V.; Abramova, T.; Godovikova, T.; Giege, R.; Silnikov, V., *Antisense Nucleic Acid Drug Devel.* 1997, **7**, 39–42.

92. Kolasa, K. A.; Morrow, J. R.; Sharma, A. P., *Inorg. Chem.* 1993, **32**, 3983–4.
93. Huesken, D.; Goodall, G.; Blommers, M. J. J.; Jahnke, W.; Hall, J.; Haener, R.; Moser, H. E., *Biochemistry* 1996, **35**, 16591–16600.
94. Daniher, A. T.; Bashkin, J. K., *Chem. Commun.*1998, 1077–1078.
95. Ramasamy, K. S.; Seifert, W., *Bioorg. Med. Chem. Lett.* 1996, **6**, 1799–1804.
96. Fukui, K.; Morimoto, M.; Segawa, H.; Tanaka, K.; Shimidzu, T., *Bioconjugate Chem.* 1996, **7**, 349–355.
97. Sarver, N.; Cantin, E. M.; Chang, P. S.; Zaia, J. A.; Ladne, P. A.; Stephens, D. A.; Rossi, J. J., *Science* 1990, **247**, 1222–1225.
98. Christoffersen, R. E.; Marr, J. J., *J. Med. Chem.* 1995, **38**, 2023–2037.
99. Eckstein, F., *NATO ASI Ser., Ser. C* 1996, **479**, 291–294.
100. Heidenreich, O.; Pieken, W.; Eckstein, F., *FASEB J.* 1993, **7**, 90–96.
101. Olsen, D. B.; Benseler, F.; Aurup, H.; Pieken, W. A.; Eckstein, F., *Biochemistry* 1991, **30**, 9735–9741.
102. Pieken, W. A.; Olsen, D. B.; Aurup, H.; Williams, D. M.; Heidenreich, O.; Benseler, F.; Eckstein, F., *Nucleic Acids Symp. Ser.* 1991, **24**, 51–53.
103. Usman, N.; Beigelman, L.; McSwiggen, J. A., *Curr. Opin. Struct. Biol.* 1996, **6**, 527–533.
104. Matulic-Adamic, J.; Karpeisky, A. M.; Gonzales, C.; Burgin, A. B., Jr.; Usman, N.; McSwiggen, J. A.; Beigelman, L., *Collect. Czech. Chem. Commun.* 1996, **61**, S271–S275.
105. Burgin, A. B., Jr.; Gonzalez, C.; Matulic-Adamic, J.; Karpeisky, A. M.; Usman, N.; McSwiggen, J. A.; Beigelman, L., *Biochemistry* 1996, **35**, 14090–14097.
106. Zlokovic, B. V.; Apuzzo, M. L., *Neurosurgery*1997, **40**, 789–803.
107. Kaplitt, M. G.; Darakchiev, B.; During, M. J., *Pediatr. Neurosurg.* 1998, **28**, 3–14.
108. Jiang, Q.; Engelhardt, J. F., *Eur. J. Hum. Genet.* 1998, **6**, 12–31.
109. Yang, K.; Clifton, G. L.; Hayes, R. L., *J. Neurotrauma* 1997, **14**, 281–297.
110. Walter, J.; High, K. A., *Adv. Vet. Med.* 1997, **40**, 119–134.
111. MW, R.; Barnes, M. N.; Rancourt, C.; Wang, M.; Grim, J.; Alvarez, R. D.; Siegal, G. P.; Curiel, D. T., *Semin. Oncol.* 1998, **25**, 397–406.

INDEX